INTRODUCTION
TO
PROJECTIVE
GEOMETRY

INTRODUCTION TO PROJECTIVE GEOMETRY

C. R. WYLIE, JR.

Professor of Mathematics
Furman University
Former Chairman of Department of Mathematics
The University of Utah

McGRAW-HILL BOOK COMPANY

New York
St. Louis
San Francisco
Dusseldorf
London
Mexico
Panama
Sydney
Toronto

This book was set in Baskerville by The Universities
Press, and printed on permanent paper and bound by
The Maple Press Company. The designer was Marsha Cohen;
the drawings were done by Harry Lazarus. The editors
were Donald K. Prentiss and Janet Wagner. John F. Harte
supervised the production.

INTRODUCTION TO PROJECTIVE GEOMETRY

Library of Congress Catalog Card Number 71-85174

72195

1234567890 MAMM 79876543210

PREFACE

The objective of this book is to provide an introduction to plane projective geometry suitable for undergraduate students majoring in mathematics or studying to become teachers of secondary school mathematics.

The material falls naturally into four main subdivisions. The first, consisting only of Chap. 1, is a historical introduction dealing with the elementary aspects of perspective, both in three dimensions, as the artists of the Middle Ages developed it, and in two dimensions, after the process of rabattement has brought the picture plane into coincidence with the object plane.

The second major portion of the book consists of Chaps. 2 to 5 and is devoted to a development of analytic projective geometry as an extension of the geometry of the euclidean plane. In Chap. 2 the concepts of ideal points and the ideal line are introduced, and the familiar cartesian plane is extended by the addition of these new elements. Chapter 3 is a presentation of the basic facts of matrix theory and linear algebra which will be needed in later chapters. In Chap. 4 the discussion of the extended plane is resumed, and such topics as the parametrization of lines and pencils, the theorems of Desargues and Pappus, perspectivities, projectivities, and involutions, cross ratio, and conics are discussed. Chapter 5 is devoted to a study of linear transformations in the extended plane, first as coordinate transformations, then as point transformations, and the six possible types of nonsingular linear transformations are identified. Then, after the notion of a group is introduced, the affine, similarity, equiareal, and euclidean subgroups of the projective group are investigated.

Chapters 6 to 9 constitute the third major subdivision of the book, which is devoted to an axiomatic development of plane projective geometry. In Chap. 6 the axioms of incidence and connection and the projectivity axiom are introduced, and various systems, both finite and infinite, are discussed as concrete representations. The theorems of Desargues and Pappus are deduced, the basic properties of projectivities are developed, and the logical interrelations of these results are explored. In Chap. 7 the properties of complete four-points and complete four-lines are investigated, and the notions of harmonic tetrads, involutory hexads, and involutions are introduced. Chapter 8 deals with conics and their projective generation, Desargues' conic theorem, Pascal's theorem, polar theory, the conic as a self-dual configuration, and projectivities on a conic. In Chap. 9, the concept of a field is discussed, the arithmetic of collinear points is developed, and an isomorphism between the points of an arbitrary line and the field of complex numbers is postulated. After the general line is thus coordinatized, both nonhomogeneous and homogeneous coordinates are introduced into the projective plane itself, equations are obtained for lines, points, and conics, and the incidence condition for a point and a line is determined.

Chapters 10 and 11, which make up the last section of the book, are concerned with the introduction of a metric into the projective plane. In Chap. 10 this is done in such a way that elliptic and hyperbolic geometry are obtained. In Chap. 11, euclidean geometry and a curious variant arising from the assumption that the circular points at infinity are real rather than complex are similarly derived. In each case this is done in a more general context than is usual. Specifically, elliptic geometry is developed not from the particular absolute conic $x_1{}^2 + x_2{}^2 + x_3{}^2 = 0$ but rather from the conic $X^T a X = 0$, where a is an arbitrary positive definite matrix; and hyperbolic geometry is developed not from the particular absolute conic $x_1{}^2 + x_2{}^2 - x_3{}^2 = 0$ but rather from the conic $X^T a X = 0$, where a is an arbitrary nonsingular indefinite matrix. In the case of euclidean geometry, this generality leads to an infinite number of distance- and angle-measurement formulas, all of which are consistent with the euclidean law of cosines, law of sines, and angle-sum theorem.

The book concludes with two appendixes, the first containing a brief review of the fundamental properties of determinants and the second containing the incidence tables for a nondesarguesian projective geometry with 10 points on each line, together with a number of examples of the curious properties of a geometry in which the theorem of Desargues does not hold.

Although algebraic techniques are used freely throughout this book, its spirit is geometric rather than algebraic. The primary emphasis, of course, is on projective geometry, but with the interests of prospective high school teachers in mind, a serious attempt has been made to clarify the relation of euclidean geometry to projective geometry and to use projective results to illuminate the familiar facts of euclidean geometry. In addition to discussions in the text itself, many of the worked examples and numerous problems have been included for this purpose.

It is the author's hope that most, if not all, of this book will be relatively easy reading for any reasonably well-prepared junior or senior student in mathematics. In an attempt to achieve this clarity of exposition, derivations and proofs have been given in more than usual detail, and a substantial number of completely worked examples and carefully drawn figures have been included. In addition there are over 800 exercises, ranging from routine applications to significant extensions of the theory. As an aid to the student, the answers to the odd-numbered exercises are given at the end of the book.

Any text represents the influence of teachers, colleagues, and students too numerous to list. However, I must acknowledge my indebtedness to the students who used a manuscript edition of this material in my class in projective geometry at the University of Utah during the academic year 1968–1969. Their comments, criticisms, and suggestions were invaluable. I must also express my deep appreciation of the contribution of my wife, Ellen, who assisted me in the tedious work of proofreading.

C. R. Wylie, Jr.

CONTENTS

one

THE ELEMENTS OF PERSPECTIVE

1.1 Introduction

The origins of projective geometry, like those of many other branches of mathematics, are to be found in man's concern with the world around him. Unlike her sister disciplines, however, projective geometry grew out of problems which were esthetic rather than practical, and those who contributed to her early development were artists rather than scientists and engineers.

Although the painters of Greece and Rome had tried with some success to give their work a three-dimensional effect, the artists of medieval Europe, preoccupied with religious themes that were mystic rather than realistic, painted in a stiff, highly stylized, starkly two-dimensional fashion. Spatial relations were ignored, backgrounds were neutral, foregrounds were usually missing, and figures whether of trees, animals, or human beings were flat and lifeless. Then toward the end of the thirteenth century, as a revived interest in the cultures of Greece and Rome heralded the dawn of the Renaissance, artists became aware of the unreality of their work and sought consciously to make it more natural and realistic. Duccio (c. 1255–1319) and Giotto (c. 1267–1337), among others, experimented with renderings of space, distance, and shape to suggest the three-dimensional relations existing among the objects they painted. Well-defined ground planes were introduced, foreshortening was attempted, and converging lines were used to give an impression of depth.

The intuitive theory of perspective developed by Duccio and Giotto culminated in the fourteenth century in the work of Lorenzetti (c. 1300–1348). Thereafter, further progress in the realistic representation of three-dimensional scenes on the two-dimensional canvas of the painter had to await the development of a mathematical theory of perspective. This came in the fifteenth century, and the men who created it, though in some cases quite competent in the mathematics of their day, were primarily artists. The first of these was Brunelleschi (1379–1446), who by 1425 had developed a system of perspective which he used in his own work and taught to other painters. The first text on perspective, a treatise by Alberti (1404–1472), appeared in 1435. Later Piero della Francesca (c. 1418–1492), a gifted mathematician as well as an outstanding painter, extended considerably the work of Alberti. Still later, both Leonardo da Vinci (1452–1519) and Albrecht Dürer (1471–1528) wrote treatises on perspective which not only presented the mathematical theory of perspective but insisted on its fundamental importance in all painting.

Reduced to its simplest terms, the theory of perspective regards the artist's canvas as a transparent screen through which he looks from a fixed vantage point at the scene he is painting. Light rays coming from each point of the scene are imagined to enter his eye, and the totality of these lines is called a **projection.** The point where any line in the projection pierces the veiwing screen is the image of the corresponding point in the scene being painted. The totality of all these image points, which, of course, becomes the painting itself, is called a **section** of the projection and conveys to the eye the same impression as the scene itself. Figures 1.1 and 1.2, which are illustrations from Dürer's text on perspective, "A Treatise on Measurement," illustrate the processes of projection and section and how the artist was supposed to make use of them.

Clearly, since any section of any projection of a given scene must convey the impression of that scene to a viewer, it follows that even though lengths, angles, and shapes may be altered, there must also be many properties which are left unchanged by these processes. From one point of view, projective geometry can be thought of as simply the study of those properties of figures which are left unchanged, or **invariant,** by projections and sections. Although this is a great oversimplification, it provides a natural

Fig. 1.1   Woodcut from Dürer's "A Treatise on Measurement."

introduction to concepts of importance in projective geometry, and we shall
devote the rest of this chapter to a study of perspective in somewhat the spirit
of the artists of the early Renaissance.

1.2 The Elements of Perspective

Let us imagine that we are viewing a scene from a fixed point, C, through a
fixed plane, ρ, and for simplicity, let us suppose that the scene is two-
dimensional and lies in a plane, σ, which is horizontal and perpendicular to
the viewing screen (Fig. 1.3). The plane σ, which contains the scene, we
shall call the **object plane.** The vertical plane ρ we shall call the **picture
plane** or **image plane.** The line joining the viewing point, C, to any point
P in the object plane is called the **line of sight** or **projecting line** of P.
The point P', in which the line of sight from P intersects the picture plane, is,
by definition, the **image** of P. A transformation such as this, in which a
point P and its image P' are always collinear with a fixed point C, is called a

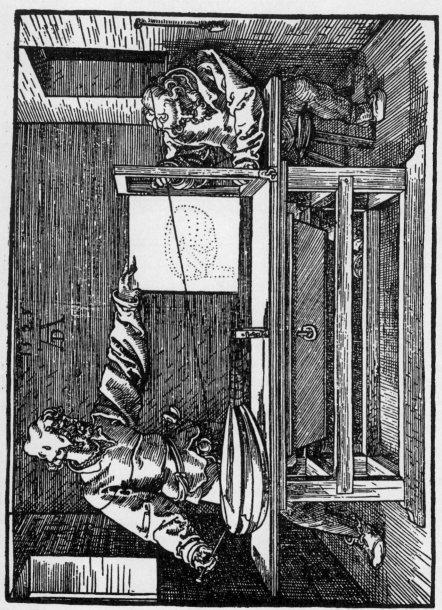

Fig. 1.2   Woodcut from Dürer's "A Treatise on Measurement."

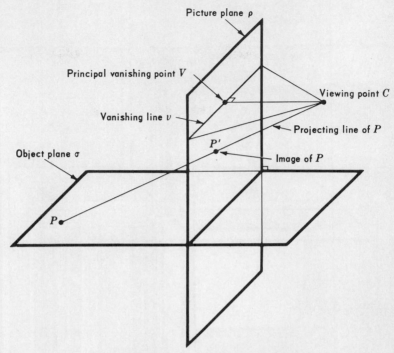

Fig. 1.3   The elements of a perspective transformation.

perspective transformation.[1] The point of intersection of the picture plane, ρ, and the line through the viewing point, C, which is perpendicular to ρ is called the **principal vanishing point,** V. The line, v, which is the intersection of the picture plane and the plane through C parallel to the plane of the scene is called the **vanishing line** or **horizon line.**

To discover the significance of the principal vanishing point, let us imagine that the scene contains a pair of parallel lines, say the rails of a railroad track, which are perpendicular to the picture plane, as shown in Fig. 1.4. By definition, the image of any point on either of these lines, say the line l_1, is found by joining it to the viewing point C and then determining where this projecting line pierces the picture plane ρ. Clearly, since the projecting lines of the various points on l_1 all pass through C and all intersect l_1, they must all lie in the plane, π_1, determined by C and l_1. The locus of the images of the points of l_1, that is, the image of l_1 itself, is therefore the line, l_1',

---

[1] Presumably, the scene in which an artist is interested always lies on the opposite side of the picture plane from the viewing point. Hence the picture itself always lies in the half of the picture plane which is on the same side of the object plane as the viewing point. However, in the mathematical discussion of perspective such restrictions are unnecessary and unnatural, and we shall assume that our transformations extend over the entire object and image planes.

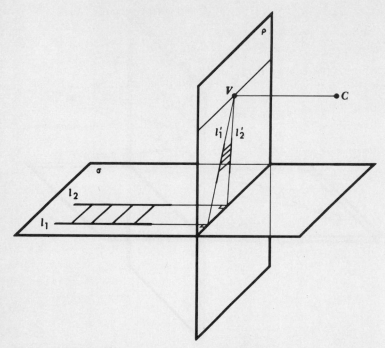

Fig. 1.4   The significance of the principal vanishing point in a perspective transformation.

in which the plane π_1 intersects the picture plane. Moreover, since l_1 is, by hypothesis, perpendicular to the picture plane ρ, it must be parallel to the line CV, which, by definition, is also perpendicular to ρ. Therefore, since parallel lines are necessarily coplanar, it follows that the line CV also lies in the plane π_1 determined by C and l_1. Thus the principal vanishing point, V, since it lies in both ρ and π_1, must lie on their intersection, l_1'. In other words, the image, l_1', of the line l_1 passes through V. Similarly, of course, the image, l_2', of l_2 also passes through V. These observations are summarized in the following theorem.

Theorem 1

All lines in the scene which are perpendicular to the picture plane appear on the picture plane as lines which pass through the principal vanishing point in the picture plane.

To appreciate the significance of the vanishing line, v, let us apply the preceding considerations to a pair of parallel lines in the scene which are not perpendicular to the picture plane, such as the lines m_1 and m_2 in Fig. 1.5.

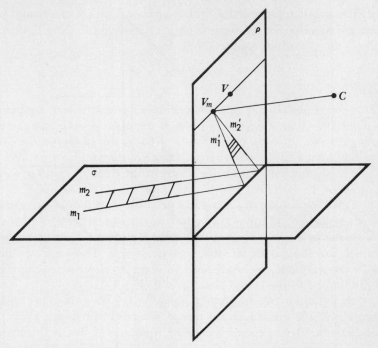

Fig. 1.5 The mapping of a general pair of parallel lines in a perspective transformation.

As before, the projecting lines of the various points of m_i all lie in the plane, π_i, determined by C and m_i. The image of m_i is therefore the line, m_i', in which the projecting plane π_i intersects the picture plane ρ. Now there is a unique line through C which is parallel to m_i and therefore in the plane π_i. Moreover, since this line is obviously horizontal, it also lies in the plane determined by C and v and must therefore intersect the vanishing line v in a point V_m. Thus the point V_m, since it lies in both π_i and ρ, must lie on their intersection, m_i'. We have thus established the following theorem.

Theorem 2

The lines of a general parallel family lying in a plane perpendicular to the picture plane appear on the picture plane as lines which pass through a unique point on the vanishing line in the picture plane.

Parallel lines, of course, never meet, but to the eye it appears that they do, and on the picture plane, in fact, their representations do indeed converge to a common point. In the picture plane, the vanishing line is simply the locus of the intersections of the images of horizontal parallel lines, and its points represent the nonexistent points in the horizontal plane to which parallel lines in the scene appear to converge as these lines recede indefinitely and "vanish" toward the horizon.

Example 1

In a three-dimensional rectangular coordinate system the viewing point, C, is the point $(0,-3,2)$, the picture plane is the plane $y = 0$, and the plane containing the scene is the plane $z = 0$. What is the image on the picture plane of the point $(2,1,0)$? What is the image of the family of parallel lines $y = x + k$ in the scene? What is the image of the circle, Γ, whose equation in the plane of the scene is $x^2 + (y - 2)^2 = 1$?

    It will be helpful in solving this problem to introduce two auxiliary coordinate systems in addition to the basic x, y, z system itself. One of these, an X, Y system in the xy plane, we shall need in order to describe configurations which are limited to the object plane, $z = 0$. The other, an X', Z' system in the xz plane, we shall need in order to describe configurations which are limited to the image plane, $y = 0$ (Fig. 1.6).

    By definition, the image of a point $P:(X,Y,0)$ in the object plane is the point $P':(X',0,Z')$ in which the line PC intersects the image plane. Now we know from analytic geometry that the equations of the line determined

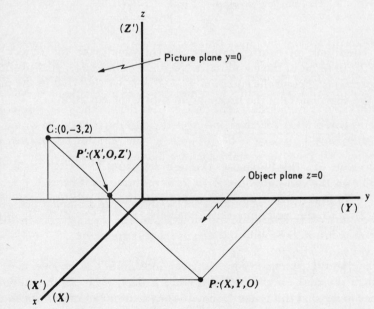

Fig. 1.6   The data for a particular perspective transformation.

by two points (x_0, y_0, z_0) and (x_1, y_1, z_1) can be written in the form

$$\frac{x - x_0}{x_1 - x_0} = \frac{y - y_0}{y_1 - y_0} = \frac{z - z_0}{z_1 - z_0}$$

Hence, taking $C:(0, -3, 2)$ as the point (x_0, y_0, z_0) and $P:(X, Y, 0)$ as the point (x_1, y_1, z_1), we have for the equations of the projecting line PC,

$$\frac{x - 0}{X - 0} = \frac{y + 3}{Y + 3} = \frac{z - 2}{0 - 2}$$

To find the coordinates $(X', 0, Z')$ of the point P' in which this line pierces the picture plane, we merely put $y = 0$ and solve these equations for x and z, getting

$$(1) \qquad x \equiv X' = \frac{3X}{Y + 3} \qquad z \equiv Z' = \frac{2Y}{Y + 3}$$

To find the image of the given point $(2, 1, 0)$, we merely let $X = 2$, $Y = 1$ in the last pair of equations. The result is the image point $X' = \frac{3}{2}$, $Z' = \frac{1}{2}$, that is, the point $(\frac{3}{2}, 0, \frac{1}{2})$.

To find the image of the family of parallel lines defined in the object plane by the equation $y = x + k$, that is, $Y = X + k$, it is convenient to solve Eqs. (1) for X and Y, getting

$$(2) \qquad X = \frac{2X'}{2 - Z'} \qquad Y = \frac{3Z'}{2 - Z'}$$

These equations, of course, express the coordinates of the object point $P:(X, Y, 0)$ in terms of the coordinates $(X', 0, Z')$ of its image, P'. Substituting the expressions (2) into the equation $Y = X + k$, we find that in the picture plane the coordinates of the image point satisfy the equation

$$\frac{3Z'}{2 - Z'} = \frac{2X'}{2 - Z'} + k \qquad \text{or} \qquad (3 + k)Z' = 2X' + 2k$$

For all values of k this line passes through the point $X' = 3$, $Z' = 2$, which is a point on the vanishing line $Z' = 2$, as required by Theorem 2.

To find the image of the circle Γ, we merely substitute the expressions (2) into the equation of Γ, namely, $x^2 + (y - 2)^2 = 1$, that is, $X^2 + (Y - 2)^2 = 1$, getting

$$\left(\frac{2X'}{2 - Z'}\right)^2 + \left(\frac{3Z'}{2 - Z'} - 2\right)^2 = 1 \qquad \text{or} \qquad (X')^2 + 6(Z')^2 - 9Z' + 3 = 0$$

It is easy to see that this is not the equation of a circle but rather the equation of an ellipse.

Exercises

1. In Example 1, does the center of the ellipse which is the image of Γ coincide with the image of the center of Γ?

2. In Example 1, what is the image of the family of parallel lines defined in the object plane by the equation $X = 3Y + k$? By the equation $Y = mX + k$?

3. In Example 1, what is the image of the parabola whose equation in the object plane is $Y = X^2$? What is the image of the parabola whose equation is $Y = 1 - X^2$?

4. Rework Example 1 if C is the point $(1,-3,2)$.

5. Rework Example 1 if the picture plane is the plane $y = 1$.

6. Rework Example 1 if C is the point $(0,-2,1)$ and the picture plane is the plane $y = 2$.

7. In a perspective transformation, are there any points which coincide with their images?

8. In a perspective transformation, is every point in the picture plane the image of some point in the object plane? Does every point in the object plane have an image in the picture plane? *Hint:* Remember that the object plane extends on both sides of the picture plane.

9. In a perspective transformation, is a line always represented by a line? Is a nonsingular conic always represented by a nonsingular conic?

10. In a perspective transformation, are there any parallel lines which are not represented in the picture plane by lines meeting on the vanishing line?

11. In a perspective transformation, is a circle ever represented by a circle?

12. In a perspective transformation, is a circle always represented by an ellipse? If not, what other possibilities are there?

13. What are the possibilities for the image of a parabola under a perspective transformation? What are the possibilities for the image of a hyperbola?

14. In a perspective transformation, is a segment always represented by a segment?

15. In a perspective transformation, if P is a point between Q and R, is it true that the image of P is between the images of Q and R?

16. If $P:(X,Y)$ is a general point in the object plane $z = 0$, and if the viewing point is an arbitrary point $C:(a,b,c)$, find the coordinates (X',Z') of the

image of P on the picture plane $y = 0$. What are the equations expressing X and Y in terms of X' and Z'?

17. In a perspective transformation, are the lengths of all segments altered in the same ratio?

18. In a perspective transformation, is the angle between two lines the same as the angle between the images of the lines?

19. In a perspective transformation, if a line is tangent to a circle, is the image of the line tangent to the image of the circle?

20. In Example 1, verify that for the four collinear points, $P_1:(0,1,0)$, $P_2:(-1,2,0)$, $P_3:(-3,4,0)$, $P_4:(-7,8,0)$ and their images P_1', P_2', P_3', P_4', the following relation holds

$$\frac{(P_1P_3)(P_2P_4)}{(P_1P_4)(P_2P_3)} = \frac{(P_1'P_3')(P_2'P_4')}{(P_1'P_4')(P_2'P_3')}$$

where (P_iP_j) denotes the distance between P_i and P_j.

21. If $P_i:(X_i,Y_i)$, $i = 1, 2, 3, 4$, are four collinear points in the object plane $z = 0$, and if $P_i':(X_i',Z_i')$ are the images of these points on the image plane $y = 0$ under a perspective transformation with center $C:(a,b,c)$, show that

$$\frac{(P_1P_3)(P_2P_4)}{(P_1P_4)(P_2P_3)} = \frac{(P_1'P_3')(P_2'P_4')}{(P_1'P_4')(P_2'P_3')}$$

22. Given a triangle in the plane $z = 0$, is it possible to find a viewing point, C, from which the triangle will appear on the plane $y = 0$ as a right triangle?

23. Answer Exercise 22 if it is required that the image of the given triangle be isosceles.

24. Answer Exercise 22 if it is required that the image of the given triangle be equilateral.

25. Discuss the perspective transformation when the object plane is not perpendicular to the picture plane.

1.3 Plane Perspective

In the last section we discussed from a mathematical point of view a restricted form of the theory of perspective originated by the artists of the early Renaissance, namely, the perspective mapping of one plane onto another. In this section we shall investigate the possibility of reducing this from a transformation in three dimensions to a transformation in two dimensions by imagining the object plane and the image plane to be the same.

Beginning with the configuration we discussed in the last section, namely, a viewing point, C, from which a given object plane is projected onto a picture plane perpendicular to it, let us rotate the picture plane about its intersection with the object plane until its upper half coincides with the half of the object plane on the opposite side of the picture plane from C.† During this process we ignore the point C, and in fact it plays no part in our later work. A general point, P, in the object plane, σ, now has an image, P', *in that same plane*, carried there by the rotation of the image plane, ρ. Of course, after the two planes coincide, a general point P and its image P' are no longer collinear with C, and P' cannot be determined from P by the steps of projection and section. However, it is still possible to describe entirely in the object plane an equivalent procedure by which the image of any point P can be constructed.

If l is the line of intersection of the object plane and the image plane, it is clear that the points of l coincide with their images both before and after the two planes are rotated into coincidence. In other words, in the transformation in the object plane they are invariant points. Now let π be the plane which passes through C and is perpendicular to both the object plane and the image plane in their unrotated positions, and let r be the line in this plane which passes through C and makes an angle of 45° with the object plane on the opposite side of the picture plane from C. Then the point, O', in which r pierces the image plane rotates into coincidence with its preimage,[1] O, as the image plane rotates (Fig. 1.7). Thus O is also an invariant point in the mapping of the object plane onto itself, and except in the special case in which r intersects l, the point O does not lie on the invariant line l.

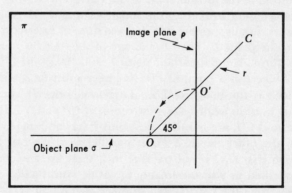

Fig. 1.7   The process of rabattement.

† This process of rotation is known in descriptive geometry and the theory of perspective as **rabattement,** from the French word *rabattre*, meaning *to lower* or *to bring down*.
[1] If a point O' is the image of a point O under a transformation of any sort, the point O is often called the **preimage** of O'.

Assuming now that the invariant point O does not lie on the invariant line l, it is clear that the image of any line in the object plane which passes through O is that same line. In fact, under a perspective transformation, a general line is transformed into a line. Hence the image of a line is uniquely determined by the images of any two of its points; and if two of its points are invariant, the line must coincide with its image. Now, as we have just seen, the point O is invariant. Moreover, with the exception of the unique line through O which is parallel to l, every line through O intersects l in a point which is invariant and, by hypothesis, distinct from O. Hence, with one possible exception, every line which passes through O contains two invariant points and therefore coincides with its image. Finally, the line, p, which passes through O and is parallel to l must also coincide with its image. In fact, the image of p must be some line on the point O, yet every line on O, except possibly p, is its own image. Hence no line except p itself can be the image of p. In other words, p is also invariant, and our argument is complete.

The fact that each line on O is its own image does not, of course, mean that each point on such a line is its own image. In fact, if this were the case, every point in the plane of our discussion would be invariant, which is clearly false. The only conclusion we can presently draw is that the image of a general point on any line through O is some other point *on that same line*. In other words, in the transformation in the object plane, a point and its image are always collinear with O. Or to put it still differently, each line on O is invariant as a whole but is not point-by-point invariant.

If we are given a point, P, in the plane σ, we now know that its image, P', is some point on the line OP. To locate P' on this line, we must be given, in addition to the location of O and l, one point, G, and its image, G'. With this information available, the image of a general point, P, can be found as follows. Let us suppose first that the line PG intersects the invariant line, l, say in the point L. Then the image of the line PG must be the line which passes through the invariant point L and the point G', which is the image of G. Since P is a point of PG, its image must lie somewhere on the line $G'L$, which is the image of PG. Therefore, since P' must also lie on the line OP, it must in fact be the intersection of OP and $G'L$ (Fig. 1.8a). Of course, if OP and $G'L$ are parallel, the point P has no image.

On the other hand, if PG is parallel to l, we can first choose a point, G_1, such that G_1G is not parallel to l, then use the construction we have just described to find the image, G'_1, of G_1, and finally determine the image of P by using the pair (G_1,G'_1) in place of the pair (G,G').

The transformation defined by the preceding construction is known as **plane perspective.** The line of invariant points, l, is known as the **axis** of the transformation, and the invariant point, O, is known as the **center** of the transformation. Clearly, a plane perspective is uniquely

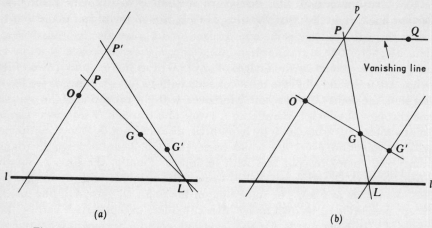

Fig. 1.8   The construction of the image of a point under a plane perspective.

determined when its axis, *l*, its center, *O*, and one additional point, *G*, and its image, *G'*, are given.

The locus of points, *P*, which have no image under a plane perspective is of considerable importance, since in many constructions it plays a more significant role than the axis of the transformation. To identify such points, we recall from the preceding discussion that a point *P* will have no image if and only if the line *G'L* in Fig. 1.8*a* is parallel to the line *OP*. Hence we begin with an arbitrary line, *p*, through *O* on which, if possible, one of the required points *P* is to be located. Then through *G'* we draw a line parallel to *p* and determine the point, *L*, in which it intersects *l*. Finally, the intersection of *LG* and the line *p* is a point *P* which has no image (Fig. 1.8*b*).

The locus of *P* is easy to determine. In fact, it follows from the construction shown in Fig. 1.8*b* that $\triangle PGO \sim \triangle LGG'$, and therefore

$$\frac{(PG)}{(LG)} = \frac{(GO)}{(GG')}$$

Since *O*, *G*, and *G'* are all fixed, it follows that $(GO)/(GG')$, and hence $(PG)/(LG)$, is a constant, independent of which line *p* is being considered. Finally, we know from elementary geometry that if *P* and *L* are points collinear with a fixed point *G* such that $(PG)/(LG)$ is a constant, and if *L* varies along a line *l*, then *P* varies along a line parallel to *l*.

For obvious reasons, the line which is the locus of points which have no images is called the **vanishing line** of a plane perspective, even though it is not the line into which the vanishing line in the picture plane is rotated (see Exercise 3). Our interest in the vanishing line of a plane perspective is based on the properties described in the following pair of theorems.

Theorem 1

Let T be a plane perspective with axis l, center O, and vanishing line v. Then the image of an arbitrary line, p, meeting l in a point, L, and v in a point, V, is the line, p', which contains L and is parallel to OV.

Proof Clearly, since L is an invariant point, the image, p', of the given line, p, must pass through L. Furthermore, if p' intersected OV, say in the point Q, then Q, being a point on the image of p which was collinear with O and V, would be the image of V. This is impossible, however, because V, being on the vanishing line of the given transformation, has no image. Hence, since p can have no point in common with OV, it must be parallel to OV, and our proof is complete.

Corollary 1

In any plane perspective, the image of the family of lines which pass through a point on the vanishing line of the perspective is a family of parallel lines.

Theorem 2

Let T be a plane perspective with center O and vanishing line v, let G' be the image of a particular point G under T, let P be an arbitrary point, and let V be the intersection of v and PG. Then the image of P under T is the point common to OP and the line through G' which is parallel to OV.

Proof Obviously, since the given point, P, is on the line PG, its image, P', must lie on the image of PG, which, by Theorem 1, is the line through G' parallel to OV. Furthermore, the image of P must also be on the line OP. Hence P' is the intersection of OP and the line on G' parallel to OV, as asserted.

Exercises

1. In a plane perspective, if P' is the image of P, is P the image of P'?

2. In a plane perspective, if G and G' are distinct points such that each is the image of the other, prove that if P' is the image of an arbitrary point P, then P is the image of P'.

3. Show how to construct the line in the object plane into which the vanishing line in the picture plane is rotated. What is the preimage of this line in the transformation in the object plane?

4. Under what conditions, if any, can a plane perspective be defined by specifying the axis, l, and the images, G' and H', of two points, G and H, neither of which is on l?

5. Is a plane perspective determined if one is given the center, O, and the images, G' and H', of two noninvariant points, G and H?

6. Show that there is always a plane perspective in which two points, G and H, have specified images, G' and H', unless $GG' \parallel HH'$.

7. If G, G', H, H', J, J' are six points such that the lines GG', HH', JJ' are concurrent, show that there is a unique plane perspective in which G' is the image of G, H' is the image of H, and J' is the image of J.

8. Using the results of Exercise 7, prove the following theorem. *If two triangles are so related that the lines joining corresponding vertices are concurrent, then the intersections of corresponding sides of the two triangles are collinear.* (This important result is known as **Desargues' theorem.**)

9. Let a plane perspective in the xy plane be defined by the axis $l: y = 0$, the center $O:(0,2)$, the point $G:(1,1)$ and its image $G':(3,-1)$. If $P:(a,b)$ is an arbitrary point, what are the coordinates of its image? What is the vanishing line of this transformation?

10. In Exercise 9, carry out a point-by-point construction of the image of the circle $x^2 + y^2 = 1$, the circle $x^2 + (y - 1)^2 = \frac{1}{4}$, the circle $x^2 + (y - 1)^2 = 1$.

11. Work Exercise 9 if the axis is the line $y = x$, the center is the point $(1,0)$, G is the point $(-1,0)$, and G' is the point $(-2,0)$.

12. Work Exercise 9 if the axis is the line $x + y = 0$, the center is the point $(1,1)$, G is the point $(-1,-1)$, and G' is the point $(2,2)$.

13. In a plane perspective in the xy plane, O is the point $(0,1)$, and l is the x axis. If the image of the point $G:(1,2)$ is the point $G':(2,3)$, what are the coordinates of the viewing point in the equivalent three-dimensional perspective?

14. Work Exercise 13 if O is the point $(1,2)$, l is still the x axis, and the image of the point $(0,3)$ is the point $(2,1)$.

15. In a plane perspective, l is the x axis, O is the point $(0,3)$, and G is the point $(0,1)$. Determine the coordinates of the image of G if the perspective is to have the property that if P' is the image of an arbitrary point P, then P is the image of P'.

16. Given two intersecting lines, show how to determine a plane perspective which will transform them into parallel lines.

17. Given the center, C, the vanishing line, v, a point G and its image G' in a plane perspective, show how to construct the axis of the perspective.

18. Prove that a plane perspective is uniquely determined when its center, its vanishing line, and the image, G', of one point, G, are given.

19. Work Exercise 9 if l is the vanishing line of the plane perspective instead of its axis.

20. In Exercise 9, if l is the vanishing line instead of the axis, carry out a point-by-point construction of the image of the circle $x^2 + y^2 = 1$, the circle $x^2 + (y - 1)^2 = \frac{1}{4}$, the circle $x^2 + (y - 1)^2 = 1$.

21. If the vanishing line of a plane perspective is the x axis, if the center of the perspective is the point $(0,-1)$, and if the image of the point $(0,\frac{1}{2})$ is the point $(0,2)$, find the equations of the transformation. What is the axis of the transformation?

22. Work Exercise 21, given that the image of the point $(0,2)$ is the point $(0,\frac{1}{2})$.

23. In Exercise 21, if the image of the point $(0,a)$ is the point $(0,b)$, determine the relation between a and b which will ensure that $(0,a)$ is also the image of $(0,b)$.

24. In Exercise 9, let L be the point in which the line through O containing a general point P and its image P' intersects l. Prove that $(OL)(PP')/(OP)(PL)$ is a constant independent of P. Is this true if L is the point in which the line OP meets the vanishing line?

25. Discuss the plane perspective transformation when the center, O, lies on the axis, l, and show, in particular, that any point and its image are still collinear with O.

1.4 Plane Constructions

In this section we shall investigate how plane perspectivities can be used in certain cases to transform a given configuration into another with special properties. Obviously, since every plane perspective transforms points into points and lines into lines, we do not expect to be able to transform a rectangle into a circle or a triangle into a pentagon, for instance. It may be possible, however, to find a plane perspective which will do such things as transform a given quadrilateral into a square or a given ellipse into a circle.

We begin by showing how to transform an arbitrary quadrilateral into a parallelogram. Let $ABCD$ be an arbitrary quadrilateral, let V_1 be the intersection of the opposite sides AB and CD, and let V_2 be the intersection

of the opposite sides *BC* and *AD* (Fig. 1.9). From Corollary 1, Theorem 1, of
the last section we know that the family of lines passing through any point on
the vanishing line of a plane perspective is transformed into a family of
parallel lines. Hence it is clear that if a perspective is set up in which V_1V_2
is the vanishing line, then regardless of the location of the axis, *l*, or the center,
O, the images of *AB* and *CD* will be parallel and the images of *BC* and *AD*
will be parallel. In fact, once *O* is chosen, the images of *AB* and *CD* are known
to be lines parallel to OV_1 and the images of *BC* and *AD* are known to be
lines parallel to OV_2. Hence when the image, *A'*, of the vertex *A* is assigned,
A'B' and *A'D'* can be drawn. Then *B'* can be located as the intersection of
A'B' and *OB*, and *D'* can be located as the intersection of *A'D'* and *OD*.
With *B'* and *D'* known, *C'* can be found immediately, and the construction
is complete.

If we desire that the quadrilateral *ABCD* be transformed into a
rectangle, rather than just a general parallelogram, the center *O* cannot be
chosen arbitrarily. In fact, since $D'C' \parallel OV_1$ and $D'A' \parallel OV_2$, it follows
(Fig. 1.9) that $\angle C'D'A' \cong \angle V_1OV_2$. Therefore, if *A'B'C'D'* is to be a rectangle, i.e., if $\angle C'D'A'$ is to be a right angle, $\angle V_1OV_2$ must also be a right
angle. Hence, since an angle inscribed in a semicircle is a right angle, the
center *O* must be chosen on the circle which has the segment $\overline{V_1V_2}$ as
diameter (Fig. 1.10).

Finally, if we wish to transform the quadrilateral *ABCD* into a
square and not just a rectangle, the center, *O*, is still further restricted and
cannot be chosen arbitrarily on the circle having $\overline{V_1V_2}$ as diameter. To

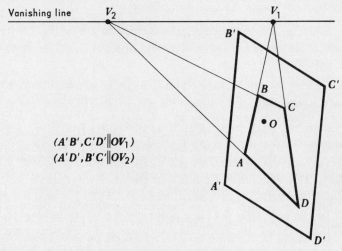

Fig. 1.9   The transformation of a general quadrilateral into a
parallelogram.

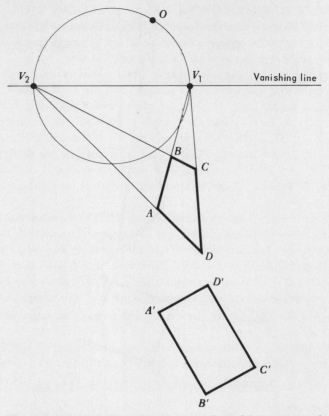

Fig. 1.10   The transformation of a quadrilateral into a rectangle.

determine where O must be located on this circle, we recall that a rectangle is a square if and only if its diagonals are perpendicular. Hence the perspective we require is one which, in addition to making $\angle C'D'A'$ a right angle, will also transform the diagonals AC and BD into lines which are perpendicular. If V_3 is the intersection of AC and the vanishing line, and if V_4 is the intersection of BD and the vanishing line, then $A'C' \parallel OV_3$ and $B'D' \parallel OV_4$. Therefore the diagonals of the image rectangle will be perpendicular if and only if O is chosen so that $\angle V_3OV_4$ is also a right angle. This will be the case only if O is a point on the circle having $\overline{V_3V_4}$ as diameter. Thus to transform $ABCD$ into a square $A'B'C'D'$, the center O of the perspective must be one of the two intersections of the circle having $\overline{V_1V_2}$ as diameter and the circle having $\overline{V_3V_4}$ as diameter (Fig. 1.11).

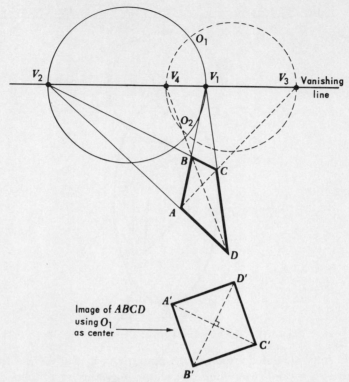

Fig. 1.11   The transformation of a quadrilateral into a square.

Plane perspective also sheds light on the relations among the various conic sections, for it is possible to transform a conic of one type into a conic of another type by a plane perspective just as well as by the equivalent three-dimensional projection and section. In doing this, it is convenient to consider the plane perspective to be defined by its center, O, the image, G', of some particular point, G, and its vanishing line, v, rather than its axis, l. As we have seen (Theorem 2, Sec. 1.3) with these data given, the image, P', of an arbitrary point, P, is found by first determining the intersection, V, of v and PG, and then locating P' as the intersection of OP and the line through G' which is parallel to OV.

Specifically, let us investigate the image of a circle, Γ, under a plane perspective. As a first possibility, let us suppose that the circle does not intersect the vanishing line. Then, since the only points in the plane which do not have images are the points of the vanishing line, it follows that every point of Γ has an image. Necessarily, then, the image of the circle is a conic

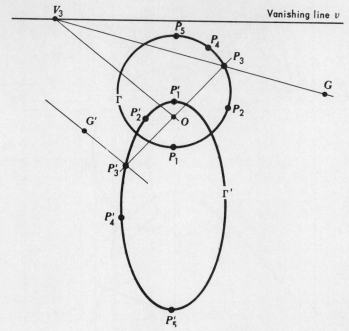

Fig. 1.12 The transformation of a circle into an ellipse.

lying in some bounded region of the plane; i.e., it is an ellipse (Fig. 1.12). Of course in special cases the image ellipse may also be a circle.

Suppose next that the circle intersects the vanishing line in two points, V_1 and V_2. These points on the circle have no images. Therefore, since the tangent to the circle at V_i meets the circle only at V_i, it is transformed into a line, parallel to OV_i, which has no point in common with the image of the circle. Every other line which passes through V_i, except the vanishing line itself, meets the circle in a second point which is not on the vanishing line and hence has an image. These lines are therefore transformed into lines parallel to OV_i each of which meets the image of the circle in a single point. The image of the circle Γ is therefore a hyperbola whose asymptotes are the images of the tangents to Γ at V_1 and V_2 (Fig. 1.13).

Intermediate between the case in which the circle Γ intersects the vanishing line in two points and the case in which it does not intersect the vanishing line is the case in which it is tangent to the vanishing line, say at the point V. Then, with the exception of the vanishing line, which has no image, the lines which pass through V are transformed into lines, parallel to OV, each of which intersects the image of Γ in a single point. The image of the circle in this case is therefore a parabola whose axis is parallel to OV (Fig. 1.14).

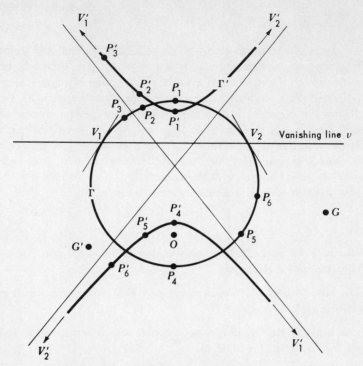

Fig. 1.13   The transformation of a circle into a hyperbola.

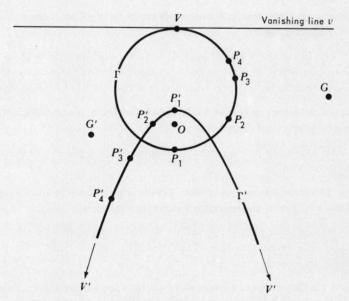

Fig. 1.14   The transformation of a circle into a parabola.

From an intuitive point of view, it appears that the points of the vanishing line are transformed into points which are infinitely far away, i.e., points which lie on a "line at infinity" (though of course there are no such points in the euclidean plane). Therefore, since hyperbolas, parabolas, and ellipses are obtained from a circle by a plane perspective in which the vanishing line meets the circle in two, one, or no (real) points, it seems intuitively plausible to say that a conic is a hyperbola, a parabola, or an ellipse according as it meets the "line at infinity" in two, one, or no (real) points. In the next chapter, and elsewhere in this book, we shall return to this idea and attempt to give it a more precise meaning.

Exercises

1. Show how to find a plane perspective which will transform a given quadrilateral into a rhombus.

2. Show how to find a plane perspective which will transform a given triangle into a right triangle.

3. Show how to find a plane perspective which will transform a given triangle into an isosceles triangle.

4. Can an arbitrary triangle be transformed by a plane perspective into an equilateral triangle?

5. Can an arbitrary triangle be transformed by a plane perspective into an isosceles right triangle?

6. Without using the a priori knowledge that such triangles do not exist in euclidean geometry, explain why it is impossible to find a plane perspective which will transform a given triangle into one which has two right angles.

7. Can an arbitrary segment be transformed by a plane perspective into a segment of prescribed length?

8. Can an arbitrary pentagon be transformed by a plane perspective into a regular pentagon?

9. Is it possible to find a plane perspective which will simultaneously transform two given triangles into isosceles triangles?

10. Given two triangles, is it possible to find a plane perspective which will simultaneously transform one of the triangles into a right triangle and the other into an isosceles triangle?

11. How can a parabola be transformed into an ellipse by a plane perspective? Into a hyperbola? Into another parabola?

12. How can a hyperbola be transformed into an ellipse by a plane perspective? Into a parabola? Into another hyperbola?

13. What special property, if any, does the image of a circle have when the center of the plane perspective is the center of the circle?

14. If a circle Γ is tangent at V to the vanishing line of a plane perspective, what line through V is transformed into the axis of the parabola which is the image of Γ? What point on Γ is the preimage of the vertex of the image parabola?

15. If a circle Γ is transformed by a plane perspective into a hyperbola, which points on Γ are the preimages of the vertices of the image hyperbola?

16. Given a plane perspective whose center is the origin, whose vanishing line is $v: y = 2$, and under which the image of the point $G:(1,1)$ is the point $G':(-1,-1)$. Carry out the point-by-point construction of the image of each of the following conics:

(a) $x^2 + y^2 = 1$ (b) $x^2 + (y-1)^2 = 1$ (c) $y = x^2$

(d) $x = y^2$ (e) $y = 1 - x^2$ (f) $xy = 1$

17. What is the locus of the vertices of right angles which are transformed into right angles by a given plane perspective?

18. If P_1, P_2, P_3, P_4 are four points no three of which are collinear, is it possible to find a plane perspective which will transform these points into four points, P_1', P_2', P_3', P_4', such that P_4' will be the point of concurrence of the altitudes of $\triangle P_1' P_2' P_3'$?

19. Answer Exercise 18 if P_4' is to be the point of concurrence of the angle bisectors of $\triangle P_1' P_2' P_3'$.

20. Answer Exercise 18 if P_4' is to be the point of concurrence of the medians of $\triangle P_1' P_2' P_3'$.

21. Can an arbitrary ellipse be transformed into a circle by a plane perspective?

1.5 Conclusion

In this chapter we have examined the historical origins of projective geometry as they are found in the work of the artists of the early Renaissance. These considerations led us to the important concepts of *projection* and *section* and *perspectivities* in two as well as in three dimensions. One significant feature revealed by our study of perspectivities was that in no case were our transformations completely one to one. More specifically, whether the perspectivity was a transformation mapping a plane into itself or a transformation mapping one plane into another, there was always a line of points in the object plane which had no images and a line of points in the image plane which had no preimages. These exceptional points seemed in some vague way to be associated with families of parallel lines, and perhaps were related to the cliché that "parallel lines meet at infinity."

In the next chapter, we shall consider the origin of projective geometry from another point of view which, though quite different from the one we have adopted in this chapter, will also confront us with the problem of exceptional elements and motivate an extension, or enlargement, of the euclidean plane in which no such exceptional elements exist.

two

THE
EXTENDED
EUCLIDEAN
PLANE

2.1 Introduction

There are several ways in which the study of projective geometry can be begun. One is the historical approach, which is based on the processes of projection and section and involves an early emphasis on constructions of various kinds. This was the theme of our introductory discussion in the last chapter. It is also possible to develop projective geometry as an extension of elementary geometry by systematically adding certain new elements to the familiar euclidean plane and then studying the properties of this enlarged system. Both this process, which might be called the *analytic* approach, and the historical development, which might be called the *synthetic* approach, emphasize the relation of projective geometry to the physical world and to preexistent systems for studying that world. In some respects this emphasis is desirable, or at least not undesirable, but on the other hand it disguises the fact that projective geometry is a branch of mathematics worthy of study in its own right, originally suggested by, but logically independent of, objects and problems in the external world.

The third approach to projective geometry is the *axiomatic* one. In it, projective geometry is developed in a purely deductive way from a set of axioms subject only to the fundamental requirement of consistency. The advantages of this method are the advantages inherent in any axiomatic development: the logical structure of the subject matter is made clear, its results are systematically arranged in a way that shows how one depends on another, and comparisons can be made between related systems having a number of axioms in common.

In this chapter and in the three which follow, we shall study the analytic approach to projective geometry, i.e., we shall attempt an enlargement of the euclidean plane which, when it is accomplished, will in effect be the classical projective plane. Our purpose in this is twofold. In the first place, it will illumine certain aspects of euclidean geometry and should therefore be helpful to those whose primary interest is the teaching of elementary geometry. Second, it will provide us with a specific model of the axiomatic system we shall introduce in Chap. 6, which will be useful in subsequent chapters for purposes of illustration and as a check on the consistency of our postulates.

2.2 Homogeneous Coordinates in the Euclidean Plane

The most characteristic feature of the euclidean plane is probably its parallel property: *through any point not on a given line there passes a unique line which is parallel to the given line*. It was this which Euclid's "defenders" tried for more than 2,000 years to prove from his other postulates, and it was the denial of this in the early years of the nineteenth century which finally revealed the existence of other geometries than Euclid's.

To a person trained only in euclidean geometry and living in a world whose accessible regions, at least, seem to be euclidean, it is difficult to see anything "unnatural" or exceptional in the fact that while most pairs of lines intersect, there are some that do not. Nonetheless, it is interesting to consider the possibility of geometries in which, without exception, two lines will always have a point in common. In particular, it is instructive to ask if the familiar euclidean plane can be enlarged into such a system by adjoining to it certain additional points to serve as the intersections of lines that were previously parallel.

To investigate this question it is convenient to begin with the coordinatized plane of elementary analytic geometry and introduce what are known as **homogeneous coordinates.**

Definition 1

If (x,y) are the rectangular coordinates of an arbitrary point, P, in the euclidean plane, E_2, and if (x_1,x_2,x_3) are any three real numbers such that $x_1/x_3 = x$ and $x_2/x_3 = y$, then the triple (x_1,x_2,x_3) is said to be a set of homogeneous coordinates[1] for P.

To emphasize the distinction between the homogeneous coordinates, (x_1,x_2,x_3), of a point in E_2 and the original rectangular coordinates, (x,y), the latter are often referred to as the **nonhomogeneous coordinates** of the point. While the nonhomogeneous coordinates of a point in E_2 are unrestricted (except, of course, that they must be real), it is clear from Definition 1 that the third homogeneous coordinate, x_3, of a point in E_2 can never be zero.

Clearly, if the homogeneous coordinates (x_1,x_2,x_3) of a point are given, its rectangular coordinates (x,y) are uniquely determined. On the other hand, if the rectangular coordinates of a point are given, the homogeneous coordinates of the point are not uniquely determined. In fact, if (x_1,x_2,x_3) are homogeneous coordinates of a point $P:(x,y)$, then for all real values of k different from zero the numbers (kx_1,kx_2,kx_3) are also homogeneous coordinates of P, since

$$\frac{kx_1}{kx_3} = \frac{x_1}{x_3} = x \qquad \text{and} \qquad \frac{kx_2}{kx_3} = \frac{x_2}{x_3} = y$$

If a polynomial equation in rectangular coordinates is transformed into homogeneous coordinates via the substitutions $x = x_1/x_3$ and $y = x_2/x_3$ and then cleared of fractions, the resulting equation will always be homogeneous; i.e., all terms in the new equation will be of the same total degree in the variables of the set $\{x_1,x_2,x_3\}$. For example, the general equation of a line in rectangular coordinates, namely,

$$l_1 x + l_2 y + l_3 = 0 \qquad l_1, l_2 \text{ not both zero}$$

becomes in homogeneous coordinates

$$l_1 \frac{x_1}{x_3} + l_2 \frac{x_2}{x_3} + l_3 = 0$$

or, clearing of fractions, $l_1 x_1 + l_2 x_2 + l_3 x_3 = 0$; and now each term is linear in the variables of the set $\{x_1,x_2,x_3\}$. Similarly, the general equation of a conic in rectangular coordinates,

(1) $$a_{11}x^2 + 2a_{12}xy + a_{22}y^2 + 2a_{13}x + 2a_{23}y + a_{33} = 0$$

[1] The homogeneous coordinates (x_1,x_2,x_3) of a point in E_2 should not be confused with the rectangular coordinates (x,y,z) of a point in euclidean three-space, E_3.

becomes in homogeneous coordinates

(2) $\quad a_{11}x_1{}^2 + 2a_{12}x_1x_2 + a_{22}x_2{}^2 + 2a_{13}x_1x_3 + 2a_{23}x_2x_3 + a_{33}x_3{}^2 = 0^\dagger$

and the left-hand side is clearly a homogeneous quadratic function of the new variables x_1, x_2, x_3. It is this property which motivated the name *homogeneous coordinates*.

In elementary analytic geometry it is customary to identify points by a pair of coordinates, (x,y), and lines by a single equation,

(3) $\qquad\qquad l_1x + l_2x + l_3 = 0 \qquad l_1, l_2$ not both zero

This asymmetry usually passes unnoticed, and in any event it causes little or no inconvenience, although it can easily be eliminated. In fact, the line defined by Eq. (3) is completely determined by the three coefficients l_1, l_2, l_3 which can thus be regarded as coordinates of the line.

Definition 2

If $l_1x + l_2y + l_3 = 0$ is the equation of a line, p, in rectangular coordinates, then the triple $[l_1,l_2,l_3]$ is said to be a set of homogeneous coordinates for p.

As a notation to distinguish between triples which are homogeneous coordinates of a line and triples which are homogeneous coordinates of a point, we shall consistently write the former in square brackets and the latter in parentheses. Thus $[1,2,3]$ denotes a line, namely, the line whose cartesian equation is $x + 2y + 3 = 0$, while $(1,2,3)$ denotes a point, namely, the point whose cartesian coordinates are $(\frac{1}{3},\frac{2}{3})$. Although a line is uniquely determined by a given set of homogeneous line-coordinates $[l_1,l_2,l_3]$, it is not true that a given line determines a unique set of homogeneous line-coordinates. In fact, since

$$l_1x + l_2y + l_3 = 0 \qquad \text{and} \qquad kl_1x + kl_2y + kl_3 = 0 \qquad k \neq 0$$

are equivalent equations, it is clear that they represent the same line. Hence, *if $[l_1,l_2,l_3]$ are homogeneous coordinates of a line, p, then $[kl_1,kl_2,kl_3]$ are also homogeneous coordinates of p provided $k \neq 0$.*

Conversely, just as a line is usually identified by an equation which is satisfied by the coordinates of every point on that line and by the co-ordinates of no other points, so it is possible to characterize a point by an equation which is satisfied by the coordinates of every line which passes through that point and by the coordinates of no other line. Specifically, if

---

† The **double-subscript** method of identifying the coefficients in the equation of the conic (1) becomes especially convenient when the equation is expressed in homogeneous co-ordinates, for now the two subscripts in any coefficient identify the two variables in the term which that coefficient multiplies.

$[l_1, l_2, l_3]$ is any line[1] which passes through the particular point with co-ordinates (x_0, y_0), then (x_0, y_0) must satisfy the equation of this line; i.e., the coordinates of the line must satisfy the equation

$$l_1 x_0 + l_2 y_0 + l_3 = 0$$

Obvious as these observations are, they are important enough to warrant explicit statement as a theorem.

Theorem 1

If $[l_1, l_2, l_3]$ are the coordinates of a line, p, then the equation of p in rectangular coordinates is $l_1 x + l_2 y + l_3 = 0$; that is, this equation is satisfied by the coordinates of every point of p and by the coordinates of no other points.

If (x, y) are the coordinates of a point, P, then the equation of P in line-coordinates is $x l_1 + y l_2 + l_3 = 0$; that is, this equation is satisfied by the coordinates of every line which passes through P and by the co-ordinates of no other lines.

When P is identified by a triple of homogeneous point-coordinates, (x_1, x_2, x_3), rather than by a pair of rectangular coordinates, (x, y), the results of Theorem 1 are described by the following corollary.

Corollary 1

If $[l_1, l_2, l_3]$ are the coordinates of a line, p, then the equation of p in homogeneous point-coordinates is

$$l_1 x_1 + l_2 x_2 + l_3 x_3 = 0$$

If (x_1, x_2, x_3) are the homogeneous coordinates of a point, P, then the equation of P in line-coordinates is

$$x_1 l_1 + x_2 l_2 + x_3 l_3 = 0$$

We are all familiar with the process of plotting the curve, or locus, defined by an equation in nonhomogeneous point-coordinates, $f(x, y) = 0$. The usual method of doing this, of course, is to assign various convenient values to x, solve for the corresponding values of y from the given equation, and then plot the points determined by the various (x, y) pairs we have thus calculated. If the equation of the locus is given in homogeneous coordinates, the process is essentially the same, for since x_3 can never be zero, and since

---

[1] For convenience, we shall often abbreviate the statement "the line whose coordinates are $[l_1, l_2, l_3]$" and say simply "the line $[l_1, l_2, l_3]$." Similarly, we shall often refer to the coordinate-triple of a point, (x_1, x_2, x_3), as though it were the point itself.

only the ratios x_1/x_3 and x_2/x_3 are significant, we can, for convenience, take x_3 to be 1, in which case x_1 and x_2 become x and y, respectively, and the problem is reduced to the familiar one in nonhomogeneous coordinates.

Now that we have introduced coordinates for lines as well as points, it is clearly possible to consider an equation in line-coordinates, say

$$F(l_1, l_2, l_3) = 0$$

and plot it very much as we plot an equation in point-coordinates. We first find sets of values $[l_1, l_2, l_3]$ which satisfy the given equation, taking into account the fact that since l_3 can have any value, including zero, we must solve for l_1 and l_2 when $l_3 = 0$ as well as when $l_3 \neq 0$, say $l_3 = 1$. Then the lines are plotted, for instance, by using their coordinates to construct their equations and then using their equations to find their intercepts. The result is a family of lines known as an **envelope.** In all but the simplest cases, an envelope will appear to consist of all the lines which are tangent to some curve; and later in our work we shall see that a conic, for example, can be studied either as the locus of its points or as the envelope of its tangents. Figure 2.1 illustrates these ideas for the locus whose equation is $x_1 + x_2 + x_3 = 0$, that is, the line $[1,1,1]$, and the envelope whose equation is $l_1 + l_2 + l_3 = 0$, that is, the point $(1,1,1)$.

Using Corollary 1, Theorem 1, it is possible to find the coordinates of a line, p, which is determined not by an equation but by two of its points, say $A:(a_1, a_2, a_3)$ and $B:(b_1, b_2, b_3)$. For if $[l_1, l_2, l_3]$ are the coordinates of p, then by Corollary 1, we must have

$$a_1 l_1 + a_2 l_2 + a_3 l_3 = 0$$
$$b_1 l_1 + b_2 l_2 + b_3 l_3 = 0$$

and it is easy to verify that the general solution of this system of equations is

$$
\begin{aligned}
l_1 &= k(a_2 b_3 - a_3 b_2) \\
l_2 &= -k(a_1 b_3 - a_3 b_1) \qquad k \text{ arbitrary} \\
l_3 &= k(a_1 b_2 - a_2 b_1)
\end{aligned}
$$

(4)

Since A and B are assumed to be distinct points, their coordinates cannot be proportional. Hence the three quantities in parentheses cannot all be zero. In fact, the first two of them cannot be zero simultaneously because, if this were the case, then multiplying the first by $-a_1$ and the second by $-a_2$ and adding, we would have

$$a_3(a_1 b_2 - a_2 b_1)$$

Since a_3 is known to be different from zero, this implies that $a_1 b_2 - a_2 b_1 = 0$. However, if this is true, then all three of the quantities in parentheses in (4) are equal to zero, which we have just seen to be impossible. Thus if $k \neq 0$,

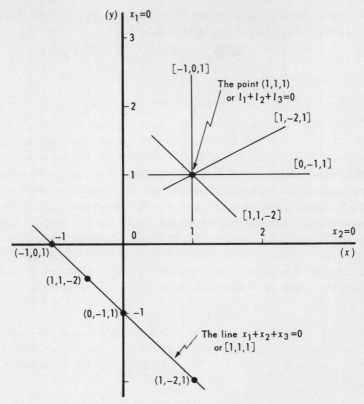

Fig. 2.1   The locus $x_1 + x_2 + x_3 = 0$ and the envelope $l_1 + l_2 + l_3 = 0$.

l_1 and l_2 cannot both be zero, and hence the expressions in (4) are the coordinates of some line. Finally, it is easy to check that this line is indeed the line AB by using Corollary 1, Theorem 1, and Eqs. (4) to show that $A:(a_1,a_2,a_3)$ and $B:(b_1,b_2,b_3)$ both lie on this line.

Since we shall often need to find the coordinates of a line defined by two points, it is interesting and useful to observe that the quantities in parentheses in (4) are precisely the expansions of the second-order determinants obtained from the array of the coordinates of A and B,

$$
\begin{array}{ccc}
a_1 & a_2 & a_3 \\
b_1 & b_2 & b_3
\end{array}
$$

by omitting the first, second, and third columns, in turn.[1]

---

[1] At present, this observation is just an informal memory device. Its full significance will become clear in the next chapter when we introduce the concept of a *matrix*.

Similarly, of course, it follows that if $P:(x_1,x_2,x_3)$ is the intersection of the lines $l:[l_1,l_2,l_3]$ and $m:[m_1,m_2,m_3]$, then

$$
\begin{aligned}
x_1 &= k(l_2m_3 - l_3m_2) \\
x_2 &= -k(l_1m_3 - l_3m_1) \qquad k \text{ arbitrary} \\
x_3 &= k(l_1m_2 - l_2m_1)
\end{aligned}
$$

(5)

Again, these expressions can be constructed from the array of the coordinates of l and m,

$$
\begin{array}{ccc}
l_1 & l_2 & l_3 \\
m_1 & m_2 & m_3
\end{array}
$$

by taking the expansions of the second-order determinants obtained by omitting the first, second, and third columns, in turn.

Summarizing what we have thus far discovered in our discussion of homogeneous coordinates, it appears that the euclidean plane, E_2, consists of:

1. A collection of objects called "points" which are in one-to-one correspondence[1] with sets of triples of the form

$$(kx_1, kx_2, kx_3) \qquad x_1, x_2, x_3, k \text{ real}$$
$$x_3 \neq 0$$
$$k \neq 0$$

2. A collection of objects called "lines" which are in one-to-one correspondence with sets of triples[2] of the form

$$[kl_1, kl_2, kl_3] \qquad l_1, l_2, l_3, k \text{ real}$$
$$l_1, l_2 \text{ not both zero}$$
$$k \neq 0$$

3. A relation, referred to equally well as "lies on," "passes through," "belongs to," or "contains," which exists between a point (x_1,x_2,x_3) and a line $[l_1,l_2,l_3]$ if and only if

$$l_1x_1 + l_2x_2 + l_3x_3 = 0$$

The relation $l_1x_1 + l_2x_2 + l_3x_3 = 0$ we shall henceforth refer to as the **incidence condition** for the point (x_1,x_2,x_3) and the line $[l_1,l_2,l_3]$.

---

[1] A **one-to-one correspondence** between two collections of objects is a correspondence in which without exception every member of either collection is associated with one and only one member of the other collection.

[2] In more sophisticated language, we can say that the points of E_2 are in one-to-one correspondence with equivalence classes of triples (x_1,x_2,x_3) in which the third element is different from zero and the lines of E_2 are in one-to-one correspondence with equivalence classes of triples $[l_1,l_2,l_3]$ in which the first two elements are not both zero, the equivalence relation in each case being proportionality (see Exercises 22 to 25).

Exercises

1. Give three sets of homogeneous coordinates for each of the points determined by the following nonhomogeneous coordinates:

(a) $(2,3)$ (b) $(-1,4)$ (c) $(-3,-5)$ (d) $(0,2)$

(e) $(\frac{2}{3},\frac{4}{3})$ (f) $(-\frac{1}{2},\frac{3}{2})$ (g) $(\frac{1}{2},-\frac{2}{3})$ (h) $(-3,0)$

2. What are the nonhomogeneous coordinates of the points determined by each of the following sets of homogeneous coordinates?

(a) $(3,2,1)$ (b) $(1,2,3)$ (c) $(0,2,-1)$ (d) $(\frac{1}{2},-\frac{3}{2},\frac{7}{2})$

(e) $(-2,0,3)$ (f) $(\frac{1}{2},\frac{1}{3},\frac{5}{6})$ (g) $(6,4,-2)$ (h) $(0,0,\frac{1}{2})$

3. Write each of the following equations in terms of homogeneous coordinates:

(a) $y = x^2$ (b) $xy = 2$ (c) $x^2 + y^2 = r^2$

(d) $2xy + 3x + y = 1$ (e) $y^2 = x^3$ (f) $y = x^3 - 2x^2$

(g) $x^2 - 2xy + 3y = 2$ (h) $xy^2 = x + y^4$

4. Write each of the following equations in terms of nonhomogeneous coordinates:

(a) $2x_1 x_3 = x_2^2$ (b) $x_1^2 + 2x_1 x_2 = x_3^2$

(c) $x_1 x_2 + x_2 x_3 + x_1 x_3 = 0$ (d) $x_1 x_2^2 = x_3^3$

(e) $x_1^3 + x_2^3 + x_3^3 = x_1 x_2 x_3$ (f) $x_1^2 x_2^3 = x_3^5$

5. What are the coordinates of the following lines?

(a) $x - 2y = 3$ (b) $3x + 2y - 1 = 0$ (c) $y = 2x$

(d) $x = 3$ (e) $y = 0$

6. Find an equation for each of the following points:

(a) $(2,-1)$ (b) $(2,3)$ (c) $(2,0)$

(d) $(1,1,2)$ (e) $(0,2,3)$

7. Plot each of the following lines:

(a) $[1,2,3]$ (b) $[1,-2,-2]$ (c) $[1,0,2]$

(d) $[0,1,0]$ (e) $[3,1,0]$

8. Plot at least three lines of each of the following envelopes:

(a) $l_1 - l_2 + l_3 = 0$ (b) $2l_1 + l_2 = 0$ (c) $l_1 + 2l_3 = 0$

(d) $l_2 = 0$ (e) $l_3 = 0$ (f) $3l_1 + 4l_2 + 2l_3 = 0$

9. Plot at least six lines of each of the following envelopes:

(a) $l_1 l_2 = l_3{}^2$ (b) $l_1 l_3 = l_2{}^2$ (c) $l_1 l_2 + l_2 l_3 = 0$

(d) $l_1 l_2{}^2 = l_3{}^3$ (e) $l_1{}^2 + l_2{}^2 = l_3{}^2$ (f) $l_1{}^2 l_3 = l_2{}^3$

10. Find a set of coordinates and an equation for the line determined by each of the following pairs of points:

(a) $(1,1,-1)$, $(2,-1,3)$

(b) $(0,1,2)$, $(2,0,1)$

(c) $(2,1,-2)$, $(4,3,2)$

(d) $(0,0,1)$, $(1,1,2)$

(e) $l_1 - l_2 = 0$, $l_2 + l_3 = 0$

(f) $l_1 + l_2 - 2l_3 = 0$, $2l_1 - l_2 - 3l_3 = 0$

11. Find a set of coordinates and an equation for the point of intersection of each of the following pairs of lines:

(a) $x_1 - x_2 + 2x_3 = 0$, $2x_1 + 3x_2 - 2x_3 = 0$

(b) $x_1 - 2x_2 - x_3 = 0$, $2x_1 + 4x_2 + 3x_3 = 0$

(c) $x_1 - x_2 = 0$, $2x_1 + x_2 = 0$

(d) $x_1 + 4x_2 = 0$, $x_3 = 0$

(e) $[1,2,1]$, $[0,2,3]$

(f) $[1,1,3]$, $[2,0,1]$

12. Express in homogeneous coordinates the formula for the distance between the points $P:(x,y)$ and $P':(x',y')$.

13. Express in homogeneous coordinates the formula for the slope of the line determined by the points $P:(x,y)$ and $P':(x',y')$.

14. Find the points, if any, which are common to the loci in each of the following pairs:

(a) $3x_1 - x_2 = 0$, $x_1{}^2 - 2x_1 x_3 - x_2 x_3 = 0$

(b) $x_1 - 2x_2 - x_3 = 0$, $x_1{}^2 + x_2{}^2 - x_3{}^2 = 0$

(c) $2x_1 - x_2 - 3x_3 = 0$, $x_1 x_2 = x_3{}^2$

(d) $x_2{}^2 - 4x_1 x_3 = 0$, $x_1{}^2 + x_1 x_2 - 8x_2 x_3 = 0$

(e) $x_1 + 2x_2 + 3x_3 = 0$, $x_1 - 2x_2 + 3x_3 = 0$

(f) $x_1 + 2x_2 + 3x_3 = 0$, $x_1 + 2x_2 - 3x_3 = 0$

15. Find the lines, if any, which are common to the envelopes in each of the following pairs:

(a) $2l_1 + l_2 - l_3 = 0$, $l_1 - 2l_2 + 2l_3 = 0$
(b) $2l_1 - l_2 + l_3 = 0$, $l_1 l_2 = l_3{}^2$
(c) $l_1 - l_2 = 0$, $l_1 l_3{}^2 = l_2{}^3$
(d) $l_2 l_3 = l_1{}^2$, $l_2 l_3 = 2l_1 l_3 - l_1{}^2$

16. Show that the coordinates of the point $(1,2,1)$ satisfy the equation $x_1{}^2 - x_2 x_3 - 2x_1 + 3x_3 = 0$ but that in general the coordinates $(k,2k,k)$ do not satisfy this equation. Explain.

17. Prove that the substitutions $x = x_1/x_3$ and $y = x_2/x_3$ will convert any polynomial equation, $f(x,y) = 0$, into one which is homogeneous.

18. Prove that the expressions in parentheses in Eqs. (4) are all zero if and only if A and B are the same point.

19. Find the equations satisfied by the coordinates of the tangents to each of the following curves:

(a) $x^2 + y^2 = 1$ (b) $y = x^2$ (c) $xy = 2$
(d) $y = x^3$ (e) $xy - y = 2$ (f) $y = x^2 - 3x + 2$
(g) $x^2 - y^2 = 1$ (h) $y^2 = x^3$

20. Show that for all values of λ and μ, except $\lambda = \mu = 0$, the line

$$[\lambda l_1 + \mu m_1, \ \lambda l_2 + \mu m_2, \ \lambda l_3 + \mu m_3]$$

passes through the intersection of the line $[l_1,l_2,l_3]$ and the line $[m_1,m_2,m_3]$.

21. Show that for all values of λ and μ, except $\lambda = \mu = 0$, the point

$$(\lambda x_1 + \mu y_1, \ \lambda x_2 + \mu y_2, \ \lambda x_3 + \mu y_3)$$

lies on the line determined by point (x_1,x_2,x_3) and point (y_1,y_2,y_3).

22. An **equivalence relation,** R, among the elements of a set, S, is a **reflexive, symmetric,** and **transitive** relation which may or may not exist between two arbitrary elements of S. If we use the notation a R b to denote that the given relation exists between the elements a and b of S, this means that R has the following properties:

1. a R a for any element a in S (reflexive).
2. For any elements a and b of S, if a R b, then b R a (symmetric).
3. For any elements a, b, and c of S, if a R b and b R c, then a R c (transitive).

If a, b, c, $\ldots$ are ordered triples, (a_1,a_2,a_3), (b_1,b_2,b_3), (c_1,c_2,c_3), $\ldots$ and if a R b means that a and b are proportional, i.e., that there exists a nonzero value, k, such that $b_1 = ka_1$, $b_2 = ka_2$, $b_3 = ka_3$, show that R is an equivalence relation in the set of triples (x_1,x_2,x_3) for which $x_3 \neq 0$ and also in the set of triples $[l_1,l_2,l_3]$ for which l_1 and l_2 are not both zero.

23. In the set of all lines in E_2, is "is parallel to" an equivalence relation? Is "is not parallel to" an equivalence relation? Is "is in the same direction as" an equivalence relation?

24. (*a*) In the set of all circles in E_2, is "is concentric with" an equivalence relation?

(*b*) In the set of all triangles in E_2, is "is similar to" an equivalence relation?

25. If an equivalence relation, R, exists in a set, S, the subset of S which consists of all the elements of S which are equivalent to a given element is called an **equivalence class.** Show that an equivalence class in S is determined equally well by any one of its elements. Show also that no two of the equivalence classes into which a set S is partitioned by a relation R can have an element in common.

2.3 Ideal Points and the Ideal Line

As a guide to our proposed enlargement of the euclidean plane, E_2, into a new system in which, without exception, every two lines will have a point in common, let us now, innocently, attempt to find the intersection of two parallel lines, say

$$l_1 x + l_2 y + l_3 = 0$$
$$l_1 x + l_2 y + l_3' = 0 \qquad l_3 \neq l_3'$$

l_1, l_2 not both zero. Naturally, we should recognize at once that these equations are inconsistent and have no solution. If we do not, then of course, we may be guilty of the absurdity of proceeding formally, writing

$$x = \frac{\begin{vmatrix} -l_3 & l_2 \\ -l_3' & l_2 \end{vmatrix}}{\begin{vmatrix} l_1 & l_2 \\ l_1 & l_2 \end{vmatrix}} = \infty \; (!) \qquad y = \frac{\begin{vmatrix} l_1 & -l_3 \\ l_1 & -l_3' \end{vmatrix}}{\begin{vmatrix} l_1 & l_2 \\ l_1 & l_2 \end{vmatrix}} = \infty \; (!)$$

and finally asserting that "parallel lines meet at infinity." In any event, this approach using nonhomogeneous coordinates tells us only what we knew by definition, namely, that parallel lines have no point in common.

If we reconsider the problem using homogeneous coordinates, our work is more illuminating. The equations of the lines now take the form

$$l_1 x_1 + l_2 x_2 + l_3 x_3 = 0$$
$$l_1 x_1 + l_2 x_2 + l_3' x_3 = 0 \qquad l_3 \neq l_3'$$

l_1, l_2 not both zero and these we can solve without difficulty. In fact, subtracting the two equations, we have

$$(l_3 - l_3')x_3 = 0$$

and hence, since $l_3 \neq l_3'$,

$$x_3 = 0$$

From this it follows immediately that $l_1x_1 + l_2x_2 = 0$, which implies that

$$x_1 = kl_2 \qquad x_2 = -kl_1 \qquad k \text{ arbitrary}$$

We must be careful not to delude ourselves here. Although we have been able to solve the homogeneous version of the equations of two parallel lines, we have *not* found a point common to the two lines because, as we saw above, even with the restriction that $k \neq 0$, the solution triple $(kl_2, -kl_1, 0)$ corresponds to no point in E_2, since its third component is equal to zero. However, *if* the euclidean plane were enlarged by adding to it additional "points" in one-to-one correspondence with sets of triples of the form

$$(kl_2, -kl_1, 0) \qquad \begin{matrix} k \neq 0 \\ l_1, l_2 \text{ not both zero} \end{matrix}$$

then lines parallel in E_2 would have a point of intersection in the extended system. Moreover, since the numbers $(kl_2, -kl_1, 0)$ satisfy the equation $l_1x_1 + l_2x_2 + l_3x_3 = 0$ for all values of l_3, it is clear that the new point which corresponds to the set of triples of the form $(kl_2, -kl_1, 0)$ lies on every line of the parallel family

$$l_1x_1 + l_2x_2 + l_3x_3 = 0 \qquad \begin{matrix} l_1, l_2 \text{ constant} \\ l_3 \text{ variable} \end{matrix}$$

and on no other lines of E_2. Hence the new points are also in one-to-one correspondence with the families of parallel lines or, equivalently, with the directions in E_2.

Finally we note that the coordinates of each of the points we have added to E_2 satisfy the equation

$$x_3 = 0$$

whose coefficients define the triple

$$[l_1, l_2, l_3] = [0, 0, 1]$$

Since its first two components are zero, this triple corresponds to no line in E_2. However, to complete our extension of E_2 we shall postulate the existence of a single new line, $x_3 = 0$, corresponding to the set of triples of the form $[0, 0, k]$, $k \neq 0$, and constituting the locus of the new points adjoined to E_2.

For obvious reasons, the points which we have added to E_2 are often called "points at infinity" and their locus, the line corresponding to the triple $[0,0,1]$ or the line whose equation is $x_3 = 0$, is called the "line at infinity." It is preferable, however, to refer to these as **ideal points** and the **ideal line,** respectively. The euclidean plane as we have now extended it by the addition of the ideal points and the ideal line thus consists of:

1. A set of objects called "points" which are in one-to-one correspondence with sets of triples of the form

$$(kx_1, kx_2, kx_3) \qquad \begin{array}{l} k \neq 0 \\ x_1, x_2, x_3 \text{ real and not all zero} \end{array}$$

2. A set of objects called "lines" which are in one-to-one correspondence with sets of triples of the form

$$[kl_1, kl_2, kl_3] \qquad \begin{array}{l} k \neq 0 \\ l_1, l_2, l_3 \text{ real and not all zero} \end{array}$$

3. A relation referred to equally well as "lies on," "passes through," "belongs to," or "contains," which exists between a point (x_1, x_2, x_3) and a line $[l_1, l_2, l_3]$ if and only if $l_1 x_1 + l_2 x_2 + l_3 x_3 = 0$.

This new system we shall refer to as the **extended euclidean plane,** E_2^+. If the coordinates (x_1, x_2, x_3) and $[l_1, l_2, l_3]$ and the proportionality constant k are allowed to take on complex as well as real values, we shall at present call the resulting system the **algebraic representation,** Π_2, although we shall see later that it is actually the classical projective plane.

In the extended euclidean plane the lines $x_1 = 0$, $x_2 = 0$, and $x_3 = 0$ are called the **axes** of the homogeneous coordinate system; these three lines, collectively, are said to form the **triangle of reference;** and the points $(1,0,0)$, $(0,1,0)$, and $(0,0,1)$ in which these lines intersect are called the **vertices** of the triangle of reference. The structure of the extended plane and its triangle of reference is suggested in Fig. 2.2.

From the description of the extended euclidean plane which we have just given, it should be clear that in this new system points and lines have completely comparable properties: points and lines correspond to precisely the same sets of triples. Not only do two points always determine a line, but, equally well, two lines *always* determine a point. The incidence condition for a point and a line is symmetric in the coordinates of the point and the coordinates of the line. In fact, algebraically speaking, the only way points and lines can be distinguished is by noting whether the corresponding coordinate triples appear in parentheses or in brackets. Thus from any theorem proved by algebraic manipulations with triples in parentheses and triples in brackets, another theorem can be obtained simply by interchanging parentheses and brackets, that is by interchanging the words "point" and "line," in the statement of the theorem and in the various steps in the proof. These observations are very important, and later on, when

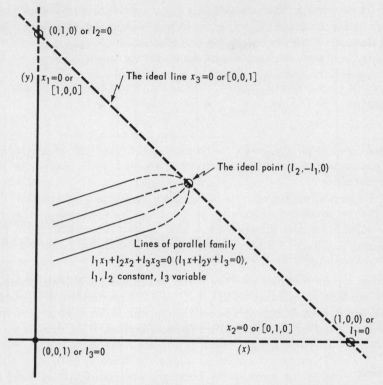

Fig. 2.2 The extended euclidean plane and its triangle of reference.

we begin our axiomatic development of projective geometry, we shall formalize them in the so-called *principle of duality*. To emphasize this principle, we shall not only refer to the line determined by two points P and Q as the line PQ but we shall also refer to the point determined by two lines p and q as the point pq. We shall also refer to the set of all points on the line PQ as the **range** of points on PQ, or simply as the range PQ, and we shall refer to the set of all lines on, or through, the point pq as the **pencil** of lines on pq, or simply as the pencil pq.

If the formula for the distance between two points $P:(x,y)$ and $P':(x',y')$ in the euclidean plane, namely,

$$d^2 = (x - x')^2 + (y - y')^2$$

is expressed in terms of homogeneous coordinates by the substitutions $x = x_1/x_3, y = x_2/x_3$ and $x' = x_1'/x_3', y' = x_2'/x_3'$, we obtain

$$d^2 = \frac{(x_1 x_3' - x_1' x_3)^2 + (x_2 x_3' - x_2' x_3)^2}{(x_3 x_3')^2}$$

Since this becomes infinite if either P or P' is an ideal point and indeterminate if both P and P' are ideal points, it is clear that in the euclidean sense, at least, distance is not defined in the extended plane.

Similarly, the euclidean formula for the angle between two lines, $l_1x + l_2y + l_3 = 0$ and $m_1x + m_2y + m_3 = 0$, namely,

$$\tan \theta = \frac{l_1m_2 - l_2m_1}{l_1m_1 + l_2m_2}$$

becomes indeterminate if either of the lines is the ideal line. Hence it is clear that the euclidean formula for measuring angles is also meaningless in the extended plane.

In passing from the euclidean plane to the extended plane we have thus lost the possibility of studying properties such as similarity, congruence, perpendicularity, and betweenness, which depend on measurement, just as we found in Chap. 1 that perspective transformations preserved neither distances nor angles. At first glance this may seem like a great price to pay for a little extra generality and the elimination of a few exceptional cases. It turns out, however, that there are many interesting and important relations among geometrical objects which do not involve any metrical considerations, and it is precisely these which are studied in projective geometry.

Exercises

1. What is the ideal point on the line defined by each of the following euclidean equations?

(a) $2x - y + 3 = 0$ (b) $3x - 4y + 5 = 0$ (c) $x + 2y + c = 0$
(d) $x = 5$ (e) $2y = 7$ (f) $x = 3y$

2. Find an equation for the line which passes through the finite point $(1,2)$ and each of the following ideal points:

(a) $(1,1,0)$ (b) $(2,3,0)$ (c) $(-3,1,0)$
(d) $(0,2,0)$ (e) $(-1,0,0)$

3. In the extended euclidean plane, find the intersection of the lines defined by each of the following pairs of euclidean equations:

(a) $x - 2y = 3, 2x - 4y = 3$ (b) $2x + 5y = 4, 2x + 5y = 7$
(c) $x - y = 1, x - y = 2$

4. Find an equation for each of the following points:

(a) $(1,2,0)$ (b) $(2,-3,0)$ (c) $(5,0,0)$ (d) $(0,-2,0)$

5. What are the ideal points on the conics whose euclidean equations are

(a) $x^2 + y^2 = 1$ (b) $x^2 - y^2 = 1$

(c) $x^2 - 3xy + 2y^2 - x + 3y = 2$ (d) $y = x^2$

(e) $4x^2 + 4xy + y^2 - 2y = 1$ (f) $x^2 + 2xy + 2y^2 - x - y = 0$

6. Prove that a conic in the euclidean plane is a hyperbola, a parabola, or an ellipse according as its representation in E_2^+ intersects the ideal line in two points, one point, or no (real) points.

7. (a) Is it possible to find a line which passes through a given ideal point and is perpendicular to a given line?

(b) Is it possible for a line to pass through two different ideal points?

8. Find the points common to the loci in each of the following pairs and identify each intersection as a point of E_2, E_2^+, or Π_2.

(a) $x_1 - 2x_3 = 0$, $x_2 x_3 - x_1^2 = 0$

(b) $x_1 - x_2 + ax_3 = 0$, $x_1^2 - x_2^2 - x_3^2 = 0$

(c) $2x_1^2 - x_2 x_3 = 0$, $x_1^2 - x_1 x_3 - x_2 x_3 = 0$

(d) $x_1 x_2 = x_3^2$, $2x_1 x_2 - 5x_1 x_3 - x_2 x_3 + 2x_3^2 = 0$

9. Find the lines common to the envelopes in each of the following pairs and identify each intersection as a line of E_2, E_2^+, or Π_2.

(a) $l_1^2 + l_2^2 - l_3^2 = 0$, $l_1 + l_2 - l_3 = 0$

(b) $l_1^2 + l_2^2 - l_3^2 = 0$, $2l_1 l_2 - l_3^2 = 0$

(c) $l_1 l_2 - l_3^2 = 0$, $l_2 l_3 - l_1^2 = 0$

(d) $l_1^2 + l_1 l_3 + 2l_2^2 = 0$, $l_1 l_3 = 2l_2^2$

10. Show that any line parallel to the axis of the parabola $y = ax^2 + bx + c$ meets the curve in one finite point and one ideal point.

11. Show that any line parallel to an asymptote of the hyperbola $x^2/a^2 - y^2/b^2 = 1$ meets the curve in one finite point and one ideal point. What are the intersections of this hyperbola and each of its asymptotes?

12. (a) Show that in Π_2 the circle defined by the euclidean equation $x^2 + y^2 + ax + by + c = 0$ contains the points $I:(1,i,0)$ and $J:(1,-i,0)$.

(b) Show that the euclidean specialization of any real conic which contains the points $I:(1,i,0)$ and $J:(1,-i,0)$ is a circle. (Because of these properties, I and J are often referred to as the **circular points at infinity**.)

13. (a) Prove that in E_2^+ the point $A:(a_1,a_2,a_3)$ and the point $B:(b_1,b_2,b_3)$ are the same point if and only if every second-order determinant in the array

$\begin{matrix} a_1 & a_2 & a_3 \\ b_1 & b_2 & b_3 \end{matrix}$ is equal to zero.

(*b*) Is the corresponding statement true for the line $l:[l_1,l_2,l_3]$ and the line $m:[m_1,m_2,m_3]$?

14. Verify that proportionality is an equivalence relation in the set of all triples of homogeneous point-coordinates and also in the set of all triples of homogeneous line-coordinates. Is this true if the triples $(0,0,0)$ and $[0,0,0]$ are included in the respective sets?

15. Discuss the problem of introducing homogeneous point-coordinates on the euclidean line, E_1. Does this process lead to the concept of an extended euclidean line, E_1^+? How?

2.4 Euclidean Specializations of the Extended Plane

In the last section we observed that the euclidean plane, E_2, is converted into the extended euclidean plane, E_2^+, by adjoining to it one new line together with the set of all points on that line. Conversely, the original euclidean plane can be recovered from the extended plane simply by removing the ideal line and its points.

Actually, the ideas of "adding to" and "removing from" are somewhat misleading, and a better appreciation of the relation between E_2 and E_2^+ is perhaps provided by the following analogy. When Columbus discovered America, it could hardly have been said that he "added" America to the Old World. What he accomplished was rather the extension of man's knowledge to a region of whose existence men had previously been unaware. Conversely, people who for one reason or another chose to ignore the discoveries of Columbus did not thereby "remove" America from the new and larger world which Columbus had revealed. Instead, they simply acted as though America did not exist and continued to live exclusively in the geographical system they had always known.

In a sense, the work of the last section added nothing to the euclidean plane but rather added to our total store of knowledge by revealing to us that there is a more extensive plane in which the familiar euclidean plane is embedded. And having recognized this, if we continue to be concerned only with euclidean geometry (as we have every right to be, if we wish), we are not thereby removing anything from the extended plane; we are simply restricting our attention to that part of the extended plane which interests us and acting as though the rest did not exist.

If, given the extended plane with its homogeneous coordinates, (x_1,x_2,x_3), we wish to return to the euclidean plane from which we derived it, we need only assign to x_3 a value different from 0, say $x_3 = 1$. Since every ideal point is characterized by the condition that $x_3 = 0$, the assumption that $x_3 = 1$ prevents the consideration of the ideal line and its points and restricts us to just the elements comprising the euclidean plane with which we started.

These observations raise the following question: Is the ideal line essentially different from the other lines in the extended plane, or can an arbitrary line in the extended plane be considered the ideal line of some euclidean plane contained in the extended plane? At present we cannot answer this question completely, but a partial answer is provided by the following considerations.

Suppose we attempt to discover in the extended plane a euclidean plane for which $x_2 = 0$ is the ideal line. To do this we must reduce the extended plane by removing from it (or refusing to consider in it) the line $x_2 = 0$ and all its points. Clearly this can be done as easily (and by the same device) as the points on the line $x_3 = 0$ were eliminated in our earlier discussion: all we need do is put $x_2 = 1$. Then interpreting

$$\frac{x_1}{x_2} = x_1 \quad \text{and} \quad \frac{x_3}{x_2} = x_3$$

as the coordinates, say x' and y', of a new rectangular coordinate system, we shall have isolated from the extended plane a euclidean plane for which $x_2 = 0$ is indeed the ideal line. Since the points of the line $x_3 = 0$ (with the exception of its intersection with $x_2 = 0$) remain in the new euclidean plane, it follows that lines which were parallel in the original euclidean plane, i.e., intersected in the extended plane in points on $x_3 = 0$ (except those which passed through the intersection of $x_3 = 0$ and $x_2 = 0$) will be intersecting lines in the new euclidean plane. On the other hand, lines in the original euclidean plane which intersected in points on the x axis, i.e., points on the line $y = 0$ or $x_2 = 0$, will be parallel in the new euclidean plane since their points of intersection will not belong to the new euclidean plane.

Similarly, the euclidean plane for which $x_1 = 0$ is the ideal line can be obtained simply by setting $x_1 = 1$ and then interpreting $x_2/x_1 = x_2$ and $x_3/x_1 = x_3$ as the coordinates, say x'' and y'', of still another rectangular coordinate system. Whether there is a euclidean plane thus corresponding to every line in the extended plane and, if so, how it can be obtained, we cannot tell until we have studied more about linear transformations in the extended plane.

Example 1

If the equations of the parallel lines

$$y = 2x + 3 \quad \text{and} \quad y = 2x + 4$$

are expressed in terms of homogeneous coordinates by the substitutions $x = x_1/x_3$ and $y = x_2/x_3$, they become, respectively,

$$2x_1 - x_2 + 3x_3 = 0 \quad \text{and} \quad 2x_1 - x_2 + 4x_3 = 0$$

If we now wish to study these lines in the new euclidean plane for which $x_2 = 0$, rather than $x_3 = 0$, is the ideal line, we must eliminate from consideration all points for which $x_2 = 0$. This can be done by simply setting $x_2 = 1$ and identifying x_1 and x_3 as the x' and y' coordinates of a new euclidean plane. It can also be done in a slightly more general fashion by dividing each of the last two equations by x_2 (which can be done if and only if $x_2 \neq 0$) and then setting $x_1/x_2 = x'$ and $x_3/x_2 = y'$. This yields the equations

$$\frac{2x_1}{x_2} - 1 + \frac{3x_3}{x_2} = 0 \quad \text{and} \quad \frac{2x_1}{x_2} - 1 + \frac{4x_3}{x_2} = 0$$

or $\qquad 2x' + 3y' - 1 = 0 \quad \text{and} \quad 2x' + 4y' - 1 = 0$

In the new euclidean plane these equations represent lines which are no longer parallel but instead intersect in the point whose new nonhomogeneous coordinates are $(\frac{1}{2},0)$. This is the case, of course, because the ideal point $(1,2,0)$, or $(\frac{1}{2},1,0)$, common to these two lines in the extended plane becomes an ordinary point in the new euclidean plane obtained by assuming $x_2 \neq 0$.

On the other hand, when the equations of the intersecting lines $y = x - 1$ and $y = 2x - 2$ in the original euclidean plane are expressed in terms of homogeneous coordinates, they become, respectively,

$$x_1 - x_2 - x_3 = 0 \quad \text{and} \quad 2x_1 - x_2 - 2x_3 = 0$$

In the new euclidean plane for which $x_2 = 0$ is the ideal line and in which $x_1/x_2 = x'$ and $x_3/x_2 = y'$, these equations become, respectively,

$$y' = x' - 1 \quad \text{and} \quad y' = x' - \tfrac{1}{2}$$

which now define parallel lines. This could have been foreseen, since in the original euclidean plane the two lines intersected in the point with nonhomogeneous coordinates $(1,0)$, which, being on the line $x_2 = 0$, does not belong to the new euclidean plane obtained by assuming $x_2 \neq 0$.

Example 2

If the equation of the parabola $y = x^2$ is expressed in terms of homogeneous coordinates, it becomes

(1) $$x_2 x_3 = x_1{}^2$$

To study this curve in the euclidean plane for which $x_1 = 0$ is the ideal line and for which the intersection of $x_2 = 0$ and $x_3 = 0$ is the origin, we must reject from consideration all points of the extended plane for which $x_1 = 0$; that is, we must assume $x_1 \neq 0$. We can then divide Eq. (1) by x_1^2 and identify x_2/x_1 and x_3/x_1 as the cartesian coordinates, say x' and y', of a new euclidean plane, getting, in turn, the equations

$$\frac{x_2}{x_1} \frac{x_3}{x_1} = 1 \qquad \text{and} \qquad x'y' = 1$$

In the new euclidean plane, the curve that was originally a parabola now appears as a hyperbola! Evidently, for a conic in the extended plane, being a parabola or a hyperbola is not an intrinsic property but depends on the particular euclidean specialization in which the conic is considered. Apparently, in the extended plane, as in life in general, things may look quite different from different points of view.

Similarly, to study the given curve in the euclidean plane for which $x_2 = 0$ is the ideal line and for which the intersection of $x_1 = 0$ and $x_3 = 0$ is the origin, we must assume $x_2 \neq 0$, divide Eq. (1) by x_2^2, and then interpret x_1/x_2 and x_3/x_2 as new cartesian coordinates, say x'' and y''. This gives us, in turn, the equations

$$\frac{x_3}{x_2} = \left(\frac{x_1}{x_2}\right)^2 \qquad \text{and} \qquad y'' = (x'')^2$$

which again describe a parabola. Later we shall see that there are still other euclidean specializations of the extended plane in which this curve appears as an ellipse.

Figure 2.3 suggests how these three euclidean representations of the curve whose equation in the extended plane is $x_2 x_3 = x_1^2$ are related to each other.

Exercises

1. What is the difference, if any, between the curve $y = x^2$ in the euclidean plane and the corresponding curve $x_2 x_3 = x_1^2$ in the extended plane?

2. Discuss the relation between the lines in each of the following pairs in the euclidean planes for which $x_1 = 0$, $x_2 = 0$, and $x_3 = 0$ are, in turn, the ideal line:

(a) $x_1 - x_2 + x_3 = 0$, $3x_1 + x_2 - x_3 = 0$
(b) $x_1 - x_2 - x_3 = 0$, $2x_1 + x_2 - 5x_3 = 0$
(c) $x_1 - x_3 = 0$, $x_1 - 2x_3 = 0$

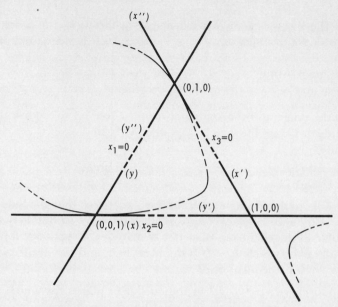

Fig. 2.3   The conic $x_2x_3 = x_1{}^2$ in the extended plane.

3. Discuss the nature of each of the following conics in the euclidean planes for which $x_1 = 0$, $x_2 = 0$, and $x_3 = 0$ are, in turn, the ideal lines:

(a) $x_1{}^2 - 2x_1x_3 + x_2{}^2 = 0$
(b) $x_2x_3 = x_1{}^2 - x_3{}^2$
(c) $x_1{}^2 - 2x_1x_3 + x_2{}^2 - 2x_2x_3 + x_3{}^2 = 0$
(d) $x_1{}^2 - x_2{}^2 - x_3{}^2 = 0$

4. Discuss the relation between the line and the conic in each of the following pairs in the euclidean planes for which $x_1 = 0$, $x_2 = 0$, and $x_3 = 0$ are, in turn, the ideal line:

(a) $2x_1 = x_2$, $x_2x_3 = 2x_1x_3 - x_1{}^2$
(b) $x_1 = x_3$, $x_1{}^2 - x_2{}^2 - x_3{}^2 = 0$
(c) $x_1 - x_2 - x_3 = 0$, $x_1{}^2 + x_2x_3 - x_3{}^2 = 0$

5. Discuss the relation between the two conics in each of the following pairs in the euclidean planes for which $x_1 = 0$, $x_2 = 0$, and $x_3 = 0$ are, in turn, the ideal line:

(a) $x_1{}^2 + x_2{}^2 - x_3{}^2 = 0$, $x_1{}^2 + x_2{}^2 - 4x_3{}^2 = 0$
(b) $x_1{}^2 + x_2x_3 - x_3{}^2 = 0$, $x_1{}^2 - x_2x_3 + x_3{}^2 = 0$
(c) $x_1{}^2 + x_2{}^2 - x_3{}^2 = 0$, $4x_1{}^2 + x_2{}^2 - 4x_3{}^2 = 0$
(d) $x_1{}^2 + x_2{}^2 - x_3{}^2 = 0$, $x_1{}^2 - x_2{}^2 + x_3{}^2 = 0$

6. What is the distance between the points in the following pairs in the euclidean planes for which $x_1 = 0$, $x_2 = 0$, and $x_3 = 0$ are, in turn, the ideal line?

(a) $(1,1,1)$ and $(-1,2,3)$ (b) $(2,1,-1)$ and $(2,1,2)$

(c) $(1,-1,1)$ and $(1,1,2)$ (d) $(1,2,3)$ and $(1,2,0)$

7. What is the tangent of the angle between the lines in the following pairs in the euclidean planes for which $x_1 = 0$, $x_2 = 0$, and $x_3 = 0$ are, in turn, the ideal line?

(a) $[1,1,0]$ and $[1,2,1]$ (b) $[0,1,0]$ and $[1,2,3]$

(c) $[2,-1,1]$ and $[3,1,-1]$ (d) $[1,2,-2]$ and $[0,0,2]$

8. Is it possible to determine l_1, l_2, l_3 and m_1, m_2, m_3 so that the lines $l_1x_1 + l_2x_2 + l_3x_3 = 0$ and $m_1x_1 + m_2x_2 + m_3x_3 = 0$ will simultaneously be perpendicular in the euclidean plane for which $x_1 = 0$ is the ideal line, the euclidean plane for which $x_2 = 0$ is the ideal line, and the euclidean plane for which $x_3 = 0$ is the ideal line?

9. Discuss the curve $x_1x_2{}^2 = x_3{}^3$ in the euclidean planes for which $x_1 = 0$, $x_2 = 0$, and $x_3 = 0$ are, in turn, the ideal line. Show by a figure similar to Fig. 2.3 how these three euclidean representations are related to each other.

10. Answer Exercise 9 for the curve $x_1x_2{}^3 = x_3{}^4$.

2.5 Conclusion

In this chapter we generalized the familiar rectangular coordinates, (x,y), of a point in the euclidean plane to a more symmetric form involving three homogeneous coordinates, (x_1,x_2,x_3), only the ratios of which were significant. We also introduced the important notions of the homogeneous coordinates of a line and the equation of a point, so that now *both* points and lines can be described *either* by a set of coordinates *or* by an equation. These investigations led us to certain coordinate triples for which there were no corresponding geometrical objects in the euclidean plane. This, in turn, motivated us to enlarge the euclidean plane by adding to it one line of points, known as *ideal points*, to which the previously exceptional coordinate triples would correspond. In this enlarged system, called the *extended euclidean plane*, E_2^+, or simply the *extended plane*, there were no longer any parallel lines since, without exception, two lines always determined a point just as two points always determine a line. In fact, we observed that in the extended plane, points and lines have properties which are algebraically indistinguishable, and from any theorem concerning points and lines a new theorem, known as the *dual* of the first, can be obtained simply by interchanging the terms "point" and "line" in its statement.

Furthermore, we found that the concepts of distance and angle, and such related properties as similarity, congruence, and perpendicularity, were no longer meaningful in the extended plane.

Finally, we discovered that the extended plane contained at least three different euclidean planes, corresponding to the choice of $x_1 = 0$, $x_2 = 0$, or $x_3 = 0$ as the ideal line. Moreover, we conjectured that any line in the extended plane could be chosen to play the role of the ideal line and that every such choice would lead to a different euclidean plane contained in the extended plane.

Not only do these considerations give us a better understanding of the familiar euclidean plane, but they also have a direct and immediate bearing on our study of projective geometry. In fact, the extended euclidean plane is nothing but the real projective plane; and when the entries in the coordinate triples of points and lines are allowed to take on complex as well as real values, the extended plane is just the classical projective plane Π_2. For this reason we shall devote the next three chapters to a study of the extended plane.

three

A LITTLE LINEAR ALGEBRA

3.1 Introduction

The work of the last chapter suggested, correctly, that the study of the extended plane involves a considerable amount of algebra. It is also true, though not yet obvious, that once we have introduced coordinates in our axiomatic development of the projective plane, algebra will likewise be a tool of great importance. In pursuing each of these themes in later chapters, matric notation and some of the simpler properties of matrices will be of great convenience in systematizing the algebraic manipulations we shall have to carry out. Accordingly, we devote this chapter to an introductory discussion of matrices, linear dependence and independence, the solution of systems of linear equations, and quadratic forms.

3.2 Matrices and Their Elementary Properties

By a **matrix** we mean simply a rectangular array of quantities, which in our work will be either real or complex numbers and which we shall refer to as **scalars.** More specifically, we have the following definition.

Definition 1

An (m,n) matrix is a rectangular array of mn quantities arranged in m rows and n columns:

$$A = \begin{Vmatrix} a_{11} & a_{12} & \cdots & a_{1n} \\ a_{21} & a_{22} & \cdots & a_{2n} \\ \cdot & \cdot & \cdots & \cdot \\ a_{m1} & a_{m2} & \cdots & a_{mn} \end{Vmatrix} = \|a_{ij}\|$$

As illustrated in Definition 1, matrices are usually named by capital letters and represented in detail by displaying some or all of the quantities, or **elements,** between double vertical bars. In presenting such a display, it is often convenient to use the **double-subscript notation** illustrated in Definition 1, in which the element in the ith row and jth column is identified by attaching the subscripts i and j, in that order, to the appropriate letter. A $(1,n)$ matrix, i.e., a matrix consisting of a single row of n elements, is called a **row matrix** or a **row vector.** An $(m,1)$ matrix, i.e., a matrix consisting of a single column containing m elements, is called a **column matrix** or a **column vector.** A matrix for which $m = n$, that is, a matrix in which the number of rows is equal to the number of columns, is called a **square matrix.** A square matrix with n rows and n columns is often called a matrix of **order** n.

Although we have not previously used the term *matrix*, we have already encountered matrices in our work. For instance, if P is a point in the extended euclidean plane, its coordinates (x_1,x_2,x_3) can be displayed either as the row matrix

$$\|x_1 \quad x_2 \quad x_3\|$$

or as the column matrix

$$\begin{Vmatrix} x_1 \\ x_2 \\ x_3 \end{Vmatrix}$$

Hence, whenever we have considered the coordinates of a point in E_2^+ we have really been dealing with a matrix. From now on we shall make this identification explicit, and we shall use the name of a point to refer equally well to the point itself and to the *column* matrix of its coordinates. Thus if $P:(x_1,x_2,x_3)$ is a point in E_2^+ or Π_2, we shall also write

$$P = \begin{Vmatrix} x_1 \\ x_2 \\ x_3 \end{Vmatrix}$$

Similarly, we shall use the name of a line to refer equally well to the line and to the column matrix of its coordinates.[1]

Another example of a matrix is provided by the array

$$\begin{array}{ccc} a_1 & a_2 & a_3 \\ b_1 & b_2 & b_3 \end{array}$$

which we introduced in Sec. 2.2 in connection with the problem of finding the coordinates of the line determined by two general points, $A:(a_1,a_2,a_3)$ and $B:(b_1,b_2,b_3)$. Now, of course, using matric notation, we would write this array in the form

$$\left\|\begin{array}{ccc} a_1 & a_2 & a_3 \\ b_1 & b_2 & b_3 \end{array}\right\|$$

Clearly, the columns of any matrix can be thought of, individually, as being column vectors. Adopting this point of view, it is sometimes convenient, for mathematical as well as typographical reasons, to assign names to the column vectors of a matrix A and use these names to abbreviate the detailed representation of A. Similarly, the rows of a matrix A can be thought of, individually, as being row vectors, and if names are assigned to these vectors we obtain another abbreviation of the detailed representation of A. Thus if A is the matrix

$$\left\|\begin{array}{cccc} a_{11} & a_{12} & \cdots & a_{1n} \\ a_{21} & a_{22} & \cdots & a_{2n} \\ \cdot & \cdot & \cdots & \cdot \\ a_{m1} & a_{m2} & \cdots & a_{mn} \end{array}\right\|$$

we can assign names to its column vectors, say,

$$C_1 = \left\|\begin{array}{c} a_{11} \\ a_{21} \\ \cdot \\ a_{m1} \end{array}\right\|, \quad C_2 = \left\|\begin{array}{c} a_{12} \\ a_{22} \\ \cdot \\ a_{m2} \end{array}\right\|, \quad \ldots, \quad C_n = \left\|\begin{array}{c} a_{1n} \\ a_{2n} \\ \cdot \\ a_{mn} \end{array}\right\|$$

and write it in the more compact form

$$(1) \qquad A = \|C_1 \quad C_2 \quad \cdots \quad C_n\|$$

or we can assign names to its row vectors, say

$$R_1 = \|a_{11} \quad a_{12} \quad \cdots \quad a_{1n}\|, \qquad R_2 = \|a_{21} \quad a_{22} \quad \cdots \quad a_{2n}\|, \qquad \ldots,$$

$$R_m = \|a_{m1} \quad a_{m2} \quad \cdots \quad a_{mn}\|$$

[1] We could, of course, have identified points and lines with the row matrices of their coordinates, but for reasons which will become clear after we have defined the multiplication of matrices, it is more convenient to identify them with the corresponding column matrices.

and write it more compactly as

(2)
$$A = \left\| \begin{array}{c} R_1 \\ R_2 \\ . \\ R_m \end{array} \right\|$$

When a matrix A is written in the form (1), we say that it has been **partitioned into columns;** when it is written in the form (2), we say that it has been **partitioned into rows.**

A matrix is said to be a **zero matrix** or a **null matrix** if and only if each of its elements is zero. All zero matrices are denoted by the symbol O.

In any square matrix, the line of elements extending from the upper left-hand corner to the lower right-hand corner is called the **principal diagonal** of the matrix. A square matrix in which every element which is not on the principal diagonal is zero is called a **diagonal matrix.** A diagonal matrix whose diagonal elements are each 1 is called a **unit matrix.** All unit matrices are denoted by the symbol I.

Definition 2

Two matrices, $A = \|a_{ij}\|$ and $B = \|b_{ij}\|$, are equal if and only if they are identical, i.e., if and only if they have the same number of rows and the same number of columns and $a_{ij} = b_{ij}$ for all values of i and j.

In general the elements of a matrix are independent of each other; hence in an arbitrary matrix A it is not to be expected that $a_{ij} = a_{ji}$. If this happens to be the case for all values of i and j, which is possible only if A is a square matrix, we say that A is a **symmetric matrix,** since elements which are *symmetrically* located with respect to the principal diagonal are now equal.

Definition 3

The (n,m) matrix obtained from a given (m,n) matrix, A, by interchanging its rows and columns is called the transpose of A.

The transpose of a matrix A is denoted by the symbol A^T. Clearly, *a matrix is symmetric if and only if it is equal to its transpose.*

The symbol for a matrix appears, at first glance, very much like the symbol for a determinant, and for a while some students have trouble distinguishing between the concept of a matrix and the concept of a determinant. Briefly, a matrix is simply a rectangular array of elements which may or may not be square, while a determinant is a rather complicated function of the elements of a necessarily square array. However, associated with every *square* matrix there is a determinant, namely, the determinant whose elements are, respectively, the elements of the given matrix.[1] If $A = \|a_{ij}\|$ is any square matrix, the **determinant of** A is denoted by either of the symbols

$$|A| \qquad \text{or} \qquad |a_{ij}|$$

Definition 4

A square matrix whose determinant is equal to zero is said to be singular.

Definition 5

A square matrix whose determinant is different from zero is said to be nonsingular.

Although the definition and elementary properties of determinants should be familiar to us from our work in college algebra, we have listed them for reference in Appendix 1.

By defining appropriately the addition and multiplication of matrices, an algebra of matrices can be developed. The following definitions are the basis for this development.

Definition 6

The sum or difference of two matrices, A and B, having the same number of rows and the same number of columns is the matrix $A \pm B$ whose elements are the sums or differences of the respective elements of A and B.

If addition is commutative for the elements of A and B, as it is, for instance, when these elements are real or complex numbers, it follows that *matric addition is commutative*; i.e.,

$$A + B = B + A$$

Neither addition nor subtraction is defined for matrices which do not have the same number of rows *and* the same number of columns.

Definition 7

The product of a matrix A and a scalar k is the matrix $kA = Ak$ whose elements are the elements of A each multiplied by the scalar k.

---

[1] The relation between determinants and square matrices is essentially the relation between the dependent and independent variables of a function.

It is important to distinguish clearly between the process of multiplying a determinant by a scalar and the process of multiplying a matrix by a scalar. To multiply a determinant by a scalar we merely multiply the elements in some *one* row or some *one* column by the scalar. On the other hand, to multiply a matrix by a scalar we must multiply *every* element in the matrix by the scalar.

Example 1

If
$$A = \begin{Vmatrix} a_{11} & a_{12} & a_{13} \\ a_{21} & a_{22} & a_{23} \\ a_{31} & a_{32} & a_{33} \end{Vmatrix}, \quad \text{then} \quad kA = \begin{Vmatrix} ka_{11} & ka_{12} & ka_{13} \\ ka_{21} & ka_{22} & ka_{23} \\ ka_{31} & ka_{32} & ka_{33} \end{Vmatrix}$$

and
$$|kA| = \begin{vmatrix} ka_{11} & ka_{12} & ka_{13} \\ ka_{21} & ka_{22} & ka_{23} \\ ka_{31} & ka_{32} & ka_{33} \end{vmatrix} = k^3 \begin{vmatrix} a_{11} & a_{12} & a_{13} \\ a_{21} & a_{22} & a_{23} \\ a_{31} & a_{32} & a_{33} \end{vmatrix} = k^3 |A|$$

On the other hand, for $k\,|A|$ we have, equivalently,

$$\begin{vmatrix} ka_{11} & ka_{12} & ka_{13} \\ a_{21} & a_{22} & a_{23} \\ a_{31} & a_{32} & a_{33} \end{vmatrix} = \begin{vmatrix} a_{11} & a_{12} & a_{13} \\ ka_{21} & ka_{22} & ka_{23} \\ a_{31} & a_{32} & a_{33} \end{vmatrix} = \ldots = \begin{vmatrix} a_{11} & a_{12} & ka_{13} \\ a_{21} & a_{22} & ka_{23} \\ a_{31} & a_{32} & ka_{33} \end{vmatrix}$$

Definition 8
The scalar product of two vectors with the same number of elements is the sum of the products of corresponding elements.

The scalar product of two vectors is also referred to as the **inner product** or **dot product** of the two vectors and is often denoted by putting a dot between the names of the two vectors.

Example 2

If
$$V_1 = \|-1 \quad 4 \quad 0 \quad 7\| \quad \text{and} \quad V_2 = \|2 \quad 3 \quad 3 \quad 1\|$$
then
$$V_1 \cdot V_2 = (-1)(2) + (4)(3) + (0)(3) + (7)(1) = 17$$

The condition from the elementary analytic geometry of E_3 that two lines be perpendicular provides a familiar example of the scalar product of two vectors. In fact, if l and m are two lines whose direction numbers are respectively (l_1, l_2, l_3) and (m_1, m_2, m_3), and if we define the vectors

$$L = \|l_1 \quad l_2 \quad l_3\| \quad \text{and} \quad M = \|m_1 \quad m_2 \quad m_3\|$$

then the well-known condition that l and m be perpendicular, namely, $l_1 m_1 + l_2 m_2 + l_3 m_3 = 0$, is simply the condition that

$$L \cdot M = 0$$

Definition 9

Two matrices, A and B, are said to be conformable in the order A, B if and only if the number of columns in A is equal to the number of rows in B.

In general, two matrices, A and B, will not be conformable either in the order A, B or in the order B, A. In fact, if A is an (m,n) matrix and B is a (p,q) matrix, A and B are conformable in the order A, B if and only if $n = p$ and are conformable in the order B, A if and only if $q = m$.

Definition 10

If A and B are two matrices which are conformable in the order A, B, then the product AB is the matrix in which the element in the ith row and jth column is the scalar product of the ith-row vector of A and the jth-column vector of B.

Example 3

$$\begin{Vmatrix} 1 & 2 \\ 3 & 4 \end{Vmatrix} \cdot \begin{Vmatrix} 5 & 6 & 7 \\ 8 & 9 & 0 \end{Vmatrix} = \begin{Vmatrix} 1 \cdot 5 + 2 \cdot 8 & 1 \cdot 6 + 2 \cdot 9 & 1 \cdot 7 + 2 \cdot 0 \\ 3 \cdot 5 + 4 \cdot 8 & 3 \cdot 6 + 4 \cdot 9 & 3 \cdot 7 + 4 \cdot 0 \end{Vmatrix}$$

$$= \begin{Vmatrix} 21 & 24 & 7 \\ 47 & 54 & 21 \end{Vmatrix}$$

It should be clear from Definition 10 and Example 3 why only conformable matrices can be multiplied. In fact, if A and B are two matrices, the scalar product of a general row vector of A and a general column vector of B can be computed if and only if these two vectors have the same number of elements. Moreover, the number of elements of any row vector of A is just the number of columns in A, and the number of elements of any column vector of B is just the number of rows of B. Hence if A and B are to be multiplied, the number of columns of A must equal the number of rows in B, which is just what is meant by saying that A and B are conformable in the order A, B. It should also be clear that when A and B are conformable in the order A, B, the number of rows in the product AB is equal to the number of rows in A, and the number of columns in AB is equal to the number of columns in B. In other words, *if A is an (m,n) matrix, and if B is an (n,q) matrix, then AB is an (m,q) matrix.*

From Definition 10, it is obvious that a matrix can be multiplied by itself if and only if it is square. Repeated products of a square matrix with itself are indicated by the usual exponential notation. Thus we write

$$AA = A^2, \qquad AAA = A^3, \qquad \ldots$$

In some respects, matric multiplication is just like the familiar multiplication of algebraic quantities; in other respects it is surprisingly different. The following theorems describe the basic properties of matric multiplication.[1]

Theorem 1

Even for matrices which are conformable in either order, multiplication is not commutative; i.e., in general

$$AB \neq BA$$

Example 4

$$\begin{Vmatrix} 1 & 2 \\ 3 & 4 \end{Vmatrix} \cdot \begin{Vmatrix} -1 & 0 \\ 4 & 2 \end{Vmatrix} = \begin{Vmatrix} 7 & 4 \\ 13 & 8 \end{Vmatrix}$$

but

$$\begin{Vmatrix} -1 & 0 \\ 4 & 2 \end{Vmatrix} \cdot \begin{Vmatrix} 1 & 2 \\ 3 & 4 \end{Vmatrix} = \begin{Vmatrix} -1 & -2 \\ 10 & 16 \end{Vmatrix}$$

Corollary 1

Both unit matrices and zero matrices commute with all suitably conformable matrices; more specifically,

$$AI = IA = A \qquad \text{and} \qquad AO = OA = O$$

Theorem 2

The vanishing of the product of two matrices does not imply that either of the matrices is a zero matrix.

Example 5

$$\begin{Vmatrix} 1 & 1 \\ 2 & 2 \end{Vmatrix} \cdot \begin{Vmatrix} -2 & 3 & 4 \\ 2 & -3 & -4 \end{Vmatrix} = \begin{Vmatrix} 0 & 0 & 0 \\ 0 & 0 & 0 \end{Vmatrix}$$

Theorem 3

For suitably conformable matrices, multiplication is distributive over addition and subtraction; that is,

$$A(B \pm C) = AB \pm AC$$

[1] Proofs of results stated here without proof may be found in C. R. Wylie, Jr., "Advanced Engineering Mathematics," 3d ed., McGraw-Hill Book Company, New York, 1966.

Example 6

$$\begin{Vmatrix} 1 & 2 \\ 3 & 4 \end{Vmatrix} \cdot \left(\begin{Vmatrix} 2 & -1 \\ 0 & 3 \end{Vmatrix} + \begin{Vmatrix} 1 & 1 \\ -1 & 2 \end{Vmatrix} \right) = \begin{Vmatrix} 1 & 2 \\ 3 & 4 \end{Vmatrix} \cdot \begin{Vmatrix} 3 & 0 \\ -1 & 5 \end{Vmatrix} = \begin{Vmatrix} 1 & 10 \\ 5 & 20 \end{Vmatrix}$$

and

$$\begin{Vmatrix} 1 & 2 \\ 3 & 4 \end{Vmatrix} \cdot \begin{Vmatrix} 2 & -1 \\ 0 & 3 \end{Vmatrix} + \begin{Vmatrix} 1 & 2 \\ 3 & 4 \end{Vmatrix} \cdot \begin{Vmatrix} 1 & 1 \\ -1 & 2 \end{Vmatrix}$$

$$= \begin{Vmatrix} 2 & 5 \\ 6 & 9 \end{Vmatrix} + \begin{Vmatrix} -1 & 5 \\ -1 & 11 \end{Vmatrix} = \begin{Vmatrix} 1 & 10 \\ 5 & 20 \end{Vmatrix}$$

Theorem 4

For suitably conformable matrices, matric multiplication is associative; i.e.,

$$A(BC) = (AB)C$$

Example 7

$$\begin{Vmatrix} 1 & 2 \\ 3 & 4 \end{Vmatrix} \cdot \left(\begin{Vmatrix} 2 & -1 & 3 \\ 0 & 1 & 2 \end{Vmatrix} \cdot \begin{Vmatrix} 1 & -1 \\ 1 & 0 \\ 2 & 4 \end{Vmatrix} \right)$$

$$= \begin{Vmatrix} 1 & 2 \\ 3 & 4 \end{Vmatrix} \cdot \begin{Vmatrix} 7 & 10 \\ 5 & 8 \end{Vmatrix} = \begin{Vmatrix} 17 & 26 \\ 41 & 62 \end{Vmatrix}$$

and

$$\left(\begin{Vmatrix} 1 & 2 \\ 3 & 4 \end{Vmatrix} \cdot \begin{Vmatrix} 2 & -1 & 3 \\ 0 & 1 & 2 \end{Vmatrix} \right) \cdot \begin{Vmatrix} 1 & -1 \\ 1 & 0 \\ 2 & 4 \end{Vmatrix}$$

$$= \begin{Vmatrix} 2 & 1 & 7 \\ 6 & 1 & 17 \end{Vmatrix} \cdot \begin{Vmatrix} 1 & -1 \\ 1 & 0 \\ 2 & 4 \end{Vmatrix} = \begin{Vmatrix} 17 & 26 \\ 41 & 62 \end{Vmatrix}$$

Theorem 5

The transpose of the product of two conformable matrices is equal to the product of the transposed matrices taken in the other order; that is,

$$(AB)^T = B^T A^T$$

Corollary 1

The transpose of the product of any number of suitably conformable matrices is equal to the product of the transposed matrices taken in the other order; i.e.,

$$(A_1 A_2 \cdots A_k)^{\bar{T}} = A_k{}^{\bar{T}} \cdots A_2{}^{\bar{T}} A_1{}^T$$

Example 8

$$\left(\left\| \begin{matrix} 1 & 2 \\ 3 & 4 \end{matrix} \right\| \cdot \left\| \begin{matrix} 2 & -1 & 0 \\ 4 & 1 & 3 \end{matrix} \right\| \right)^T = \left\| \begin{matrix} 10 & 1 & 6 \\ 22 & 1 & 12 \end{matrix} \right\|^T = \left\| \begin{matrix} 10 & 22 \\ 1 & 1 \\ 6 & 12 \end{matrix} \right\|$$

and

$$\left\| \begin{matrix} 2 & -1 & 0 \\ 4 & 1 & 3 \end{matrix} \right\|^T \cdot \left\| \begin{matrix} 1 & 2 \\ 3 & 4 \end{matrix} \right\|^T = \left\| \begin{matrix} 2 & 4 \\ -1 & 1 \\ 0 & 3 \end{matrix} \right\| \cdot \left\| \begin{matrix} 1 & 3 \\ 2 & 4 \end{matrix} \right\| = \left\| \begin{matrix} 10 & 22 \\ 1 & 1 \\ 6 & 12 \end{matrix} \right\|$$

Theorem 6

If A and B are square matrices of the same order, the determinant of the product AB is equal to the product of the determinant of A and the determinant of B; i.e.,

$$|AB| = |A| \cdot |B|$$

Example 9

If $A = \left\| \begin{matrix} 1 & 2 \\ 3 & 4 \end{matrix} \right\|$ and $B = \left\| \begin{matrix} -2 & 3 \\ 5 & 1 \end{matrix} \right\|$, then $AB = \left\| \begin{matrix} 8 & 5 \\ 14 & 13 \end{matrix} \right\|$ and

$|AB| = \left| \begin{matrix} 8 & 5 \\ 14 & 13 \end{matrix} \right| = 34$. On the other hand, $|A| = -2$, $|B| = -17$, and $|A| \cdot |B| = (-2)(-17) = 34$.

Exercises

1. If $A = \left\| \begin{matrix} 1 & 2 & 4 \\ -1 & 0 & 3 \end{matrix} \right\|$ and $B = \left\| \begin{matrix} 1 & 2 \\ 2 & 3 \\ 3 & 4 \end{matrix} \right\|$, compute all two- and three-factor products which can be formed from these three matrices.

2. If $V_1 = \| 1 \quad 2 \quad 3 \|$ and $V_2 = \left\| \begin{matrix} 3 \\ -2 \\ 1 \end{matrix} \right\|$, what is $V_1 V_2$? What is $V_2 V_1$? What is $V_1^T V_2^T$? What is $V_2^T V_1^T$?

3. If $A = \left\| \begin{matrix} 1 & 2 \\ -1 & 3 \end{matrix} \right\|$, compute A^2 and verify that

$$A^2 - 3A + 2I = (A - I)(A - 2I) = (A - 2I)(A - I)$$

4. In Exercise 3, verify that $A^2 - 4A + 5I = O$ and use this fact to compute A^3, A^4, and A^5 by first expressing them in terms of the matrices A and I. Check these results by calculating A^3, A^4, and A^5 directly from A.

5. Work Exercise 3 if:

$$(a)\ A = \begin{Vmatrix} 1 & -2 \\ 3 & 2 \end{Vmatrix} \qquad (b)\ A = \begin{Vmatrix} 1 & 1 & 0 \\ 0 & 1 & 2 \\ -1 & 3 & 0 \end{Vmatrix}$$

6. Show that $\begin{Vmatrix} \cos\theta & \sin\theta \\ -\sin\theta & \cos\theta \end{Vmatrix}^n = \begin{Vmatrix} \cos n\theta & \sin n\theta \\ -\sin n\theta & \cos n\theta \end{Vmatrix}$

7. Show that $\begin{Vmatrix} \cosh\theta & \sinh\theta \\ \sinh\theta & \cosh\theta \end{Vmatrix}^n = \begin{Vmatrix} \cosh n\theta & \sinh n\theta \\ \sinh n\theta & \cosh n\theta \end{Vmatrix}$

8. If $A = \begin{Vmatrix} 1 & 2 \\ 2 & 4 \end{Vmatrix}$, find a nonzero (2,2) matrix X such that AX is a zero matrix. Is $XA = O$? If $A = \begin{Vmatrix} 1 & 2 \\ 2 & 3 \end{Vmatrix}$, is it possible to find a nonzero (2,2) matrix X such that AX is a zero matrix?

9. If A and B are symmetric matrices of the same order, under what conditions, if any, is AB also symmetric?

10. If A and B are square matrices which commute, i.e., square matrices such that $AB = BA$, prove that A^2 and B^2 also commute. Is this true for general positive integral powers of A and B?

11. Prove that $(A + B)^T = A^T + B^T$.

12. If

$$A = \begin{Vmatrix} 1 & 2 & 3 \\ 2 & 3 & 4 \\ 3 & 4 & -1 \end{Vmatrix} \quad \text{and} \quad B = \begin{Vmatrix} 1 & -1 & 1 \\ 2 & 2 & -2 \\ 0 & 1 & 3 \end{Vmatrix}$$

verify that $|AB| = |A| \cdot |B|$. What is $|BA|$? Is it always true that $|BA| = |AB|$?

13. If $X = \begin{Vmatrix} x_1 \\ x_2 \\ x_3 \end{Vmatrix}$ and $A = \begin{Vmatrix} 1 & -1 & 2 \\ -1 & -1 & 3 \\ 2 & 3 & 4 \end{Vmatrix}$, what is $X^T A X$?

14. Determine the most general (2,2) matrix X such that $X^2 = I$.

15. If $A = \begin{Vmatrix} 1 & 2 \\ 3 & 4 \end{Vmatrix}$, $B = \begin{Vmatrix} 2 & 1 \\ 1 & -1 \end{Vmatrix}$, $C = \begin{Vmatrix} 1 & 2 \\ 2 & 4 \end{Vmatrix}$, and $X = \begin{Vmatrix} x_1 & x_2 \\ x_3 & x_4 \end{Vmatrix}$,

solve each of the following equations:

(a) $AX = B$ (b) $BX = A$ (c) $AX = C$ (d) $BX = C$

Is it possible to solve the equation $CX = A$?

3.3 Adjoints and Inverses

In the last section we began the development of the algebra of matrices by defining the addition, subtraction, and multiplication of matrices. To complete our discussion, it would seem that we should now define the division of matrices, but this is not the case. Instead, we shall introduce a related concept suggested by the following observations from elementary algebra.

In elementary algebra, it is a familiar fact that any quantity Q which is not equal to zero has a *reciprocal*, or *multiplicative inverse*, $1/Q = Q^{-1}$, with the property that

$$(1) \qquad\qquad Q^{-1}Q = QQ^{-1} = 1$$

Division, which we sometimes consider, inaccurately, to be a process independent of multiplication, is actually nothing but multiplication involving the reciprocal of the divisor as one factor. In matric algebra we do not define division as such, but in an important class of cases, namely, the class of nonsingular square matrices, we do define the reciprocal, or inverse, of a matrix. Then using inverses, we are able through multiplication to accomplish all that we might properly expect to do by division.

Preparatory to defining the **inverse** of a nonsingular square matrix A, we must first define the **cofactor matrix** and the **adjoint matrix** of A.

Definition 1

If $A = \|a_{ij}\|$ is a square matrix, and if A_{ij} is the cofactor[1] of the element a_{ij} in the determinant of A, then the matrix $\|A_{ij}\|$ obtained by replacing each element of A by its cofactor in $|A|$ is called the cofactor matrix of A.

Definition 2

The transpose of the cofactor matrix of a square matrix A is called the adjoint of A.

The adjoint of a square matrix A is indicated by the symbol adj A.

Definition 3

The reciprocal, or inverse, A^{-1}, of a nonsingular square matrix A is the adjoint of A divided by the determinant of A; that is,

$$A^{-1} = \frac{\text{adj } A}{|A|} = \frac{\|A_{ij}\|^T}{|A|} = \frac{\|A_{ji}\|}{|A|}$$

[1] See Appendix 1 for the definition of the cofactor of an element of a determinant.

Clearly, although every square matrix has an adjoint, only nonsingular square matrices have inverses, since the definition of the inverse of a matrix A requires that the determinant of A be different from zero.

The importance of the inverse of a matrix stems from the property described in the following theorem, whose resemblance to the scalar relation (1) is obvious.

Theorem 1

The product of a nonsingular matrix A and its inverse in either order is a unit matrix; that is,

$$AA^{-1} = A^{-1}A = I$$

Proof Let A be a nonsingular matrix, and consider first the product

$$A^{-1}A = \frac{1}{|A|} \begin{Vmatrix} A_{11} & A_{21} & \cdots & A_{n1} \\ \cdot & \cdot & \cdots & \cdot \\ A_{1i} & A_{2i} & \cdots & A_{ni} \\ \cdot & \cdot & \cdots & \cdot \\ A_{1n} & A_{2n} & \cdots & A_{nn} \end{Vmatrix} \cdot \begin{Vmatrix} a_{11} & \cdots & a_{1j} & \cdots & a_{1n} \\ a_{21} & \cdots & a_{2j} & \cdots & a_{2n} \\ \cdot & \cdots & \cdot & \cdots & \cdot \\ a_{n1} & \cdots & a_{nj} & \cdots & a_{nn} \end{Vmatrix}$$

From the definition of matric multiplication, it is clear that the element in the ith row and jth column of the product of the two matrices on the right is $A_{1i}a_{1j} + A_{2i}a_{2j} + \cdots + A_{ni}a_{nj}$. If $i = j$, this sum is equal to $|A|$ since it is just the expansion of the determinant of A in terms of the elements in the jth column of A. If $i \neq j$, this sum consists of the elements of one column of $|A|$ multiplied by the cofactors of the corresponding elements of a *different* column and is therefore equal to zero by Theorem 9, Appendix 1. Thus

$$A^{-1}A = \frac{1}{|A|} \begin{Vmatrix} |A| & 0 & \cdots & 0 \\ 0 & |A| & \cdots & 0 \\ \cdot & \cdot & \cdots & \cdot \\ 0 & 0 & \cdots & |A| \end{Vmatrix} = \begin{Vmatrix} 1 & 0 & \cdots & 0 \\ 0 & 1 & \cdots & 0 \\ \cdot & \cdot & \cdots & \cdot \\ 0 & 0 & \cdots & 1 \end{Vmatrix} = I$$

as asserted. The proof that $AA^{-1} = I$ follows in exactly the same fashion.

Corollary 1

For any square matrix A,

$$(\text{adj } A)A = A(\text{adj } A) = |A|\, I$$

Corollary 2

If A is a nonsingular matrix, then A^{-1} is also nonsingular and $|A^{-1}| = 1/|A|$.

Example 1

If $A = \begin{Vmatrix} 1 & 2 & -2 \\ 3 & 3 & -1 \\ 2 & 1 & 0 \end{Vmatrix}$, then the determinant of A is $\begin{vmatrix} 1 & 2 & -2 \\ 3 & 3 & -1 \\ 2 & 1 & 0 \end{vmatrix} = 3.$

The cofactor matrix of A is

$$\begin{Vmatrix} \begin{vmatrix} 3 & -1 \\ 1 & 0 \end{vmatrix} & -\begin{vmatrix} 3 & -1 \\ 2 & 0 \end{vmatrix} & \begin{vmatrix} 3 & 3 \\ 2 & 1 \end{vmatrix} \\ -\begin{vmatrix} 2 & -2 \\ 1 & 0 \end{vmatrix} & \begin{vmatrix} 1 & -2 \\ 2 & 0 \end{vmatrix} & -\begin{vmatrix} 1 & 2 \\ 2 & 1 \end{vmatrix} \\ \begin{vmatrix} 2 & -2 \\ 3 & -1 \end{vmatrix} & -\begin{vmatrix} 1 & -2 \\ 3 & -1 \end{vmatrix} & \begin{vmatrix} 1 & 2 \\ 3 & 3 \end{vmatrix} \end{Vmatrix} = \begin{Vmatrix} 1 & -2 & -3 \\ -2 & 4 & 3 \\ 4 & -5 & -3 \end{Vmatrix}$$

The adjoint of A is its transposed cofactor matrix; that is,

$$\operatorname{adj} A = \begin{Vmatrix} 1 & -2 & 4 \\ -2 & 4 & -5 \\ -3 & 3 & -3 \end{Vmatrix}$$

The inverse of A is therefore

$$A^{-1} = \frac{\operatorname{adj} A}{|A|} = \tfrac{1}{3} \begin{Vmatrix} 1 & -2 & 4 \\ -2 & 4 & -5 \\ -3 & 3 & -3 \end{Vmatrix}$$

and

$$A^{-1}A = \tfrac{1}{3} \begin{Vmatrix} 1 & -2 & 4 \\ -2 & 4 & -5 \\ -3 & 3 & -3 \end{Vmatrix} \cdot \begin{Vmatrix} 1 & 2 & -2 \\ 3 & 3 & -1 \\ 2 & 1 & 0 \end{Vmatrix}$$

$$= \tfrac{1}{3} \begin{Vmatrix} 3 & 0 & 0 \\ 0 & 3 & 0 \\ 0 & 0 & 3 \end{Vmatrix} = \begin{Vmatrix} 1 & 0 & 0 \\ 0 & 1 & 0 \\ 0 & 0 & 1 \end{Vmatrix}$$

From Theorem 1 it is clear that if A is a nonsingular matrix, then $X = A^{-1}$ is a solution of each of the matric equations

$$AX = I \quad \text{and} \quad XA = I$$

Actually, as the next theorem assures us, $X = A^{-1}$ is the *only* solution of each of these equations.

Theorem 2

If A is a nonsingular matrix, then $X = A^{-1}$ is the unique solution of each of the equations

$$AX = I \quad \text{and} \quad XA = I$$

Proof Consider first the equation $AX = I$, and suppose that both X_1 and X_2 satisfy this equation. This means that $AX_1 = I$ and $AX_2 = I$, from which we conclude that

$$AX_1 = AX_2$$

Now, by hypothesis, A is nonsingular, and therefore A^{-1} exists. Hence, multiplying the last equation through on the left by A^{-1}, we have

$$
\begin{aligned}
A^{-1}(AX_1) &= A^{-1}(AX_2) \\
(A^{-1}A)X_1 &= (A^{-1}A)X_2 && \text{by Theorem 4, Sec. 3.2} \\
IX_1 &= IX_2 && \text{by Theorem 1} \\
X_1 &= X_2 && \text{by Corollary 1, Theorem 1, Sec. 3.2}
\end{aligned}
$$

This proves that the equation $AX = I$ has just one solution, and from Theorem 1 it follows that this solution is $X = A^{-1}$. A similar argument proves that $X = A^{-1}$ is also the unique solution of the equation $XA = I$.

Using the last theorem, we can now prove several other useful results about the inverse of a matrix.

Theorem 3

If A is a nonsingular matrix, then $(A^{-1})^{-1} = A$.

Proof By Theorem 2, the inverse of A^{-1} is the unique solution of the equation $A^{-1}X = I$. Moreover, by Theorem 1, it is clear that $X = A$ satisfies this equation. Hence, since this equation has only one solution, that solution must be $X = A$. In other words, the inverse of A^{-1}, that is, $(A^{-1})^{-1}$, is A, as asserted.

Theorem 4

If A and B are conformable nonsingular matrices, then $(AB)^{-1} = B^{-1}A^{-1}$.

Proof By Theorem 2, the inverse of the matrix AB is the unique solution of the equation $(AB)X = I$. Moreover, it is clear that $B^{-1}A^{-1}$ satisfies this equation since

$$
\begin{aligned}
(AB)(B^{-1}A^{-1}) &= A(BB^{-1})A^{-1} && \text{by Theorem 4, Sec. 3.2} \\
&= AIA^{-1} && \text{by Theorem 1} \\
&= AA^{-1} && \text{by Corollary 1, Theorem 1, Sec. 3.2} \\
&= I && \text{by Theorem 1}
\end{aligned}
$$

Therefore the inverse of AB, that is, $(AB)^{-1}$, is equal to $B^{-1}A^{-1}$, as asserted.

Corollary 1

If $A_1, A_2, \ldots, A_k$ are conformable nonsingular matrices, then

$$(A_1A_2 \cdots A_k)^{-1} = A_k^{-1} \cdots A_2^{-1}A_1^{-1}$$

Once the multiplication of matrices is defined, it becomes possible to define positive integral powers of any square matrix, A, by writing, as usual

$$AA = A^2, \qquad AAA = A^3, \qquad \cdots$$

Similarly, once the inverse of a nonsingular matrix is defined, negative integral powers of *nonsingular* matrices can be defined by writing

$$A^{-n} = (A^{-1})^n$$

Negative powers of singular matrices are not defined.

With these definitions and the understanding that $A^0 = I$, it is clear that the familiar laws of exponents

$$A^r A^s = A^{r+s} \qquad \text{and} \qquad (A^r)^s = A^{rs}$$

hold for all square matrices, singular or not, if r and s are nonnegative integers and hold for all integral values of r and s if A is nonsingular.

As our final result in this section, we have the following theorem.

Theorem 5

If $A = \|a_{ij}\|$ is a nonsingular matrix, and if α_{ji} is the cofactor of the element in the jth row and ith column in the determinant of A^{-1}, then

$$|A|\, \alpha_{ji} = a_{ij}$$

Proof By Theorem 3, the inverse of A^{-1} is A. On the other hand, by definition the inverse of A^{-1} is $1/|A^{-1}|$ times the transposed cofactor matrix of A^{-1}. Therefore, equating corresponding elements in the two versions of $(A^{-1})^{-1}$, we have

$$a_{ij} = \frac{\alpha_{ji}}{|A^{-1}|}$$

Finally, since $|A^{-1}| = 1/|A|$, we have

$$a_{ij} = |A|\, \alpha_{ji}$$

as asserted.

If we apply Theorem 5 to the special case of a (3,3) matrix and write the cofactor α_{ji} at length, we have the following useful result.

Corollary 1

If $A = \|a_{ij}\|$ is a nonsingular (3,3) matrix, if A_{ij} is the cofactor of a_{ij} in $|A|$, and if (i,j,k) is any permutation of the numbers 1, 2, 3, then

$$A_{jj}A_{kk} - A_{jk}A_{kj} = |A|\, a_{ii}$$
$$A_{jk}A_{ki} - A_{ji}A_{kk} = |A|\, a_{ij}$$

Exercises

1. Find the adjoint of each of the following matrices and, when it exists, find the inverse:

(a) $\begin{Vmatrix} 1 & 2 \\ 2 & 3 \end{Vmatrix}$
(b) $\begin{Vmatrix} 1 & 2 \\ 2 & 4 \end{Vmatrix}$
(c) $\begin{Vmatrix} 2 & 3 \\ -4 & 0 \end{Vmatrix}$
(d) $\begin{Vmatrix} 2 & -1 \\ 1 & 1 \end{Vmatrix}$

2. Find the adjoint of each of the following matrices and, when it exists, find the inverse:

(a) $\begin{Vmatrix} 1 & 1 & -1 \\ -1 & 0 & 2 \\ 2 & 5 & 3 \end{Vmatrix}$
(b) $\begin{Vmatrix} 1 & 1 & 1 \\ 1 & 2 & 3 \\ 3 & 2 & 1 \end{Vmatrix}$
(c) $\begin{Vmatrix} 1 & 0 & 2 \\ 3 & 1 & -1 \\ 2 & 4 & 3 \end{Vmatrix}$

3. For what values of λ, if any, do the following matrices have inverses?

(a) $\begin{Vmatrix} \lambda - 1 & 2 \\ 3 & \lambda \end{Vmatrix}$
(b) $\begin{Vmatrix} \lambda - 1 & \lambda - 2 \\ \lambda - 3 & \lambda - 4 \end{Vmatrix}$
(c) $\begin{Vmatrix} 3 & 1 & 0 \\ -4 & 2 & 5 \\ \lambda^2 & \lambda & 1 \end{Vmatrix}$

(d) $\begin{Vmatrix} \lambda - 1 & \lambda & \lambda + 1 \\ 2 & -1 & 3 \\ \lambda + 3 & \lambda - 2 & \lambda + 7 \end{Vmatrix}$
(e) $\begin{Vmatrix} 2 - \lambda & -1 & 2\lambda \\ 1 & 2 & 3 \\ 2 & 2 & 1 \end{Vmatrix}$

4. For each of the following pairs of matrices, verify that $(AB)^{-1} = B^{-1}A^{-1}$:

(a) $A = \begin{Vmatrix} 1 & 1 \\ 1 & 2 \end{Vmatrix}, \qquad B = \begin{Vmatrix} 1 & 2 \\ 2 & 5 \end{Vmatrix}$

(b) $A = \begin{Vmatrix} 2 & 0 \\ -1 & 1 \end{Vmatrix}, \qquad B = \begin{Vmatrix} 0 & -1 \\ 2 & 3 \end{Vmatrix}$

(c) $A = \begin{Vmatrix} 1 & 0 & -1 \\ 0 & 2 & 0 \\ 1 & 1 & 3 \end{Vmatrix}, \qquad B = \begin{Vmatrix} 1 & 1 & 1 \\ 1 & 2 & 1 \\ 1 & 3 & 2 \end{Vmatrix}$

(d) $A = \begin{Vmatrix} 1 & -1 & 0 \\ 2 & 0 & 1 \\ 1 & 3 & 1 \end{Vmatrix}, \qquad B = \begin{Vmatrix} 2 & 1 & 1 \\ 1 & 2 & 2 \\ 1 & 2 & 4 \end{Vmatrix}$

5. Verify the relations $|A^{-1}| = 1/|A|$ and $(A^{-1})^{-1} = A$ for each of the following matrices:

(a) $\begin{Vmatrix} 1 & 1 \\ 2 & 3 \end{Vmatrix}$
(b) $\begin{Vmatrix} 2 & 3 \\ 4 & 5 \end{Vmatrix}$
(c) $\begin{Vmatrix} 1 & 1 & 0 \\ 2 & 2 & 1 \\ 1 & 2 & 3 \end{Vmatrix}$
(d) $\begin{Vmatrix} 1 & 2 & 3 \\ 1 & 0 & 1 \\ -1 & 2 & 0 \end{Vmatrix}$

6. Prove Corollary 1, Theorem 4.

7. If A is a nonsingular matrix which commutes with a matrix B, prove that A^{-1} commutes with B. If B is also nonsingular, do A^{-1} and B^{-1} commute?

8. (a) If A is a nonsingular matrix, show that the determinant of the adjoint of A is equal to the $(n-1)$st power of the determinant of A.

(b) If A is a nonsingular matrix, show that the adjoint of the adjoint of A is equal to A times the $(n-2)$nd power of the determinant of A.

9. Show by an example that $AB = AC$ does not imply that $B = C$. Are there any conditions under which it is possible to draw the conclusion that $B = C$ from the fact that $AB = AC$?

10. Under what conditions, if any, is $(AB)^2 = A^2B^2$?

3.4 Linear Transformations

By a **linear transformation** we mean a relation of the form

$$T_a: \begin{array}{l} y_1 = a_{11}x_1 + a_{12}x_2 + \cdots + a_{1n}x_n \\ y_2 = a_{21}x_1 + a_{22}x_2 + \cdots + a_{2n}x_n \\ \quad \cdot \qquad \cdot \qquad \cdot \qquad \cdots \qquad \cdot \\ y_n = a_{n1}x_1 + a_{n2}x_2 + \cdots + a_{nn}x_n \end{array}$$

connecting the variables $(x_1, x_2, \ldots, x_n)$ and the variables $(y_1, y_2, \ldots, y_n)$ in which the a_{ij}'s are coefficients independent of the x's and the y's. For example, if $n = 2$, the transformation T_a might be one sending a point in the euclidean plane with cartesian coordinates (x_1, x_2) into the point with cartesian coordinates (y_1, y_2). Similarly, if $n = 3$, T_a might be a transformation in euclidean three-space sending a point with cartesian coordinates (x_1, x_2, x_3) into the point with cartesian coordinates (y_1, y_2, y_3), or it might be a transformation in the extended euclidean plane sending a point with homogeneous coordinates (x_1, x_2, x_3) into the point with homogeneous coordinates (y_1, y_2, y_3).

Using the ideas of matric multiplication and equality, it is possible to express the transformation T_a in a very compact form. To do this, let us define the variable column vectors

$$X = \begin{Vmatrix} x_1 \\ x_2 \\ \cdot \\ x_n \end{Vmatrix} \quad \text{and} \quad Y = \begin{Vmatrix} y_1 \\ y_2 \\ \cdot \\ y_n \end{Vmatrix}$$

and the coefficient matrix

$$A = \begin{Vmatrix} a_{11} & a_{12} & \cdots & a_{1n} \\ a_{21} & a_{22} & \cdots & a_{2n} \\ \cdot & \cdot & \cdots & \cdot \\ a_{n1} & a_{n2} & \cdots & a_{nn} \end{Vmatrix}$$

Now A is an (n,n) matrix, and X is an $(n,1)$ matrix. Hence A and X are conformable in the order A, X and can therefore be multiplied, giving us the $(n,1)$, or column, matrix

$$AX = \begin{Vmatrix} a_{11} & a_{12} & \cdots & a_{1n} \\ a_{21} & a_{22} & \cdots & a_{2n} \\ \cdot & \cdot & \cdots & \cdot \\ a_{n1} & a_{n2} & \cdots & a_{nn} \end{Vmatrix} \cdot \begin{Vmatrix} x_1 \\ x_2 \\ \cdot \\ x_n \end{Vmatrix} = \begin{Vmatrix} (a_{11}x_1 + a_{12}x_2 + \cdots + a_{1n}x_n) \\ (a_{21}x_1 + a_{22}x_2 + \cdots + a_{2n}x_n) \\ \cdots\cdots\cdots\cdots\cdots\cdots\cdots \\ (a_{n1}x_1 + a_{n2}x_2 + \cdots + a_{nn}x_n) \end{Vmatrix}$$

Furthermore, since matrices are equal if and only if elements in corresponding positions in the two matrices are equal, it follows that the condition

$$Y = AX \qquad \text{that is,} \qquad \begin{Vmatrix} y_1 \\ y_2 \\ \cdot \\ y_n \end{Vmatrix} = \begin{Vmatrix} (a_{11}x_1 + a_{12}x_2 + \cdots + a_{1n}x_n) \\ (a_{21}x_1 + a_{22}x_2 + \cdots + a_{2n}x_n) \\ \cdots\cdots\cdots\cdots\cdots\cdots\cdots \\ (a_{n1}x_1 + a_{n2}x_2 + \cdots + a_{nn}x_n) \end{Vmatrix}$$

is precisely equivalent to the n scalar equations which collectively constitute the original definition of T_a. Thus, using the matrices X, Y, and A, we have shown that regardless of the value of n, T_a can be written in the form

$$T_a: Y = AX$$

The economy of the matric notation in this connection is obvious.

If we consider the equations of T_a when $n = 3$ to be a transformation in E_2^+ or Π_2, the image of any point $P:(a_1,a_2,a_3)$ can be found by substituting a_i for x_i, that is, by substituting the matrix $P = \begin{Vmatrix} a_1 \\ a_2 \\ a_3 \end{Vmatrix}$ for the variable matrix $X = \begin{Vmatrix} x_1 \\ x_2 \\ x_3 \end{Vmatrix}$. This would not be possible if the coordinate matrix of the point P had been defined to be the *row* matrix $\| a_1 \ \ a_2 \ \ a_3 \|$, because the $(3,3)$ matrix A and the $(1,3)$ matrix $P = \| a_1 \ \ a_2 \ \ a_3 \|$ are not conformable in the order A, P. It was for this reason that in Sec. 3.2 we defined the coordinate matrix of a general point, P, in E_2^+ or Π_2 to be a column matrix rather than a row matrix.

In studying geometry, one is frequently concerned with the effect of performing two or more transformations in succession; i.e., one is interested in determining the **composition** of two transformations:

Definition 1

The transformation, T_{ba}, which results when a transformation T_a is followed by a transformation T_b is known as the composition of T_a and T_b.

To investigate this problem in the case of linear transformations, let us suppose that we have a transformation $T_a: Y = AX$ which transforms a vector X into a vector Y and a second transformation $T_b\colon Z = BY$ which transforms a vector Y into a vector Z. The net result of applying first T_a and then T_b is to transform the vector X into the vector Z, and it is a matter of some importance to find the equations of the equivalent transformation which connects X directly with Z. This can be done, of course, by eliminating the variables $(y_1, y_2, \ldots, y_n)$ between the equations of T_a and T_b, but to do this, do we use the matric or the scalar form of the equations of the transformations?

If we use the matric form of the equations, the elimination is very simple, for since

$$Z = BY \qquad \text{and} \qquad Y = AX$$

we have, on substituting for Y,

$$Z = B(AX) = (BA)X$$

which asserts that the composition of the two transformations is a linear transformation whose coefficient matrix is the product of the coefficient matrices of the component transformations in the order B, A. It is not difficult to show that this is always the same as the result obtained by eliminating the y's from the scalar equations of the transformation. For brevity, however, we shall do this only for the special case $n = 2$. When $n = 2$, we have, specifically,

$$T_a\colon \begin{array}{l} y_1 = a_{11}x_1 + a_{12}x_2 \\ y_2 = a_{21}x_1 + a_{22}x_2 \end{array} \qquad \text{and} \qquad T_b\colon \begin{array}{l} z_1 = b_{11}y_1 + b_{12}y_2 \\ z_2 = b_{21}y_1 + b_{22}y_2 \end{array}$$

Now, eliminating y_1 and y_2 by substituting from the equations of T_a into the the equations of T_b, we have

$$T_{ba}\colon \begin{array}{l} z_1 = b_{11}(a_{11}x_1 + a_{12}x_2) + b_{12}(a_{21}x_1 + a_{22}x_2) \\ z_2 = b_{21}(a_{11}x_1 + a_{12}x_2) + b_{22}(a_{21}x_1 + a_{22}x_2) \end{array}$$

or, removing parentheses and collecting terms,

$$T_{ba}\colon \begin{array}{l} z_1 = (b_{11}a_{11} + b_{12}a_{21})x_1 + (b_{11}a_{12} + b_{12}a_{22})x_2 \\ z_2 = (b_{21}a_{11} + b_{22}a_{21})x_1 + (b_{21}a_{12} + b_{22}a_{22})x_2 \end{array}$$

This is a linear transformation which can be written in matric notation in the form

$$\left\| \begin{array}{c} z_1 \\ z_2 \end{array} \right\| = \left\| \begin{array}{cc} (b_{11}a_{11} + b_{12}a_{21}) & (b_{11}a_{12} + b_{12}a_{22}) \\ (b_{21}a_{11} + b_{22}a_{21}) & (b_{21}a_{12} + b_{22}a_{22}) \end{array} \right\| \cdot \left\| \begin{array}{c} x_1 \\ x_2 \end{array} \right\|$$

Moreover, the coefficient matrix is clearly just the matric product

$$BA = \left\| \begin{array}{cc} b_{11} & b_{12} \\ b_{21} & b_{22} \end{array} \right\| \cdot \left\| \begin{array}{cc} a_{11} & a_{12} \\ a_{21} & a_{22} \end{array} \right\|$$

Thus, for the case $n = 2$ we have proved the following important result.

Theorem 1

The result of following a linear transformation T_a: $Y = AX$ with the linear transformation T_b: $Z = BY$ is the linear transformation T_{ba}: $Z = BAX$, whose coefficient matrix is the product, BA, of the coefficient matrices of T_b and T_a.

Another problem which is often encountered in working with linear transformations is the following: given a transformation T_a: $Y = AX$ expressing Y in terms of X, to find the inverse transformation which expresses X in terms of Y. If the problem involved a single *scalar* equation, $y = ax$, the solution would be immediate: we can solve for x in terms of y if and only if a is different from zero, and when this is the case, we have

$$x = \frac{y}{a} = a^{-1}y$$

In the matric case, $Y = AX$, the situation is almost the same. If the determinant of the coefficients, $|A|$, is different from zero, then A^{-1} exists and we can multiply the equation $Y = AX$ through on the left by A^{-1}, getting

$$A^{-1}Y = A^{-1}AX = (A^{-1}A)X = IX = X$$

We have thus established a further important result.

Theorem 2

If $|A| \neq 0$, the linear transformation T_a: $Y = AX$ can be solved for X in terms of Y, and when this is the case, the inverse transformation is T_a^{-1}: $X = A^{-1}Y$.

Example 1

What is the locus of the images of the points of the conic Γ: $x_1x_3 = ax_2{}^2$, $a \neq 0$, under the transformation

$$T: \quad \begin{aligned} y_1 &= -x_1 \\ y_2 &= 2x_1 + x_2 \\ y_3 &= 2x_1 + 2x_2 - x_3 \end{aligned}$$

The most obvious way to attack this problem is to identify a general point on Γ, find its image under the transformation T, and then attempt to discover an equation satisfied by the coordinates of all such image points. To do this, let us first assign to x_1 the arbitrary value s and to x_2 the arbitrary value t. Then except when $s = 0$, that is, except for the point

(0,0,1), the coordinates of an arbitrary point on Γ are of the form

$$\left(s,t,\frac{at^2}{s}\right)$$

Moreover, if we multiply by the proportionality constant s, we obtain (s^2,st,at^2), and this family of coordinate-triples includes the triple $(0,0,1)$ corresponding to the previously excluded value $s = 0$. Under T, the image of the point (s^2,st,at^2) becomes

(1)
$$\begin{aligned} y_1 &= -s^2 \\ y_2 &= 2s^2 + st \\ y_3 &= 2s^2 + 2st - at^2 \end{aligned}$$

To obtain an equation satisfied identically by the coordinates (y_1,y_2,y_3), we must eliminate the parameters s and t from these expressions. To do this, we solve for s^2 from the first equation, getting

$$s^2 = -y_1$$

Then from the second equation we find that

$$st = 2y_1 + y_2$$

and from the third equation we obtain

$$t^2 = \frac{2y_1 + 2y_2 - y_3}{a}$$

Finally, since $(st)^2 = s^2 t^2$, we have

$$(2y_1 + y_2)^2 = (-y_1)\left(\frac{2y_1 + 2y_2 - y_3}{a}\right)$$

or

(2)
$$(4a + 2)y_1{}^2 + (4a + 2)y_1 y_2 + ay_2{}^2 - y_1 y_3 = 0$$

Conversely, if (y_1,y_2,y_3) satisfy Eq. (2), then the above steps can be reversed, Eqs. (1) can be obtained, and we can conclude that (y_1,y_2,y_3) are the coordinates of the image of some point on Γ. Thus the locus of the images of the points of the conic Γ is also a conic. Furthermore, if $a = -\frac{1}{2}$, the image of Γ is Γ itself; in other words, the conic $2x_1 x_3 + x_2{}^2 = 0$ is invariant under the transformation T.

Another way of solving this problem, which is equivalent to what we have just done but more straightforward, is to find the inverse of T, that is, the equations giving the x's in terms of the y's, and then substitute these expressions into the equation of Γ: From the equations of T we find easily that its inverse is

$$T^{-1}: \quad \begin{aligned} x_1 &= -y_1 \\ x_2 &= 2y_1 + y_2 \\ x_3 &= 2y_1 + 2y_2 - y_3 \end{aligned}$$

Hence, substituting these into the equations of Γ, we obtain

$$-y_1(2y_1 + 2y_2 - y_3) = a(2y_1 + y_2)^2$$

which simplifies at once to Eq. (2).

Incidentally, in this case the matrix of T^{-1} is the same as the matrix of T; that is, the matrix

$$\begin{Vmatrix} -1 & 0 & 0 \\ 2 & 1 & 0 \\ 2 & 2 & -1 \end{Vmatrix}$$

is its own inverse. Linear transformations whose matrices are their own inverses will play an important role in much of our later work.

Exercises

1. If $A = \begin{Vmatrix} 1 & 1 & 0 \\ 1 & 2 & -1 \\ 3 & 1 & 1 \end{Vmatrix}$ and $B = \begin{Vmatrix} 3 & 0 & -1 \\ 2 & 1 & 2 \\ 0 & 1 & 3 \end{Vmatrix}$, use the scalar forms of the transformations $T_a\colon Y = AX$ and $T_b\colon Z = BY$ to verify that the matrix of the composition of T_a and T_b is BA.

2. In Exercise 1, what is the composition of T_b and T_a? What is the inverse of T_a? What is the inverse of T_b?

3. If $A = \begin{Vmatrix} 0 & 1 & -1 \\ 2 & 1 & -2 \\ 1 & 1 & -2 \end{Vmatrix}$, what is the composition of the transformations $Y = AX$ and $Z = AY$? What is the inverse of the transformation $Y = AX$?

4. Under the transformation $Y = AX$ of Exercise 3, show that:
(a) A point and its image are always collinear with the point $(1,2,1)$.
(b) Every point on the line $x_1 + x_2 - x_3 = 0$ is transformed into itself.
(c) The image of every point on the conic $x_1{}^2 - 2x_1x_2 + x_3{}^2 = 0$ is also a point of this conic.

5. Show that any values of x_1, x_2, y_1, y_2 which satisfy the matric equation

$$\begin{Vmatrix} y_1 \\ y_2 \end{Vmatrix} = k \begin{Vmatrix} b & d \\ -a & -c \end{Vmatrix} \cdot \begin{Vmatrix} x_1 \\ x_2 \end{Vmatrix}$$

will also satisfy the bilinear equation $ax_1y_1 + bx_1y_2 + cx_2y_1 + dx_2y_2 = 0$. Under what conditions, if any, is the converse true?

6. (a) If A is a nonsingular (3,3) matrix, show that in E_2^+ the transformation $Y' = AY$ always transforms a line into a line; i.e., show that the locus of the images of the points of an arbitrary line is a line.

(*b*) What is the image of the line $x_1 - x_2 + 2x_3 = 0$ under the transformation T_a of Exercise 1 ?

7. (*a*) If A is a nonsingular $(3,3)$ matrix, show that in E_2^+ the transformation $Y = AX$ always transforms a conic into a conic; i.e., show that the locus of the images of the points of an arbitrary conic is a conic.

(*b*) What is the image of the conic $x_1 x_2 = x_3^2$ under the transformation T_b of Exercise 1 ?

8. Let $P:(p_1,p_2,p_3)$ and $Q:(q_1,q_2,q_3)$ be two points on an arbitrary line l in E_2^+, and let P' and Q' be the images of these points under the transformation T_a of Exercise 1. Find the equations expressing the coordinates of the image line, $l' = P'Q'$, in terms of the coordinates of l. How is the matrix of this transformation of line-coordinates related to the matrix of T_a ?

9. Work Exercise 8 using the transformation T_b of Exercise.1.

10. Give a proof of Theorem 1 which will be valid for all positive integral values of n.

3.5 Linear Dependence and Independence

Many problems in geometry involve the question of whether or not special relations exist between the elements of a given configuration. For instance, a construction may furnish us with three points, and we may wish to know the conditions, if any, under which the points will be collinear. Or we may wish to know whether three lines, determined in some particular way, can ever be concurrent. When a coordinate system is available for the description of geometric objects, questions such as these are often equivalent to the determination of whether related algebraic objects, such as coordinate-vectors or -equations, are linearly dependent or independent. Accordingly, we shall devote this section to a brief discussion of the concepts of linear dependence and linear independence, preparatory to resuming our study of the extended euclidean plane in the next chapter, where these ideas will be of great utility.

Definition 1

The quantities $Q_1, Q_2, \ldots, Q_n$ are said to be linearly dependent if and only if there exists a set of constants, $c_1, c_2, \ldots, c_n$, at least one of which is different from zero, such that the equation $c_1 Q_1 + c_2 Q_2 + \cdots + c_n Q_n = 0$ holds identically.

Definition 2

The quantities $Q_1, Q_2, \ldots, Q_n$ are said to be linearly independent if and only if they are not linearly dependent; i.e., if and only if the only equation of the form $c_1Q_1 + c_2Q_2 + \cdots + c_nQ_n = 0$ which they satisfy identically has $c_1 = c_2 = \cdots = c_n = 0$.

An expression of the form $c_1Q_1 + c_2Q_2 + \cdots + c_nQ_n$ in which at least one of the c's is different from zero is said to be a **linear combination** of the Q's.

The word *dependent* in the phrase *linearly dependent* suggests that if the quantities $Q_1, Q_2, \ldots, Q_n$ are linearly dependent, then at least one of the Q's depends upon, or can be expressed in terms of, the others. This inference is made more precise in the following theorem.

Theorem 1

If the quantities $Q_1, Q_2, \ldots, Q_n$ are linearly dependent, then at least one of the quantities (but not necessarily each one) can be expressed as a linear combination of the remaining ones.

Proof Since $Q_1, Q_2, \ldots, Q_n$ are linearly dependent, they necessarily satisfy a linear equation of the form $c_1Q_1 + c_2Q_2 + \cdots + c_nQ_n = 0$ in which at least one of the c's, say c_i, is different from zero. This being the case, we can solve for Q_i by dividing by c_i, getting

$$Q_i = -\frac{c_1}{c_i}Q_1 - \frac{c_2}{c_i}Q_2 - \cdots - \frac{c_n}{c_i}Q_n$$

which expresses Q_i as a linear combination of the remaining Q's. Since some, though not all, of the c's may be zero, it is clear that we may not be able to solve for each of the Q's in this fashion.

Any member of a set of quantities, $Q_1, Q_2, \ldots, Q_n$, which can be expressed as a linear combination of other members of the set is said to be **linearly dependent** upon those quantities.

Example 1

Show that the vectors $V_1 = \begin{Vmatrix} 1 \\ 0 \\ 0 \end{Vmatrix}$, $V_2 = \begin{Vmatrix} 0 \\ 1 \\ 0 \end{Vmatrix}$, $V_3 = \begin{Vmatrix} 0 \\ 0 \\ 1 \end{Vmatrix}$ are linearly independent.

To show that V_1, V_2, V_3 are linearly independent, let us consider the possibility that they satisfy a linear equation of the form $c_1V_1 + c_2V_2 + c_3V_3 = O$, where, of course, the symbol O denotes the zero element consistent

with the nature of the given quantities, that is, O denotes the zero *vector*
$\begin{Vmatrix} 0 \\ 0 \\ 0 \end{Vmatrix}$. Using the properties of matric multiplication and addition, the assumed relation can be written

$$c_1 V_1 + c_2 V_2 + c_3 V_3 = c_1 \begin{Vmatrix} 1 \\ 0 \\ 0 \end{Vmatrix} + c_2 \begin{Vmatrix} 0 \\ 1 \\ 0 \end{Vmatrix} + c_3 \begin{Vmatrix} 0 \\ 0 \\ 1 \end{Vmatrix}$$

$$= \begin{Vmatrix} c_1 \\ 0 \\ 0 \end{Vmatrix} + \begin{Vmatrix} 0 \\ c_2 \\ 0 \end{Vmatrix} + \begin{Vmatrix} 0 \\ 0 \\ c_3 \end{Vmatrix} = \begin{Vmatrix} c_1 \\ c_2 \\ c_3 \end{Vmatrix} = \begin{Vmatrix} 0 \\ 0 \\ 0 \end{Vmatrix}$$

which, from the definition of matric equality, implies that $c_1 = c_2 = c_3 = 0$. Thus, since the only linear relation which V_1, V_2, and V_3 can satisfy has $c_1 = c_2 = c_3 = 0$, it follows from Definition 2 that they are linearly independent.

Example 2

Show that the vectors $V_1 = \begin{Vmatrix} 1 \\ 2 \\ 1 \end{Vmatrix}$, $V_2 = \begin{Vmatrix} 2 \\ -1 \\ 3 \end{Vmatrix}$, $V_3 = \begin{Vmatrix} 3 \\ -4 \\ 5 \end{Vmatrix}$ are linearly

dependent, and express each as a linear combination of the others.

Again, as in Example 1, let us try to determine values of c_1, c_2, c_3 such that $c_1 V_1 + c_2 V_2 + c_3 V_3 = O$. In this case we have

$$c_1 V_1 + c_2 V_2 + c_3 V_3 = c_1 \begin{Vmatrix} 1 \\ 2 \\ 1 \end{Vmatrix} + c_2 \begin{Vmatrix} 2 \\ -1 \\ 3 \end{Vmatrix} + c_3 \begin{Vmatrix} 3 \\ -4 \\ 5 \end{Vmatrix}$$

$$= \begin{Vmatrix} (c_1 + 2c_2 + 3c_3) \\ (2c_1 - c_2 - 4c_3) \\ (c_1 + 3c_2 + 5c_3) \end{Vmatrix} = \begin{Vmatrix} 0 \\ 0 \\ 0 \end{Vmatrix}$$

which will be true if and only if, simultaneously,

$$c_1 + 2c_2 + 3c_3 = 0$$
$$2c_1 - c_2 - 4c_3 = 0$$
$$c_1 + 3c_2 + 5c_3 = 0$$

This system of equations can easily be solved by elimination. For instance, if we subtract the first equation from the third, we obtain

$$c_2 + 2c_3 = 0 \qquad \text{or} \qquad c_2 = -2c_3$$

Then substituting this into the first equation, we have

$$c_1 + 2(-2c_3) + 3c_3 = 0 \qquad \text{or} \qquad c_1 = c_3$$

Finally, assigning c_3 the arbitrary value k, we find that each equation is satisfied by the expressions $c_1 = k$, $c_2 = -2k$, $c_3 = k$ for all values of k. Thus V_1, V_2, V_3 satisfy the equation

$$V_1 - 2V_2 + V_3 = O$$

which proves that they are linearly dependent. From this equation we have, of course,

(1)
$$\begin{aligned}
V_1 &= 2V_2 - V_3 \\
V_2 &= \tfrac{1}{2}V_1 + \tfrac{1}{2}V_3 \\
V_3 &= -V_1 + 2V_2
\end{aligned}$$

Since we have already agreed to identify a point and the column vector of its homogeneous coordinates, we can think of the vectors V_1, V_2, V_3 as being points in E_2^+ or Π_2. With this interpretation, we can restate the results of the last example and say that the *points* V_1, V_2, V_3 are linearly dependent. Similarly, we can think of Eqs. (1) as expressing relations between points as well as between matrices. In the future we shall often speak, in this sense, of the linear dependence of points and of lines and write relations like Eqs. (1) between points and between lines.

Example 3

Determine for what values of a and b, if any, the expressions $x_1 + x_2$, $x_2 + ax_3$, $bx_1 + x_2 - x_3$ are linearly dependent.

If these three expressions are to be linearly dependent, they must satisfy identically some linear equation of the form

$$c_1(x_1 + x_2) + c_2(x_2 + ax_3) + c_3(bx_1 + x_2 - x_3) = 0$$

or

(2)
$$(c_1 + bc_3)x_1 + (c_1 + c_2 + c_3)x_2 + (ac_2 - c_3)x_3 = 0$$

in which at least one of the c's is different from zero. In particular, evaluating Eq. (2) for each of the sets $(x_1, x_2, x_3) = (1,0,0)$, $(0,1,0)$, $(0,0,1)$, we find that the c's must satisfy the three equations

(3)
$$c_1 + bc_3 = 0$$

(4)
$$c_1 + c_2 + c_3 = 0$$

(5)
$$ac_2 - c_3 = 0$$

From (5) we have

(6) $$c_3 = ac_2$$

and then from (3) we have

(7) $$c_1 = -bc_3 = -abc_2$$

These values will satisfy Eq. (4) if and only if

$$-abc_2 + c_2 + ac_2 = 0 \qquad \text{or} \qquad (1 + a - ab)c_2 = 0$$

Clearly, this will be true if $1 + a - ab = 0$ or if $c_2 = 0$. However, if $c_2 = 0$, then, from (6) and (7), both c_1 and c_3 must be zero, and there is no nonzero c, as required for linear dependence. Hence we conclude that if the given expressions are to be linearly dependent, then it is necessary that $1 + a - ab = 0$. That this condition is also sufficient follows immediately from Eq. (2) since if $c_1 = -abc_2$, $c_3 = ac_2$ (c_2 arbitrary), *and* $1 + a - ab = 0$, the coefficient of each of the x's vanishes. Thus if $1 + a - ab = 0$, there are nonzero values of c_1, c_2, c_3 such that Eq. (2) is satisfied no matter what values are substituted for the x's, which is sufficient to show that the given expressions are linearly dependent.

The geometric interpretation of the preceding discussion is quite interesting. If we equate each of the given expressions to zero and consider the three lines thus defined,

$$l_1: x_1 + x_2 = 0 \qquad l_2: x_2 + ax_3 = 0 \qquad l_3: bx_1 + x_2 - x_3 = 0$$

it is clear that l_1 and l_2 intersect in the point $(a, -a, 1)$. This point will lie on l_3, that is, the three lines will be concurrent, if and only if $ba - a - 1 = 0$, which is just the condition that the expressions defining l_1, l_2, and l_3 be linearly dependent.

Exercises

1. Show that the matrices $\begin{Vmatrix} 0 \\ 1 \\ 1 \end{Vmatrix}$, $\begin{Vmatrix} 1 \\ 0 \\ 1 \end{Vmatrix}$, $\begin{Vmatrix} 1 \\ 1 \\ 0 \end{Vmatrix}$ are linearly independent.

2. Determine whether the matrices $\begin{Vmatrix} 1 \\ 2 \\ 3 \end{Vmatrix}$, $\begin{Vmatrix} 3 \\ 2 \\ 1 \end{Vmatrix}$, $\begin{Vmatrix} 2 \\ 1 \\ 3 \end{Vmatrix}$ are linearly dependent or linearly independent.

3. Show that the matrices $\begin{Vmatrix} 1 & 2 \\ 3 & 4 \end{Vmatrix}$, $\begin{Vmatrix} 2 & -1 \\ 0 & 4 \end{Vmatrix}$, $\begin{Vmatrix} 0 & 5 \\ 6 & 5 \end{Vmatrix}$ are linearly dependent.

4. What conditions must a, b, c, d satisfy in order that the matrices

$$\begin{Vmatrix} 1 & 0 \\ 1 & 2 \end{Vmatrix}, \qquad \begin{Vmatrix} 0 & -1 \\ 2 & 1 \end{Vmatrix}, \qquad \begin{Vmatrix} a & b \\ c & d \end{Vmatrix}$$

be linearly dependent?

5. Show that 1, x, and x^2 are linearly independent.

6. Show that the expressions $x_1^2 - x_2x_3$, $x_2^2 - x_1x_3$, $x_3^2 - x_1x_2$ are linearly independent.

7. Determine whether the expressions $x_1 + x_2$, $x_2 - x_3$, $x_1 - x_2 - x_3$ are linearly dependent or linearly independent.

8. Are $\sin x$ and $\cos x$ linearly dependent?

9. (a) Are $\sin^2 x$ and $\cos^2 x$ linearly dependent?
   (b) Are $\sin^2 x$, $\cos^2 x$, and $\cos 2x$ linearly dependent?
   (c) Are $\sin^2 x$, $\cos^2 x$, and $\sin 2x$ linearly dependent?

10. Determine a, b, c so that the expressions $x^2 + y^2 - 1$, $x^2 + y^2 - 2x - 1$, and $x^2 + y^2 - ax - by - c$ will be linearly dependent. Show further that the linear dependence of these expressions is the necessary and sufficient condition that the circles $x^2 + y^2 - 1 = 0$, $x^2 + y^2 - 2x - 1 = 0$, and $x^2 + y^2 - ax - by - c = 0$ should have two points in common.

11. If $P:(x_1,x_2,x_3)$ is any point on the line determined by the points $A:(1,4,-2)$ and $B:(2,-2,3)$, show that the matrices

$$ P = \left\| \begin{matrix} x_1 \\ x_2 \\ x_3 \end{matrix} \right\| \qquad A = \left\| \begin{matrix} 1 \\ 4 \\ -2 \end{matrix} \right\| \qquad B = \left\| \begin{matrix} 2 \\ -2 \\ 3 \end{matrix} \right\| $$

are linearly dependent, and show further that P can be expressed as a linear combination of A and B.

12. Work Exercise 11 for the pairs of points:
(a) $A:(2,0,3)$, $B:(-1,3,1)$ (b) $A:(3,1,2)$, $B:(-2,1,0)$

13. If the point $P:(x_1,x_2,x_3)$ is not on the line determined by the points $A:(1,1,0)$ and $B:(1,-1,2)$, show that the matrices

$$ P = \left\| \begin{matrix} x_1 \\ x_2 \\ x_3 \end{matrix} \right\| \qquad A = \left\| \begin{matrix} 1 \\ 1 \\ 0 \end{matrix} \right\| \qquad \text{and} \qquad B = \left\| \begin{matrix} 1 \\ -1 \\ 2 \end{matrix} \right\| $$

are linearly independent.

14. Work Exercise 13 for the pairs of points:
(a) $A:(1,0,2)$, $B:(1,1,1)$ (b) $A:(0,2,1)$, $B:(3,0,-1)$

15. (a) Show that if 0 is included in a set of quantities, the members of the set are always linearly dependent.

(*b*) Show that if the quantities $Q_1, Q_2, \ldots, Q_n$ are linearly independent, the members of every subset of the Q's are also linearly independent. Is the converse true?

(*c*) Show that a single nonzero quantity is always linearly independent.

3.6 Systems of Linear Equations

From the work of the last section, it is clear that to answer questions concerning linear dependence and independence it is usually necessary to consider the existence and nature of solutions of systems of linear equations. For this reason, we shall devote this section to a review of the elementary properties and solution processes for systems of simultaneous linear equations.

By a **system of simultaneous linear equations** we mean a set of equations of the form

(1)
$$\begin{aligned}
a_{11}x_1 + a_{12}x_2 + \cdots + a_{1n}x_n &= b_1 \\
a_{21}x_1 + a_{22}x_2 + \cdots + a_{2n}x_n &= b_2 \\
&\;\;\vdots \\
a_{m1}x_1 + a_{m2}x_2 + \cdots + a_{mn}x_n &= b_m
\end{aligned}$$

in which the number of equations, m, is not necessarily equal to the number of unknowns, n. If each of the m quantities $b_1, b_2, \ldots, b_m$ is equal to zero, the system (1) is said to be **homogeneous;** if at least one of the b's is different from zero, the system is said to be **nonhomogeneous.** Clearly, if we define the matrices

$$A = \begin{Vmatrix} a_{11} & a_{12} & \cdots & a_{1n} \\ a_{21} & a_{22} & \cdots & a_{2n} \\ \cdot & \cdot & \cdots & \cdot \\ a_{m1} & a_{m2} & \cdots & a_{mn} \end{Vmatrix} \qquad X = \begin{Vmatrix} x_1 \\ x_2 \\ \cdot \\ x_n \end{Vmatrix} \qquad B = \begin{Vmatrix} b_1 \\ b_2 \\ \cdot \\ b_m \end{Vmatrix}$$

the system can be written in the compact matric form

$$AX = B$$

In this form, the matrix A is called the **coefficient matrix** of the system, and the matrix $\|A \;\; B\|$, obtained by adjoining the column matrix B to the coefficient matrix A as an $(n + 1)$st column, is known as the **augmented matrix** of the system.

One of the most fundamental properties of the homogeneous system $AX = O$ is contained in the following theorem.

Theorem 1

If $X = X_1$ and $X = X_2$ are two solution vectors of the homogeneous system $AX = O$, then for all values of the scalar constants c_1 and c_2 the vector $X = c_1X_1 + c_2X_2$ is also a solution.

Proof   To determine whether $X = c_1 X_1 + c_2 X_2$ is a solution of the equation $AX = O$, we substitute it into the equation, getting, by familiar laws of matrices,

$$(2) \qquad A(c_1 X_1 + c_2 X_2) = c_1 A X_1 + c_2 A X_2 \overset{?}{=} O$$

Now the matric products AX_1 and AX_2 are both zero, since by hypothesis both X_1 and X_2 are solutions of the equation $AX = O$. Hence the second expression in (2) is zero for all values of c_1 and c_2, which proves that $X = c_1 X_1 + c_2 X_2$ is a solution of $AX = O$, as asserted.

Since any vector $X = c_1 X_1 + c_2 X_2$ is linearly dependent upon the vectors X_1 and X_2, another way to state the last theorem is to say that *any vector linearly dependent upon solution vectors of the homogeneous system $AX = O$ is also a solution.*

Before we can state the fundamental theorem concerning the existence of solutions of the general system of equations $AX = B$, we must introduce the notion of the **rank** of a matrix.

Definition 1

The array of elements common to any k rows and any l columns of a matrix A is called a (k,l) submatrix of A.

Definition 2

The rank of a matrix A is the largest value of r for which there exists an (r,r) submatrix of A whose determinant is different from zero.

The rank of a matrix A, as defined in Definition 2, is sometimes referred to more specifically as the **determinant rank** of A.

Example 1

The matrix $A = \begin{Vmatrix} 1 & 2 & 2 & 3 \\ 4 & 3 & -1 & 0 \\ 2 & -1 & -5 & -6 \end{Vmatrix}$ is of rank 2, since the determinant of each of the (3,3) submatrices

$$\begin{Vmatrix} 2 & 2 & 3 \\ 3 & -1 & 0 \\ -1 & -5 & -6 \end{Vmatrix} \qquad \begin{Vmatrix} 1 & 2 & 3 \\ 4 & -1 & 0 \\ 2 & -5 & -6 \end{Vmatrix}$$

$$\begin{Vmatrix} 1 & 2 & 3 \\ 4 & 3 & 0 \\ 2 & -1 & -6 \end{Vmatrix} \qquad \begin{Vmatrix} 1 & 2 & 2 \\ 4 & 3 & -1 \\ 2 & -1 & -5 \end{Vmatrix}$$

is zero, while at least one (2,2) submatrix, for instance, the matrix $\left\| \begin{matrix} 1 & 2 \\ 4 & 3 \end{matrix} \right\|$ in the upper left-hand corner, has a nonzero determinant.

We are now in a position to state the fundamental theorem on the existence of solutions of the general system $AX = B$.

Theorem 2

A system of m linear equations in n unknowns, $AX = B$, has a solution if and only if the coefficient matrix A and the augmented matrix $\|A \quad B\|$ have the same rank.

In general, the ranks of the coefficient matrix and the augmented matrix are not computed as a first step in the solution of a nonhomogeneous system $AX = B$. Instead, a straightforward solution process is undertaken which will give the general solution of the system if there is one and reveal with certainty the nonexistence of solutions if such is the case. The next example illustrates the general method of solution, which is known as the **Gauss reduction.**

Example 2

Determine whether or not the system of equations

$$
\begin{aligned}
x_1 + 2x_2 + x_3 - x_4 &= 1 \\
x_1 + 3x_2 + x_3 - 2x_4 &= 5 \\
x_1 + 4x_2 - 2x_3 + 6x_4 &= 3 \\
x_1 + 5x_2 - 4x_3 + 11x_4 &= 1
\end{aligned}
$$

has a solution, and find the general solution if one exists.

We begin by assuming that the given equations do hold simultaneously. Then, subtracting the first equation from each of the others, in order to eliminate x_1 from every equation after the first, we obtain the equivalent system

$$
\begin{aligned}
x_1 + 2x_2 + x_3 - x_4 &= 1 \\
x_2 \qquad\quad - x_4 &= 4 \\
2x_2 - 3x_3 + 7x_4 &= 2 \\
3x_2 - 5x_3 + 12x_4 &= 0
\end{aligned}
$$

Next we subtract twice the new second equation from the new third equation, and three times the new second equation from the new fourth equation, in order to eliminate x_2 from the last two equations. This gives us the equivalent system

$$\begin{aligned}
x_1 + 2x_2 + x_3 - x_4 &= 1 \\
x_2 \qquad - x_4 &= 4 \\
-3x_3 + 9x_4 &= -6 \\
-5x_3 + 15x_4 &= -12
\end{aligned}$$

Finally, multiplying the third equation in the last set by $\frac{5}{3}$ and subtracting the result from the last equation, in order to eliminate x_3 from the last equation, we obtain the system

$$\begin{aligned}
x_1 + 2x_2 + x_3 - x_4 &= 1 \\
x_2 \qquad - x_4 &= 4 \\
-3x_3 + 9x_4 &= -6 \\
0 &= -2 \ (!)
\end{aligned}$$

The assumption that the original system was solvable has thus led us to a contradiction, which proves that the original system is inconsistent, i.e., has no solution. According to Theorem 2, the fact that the given system of equations has no solution implies that the rank of the coefficient matrix A and the rank of the augmented matrix $\|A \quad B\|$ are not equal. Indeed, in this case it is not difficult to show that the rank of A is 3 while the rank of $\|A \quad B\|$ is 4.

For the homogeneous system $AX = O$, it is clear that the rank of the coefficient matrix and the rank of the augmented matrix are always the same, since every submatrix in A is also in $\|A \quad B\|$ while every submatrix which is in $\|A \quad B\|$ but not in A is simply some submatrix of A augmented with a final column of zeros. Thus, according to Theorem 2, the homogeneous system $AX = O$ always has at least one solution. This, of course, is obvious, since any homogeneous system is clearly satisfied by the solution $x_1 = x_2 = \cdots = x_n = 0$. This solution is usually of no interest to us, however, since, in particular, coordinate-triples of the form $(0,0,0)$ or $[0,0,0]$ correspond to no geometrical objects. For this reason, we shall call the obvious solution $X = O$ of the system $AX = O$ a **trivial solution.** The next theorem gives us information about the existence of nontrivial solutions of the homogeneous system.

Theorem 3

If n is the number of unknowns in the homogeneous system $AX = O$, and if r is the rank of the coefficient matrix, A, then the maximum number of linearly independent solution vectors of the system is $n - r$.

Example 3

Solve the system of equations

$$x_1 + 2x_2 - 2x_3 + 3x_4 = 0$$
$$x_1 + 3x_2 - x_3 + 2x_4 = 0$$
$$x_1 + x_2 - 3x_3 + 4x_4 = 0$$
$$x_1 + 9x_2 + 5x_3 - 4x_4 = 0$$

Subtracting the first equation from each of the others in order to eliminate x_1 from the last three equations, we obtain the system

$$x_1 + 2x_2 - 2x_3 + 3x_4 = 0$$
$$x_2 + x_3 - x_4 = 0$$
$$- x_2 - x_3 + x_4 = 0$$
$$7x_2 + 7x_3 - 7x_4 = 0$$

Now if we add the new second equation to the new third equation, and subtract seven times the new second equation from the new fourth equation, we obtain

$$x_1 + 2x_2 - 2x_3 + 3x_4 = 0$$
$$x_2 + x_3 - x_4 = 0$$
$$0 = 0$$
$$0 = 0$$

This system, which of course is equivalent to the original system, i.e., is satisfied by precisely the same set of values that satisfy the original system, can be solved at once by assigning arbitrary values to x_3 and x_4, say $x_3 = \lambda$ and $x_4 = \mu$, and then solving for x_1 and x_2 in terms of λ and μ. The result is

$$x_1 = 4\lambda - 5\mu$$
$$x_2 = -\lambda + \mu$$
$$x_3 = \lambda$$
$$x_4 = \mu$$

or in matric form

$$\begin{Vmatrix} x_1 \\ x_2 \\ x_3 \\ x_4 \end{Vmatrix} = \lambda \begin{Vmatrix} 4 \\ -1 \\ 1 \\ 0 \end{Vmatrix} + \mu \begin{Vmatrix} -5 \\ 1 \\ 0 \\ 1 \end{Vmatrix}$$

The two particular solution vectors $X_1 = \begin{Vmatrix} 4 \\ -1 \\ 1 \\ 0 \end{Vmatrix}$ and $X_2 = \begin{Vmatrix} -5 \\ 1 \\ 0 \\ 1 \end{Vmatrix}$, corresponding, respectively, to the values $\lambda = 1$, $\mu = 0$, and $\lambda = 0$, $\mu = 1$, are easily seen to be linearly independent, and what we have shown is that any solution of the given system can be expressed as a linear combination of these two linearly independent solutions.

The Gauss reduction provides a straightforward procedure for solving any system of the form $AX = B$. However, in certain important particular cases there are other methods of solution which are often more convenient. Specifically, for systems in which the number of equations is equal to the number of unknowns, we have the results described in the next three theorems, the first of which, at least, should be familiar to us from college algebra.

Theorem 4   Cramer's rule

If the coefficient matrix A of a system of n equations in n unknowns, $AX = B$, is nonsingular, then the system has the unique solution

$$x_1 = \frac{|D_1|}{|A|}, \qquad x_2 = \frac{|D_2|}{|A|}, \qquad \cdots, \qquad x_n = \frac{|D_n|}{|A|}$$

where D_i is the matrix obtained from A by replacing the ith column of A by the column vector B.

Example 4

Without solving for x_1 or x_3, find the value of x_2 from the equations

$$\begin{aligned}
x_1 + 2x_2 + x_3 &= 2 \\
2x_1 - 5x_2 + x_3 &= -3 \\
3x_1 - x_2 + 2x_3 &= 0
\end{aligned}$$

Using Cramer's rule, we have for the value of x_2 the quotient $|D_2|/|A|$, or

$$x_2 = \frac{\begin{vmatrix} 1 & 2 & 1 \\ 2 & -3 & 1 \\ 3 & 0 & 2 \end{vmatrix}}{\begin{vmatrix} 1 & 2 & 1 \\ 2 & -5 & 1 \\ 3 & -1 & 2 \end{vmatrix}} = \frac{1}{2}$$

For a system of homogeneous linear equations, $AX = O$, the rank of the coefficient matrix and the rank of the augmented matrix are obviously equal. Hence, by Theorem 2, the system $AX = O$ always has at least one solution, which clearly is the trivial solution $x_1 = x_2 = \cdots = x_n = 0$. If A is an (n,n) matrix of rank n, then, by Theorem 3, the only solution of $AX = O$

is the trivial solution. On the other hand, if A is an (n,n) matrix whose rank is less than n, then Theorem 3 guarantees that $AX = O$ has at least one nontrivial solution in addition to the obvious trivial solution. These observations are summarized in the following exceedingly important theorem.

Theorem 5

A homogeneous system of n linear equations in n unknowns has a nontrivial solution, i.e., a solution other than $x_1 = x_2 = \cdots = x_n = 0$, if and only if the determinant of the coefficients is equal to zero.

In particular, when the rank of the coefficient matrix of a homogeneous system of n linear equations in n unknowns is equal to $n - 1$, we have the following useful result.

Theorem 6

If the coefficient matrix of a homogeneous system of n linear equations in n unknowns, $AX = O$, is of rank $n - 1$, and if the submatrix obtained from A by omitting the ith row of A is also of rank $n - 1$, then the general solution of the given system is

$$x_1 = cA_{i1}, \qquad x_2 = cA_{i2}, \qquad \ldots, \qquad x_n = cA_{in}$$

where c is an arbitrary constant and A_{ij} is the cofactor of the element in the ith row and jth column of $|A|$.

Proof Since the rank of the $(n - 1, n)$ matrix remaining when the ith row is deleted from A is $n - 1$, it follows that at least one of the cofactors A_{ij} of the elements in the ith row of $|A|$ is different from zero. Hence not all the values $x_j = cA_{ij}$ are equal to zero, and the asserted solution is nontrivial. To verify that it is indeed a solution, we merely substitute it into the kth equation of the system, namely,

$$a_{k1}x_1 + a_{k2}x_2 + \cdots + a_{kn}x_n = 0$$

and check that it is satisfied for all values of k from $k = 1$ to $k = n$:

$$a_{k1}(cA_{i1}) + a_{k2}(cA_{i2}) + \cdots + a_{kn}(cA_{in}) \overset{?}{=} 0$$

or, factoring out the constant c,

$$c(a_{k1}A_{i1} + a_{k2}A_{i2} + \cdots + a_{kn}A_{in}) \overset{?}{=} 0$$

Now if $k \neq i$, the expression in parentheses is simply the sum of the products of the elements in the kth row of $|A|$ and the cofactors of the corresponding elements in the ith row of $|A|$ and, by Theorem 9, Appendix 1, is equal to zero. If $k = i$, the quantity in parentheses is the sum of the products of the elements in the ith row of $|A|$ and their respective cofactors and is equal to

$|A|$, from the definition of a determinant. However, $|A| = 0$ by hypothesis, and hence in this case also the expression is zero. Thus for all values of c, the given expressions satisfy each equation of the system, and the theorem is proved.

As an obvious consequence of the last theorem, we have the following useful result.

Corollary 1

If the coefficient matrix, A, of a homogeneous system of $n - 1$ linear equations in n unknowns, $AX = O$, is of rank $n - 1$, then the general solution of the system is

$$x_1 = c\,|M_1|, \qquad x_2 = -c\,|M_2|, \qquad \ldots, \qquad x_n = (-1)^{n+1}c\,|M_n|$$

where M_j is the $(n - 1, n - 1)$ submatrix obtained from A by deleting the jth column of A.

It is this corollary which justifies the memory device which we suggested in Sec. 2.3 for calculating the coordinates of the point of intersection of two lines and the coordinates of the line determined by two points. Theorems 5 and 6 will be used again and again in the work ahead of us in this book.

Example 5

Find the general solution of the system of equations:

$$\begin{aligned}
x_1 + 2x_2 - x_3 + x_4 &= 0 \\
2x_1 - x_2 + 3x_3 + 2x_4 &= 0 \\
2x_1 + 3x_2 + x_3 - 3x_4 &= 0
\end{aligned}$$

The coefficient matrix of this $(3,4)$ system, namely,

$$A = \begin{Vmatrix} 1 & 2 & -1 & 1 \\ 2 & -1 & 3 & 2 \\ 2 & 3 & 1 & -3 \end{Vmatrix}$$

is clearly of rank 3 since, specifically, the determinant of the $(3,3)$ submatrix obtained by omitting the first column of A is equal to -35, which is different from zero. Hence, using Corollary 1, Theorem 6, we have for all values of c

$$x_1 = c \begin{vmatrix} 2 & -1 & 1 \\ -1 & 3 & 2 \\ 3 & 1 & -3 \end{vmatrix} = -35c \qquad x_2 = -c \begin{vmatrix} 1 & -1 & 1 \\ 2 & 3 & 2 \\ 2 & 1 & -3 \end{vmatrix} = 25c$$

$$x_3 = c \begin{vmatrix} 1 & 2 & 1 \\ 2 & -1 & 2 \\ 2 & 3 & -3 \end{vmatrix} = 25c \qquad x_4 = -c \begin{vmatrix} 1 & 2 & -1 \\ 2 & -1 & 3 \\ 2 & 3 & 1 \end{vmatrix} = 10c$$

If we replace $5c$ by a new arbitrary constant, k, the solution assumes the somewhat simpler form $x_1 = -7k$, $x_2 = 5k$, $x_3 = 5k$, $x_4 = 2k$.

Example 6

Determine for what values of λ, if any, the system

$$
\begin{aligned}
(1 - \lambda)x_1 + && x_2 + && x_3 &= 0 \\
2x_1 + && (2 - \lambda)x_2 + && x_3 &= 0 \\
4x_1 + && 2x_2 + && (3 + \lambda)x_3 &= 0
\end{aligned}
$$

has a nontrivial solution, and find the solutions corresponding to each of these values.

According to Theorem 5, the given system will have a nontrivial solution if and only if the determinant of the coefficients is equal to zero; i.e., if and only if

$$
\begin{vmatrix}
(1 - \lambda) & 1 & 1 \\
2 & (2 - \lambda) & 1 \\
4 & 2 & (3 + \lambda)
\end{vmatrix} = 0
$$

Expanding this and collecting terms, we obtain the polynomial equation

$$
\lambda^3 - 3\lambda - 2 = 0
$$

The roots of this equation are easily found to be $\lambda = -1, -1, 2$; and for these values of λ, and for no others, the given system has a nontrivial solution.

When $\lambda = 2$, the system becomes

$$
\begin{aligned}
-x_1 + x_2 + x_3 &= 0 \\
2x_1 + x_3 &= 0 \\
4x_1 + 2x_2 + 5x_3 &= 0
\end{aligned}
$$

Although it is not difficult to solve these equations by elimination, Theorem 6 provides a more systematic method. Since the rank of the matrix of coefficients in the first two equations is 2, the solution can be read from the 2×2 determinants in this matrix:

$$
\begin{Vmatrix}
-1 & 1 & 1 \\
2 & 0 & 1
\end{Vmatrix}
$$

The resulting solution is $x_1 = k$, $x_2 = 3k$, $x_3 = -2k$.

When $\lambda = -1$, the system becomes

$$
\begin{aligned}
2x_1 + x_2 + x_3 &= 0 \\
2x_1 + x_2 + x_3 &= 0 \\
4x_1 + 2x_2 + 2x_3 &= 0
\end{aligned}
$$

In this case the rank of the coefficient matrix is 1, the three equations are equivalent, and the solution can be read from any one of them. Letting

$x_1 = \lambda$ and $x_2 = \mu$, and solving for x_3, we find

$$x_1 = \lambda$$
$$x_2 = \mu$$
$$x_3 = -2\lambda - \mu$$

or $X = \lambda X_1 + \mu X_2$, where X_1 and X_2 are, respectively, the particular solu-

tion vectors $\begin{Vmatrix} 1 \\ 0 \\ -2 \end{Vmatrix}$ and $\begin{Vmatrix} 0 \\ 1 \\ -1 \end{Vmatrix}$. These are clearly independent vectors, and

the general solution for $\lambda = -1$ thus consists of all vectors linearly dependent upon the two independent vectors X_1 and X_2, as required by Theorem 3, with $n = 3$ and $r = 1$, and Theorem 1.

Exercises

1. What is the rank of each of the following matrices?

(a) $\begin{Vmatrix} 1 & 2 & 3 \\ 2 & 1 & 1 \\ 1 & 1 & 2 \end{Vmatrix}$ (b) $\begin{Vmatrix} 5 & -1 & 1 \\ 7 & 1 & 3 \\ 2 & 2 & 2 \end{Vmatrix}$ (c) $\begin{Vmatrix} 6 & 12 & -9 \\ -2 & -4 & 3 \\ 4 & 8 & -6 \end{Vmatrix}$

2. Using the Gauss reduction, determine which of the following systems have solutions, and find the solutions when they exist:

(a)
$$x + 2y + 4z - w = 3$$
$$3x + 4y + 5z - w = 7$$
$$x + 3y + 4z + 5w = 4$$

(b)
$$x + 2y + 4z + w = 3$$
$$2x - y + z + 3w = 7$$
$$-4x + 7y + 5z - 7w = 4$$

(c)
$$x + 2y + 4z + w = 0$$
$$2x - y + z + 3w = 0$$
$$-4x + 7y + 5z - 7w = 0$$

(d)
$$x + 4y - z + 2w = 0$$
$$3x + 5y - 2z - w = 0$$
$$x + 5y + 2z + w = 0$$

3. Using Cramer's rule, solve each of the following systems:

(a)
$$x - y - 2z = 0$$
$$-x + 2y + 2z = 3$$
$$3x + y - z = 1$$

(b)
$$x + 3y + z = 1$$
$$-x + 2y + 4z = 2$$
$$3x - 2y - 6z = 0$$

(c)
$$2x - y + 2z = 1$$
$$x - y - 3z = 1$$
$$4x - 2y + 3z = 2$$

4. Using Corollary 1, Theorem 6, solve each of the following systems:

(a)
$$x_1 + 2x_2 - 3x_3 = 0$$
$$2x_1 + 2x_2 + x_3 = 0$$

(b)
$$2x_1 - 4x_2 + x_3 = 0$$
$$-x_1 + 2x_2 + x_3 = 0$$

(c)
$$x_1 + x_2 - x_3 + x_4 = 0$$
$$x_1 + 2x_2 + x_3 + 3x_4 = 0$$
$$2x_1 - x_2 + 3x_3 - x_4 = 0$$

(d)
$$2x_1 - x_2 - x_3 = 0$$
$$3x_2 - 2x_3 + x_4 = 0$$
$$x_1 + x_2 - x_3 + 3x_4 = 0$$

5. (*a*) Prove that for any system of the form $AX = B$, the rank of the coefficient matrix A is always equal to or less than the rank of the augmented matrix $\|A \quad B\|$.

(*b*) Prove Corollary 1, Theorem 6.

For each of the following systems, determine for what values of λ the system has a nontrivial solution, and find the solutions corresponding to these values.

6. $(1 - \lambda)x_1 + \qquad 2x_2 = 0$
$\qquad 3x_1 + (2 - \lambda)x_2 = 0$

7. $(2 + \lambda)x_1 + \quad (1 - \lambda)x_2 = 0$
$\quad (6 - \lambda)x_1 - (5 - 2\lambda)x_2 = 0$

8. $-\lambda x_1 \qquad\qquad + \quad x_3 = 0$
$\quad x_1 - (1 + \lambda)x_2 + \quad x_3 = 0$
$\quad x_1 \qquad\qquad - \lambda x_3 = 0$

9. $(1 - \lambda)x_1 \qquad\qquad + \qquad x_3 = 0$
$\quad x_1 + (2 - \lambda)x_2 + \qquad 2x_3 = 0$
$\quad 3x_1 + \qquad 3x_2 + (2 - \lambda)x_3 = 0$

10. $(1 - \lambda)x_1 \qquad\qquad + \qquad x_3 = 0$
$\quad x_1 + (2 - \lambda)x_2 + \qquad 2x_3 = 0$
$\quad 3x_1 + \qquad 3x_2 + (2 - \lambda)x_3 = 0$

3.7 Quadratic Forms and Bilinear Forms

In the analytical study of conics, which we shall begin in the next chapter, the notation and elementary properties of the mathematical objects known as quadratic forms and bilinear forms will be of considerable use to us. We shall therefore conclude this preparatory chapter with a discussion of these concepts.

By a **quadratic form** we mean a homogeneous second-degree expression in n variables of the form

$$Q(x) = a_{11}x_1^2 + 2a_{12}x_1x_2 + \cdots + 2a_{1n}x_1x_n$$
$$+ \quad a_{22}x_2^2 \quad + \cdots + 2a_{2n}x_2x_n$$
$$+ \cdots + \quad \cdot$$
$$+ \quad a_{nn}x_n^2$$

In order that matric notation may be applied to quadratic forms, it is customary to separate each of the cross products into two equal terms and introduce additional coefficients for the new terms by the definition $a_{ji} = a_{ij}$. When this has been done, $Q(x)$ can be expressed in the more symmetric form

$$Q(x) = \quad a_{11}x_1^2 \quad + a_{12}x_1x_2 + \cdots + a_{1n}x_1x_n$$
$$+ a_{21}x_2x_1 + a_{22}x_2^2 \quad + \cdots + a_{2n}x_2x_n$$
$$+ \cdots\cdots\cdots\cdots\cdots\cdots\cdots\cdots \qquad a_{ji} = a_{ij}$$
$$+ a_{n1}x_nx_1 + a_{n2}x_nx_2 + \cdots + a_{nn}x_n^2$$

If we now define the matrices

$$X = \begin{Vmatrix} x_1 \\ x_2 \\ \cdot \\ x_n \end{Vmatrix} \quad \text{and} \quad A = \begin{Vmatrix} a_{11} & a_{12} & \cdots & a_{1n} \\ a_{21} & a_{22} & \cdots & a_{2n} \\ \cdot & \cdot & \cdots & \cdot \\ a_{n1} & a_{n2} & \cdots & a_{nn} \end{Vmatrix} \quad a_{ji} = a_{ij}$$

it is clear from the definition of matric multiplication that $Q(x)$ can be written in the compact form

$$Q(x) = X^T A X$$

where, since $a_{ji} = a_{ij}$, the **matrix of the quadratic form**, A, is necessarily symmetric.

In our study of conics in E_2^+ and Π_2, and later in our development of noneuclidean geometry, we shall be concerned with quadratic forms in which the coefficients, a_{ij}, are real. Such forms are ordinarily classified according to the values they can take on for real values of the x's.

If a quadratic form with real coefficients has the property that it is equal to or greater than zero for all real values of its variables, it is said to be **positive**. A positive form which is zero *only* for the values $x_1 = x_2 = \cdots = x_n = 0$ is said to be **positive definite**. A positive form which is zero for real values other than $x_1 = x_2 = \cdots = x_n = 0$ is said to be **positive semidefinite**. A quadratic form with real coefficients which is equal to or less than zero for all real values of its variables is said to be **negative**. A negative form which is zero *only* for the values $x_1 = x_2 = \cdots = x_n = 0$ is said to be **negative definite**. A negative form which is zero for real values other than $x_1 = x_2 = \cdots = x_n = 0$ is said to be **negative semidefinite**. Clearly, a negative definite or semidefinite form can be converted into a positive form of corresponding type by multiplying it by -1. A quadratic form which can take on both positive and negative values for real values of its variables is said to be **indefinite**. Examples of quadratic forms of each type are given in Table 3.1.

Table 3.1

Type of quadratic form	Example
Positive definite	$x_1{}^2 + x_2{}^2$
Negative definite	$-(x_1{}^2 + x_2{}^2)$
Positive semidefinite	$(x_1 - x_2)^2$
Negative semidefinite	$-(x_1 - x_2)^2$
Indefinite	$x_1{}^2 - x_2{}^2$

The matrix, A, of a quadratic form, $Q(x) = X^T AX$, is said to be **positive** or **negative definite, semidefinite,** or **indefinite** according to the nature of $Q(x)$. Correspondingly, $Q(x)$ is said to be **singular** or **nonsingular** according as its matrix, A, is singular or nonsingular, i.e., according as $|A|$ is equal to zero or different from zero.

A quadratic form which is definite is necessarily nonsingular. In fact, if we suppose that $Q(x)$ is both definite and singular, we are led at once to a contradiction, as follows. Let us first write $Q(x)$ in the partially factored form

$$Q(x) = \quad (a_{11}x_1 + a_{12}x_2 + \cdots + a_{1n}x_n)x_1$$
$$+ (a_{21}x_1 + a_{22}x_2 + \cdots + a_{2n}x_n)x_2$$
$$+ \cdots\cdots\cdots\cdots\cdots\cdots\cdots\cdots\cdots$$
$$+ (a_{n1}x_1 + a_{n2}x_2 + \cdots + a_{nn}x_n)x_n$$

Then, assuming that $|A| = 0$, Theorem 5, Sec. 3.6, assures us that there is a nontrivial solution to the system of equations obtained by equating to zero the expressions in parentheses in $Q(x)$. Finally, we observe that for the values of the x's in this (nontrivial) solution, $Q(x)$ itself is equal to zero, contrary to the hypothesis that it is definite.

The converse of the preceding observation is not true. In other words, a nonsingular quadratic form is not necessarily definite. For example, the form $x_1{}^2 - 4x_1x_2 + 3x_2{}^2 + 2x_3{}^2$ is nonsingular since the determinant of its matrix, namely,

$$\begin{vmatrix} 1 & -2 & 0 \\ -2 & 3 & 0 \\ 0 & 0 & 2 \end{vmatrix} = -2$$

is different from zero, yet it is not definite since it is zero for the nontrivial set of values $x_1 = 3$, $x_2 = 1$, $x_3 = 0$.

If the variables in a quadratic form are homogeneous coordinates, the locus obtained by equating to zero a form $Q(x)$ which is definite contains no real points, since the only real values which can satisfy the equation $Q(x) = 0$ are $x_1 = x_2 = \cdots = x_n = 0$, and these correspond to no geometric object (at least if $n = 3$). On the other hand, the locus obtained by equating to zero a form which is semidefinite or indefinite does contain real points, since forms of either of these types can take on the value zero for real values of $x_1, x_2, \ldots, x_n$ which are not all zero. For this reason (and for many others) it is desirable to have a test for the definiteness of quadratic forms. Such a criterion is provided by the following theorems, for whose proof we must refer to texts on higher algebra.[1]

---

[1] See, for instance, W. L. Ferrar, "Algebra," pp. 138–141, Oxford Book Company, Inc., New York, 1941.

Theorem 1

A necessary and sufficient condition that the real quadratic form $Q(x) = X^T A X$ be positive definite is that the quantities

$$a_{11}, \quad \begin{vmatrix} a_{11} & a_{12} \\ a_{21} & a_{22} \end{vmatrix}, \quad \begin{vmatrix} a_{11} & a_{12} & a_{13} \\ a_{21} & a_{22} & a_{23} \\ a_{31} & a_{32} & a_{33} \end{vmatrix}, \quad \dots, \quad |A| = \begin{vmatrix} a_{11} & a_{12} & \cdots & a_{1n} \\ a_{21} & a_{22} & \cdots & a_{2n} \\ \cdot & \cdot & \cdots & \cdot \\ a_{n1} & a_{n2} & \cdots & a_{nn} \end{vmatrix}$$

should all be positive.

Theorem 2

A necessary and sufficient condition that the real quadratic form $Q(x) = X^T A X$ be negative definite is that the quantities

$$a_{11}, \quad \begin{vmatrix} a_{11} & a_{12} \\ a_{21} & a_{22} \end{vmatrix}, \quad \begin{vmatrix} a_{11} & a_{12} & a_{13} \\ a_{21} & a_{22} & a_{23} \\ a_{31} & a_{32} & a_{33} \end{vmatrix}, \quad \dots, \quad |A| = \begin{vmatrix} a_{11} & a_{12} & \cdots & a_{1n} \\ a_{21} & a_{22} & \cdots & a_{2n} \\ \cdot & \cdot & \cdots & \cdot \\ a_{n1} & a_{n2} & \cdots & a_{nn} \end{vmatrix}$$

should alternate in sign, with a_{11} negative.

The quantities which appear in the statements of Theorems 1 and 2 are all determinants of square submatrices of A whose principal diagonals lie along the principal diagonal of A. Such submatrices, whether or not their diagonal elements are consecutive elements on the principal diagonal of A, are known as **principal submatrices** of A, and their determinants are known as **principal minors** of A. Clearly, by first permuting the names of the variables in $X^T A X$ and then applying Theorems 1 and 2, equivalent conditions for definiteness involving other principal minors of A can be obtained. Considering all possible permutations of the names of the variables, we obtain in this fashion the following somewhat more general theorems.

Theorem 3

A necessary and sufficient condition that the real quadratic form $X^T A X$ be positive definite is that every principal minor of A be positive.

Theorem 4

A necessary and sufficient condition that the real quadratic form $X^T A X$ be negative definite is that every principal minor of A of odd order be negative and every principal minor of A of even order be positive.

Example 1

The quadratic form

$$\|x_1 \quad x_2 \quad x_3\| \cdot \begin{Vmatrix} 5 & -2 & 1 \\ -2 & 4 & -6 \\ 1 & -6 & 11 \end{Vmatrix} \cdot \begin{Vmatrix} x_1 \\ x_2 \\ x_3 \end{Vmatrix} = \begin{cases} 5x_1{}^2 & - 2x_1 x_2 & + & x_1 x_3 \\ - 2x_2 x_1 & + 4x_2{}^2 & - & 6x_2 x_3 \\ + x_3 x_1 & - 6x_3 x_2 & + & 11x_3{}^2 \end{cases}$$

is positive definite since the three quantities

$$5, \qquad \begin{vmatrix} 5 & -2 \\ -2 & 4 \end{vmatrix} = 16, \qquad \text{and} \qquad \begin{vmatrix} 5 & -2 & 1 \\ -2 & 4 & -6 \\ 1 & -6 & 11 \end{vmatrix} = 16$$

are all positive. In fact, it is easy to verify that the given form can be written equivalently as

$$(x_1 - 2x_2 + 3x_3)^2 + (2x_1 - x_3)^2 + x_3{}^2$$

Since this is a sum of squares, the only real values for which it can vanish are those which satisfy the three equations

$$x_1 - 2x_2 + 3x_3 = 0 \qquad 2x_1 - x_3 = 0 \qquad \text{and} \qquad x_3 = 0$$

that is, $x_1 = x_2 = x_3 = 0$.

On the other hand, the quadratic form

$$\| x_1 \quad x_2 \quad x_3 \| \cdot \begin{Vmatrix} -3 & -2 & 5 \\ -2 & 4 & -6 \\ 5 & -6 & 8 \end{Vmatrix} \cdot \begin{Vmatrix} x_1 \\ x_2 \\ x_3 \end{Vmatrix} = \begin{cases} -3x_1{}^2 - 2x_1x_2 + 5x_1x_3 \\ -2x_2x_1 + 4x_2{}^2 - 6x_2x_3 \\ +5x_3x_1 - 6x_3x_2 + 8x_3{}^2 \end{cases}$$

is not definite, since the three quantities

$$-3, \qquad \begin{vmatrix} -3 & -2 \\ -2 & 4 \end{vmatrix} = -16, \qquad \text{and} \qquad \begin{vmatrix} -3 & -2 & 5 \\ -2 & 4 & -6 \\ 5 & -6 & 8 \end{vmatrix} = 0$$

do not satisfy either the conditions of Theorem 1 or Theorem 2. In fact, since the determinant of its coefficient matrix is equal to zero, this quadratic form is singular, which is sufficient to prove that it is not definite. More specifically, since this form takes on the value 3 when $x_1 = x_2 = x_3 = 1$ and takes on the value -1 when $x_1 = 1$, $x_2 = 2$, $x_3 = 1$, it is actually indefinite.

Closely associated with the quadratic form $X^T A X$ is the expression $Y^T A X$, where Y is the column matrix

$$\begin{Vmatrix} y_1 \\ y_2 \\ \cdot \\ y_n \end{Vmatrix}$$

Written at length, this expression is

$$\| y_1 \quad y_2 \quad \cdots \quad y_n \| \cdot \begin{Vmatrix} a_{11} & a_{12} & \cdots & a_{1n} \\ a_{21} & a_{22} & \cdots & a_{2n} \\ \cdot & \cdot & \cdots & \cdot \\ a_{n1} & a_{n2} & \cdots & a_{nn} \end{Vmatrix} \cdot \begin{Vmatrix} x_1 \\ x_2 \\ \cdot \\ x_n \end{Vmatrix}$$

$$= \begin{cases} (a_{11}y_1 + a_{21}y_2 + \cdots + a_{n1}y_n)x_1 \\ + (a_{12}y_1 + a_{22}y_2 + \cdots + a_{n2}y_n)x_2 \\ + \cdots \cdots \cdots \cdots \cdots \cdots \\ + (a_{1n}y_1 + a_{2n}y_2 + \cdots + a_{nn}y_n)x_n \end{cases}$$

Since this is linear in the variables $(x_1, x_2, \ldots, x_n)$ and also in the variables $(y_1, y_2, \ldots, y_n)$ it is known as a **bilinear form.** When it is desired to emphasize its relation to the quadratic form $X^T A X$, the bilinear form $Y^T A X$ is usually referred to as the **polar of** Y **with respect to** $X^T A X$.

Given a quadratic form $X^T A X$, it is often necessary to consider its evaluation for a vector X which is a linear combination of two other vectors, say $X = \lambda P + \mu R$. Substituting this expression for X, we have

$$X^T A X = (\lambda P + \mu R)^T A (\lambda P + \mu R)$$

Then, using the fact that the transpose of a sum of matrices is equal to the sum of the transposes and the fact that matric multiplication is distributive over addition, we have, further,

$$(1) \qquad \lambda^2 P^T A P + \lambda \mu P^T A R + \lambda \mu R^T A P + \mu^2 R^T A R$$

Now any quadratic or bilinear form, being the product of three matrices whose symbols are, respectively, $(1,n)$, (n,n), and $(n,1)$, is a $(1,1)$ matrix, i.e., a scalar, and is therefore equal to its transpose. Hence, in particular, the bilinear form $R^T A P$ in the third term in (1) is equal to its transpose, and we have

$$R^T A P = (R^T A P)^T$$

From this, using Corollary 1, Theorem 5, Sec. 3.2, together with the fact that the transpose of the transpose of a matrix is just the original matrix and the fact that A is symmetric and hence equal to its transpose, we have

$$(R^T A P)^T = P^T A^T (R^T)^T = P^T A R$$

Thus the second and third terms in (1) combine, and we have finally

$$(2) \qquad (\lambda P + \mu R)^T A (\lambda P + \mu R) = \lambda^2 P^T A P + 2\lambda \mu P^T A R + \mu^2 R^T A R$$

We shall make frequent use of this formula in the work ahead of us.

Exercises

1. Determine the type of each of the following quadratic forms:

(a) $x_1^2 + 4x_1 x_2 + 3x_2^2$

(b) $x_1^2 + 4x_1 x_2 + 5x_2^2$

(c) $x_1^2 + 4x_1 x_2 + 5x_2^2 + 2x_1 x_3 + 4x_2 x_3 + 2x_3^2$

(d) $-x_1^2 + 4x_1 x_2 - 5x_2^2 + 4x_1 x_3 + 6x_2 x_3 - 2x_3^2$

(e) $x_1^2 + 4x_1 x_2 + 5x_2^2 + 2x_1 x_3 + 6x_2 x_3 + 2x_3^2$

2. If $A = \begin{Vmatrix} 1 & -1 & 1 \\ -1 & 1 & 3 \\ 1 & 3 & -4 \end{Vmatrix}$, $P = \begin{Vmatrix} 2 \\ 1 \\ 1 \end{Vmatrix}$, and $R = \begin{Vmatrix} -1 \\ 0 \\ 1 \end{Vmatrix}$, determine for

what vector or vectors, X, of the family $\lambda P + \mu R$ the quadratic form $X^T A X$ is equal to zero.

3. Work Exercise 2 if $A = \begin{Vmatrix} 3 & -1 & 2 \\ -1 & -1 & -1 \\ 2 & -1 & 2 \end{Vmatrix}$, $P = \begin{Vmatrix} 2 \\ 2 \\ -1 \end{Vmatrix}$, $R = \begin{Vmatrix} 1 \\ 1 \\ 1 \end{Vmatrix}$

4. Work Exercise 2 if $A = \begin{Vmatrix} 6 & -2 & 3 \\ -2 & -2 & -1 \\ 3 & -1 & 0 \end{Vmatrix}$, $P = \begin{Vmatrix} 1 \\ -3 \\ 2 \end{Vmatrix}$, $R = \begin{Vmatrix} 0 \\ 3 \\ 1 \end{Vmatrix}$

5. Show that if Y is linearly dependent upon the vectors Y_1 and Y_2, then the polar of Y with respect to the quadratic form $X^T A X$ is linearly dependent upon the polars of Y_1 and Y_2 with respect to $X^T A X$.

6. If $Y = \begin{Vmatrix} y_1 \\ y_2 \\ y_3 \end{Vmatrix}$ and $X^T A X$ is a quadratic form in the variables (x_1, x_2, x_3),

show that the polar of Y with respect to $X^T A X$ can be obtained by applying the operator

$$\tfrac{1}{2}\left(y_1 \frac{\partial}{\partial x_1} + y_2 \frac{\partial}{\partial x_2} + y_3 \frac{\partial}{\partial x_3}\right)$$

to $X^T A X$. (For this reason

$$y_1 \frac{\partial}{\partial x_1} + y_2 \frac{\partial}{\partial x_2} + y_3 \frac{\partial}{\partial x_3}$$

is called the **polar operator.**)

7. (a) If Q_1 and Q_2 are quadratic forms, show that any linear combination of Q_1 and Q_2 is also a quadratic form.
(b) If Q is a quadratic form which is linearly dependent upon the quadratic forms Q_1 and Q_2, show that the polar of a vector Y with respect to Q is linearly dependent upon the polars of Y with respect to Q_1 and Q_2.

8. If $A = \begin{Vmatrix} 1 & 0 & 0 \\ 0 & 0 & -1 \\ 0 & -1 & 0 \end{Vmatrix}$ and $B = \begin{Vmatrix} 0 & 1 & 1 \\ 1 & 0 & -1 \\ 1 & -1 & 0 \end{Vmatrix}$, determine whether

or not there are any singular quadratic forms in the family $\lambda X^T A X + \mu X^T B X$.

9. Work Exercise 8 if $A = \begin{Vmatrix} 0 & -1 & 0 \\ -1 & 0 & 0 \\ 0 & 0 & 1 \end{Vmatrix}$ and $B = \begin{Vmatrix} 0 & 5 & 1 \\ 5 & 0 & -3 \\ 1 & -3 & 0 \end{Vmatrix}$

10. (a) If the quadratic form $X^T A X$ is indefinite, prove that there is at least one nontrivial vector X for which $X^T A X = 0$.

(*b*) If P is a vector for which an indefinite quadratic form $Q = X^T A X$ is equal to zero, and if ϵ is an arbitrary (small) positive number, show that there are vectors each of whose components differ from the corresponding components of P by less than ϵ for which Q is positive, and also such vectors for which Q is negative.

3.8 Conclusion

In this section we have introduced the vocabulary, notation, and elementary properties of matrices, and we have discussed such related ideas as linear dependence and independence, the solution of systems of linear equations, and quadratic and bilinear forms. Our work has been largely preparatory, and except for a brief discussion of linear transformations and a passing reference to concurrent lines in one of our examples, we have given no geometric applications. However, in the analytical work ahead of us we shall make extensive use of matric ideas: the notion of linear dependence will enter into our parametric treatment of the straight line and our study of projectivities between lines; our study of the conic will be essentially the study of quadratic forms in three variables; bilinear forms will be central in our discussion of the properties of tangents to conics; the concept of the rank of a matrix will be important not only in our discussion of singular conics but also in our investigation of linear transformations; and all these ideas, together with a few more specialized ones which we shall introduce when needed, will be fundamental in our development of noneuclidean geometry.

four

FURTHER PROPERTIES OF THE EXTENDED PLANE

4.1 Introduction

In this chapter we shall resume our investigation of the extended euclidean plane, E_2^+, and the algebraic representation, Π_2. After first developing algebraic criteria for the collinearity of points and the concurrence of lines, we shall use these ideas to prove the theorems of Desargues and Pappus. Then we shall define and investigate perspectivities and projectivities between lines and the projective invariant known as the cross ratio of four points or four lines. Finally, we shall study conics and their polar properties.

  Because the equations involved are all linear, the study of points and lines is the same in E_2^+ and in Π_2, and all our results will be valid in both systems. On the other hand, since the study of conics involves second-degree equations, whose roots need not be real even though their coefficients are, the properties of conics in the two systems are somewhat different. In studying conics, we shall consistently work in Π_2 and merely note, where appropriate, how our results would be modified in E_2^+.

4.2 Elementary Properties of Points and Lines

For many purposes it is convenient to be able to tell quickly whether or not three points, (a_1,a_2,a_3), (b_1,b_2,b_3), (c_1,c_2,c_3), in E_2^+ or Π_2 are collinear. Clearly this will be the case if and only if there exists a line $[l_1,l_2,l_3]$ which contains each of the given points. Furthermore, in view of the incidence condition of Sec. 2.3, this will be the case if and only if, simultaneously,

$$a_1 l_1 + a_2 l_2 + a_3 l_3 = 0$$
$$b_1 l_1 + b_2 l_2 + b_3 l_3 = 0$$
$$c_1 l_1 + c_2 l_2 + c_3 l_3 = 0$$

Finally, according to Theorem 5, Sec. 3.6, this system of homogeneous linear equations will be satisfied by a nontrivial triple of numbers $[l_1,l_2,l_3]$ if and only if the determinant of the coefficients of the system is equal to zero; i.e., if and only if

(1)
$$\begin{vmatrix} a_1 & a_2 & a_3 \\ b_1 & b_2 & b_3 \\ c_1 & c_2 & c_3 \end{vmatrix} = 0$$

Thus we have established the following important result.

Theorem 1

Three points in either E_2^+ or Π_2 are collinear if and only if the determinant whose rows consist, respectively, of the coordinates of the points is equal to zero.

Since the value of a determinant is not changed if the rows and columns of the determinant are interchanged, Theorem 1 can be restated in the following form.

Corollary 1

Three points in either E_2^+ or Π_2 are collinear if and only if the determinant whose columns are the coordinate-vectors of the points is equal to zero.

Because of the symmetry of the relations between points and lines in both E_2^+ and Π_2, it is clear that corresponding to Theorem 1 there must be a theorem giving an analogous condition that three lines should pass through the same point. Theoretically, a proof of this theorem is unnecessary, but to emphasize this fact we shall present a proof, noting that it is algebraically identical to the argument in the preceding paragraph, with triples in brackets replacing triples in parentheses, and vice versa.

Three lines, $[l_1, l_2, l_3]$, $[m_1, m_2, m_3]$, $[n_1, n_2, n_3]$, will be concurrent if and only if there exists a point (a_1, a_2, a_3) which lies on each of the lines. Furthermore, in view of the incidence condition of Sec. 2.3, this will be the case if and only if, simultaneously,

$$l_1 a_1 + l_2 a_2 + l_3 a_3 = 0$$
$$m_1 a_1 + m_2 a_2 + m_3 a_3 = 0$$
$$n_1 a_1 + n_2 a_2 + n_3 a_3 = 0$$

Finally, according to Theorem 5, Sec. 3.6, this system of homogeneous linear equations will be satisfied by a nontrivial triple of numbers (a_1, a_2, a_3) if and only if the determinant of the coefficients of this system is equal to zero; i.e., if and only if

(2)
$$\begin{vmatrix} l_1 & l_2 & l_3 \\ m_1 & m_2 & m_3 \\ n_1 & n_2 & n_3 \end{vmatrix} = 0$$

Thus we have established the dual of Theorem 1:

Theorem 2

Three lines in either E_2^+ or Π_2 are concurrent if and only if the determinant whose rows consist, respectively, of the coordinates of the lines is equal to zero.

The condition of Theorem 2 can, of course, be restated in a form analogous to Corollary 1, Theorem 1.

Corollary 1

Three lines in either E_2^+ or Π_2 are concurrent if and only if the determinant whose columns are the coordinate-vectors of the lines is equal to zero.

Also, since the coordinates of a line are, by definition, the coefficients in the equation of the line, the condition of Theorem 2 can be restated in the following form.

Corollary 2

Three lines, $l_1 x_1 + l_2 x_2 + l_3 x_3 = 0$, $m_1 x_1 + m_2 x_2 + m_3 x_3 = 0$, and $n_1 x_1 + n_2 x_2 + n_3 x_3 = 0$, are concurrent if and only if the determinant of the coefficients $\begin{vmatrix} l_1 & l_2 & l_3 \\ m_1 & m_2 & m_3 \\ n_1 & n_2 & n_3 \end{vmatrix}$ is equal to zero.

Using Theorem 1, we can now prove the following important result.

Theorem 3

A point $C = \left\| \begin{matrix} c_1 \\ c_2 \\ c_3 \end{matrix} \right\|$ is collinear with the distinct points $A = \left\| \begin{matrix} a_1 \\ a_2 \\ a_3 \end{matrix} \right\|$ and $B = \left\| \begin{matrix} b_1 \\ b_2 \\ b_3 \end{matrix} \right\|$ if and only if there exist numbers λ and μ such that

$$c_i = \lambda a_i + \mu b_i \qquad i = 1, 2, 3$$

i.e., if and only if $C = \lambda A + \mu B$.

Proof To prove the sufficiency of the given condition, we observe that if $c_i = \lambda a_i + \mu b_i$, $i = 1, 2, 3$, then the determinant

$$\begin{vmatrix} a_1 & b_1 & c_1 \\ a_2 & b_2 & c_2 \\ a_3 & b_3 & c_3 \end{vmatrix} = \begin{vmatrix} a_1 & b_1 & \lambda a_1 + \mu b_1 \\ a_2 & b_2 & \lambda a_2 + \mu b_2 \\ a_3 & b_3 & \lambda a_3 + \mu b_3 \end{vmatrix}$$

is equal to zero, since its last column can be reduced to zero by subtracting from it λ times the first column and μ times the second. Hence, by Corollary 1, Theorem 1, the three points A, B, C are collinear, as asserted.

To prove the necessity of the condition, let us suppose that the three points are collinear and attempt to find numbers α, β, γ such that

(3)
$$\begin{aligned} \alpha a_1 + \beta b_1 + \gamma c_1 = 0 \\ \alpha a_2 + \beta b_2 + \gamma c_2 = 0 \\ \alpha a_3 + \beta b_3 + \gamma c_3 = 0 \end{aligned}$$

By Corollary 1, Theorem 1, since A, B, C are assumed to be collinear, the determinant of the coefficients of this system of equations is equal to zero. Therefore, by Theorem 5, Sec. 3.6, this system is satisfied by numbers α, β, γ not all of which are zero. Moreover, in any such solution the number γ must be different from zero, because if γ were zero, it would follow from (3) that (a_1, a_2, a_3) and (b_1, b_2, b_3) were proportional triples, contrary to the hypothesis that A and B are distinct points. Thus, since $\gamma \neq 0$, we can solve Eqs. (3) for c_1, c_2, c_3, getting

$$c_1 = -\frac{\alpha}{\gamma} a_1 - \frac{\beta}{\gamma} b_1$$

$$c_2 = -\frac{\alpha}{\gamma} a_2 - \frac{\beta}{\gamma} b_2 \qquad \alpha, \beta \text{ not both zero}$$

$$c_3 = -\frac{\alpha}{\gamma} a_3 - \frac{\beta}{\gamma} b_3$$

If we now define λ and μ by the equations $\lambda = -\alpha/\gamma$ and $\mu = -\beta/\gamma$, we have succeeded in expressing the coordinates of C in the form prescribed by the theorem. Finally, since A and B are distinct points, their coordinates cannot be proportional. Hence at least one of the second-order determinants from the first two columns in the matrix of the coefficients of (3) must be different from zero. The rank of this matrix is therefore 2, and hence, by Theorem 3, Sec. 3.6, the solution for α, β, γ is unique to within an irrelevant constant of proportionality. The last assertion of the theorem follows, of course, since the equations

$$c_i = \lambda a_i + \mu b_i \qquad i = 1, 2, 3$$

are just the scalar equivalents of the matric equation $C = \lambda A + \mu B$. Obviously λ and μ cannot both be zero since this would imply $c_1 = c_2 = c_3 = 0$, which is impossible.

In Theorem 3, the points A and B are assumed to be distinct, but C can be any point collinear with A and B, including A and B themselves. Thus Theorem 3 establishes a one-to-one correspondence between the points of the line AB and the sets of ordered pairs

$$(k\lambda, k\mu) \qquad k \neq 0$$
$$\lambda, \mu \text{ not both zero}$$

Such a correspondence is called a **parametrization** of the range of points on the line AB, and we say that AB has been **parametrized** in terms of A and B as **base points,** with λ and μ as **parameters.** In particular, in such a parametrization:

A corresponds to the set of ordered pairs of the form $(k,0)$, $k \neq 0$.
B corresponds to the set of ordered pairs of the form $(0,k)$, $k \neq 0$.

Accompanying Theorem 3 we have the following theorem, which makes the corresponding assertion about lines. Its proof, which is algebraically identical to the proof of Theorem 3, we shall leave as an exercise.

Theorem 4

A line $n = \left\| \begin{array}{c} n_1 \\ n_2 \\ n_3 \end{array} \right\|$ is concurrent with the distinct lines $l = \left\| \begin{array}{c} l_1 \\ l_2 \\ l_3 \end{array} \right\|$ and

$m = \left\| \begin{array}{c} m_1 \\ m_2 \\ m_3 \end{array} \right\|$ if and only if there exist numbers λ and μ, such that

$$n_i = \lambda l_i + \mu m_i \qquad i = 1, 2, 3$$

that is, if and only if $n = \lambda l + \mu m$.

Theorem 4 establishes a one-to-one correspondence between the lines which pass through the intersection of l and m and the sets of ordered pairs

$$[k\lambda, k\mu] \qquad k \neq 0$$
$$\lambda, \mu \text{ not both zero}$$

Such a correspondence is called a **parametrization** of the pencil of lines on the point lm, and we say that lm has been **parametrized** in terms of l and m as **base lines,** with λ and μ as **parameters.** In particular, in such a parametrization:

l corresponds to the set of ordered pairs of the form $[k,0]$, $k \neq 0$.
m corresponds to the set of ordered pairs of the form $[0,k]$, $k \neq 0$.

Example 1

Given the points $A:(2,1,-1)$, $B:(1,-1,0)$, $C:(3,1,2)$, $D:(2,1,1)$. Using the parametrization of AB in terms of A and B as base points and the parametrization of CD in terms of C and D as base points, find the intersection of AB and CD.

Let $P:(p_1,p_2,p_3)$ be the required intersection. Then, since P is a point of the line AB, it follows from Theorem 3 that multipliers α and β exist such that $P = \alpha A + \beta B$ or, specifically,

$$p_1 = \alpha(2) \quad + \beta(1)$$
$$p_2 = \alpha(1) \quad + \beta(-1)$$
$$p_3 = \alpha(-1) + \beta(0)$$

Similarly, since P is also a point of the line CD, multipliers γ and δ exist such that $P = \gamma C + \delta D$, or

$$p_1 = \gamma(3) + \delta(2)$$
$$p_2 = \gamma(1) + \delta(1)$$
$$p_3 = \gamma(2) + \delta(1)$$

Therefore, since all coordinate-vectors of P must be proportional, it follows that

$$2\alpha + \beta = k(3\gamma + 2\delta)$$
$$\alpha - \beta = k(\ \gamma + \ \delta)$$
$$-\alpha \quad\ \ = k(2\gamma + \ \delta)$$

Using Corollary 1, Theorem 6, Sec. 3.6, the solution of this system of homogeneous equations is readily found to be

$$\alpha = -k \qquad \beta = k \qquad \gamma = 3 \qquad \delta = -5$$

Thus, as a point of AB, the intersection P has parameters $(-1,1)$, and as a point of CD it has parameters $(3,-5)$. Using either pair of parameters on the corresponding line, we find the coordinates of P to be $(1,2,-1)$.

Example 2

A line l is initially parametrized in terms of the base points A and B with parameters (λ,μ). If a second parametrization with base points $A' = \alpha_1 A + \alpha_2 B$ and $B' = \beta_1 A + \beta_2 B$ and parameters (λ',μ') is introduced on l (Fig. 4.1), find the relations between the parameters (λ,μ) and (λ',μ') of a general point on l.

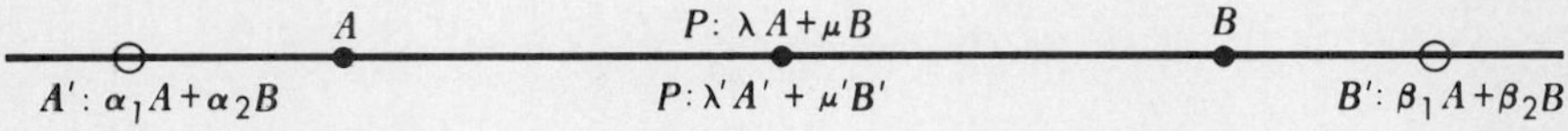

Fig. 4.1   A line parametrized in terms of two different sets of base points.

In the second parametrization of l, the coordinate-vector, P, of a general point on l is given by the expression $P = \lambda'A' + \mu'B'$. Hence, substituting for A' and B', we have

$$P = \lambda'(\alpha_1 A + \alpha_2 B) + \mu'(\beta_1 A + \beta_2 B) = (\alpha_1\lambda' + \beta_1\mu')A + (\alpha_2\lambda' + \beta_2\mu')B$$

Therefore, since we also have $P = \lambda A + \mu B$, it follows that, to within an irrelevant factor of proportionality, we can take

$$\lambda = \alpha_1\lambda' + \beta_1\mu' \qquad \text{and} \qquad \mu = \alpha_2\lambda' + \beta_2\mu'$$

Using matric notation, these two scalar equations can be written in the compact form

(4) $\qquad \Lambda = M\Lambda' \quad \text{where} \quad \Lambda = \begin{Vmatrix} \lambda \\ \mu \end{Vmatrix} \quad M = \begin{Vmatrix} \alpha_1 & \beta_1 \\ \alpha_2 & \beta_2 \end{Vmatrix} \quad \Lambda' = \begin{Vmatrix} \lambda' \\ \mu' \end{Vmatrix}$

It is interesting to note that the matrix M is the transpose of the matrix of coefficients of the pair of equations which express the new base points, A' and B', in terms of the original base points, A and B; that is, M is the matrix whose columns are, respectively, the parameters of A' and B' in terms of the original base points. Since A' and B' are distinct points, it follows that $|M|$ is different from zero. Hence M^{-1} exists, and Eq. (4) can be solved for Λ' in terms of Λ:

(5) $\qquad\qquad\qquad\qquad \Lambda' = M^{-1}\Lambda$

The next two theorems extend Theorems 3 and 4 by describing how E_2^+ and Π_2 themselves can be parametrized.

Theorem 5

If $A = \begin{Vmatrix} a_1 \\ a_2 \\ a_3 \end{Vmatrix}$, $B = \begin{Vmatrix} b_1 \\ b_2 \\ b_3 \end{Vmatrix}$, and $C = \begin{Vmatrix} c_1 \\ c_2 \\ c_3 \end{Vmatrix}$ are three noncollinear

points, then any point $D = \begin{Vmatrix} d_1 \\ d_2 \\ d_3 \end{Vmatrix}$ is linearly dependent upon A, B, and C;

that is, there exist numbers (λ,μ,ν), such that $D = \lambda A + \mu B + \nu C$.

Proof   Let $A:(a_1,a_2,a_3)$, $B:(b_1,b_2,b_3)$, $C:(c_1,c_2,c_3)$ be three noncollinear points, and let $D:(d_1,d_2,d_3)$ be any point. Then in scalar form, the assertion of the theorem is that numbers (λ,μ,ν) can be found such that

(6)
$$\begin{aligned} d_1 &= \lambda a_1 + \mu b_1 + \nu c_1 \\ d_2 &= \lambda a_2 + \mu b_2 + \nu c_2 \\ d_3 &= \lambda a_3 + \mu b_3 + \nu c_3 \end{aligned}$$

Now by hypothesis, A, B, and C are noncollinear points. Hence, by Corollary 1, Theorem 1, the determinant of the coefficients in (6) is different from zero. Moreover, at least one of the numbers (d_1,d_2,d_3) is different from zero since these are the coordinates of the point D. Therefore (6) is a system of non-homogeneous linear equations whose coefficient matrix is nonsingular. It follows, then, from Theorem 4, Sec. 3.6, that there exists a unique set of values (λ,μ,ν), which satisfy these equations, and the theorem is proved. Clearly λ, μ, ν cannot all be zero since this would imply $d_1 = d_2 = d_3 = 0$, which is impossible.

It might appear from a careless reading of the proof of Theorem 5 that there is a unique set of values (λ,μ,ν) which serve to express the point D as a linear combination of the points A, B, C. It is true that the system (6) has a unique solution, but this solution pertains only to a particular representation of D. Actually D is represented equally well by any vector of the

family $k \begin{Vmatrix} d_1 \\ d_2 \\ d_3 \end{Vmatrix} = \begin{Vmatrix} kd_1 \\ kd_2 \\ kd_3 \end{Vmatrix}$; hence if (λ,μ,ν) is one set of constants which can be

used to express D in terms of A, B, and C, then for any nonzero value of k, $(k\lambda,k\mu,k\nu)$ is another such set, and these are the only ones. These observations are summarized in the following corollary of Theorem 5.

Corollary 1

If A, B, C are three noncollinear points, and if an arbitrary point D is expressed in each of the forms

$$D = \lambda A + \mu B + \nu C \qquad \text{and} \qquad D = \lambda' A + \mu' B + \nu' C$$

then $\lambda' = k\lambda$, $\mu' = k\mu$, $\nu' = k\nu$.

The dual results for lines can be established in the same fashion:

Theorem 6

If $l = \left\| \begin{matrix} l_1 \\ l_2 \\ l_3 \end{matrix} \right\|$, $m = \left\| \begin{matrix} m_1 \\ m_2 \\ m_3 \end{matrix} \right\|$, and $n = \left\| \begin{matrix} n_1 \\ n_2 \\ n_3 \end{matrix} \right\|$ are three nonconcurrent

lines, then any line $p = \left\| \begin{matrix} p_1 \\ p_2 \\ p_3 \end{matrix} \right\|$ is linearly dependent upon l, m, and n; that is,

there exist numbers (λ, μ, ν), such that $p = \lambda l + \mu m + \nu n$.

Corollary 1

If l, m, n are three nonconcurrent lines, and if an arbitrary line p is expressed in each of the forms

$$p = \lambda l + \mu m + \nu n \qquad \text{and} \qquad p = \lambda' l + \mu' m + \nu' n$$

then $\lambda' = k\lambda$, $\mu' = k\mu$, $\nu' = k\nu$.

Exercises

1. Verify that the points in each of the following sets are collinear, and express the coordinates of each point as linear combinations of the coordinates of the other two points:

(a) $(-2,2,1)$, $(2,3,2)$, $(4,1,1)$ (b) $(5,-2,2)$, $(2,-1,0)$, $(1,0,2)$

(c) $(-1,1,0)$, $(1,2,3)$, $(-2,5,3)$ (d) $(1,-1,5)$, $(0,2,1)$, $(1,-3,4)$

2. Verify that the lines in each of the following sets are concurrent, and express the coordinates of each line as linear combinations of the coordinates of the other two lines:

(a) $[1,3,3]$, $[-1,2,1]$, $[3,-1,1]$ (b) $[-1,0,2]$, $[2,-1,0]$, $[1,-2,6]$

(c) $[1,1,-2]$, $[1,2,1]$, $[1,0,-5]$ (d) $[-4,3,1]$, $[2,1,1]$, $[-2,3,5]$

(e) $x_1 + 2x_2 + 3x_3 = 0$, $-2x_1 + 5x_2 + 3x_3 = 0$, $x_1 - x_2 = 0$

(f) $2x_1 - x_2 + x_3 = 0$, $-2x_1 + 3x_2 + 5x_3 = 0$, $-4x_1 + 3x_2 + x_3 = 0$

3. Verify that the first three points in each of the following sets are noncollinear, and express the coordinates of the fourth point as linear combinations of the coordinates of the first three:

(a) $(0,1,1)$, $(1,0,-1)$, $(2,1,0)$; $(1,2,3)$

(b) $(1,1,1)$, $(1,1,-1)$, $(1,2,-1)$; $(-1,0,4)$

(c) $(1,2,3)$, $(2,3,4)$, $(2,0,1)$; $(0,1,1)$

(d) $(1,2,1)$, $(2,0,1)$, $(0,2,-1)$; $(3,2,-2)$

4. If, for convenience, the incidence condition $l_1x_1 + l_2x_2 + l_3x_3 = 0$ is abbreviated to $(lx) = 0$, show that:

(a) The equation of the line which passes through the point (a_1,a_2,a_3) and the intersection of the distinct lines $(ux) = 0$ and $(vx) = 0$ is $(va)(ux) - (ua)(vx) = 0$.

(b) The equation of the point of intersection of the line $[\alpha_1,\alpha_2,\alpha_3]$ and the line determined by the points $(pl) = 0$ and $(ql) = 0$ is $(q\alpha)(pl) - (p\alpha)(ql) = 0$.

5. Given $A:(1,1,0)$, $l: x_1 + x_2 + x_3 = 0$, and $Y:(y_1,y_2,y_3)$. If B is the intersection of l and the line AY, find the parameters (λ,μ) of Y with respect to A and B.

6. Given $A:(1,0,-1)$, $l: x_1 - x_2 = 0$, $m: x_2 + x_3 = 0$, and $Y:(y_1,y_2,y_3)$. If B and C are, respectively, the intersections of l and m with the line AY, find the parameters (λ,μ) of Y with respect to B and C.

7. (a) Write out a proof of Theorem 4.
(b) Write out a proof of Theorem 6.

8. (a) What modifications, if any, must be made in Theorems 1 and 2 in order that they shall be valid in the euclidean plane?
(b) What modifications, if any, must be made in Theorems 3 and 4 in order that they shall be valid in the euclidean plane?
(c) What modification, if any, must be made in Theorems 5 and 6 in order that they shall be valid in the euclidean plane?

9. (a) Is Theorem 5 true if A, B, and C are collinear? Why?
(b) Prove Theorem 5, using Theorem 6, Sec. 3.6, instead of Cramer's rule.

10. In the parametrization described by Theorem 5, what is the locus of points for which $\lambda = 0$? for which $\mu = 0$? For which $\nu = 0$? For which $\lambda = 0$ and $\mu = 0$? For which $\lambda = 0$ and $\nu = 0$? For which $\mu = 0$ and $\nu = 0$? Can λ, μ, ν be considered homogeneous point-coordinates?

11. With the notation introduced in Exercise 4, prove the following result. If $(ax) = 0$, $(bx) = 0$, $(cx) = 0$ are three nonconcurrent lines, and if $U:(u_1,u_2,u_3)$, $V:(v_1,v_2,v_3)$, $W:(w_1,w_2,w_3)$ are three points, then U, V, W are collinear if and only if

$$\begin{vmatrix} (au) & (av) & (aw) \\ (bu) & (bv) & (bw) \\ (cu) & (cv) & (cw) \end{vmatrix} = 0$$

4.3 The Theorems of Desargues and Pappus

Using the results of the preceding section, we can now prove the famous theorems of Desargues and Pappus.

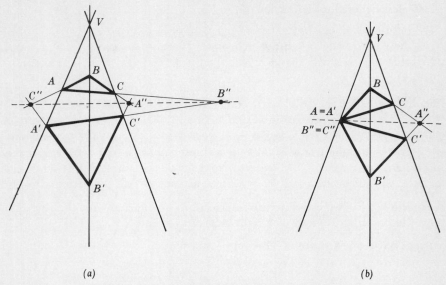

(a)   (b)

Fig. 4.2   Figure illustrating Desargues' theorem.

Theorem 1   Desargues'[1] theorem

If the lines determined by corresponding vertices of two triangles are concurrent, then the intersections of corresponding sides[2] of the two triangles are collinear.

Proof   Let A, B, C and A', B', C' be the vertices of two triangles, and let the lines AA', BB', CC' be concurrent in a point V. Let C'' be the intersection of the lines AB and $A'B'$, let B'' be the intersection of the lines AC and $A'C'$, and let A'' be the intersection of the lines BC and $B'C'$ (Fig. 4.2a). We observe first that if a pair of corresponding vertices, say A and A', coincide, then the assertion of the theorem is obviously true (Fig. 4.2b). Hence we need consider only the general case in which $A \neq A'$, $B \neq B'$, $C \neq C'$. Under these conditions, the coordinate-vector of V must, by Theorem 3, Sec. 4.2, be expressible equally well as a linear combination of the coordinate-vectors of A and

---

[1] Named for the French mathematician Gérard Desargues (1593–1662).

[2] Since segments are defined in terms of the relation of betweenness, which in turn depends upon the concept of distance, and since, as we observed in Sec. 2.4, distance is not defined in either E_2^+ or Π_2, it follows that the notion of a segment is meaningless in both E_2^+ and Π_2. For this reason, when we speak of a triangle in E_2^+ or Π_2 we do not mean a figure consisting of three noncollinear points and three *segments* determined by these points, but rather a figure consisting of three noncollinear points and the three lines these points determine. Thus in both E_2^+ and Π_2, when we speak of the sides of a triangle we mean, necessarily, the *lines* determined by the vertices of the triangle. Triangles are thus self-dual figures; and we shall extend our notation accordingly and denote by the symbol $\triangle abc$ the triangle whose sides are the lines a, b, c just as we have in the past used the symbol $\triangle ABC$ to denote the triangle whose vertices are the points A, B, C.

A', B and B', or C and C'; that is,

(1) $$V = \lambda_a A + \lambda_a' A' = \lambda_b B + \lambda_b' B' = \lambda_c C + \lambda_c' C'$$

From the last equality in this chain, we have, by transposing,

(2) $$\lambda_b B - \lambda_c C = -\lambda_b' B' + \lambda_c' C'$$

Furthermore, $\lambda_b B - \lambda_c C$ is not a null vector; for if this were the case, then B and C would be proportional and hence would be coordinate-vectors of the same point, contrary to hypothesis. Therefore $\lambda_b B - \lambda_c C$ is the coordinate-vector of some point, and since it is a linear combination of B and C, this point is on the line BC. Moreover, from (2) it is clear that this point can equally well be written $-\lambda_b' B' + \lambda_c' C'$, which shows that it is also on the line $B'C'$. But the only point which is simultaneously on the line BC and the line $B'C'$ is their intersection, A''; that is, either member of Eq. (2) is a coordinate-vector of A''. Similarly, by working with the other two equalities in the chain (1), we find

(3) $$B'' = \lambda_a A - \lambda_c C = -\lambda_a' A' + \lambda_c' C'$$

(4) $$C'' = \lambda_a A - \lambda_b B = -\lambda_a' A' + \lambda_b' B'$$

To prove the theorem, we must now show that A'', B'', C'' are collinear. To do this, it is sufficient, by Theorem 3, Sec. 4.2, to prove that A'' is a linear combination of B'' and C''; and this follows at once, since, using the first members of (2), (3), and (4), we have, by inspection,

$$A'' = B'' - C''$$

The dual of Theorem 1, which is also its converse, can be proved in exactly the same fashion or justified without further proof by the principle of duality:

Theorem 2

If the intersections of corresponding sides of two triangles are collinear, then the lines determined by corresponding vertices of the two triangles are concurrent.

The relations existing between the vertices of the two triangles and the sides of the two triangles in Desargues' theorem are examples of two relations of sufficient importance to be stated explicitly.

Definition 1

Two sets of points are said to be centrally perspective from the center V if the sets are in one-to-one correspondence and if the lines determined by corresponding points all pass through V.

Definition 2

Two sets of lines are said to be axially perspective on the axis v if the sets are in one-to-one correspondence and if the points of intersection of corresponding lines all lie on v.

Using the notions of central and axial perspectivity, Desargues' theorem and its dual can be restated in the following concise forms: *if two triangles are centrally perspective, they are also axially perspective*, and *if two triangles are axially perspective, they are also centrally perspective*.

We must remember, of course, that our proof of Desargues' theorem has established it only in the extended euclidean plane, E_2^+, and in the algebraic representation, Π_2. Later on, in our axiomatic development of projective geometry, we shall see that there are projective planes in which the theorem of Desargues is false. An example of such a system is given in Appendix 2.

Example 1

Given that the line determined by the midpoints of two sides of a triangle in E_2 is parallel to the third side, prove that the medians of a triangle are concurrent.

Let $\triangle ABC$ be an arbitrary triangle in the euclidean plane, and let A', B', C' be, respectively, the midpoints of the segments $\overline{BC}$, $\overline{CA}$, $\overline{AB}$ (Fig. 4.3). Then, as is well known, $AB \parallel A'B'$, $BC \parallel B'C'$, and $CA \parallel C'A'$. Interpreting these relations in the extended euclidean plane, we see that $\triangle ABC$ and $\triangle A'B'C'$ are axially perspective, since their corresponding sides, being parallel, necessarily intersect on the ideal line. Therefore, by the

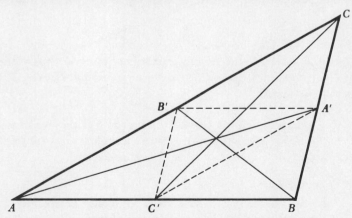

Fig. 4.3   Figure illustrating the concurrence of the medians of a triangle.

converse of Desargues' theorem, $\triangle ABC$ and $\triangle A'B'C'$ are also centrally perspective; that is, AA', BB', and CC' are concurrent lines. But these lines are, by definition, the medians of $\triangle ABC$; hence the assertion of the example is established.

Theorem 3   The theorem of Pappus[1]

Let A, B, C be three points on a line l and let A', B', C' be three points on a second line l', none of these points coinciding with the point which is common to l and l'. Then the intersection of BC' and $B'C$, the intersection of AC' and $A'C$, and the intersection of AB' and $A'B$ are collinear.

Proof   Let l and l' be two lines intersecting in the point P. Let A, B, C be three points on l, each distinct from P; let A', B', C' be three points on l', each distinct from P; and let:

A'' be the intersection of BC' and $B'C$
B'' be the intersection of AC' and $A'C$
C'' be the intersection of AB' and $A'B$

(Fig. 4.4). On l let us choose a point Q, distinct from P, to be used with P as a base point for the parametrization of l. Then, by Theorem 3, Sec. 4.2,

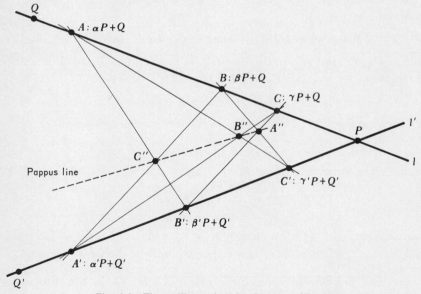

Fig. 4.4   Figure illustrating the theorem of Pappus.

[1] Pappus was a Greek mathematician of the fourth century A.D.

we can write

$$A = \alpha_p P + \alpha_q Q \qquad B = \beta_p P + \beta_q Q \qquad C = \gamma_p P + \gamma_q Q$$

Moreover, since P is distinct from A, B, and C, neither α_q, β_q, nor γ_q can be zero. Hence, since only the ratios of the parameters are significant, it is no specialization to write

(5) $$A = \alpha P + Q \qquad B = \beta P + Q \qquad C = \gamma P + Q$$

where $\alpha = \alpha_p/\alpha_q$, $\beta = \beta_p/\beta_q$, $\gamma = \gamma_p/\gamma_q$. Similarly, if we parametrize l' in terms of P and a second point, Q', distinct from P, we have

(6) $$A' = \alpha' P + Q' \qquad B' = \beta' P + Q' \qquad C' = \gamma' P + Q'$$

Now A'' is, by definition, the intersection of BC' and $B'C$. Hence it must be expressible as a linear combination of B and C', say $\lambda B + \mu' C'$, and also as a linear combination of B' and C, say $\lambda' B' + \mu C$. Substituting for B, B', C, and C' from (5) and (6) into these expressions and collecting terms, we find that A'' can be expressed in either of the forms

(7) $$A'' = (\lambda \beta + \mu' \gamma') P + \lambda Q + \mu' Q'$$
(8) $$A'' = (\lambda' \beta' + \mu \gamma) P + \mu Q + \lambda' Q'$$

Since these equations express the same point, A'', as linear combinations of the same noncollinear base points, P, Q, and Q', it follows from Corollary 1, Theorem 5, Sec. 4.2, that

$$\lambda \beta + \mu' \gamma' = k(\lambda' \beta' + \mu \gamma) \qquad \lambda = k\mu \qquad \mu' = k\lambda'$$

Hence, using the last two of these equations to eliminate λ and μ' from the first equation, we have after collecting terms on λ' and μ,

$$\mu(\beta - \gamma) = \lambda'(\beta' - \gamma')$$

Therefore, taking $\lambda' = \beta - \gamma$ and $\mu = \beta' - \gamma'$, we have from (8)

(9) $$A'' = (\beta \beta' - \gamma \gamma') P + (\beta' - \gamma') Q + (\beta - \gamma) Q'$$

In exactly the same fashion, we find

(10) $$B'' = (\gamma \gamma' - \alpha \alpha') P + (\gamma' - \alpha') Q + (\gamma - \alpha) Q'$$
(11) $$C'' = (\alpha \alpha' - \beta \beta') P + (\alpha' - \beta') Q + (\alpha - \beta) Q'$$

From (9), (10), and (11), it is clear that $A'' = -B'' - C''$. Hence, by Theorem 3, Sec. 4.2, A'', B'', and C'' are collinear, and the theorem is proved.

The line which contains the points A'', B'', C'' in the theorem of Pappus is often called the **pappus line** of the two collinear triples (A,B,C) and (A',B',C').

Exercises

1. Verify Desargues' theorem for the following pairs of triangles:

(a) $A:(1,1,1)$, $B:(1,-1,1)$, $C:(1,-2,0)$; $A':(1,1,2)$, $B':(1,-1,0)$, $C':(1,-2,2)$

(b) $A:(1,1,2)$, $B:(2,0,1)$, $C:(1,2,1)$; $A':(2,2,1)$, $B':(3,1,2)$, $C':(0,1,0)$

(c) $A:(1,1,2)$, $B:(3,1,2)$, $C:(0,1,0)$; $A':(2,2,1)$, $B':(2,0,1)$, $C':(1,2,1)$

(d) $A:(1,1,2)$, $B:(1,-1,1)$, $C:(1,1,-2)$; $A':(0,1,1)$, $B':(2,-2,1)$, $C':(1,0,-1)$

2. Verify the theorem of Pappus for the following pairs of collinear point-triples:

(a) $A:(1,-1,0)$, $B:(1,1,2)$, $C:(0,1,1)$; $A':(1,-1,1)$, $B':(2,-2,1)$, $C':(-1,1,1)$

(b) $A:(1,0,-1)$, $B:(1,1,-2)$, $C:(1,2,-3)$; $A':(1,1,0)$, $B':(1,3,1)$, $C':(0,2,1)$

(c) $A:(2,1,1)$, $B:(1,1,0)$, $C:(3,1,2)$; $A':(2,0,-1)$, $B':(1,1,0)$, $C':(0,2,1)$

(d) $A:(2,1,-1)$, $B:(0,1,1)$, $C:(1,2,1)$; $A':(1,1,1)$, $B':(0,2,1)$, $C':(1,3,2)$

3. What is the dual of the theorem of Pappus? Construct a figure illustrating this dual theorem.

4. Using Desargues' theorem, prove the following special case of the theorem of Pappus. If A, B, C are three points on a line l, and if A', B', C' are three points on a second line l', none of these points coinciding with the point of intersection of l and l', and if the triple (A,B,C) and the triple (A',B',C') are centrally perspective, then the pappus line of the two triples is concurrent with l and l'.

5. Show how the result of Exercise 4 can be used to solve the following problem. Given two lines, l and l', whose intersection is inaccessible, and a point Q which does not lie on either l or l', construct the line through Q which is concurrent with l and l'.

6. If $\triangle A_1B_1C_1$, $\triangle A_2B_2C_2$, and $\triangle A_3B_3C_3$ are perspective from a point V, show that the axes of perspectivity for the triangles taken two at a time are concurrent.

7. State Desargues' theorem in a form which is valid in the euclidean plane if:

(a) The two triangles are perspective from a finite point.

(b) The two triangles are perspective from an ideal point.

8. State the theorem of Pappus in a form which is valid in the euclidean plane.

9. Since the assertion of Example 1 refers to a configuration in the euclidean plane whereas our proof was constructed in the extended euclidean plane, it is necessary that we show that the point common to the medians in the extended plane is actually a point of the euclidean plane. Complete the solution of Example 1 by doing this.

10. Using the symbol $(lx) = 0$ as an abbreviation for the incidence condition $l_1x_1 + l_2x_2 + l_3x_3 = 0$, as in Exercise 4, Sec. 4.2, let the sides of $\triangle ABC$ be the lines whose equations are $BC: (ax) = 0$, $AC: (bx) = 0$, $AB: (cx) = 0$, and let the vertices of $\triangle A'B'C'$ be the points whose equations are $A': (\alpha l) = 0$, $B': (\beta l) = 0$, $C': (\gamma l) = 0$. Then:

(a) Show that the equations of the lines AA', BB', CC' are, respectively, $AA': (c\alpha)(bx) - (b\alpha)(cx) = 0, BB': (a\beta)(cx) - (c\beta)(ax) = 0, CC': (b\gamma)(ax) - (a\gamma)(bx) = 0$.

(b) Show that $\triangle ABC$ and $\triangle A'B'C'$ are centrally perspective if and only if $(a\beta)(b\gamma)(c\alpha) = (b\alpha)(c\beta)(a\gamma)$.

11. With the notation and data of Exercise 10, find the equations of the points of intersection of the pairs of corresponding sides, $(BC,B'C')$, $(AC,A'C')$, $(AB,A'B')$, of $\triangle ABC$ and $\triangle A'B'C'$, and thence derive a necessary and sufficient condition that $\triangle ABC$ and $\triangle A'B'C'$ be axially perspective. Combining this result with the result of part (b) of Exercise 10, obtain another proof of Desargues' theorem.

12. Using the result of part (b), Exercise 10, show that if $\triangle ABC$ is centrally perspective with $\triangle A'B'C'$ and $\triangle B'C'A'$, then it is also centrally perspective with $\triangle C'A'B'$. Draw a figure to illustrate this fact.

13. If two quadrilaterals are centrally perspective, are they also axially perspective?

14. Using the abbreviated notation introduced in Exercise 10, let the equations of the sides of $\triangle ABC$ be $BC: (ax) = 0$, $AC: (bx) = 0$, $AB: (cx) = 0$, and let the equations of the vertices be $A: (\alpha l) = 0, B: (\beta l) = 0, C: (\gamma l) = 0$. Let the pencil on A be parametrized in terms of AB and AC as base lines, let the pencil on B be parametrized in terms of BC and BA as base lines, let the pencil on C be parametrized in terms of CA and CB as base lines, and let a', b', c' be, respectively, the lines on A, B, C whose parameters are $[1,-1]$, $[1,-1]$, $[1,1]$. If A'' is the intersection of a' and BC, if B'' is the intersection of b' and AC, and if C'' is the intersection of c' and AB, prove that A'', B'', C'' are collinear, and thence prove that $\triangle ABC$ and $\triangle a'b'c'$ are centrally perspective.

15. Using the results of Exercise 14, prove the following euclidean theorem: *the points in which the bisectors of two interior angles of a triangle meet the opposite sides and the point in which the bisector of the external angle at the remaining vertex intersects the remaining side are collinear.*

4.4 Perspectivities

In the last section, after our discussion of Desargues' theorem, we introduced the notion of centrally perspective sets of points. The case in which each of

two centrally perspective sets consists of collinear points is especially important because it involves precisely the configuration obtained by projecting a
set of points on one line onto a set of points on a second line. Such a relation
between points of two ranges on distinct lines is known as a **perspectivity,**
and is fundamental in the study of projective geometry.

Definition 1

If two sets of points on distinct lines are in one-to-one correspondence, and if the lines joining corresponding points all pass through a
fixed point, V, the relation between the sets is called a perspectivity and the
point V is called its center.

The fact that a set of points $(A,B,C,...)$ on a line l is perspective
from a point V with a set of points $(A',B',C',...)$ on a second line l' is indicated
by writing

$$l(A,B,C,...) \overset{V}{\wedge} l'(A',B',C',...)$$

Occasionally, when it is unnecessary to indicate explicitly the correspondence
between the two sets, this notation is abbreviated to

$$l \overset{V}{\wedge} l'$$

Also, when the points A, B, C, . . . are known to be determined by l and the
respective lines a, b, c, . . . , and when the points A', B', C', . . . are known to
be determined by l' and the respective lines a', b', c', . . . , we shall sometimes
describe the perspectivity $l(A,B,C,...) \overset{V}{\wedge} l'(A',B',C',...)$ by writing

$$l(a,b,c,...) \overset{V}{\wedge} l'(a',b',c',...)$$

Clearly, a perspectivity between two ranges is uniquely determined
when the center of the perspectivity is given. Moreover, in any perspectivity
between the ranges on l and l', the point of intersection, S, of l and l' is
self-corresponding in the one-to-one correspondence between the two ranges
(Fig. 4.5a).

Dually, of course, we have the notion of a perspectivity between
two sets of lines on distinct points:

Definition 2

If two sets of lines on distinct points are in one-to-one correspondence, and if the points of intersection of corresponding lines all lie on a fixed
line, v, the relation between the sets is called a perspectivity and the line v is
called its axis.

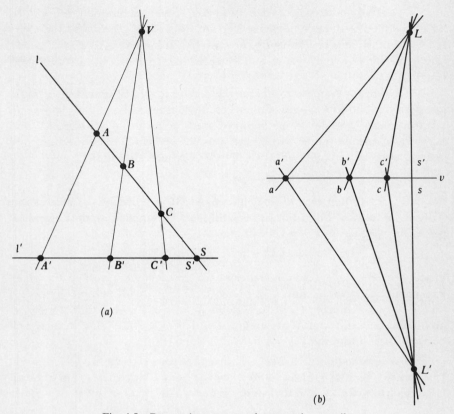

(a)

(b)

Fig. 4.5 Perspective ranges and perspective pencils.

The fact that a set of lines $(a,b,c,...)$ on a point L is perspective on a line v with a set of lines $(a',b',c',...)$ on a second point L' is indicated by writing

$$L(a,b,c,...) \overset{v}{\barwedge} L'(a',b',c',...)$$

or, in abbreviated form, simply

$$L \overset{v}{\barwedge} L'$$

Also, when the lines a, b, c, . . . are known to be determined by L and the respective points A, B, C, . . . , and when the lines a', b', c', . . . are known to be determined by L' and the respective points A', B', C', . . . , we shall sometimes describe the perspectivity $L(a,b,c,...) \overset{v}{\barwedge} L'(a',b',c,'...)$ by writing

$$L(A,B,C,...) \overset{v}{\barwedge} L'(A',B',C',...)$$

Clearly, a perspectivity between two pencils is uniquely determined when the axis of the perspectivity is given. Moreover, in any perspectivity between the pencils on L and L', the line, s, determined by L and L' is self-corresponding in the one-to-one correspondence between the two pencils (Fig. 4.5b).

We now propose to obtain the algebraic description of the general perspectivity between ranges of points on distinct lines. To do this, let l be an arbitrary line, and let it be parametrized in terms of $A:(a_1,a_2,a_3)$ and $B:(b_1,b_2,b_3)$ as base points with λ and μ as parameters. The coordinates of a general point $X:(x_1,x_2,x_3)$ on l can then be written in the form

$$x_i = \lambda a_i + \mu b_i \qquad i = 1, 2, 3$$

Similarly, let l' be a second line, parametrized in terms of the base points $A':(a_1',a_2',a_3')$ and $B':(b_1',b_2',b_3')$ with λ' and μ' as parameters, so that a general point $X':(x_1',x_2',x_3')$ on l' has coordinates

$$x_i' = \lambda' a_i' + \mu' b_i' \qquad i = 1, 2, 3$$

Finally, let $V:(v_1,v_2,v_3)$ be the center of the perspectivity. Now in the given perspectivity between l and l', a point $X':(x_1',x_2',x_3')$ with parameters (λ',μ') on l' will be the image of a general point $X:(x_1,x_2,x_3)$ with parameters (λ,μ) on l if and only if X and X' are collinear with V (Fig. 4.6). This, in turn, will be the case if and only if

$$(1) \quad \begin{vmatrix} v_1 & v_2 & v_3 \\ x_1 & x_2 & x_3 \\ x_1' & x_2' & x_3' \end{vmatrix} = \begin{vmatrix} v_1 & v_2 & v_3 \\ \lambda a_1 + \mu b_1 & \lambda a_2 + \mu b_2 & \lambda a_3 + \mu b_3 \\ \lambda' a_1' + \mu' b_1' & \lambda' a_2' + \mu' b_2' & \lambda' a_3' + \mu' b_3' \end{vmatrix} = 0$$

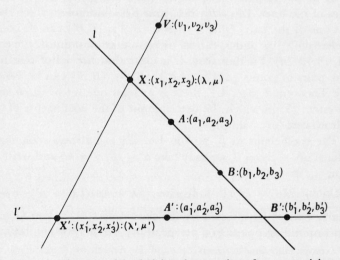

Fig. 4.6   Figure used in deriving the equation of a perspectivity.

If the last determinant is expanded,[1] this condition becomes

(2) $$\alpha\lambda\lambda' + \beta\lambda\mu' + \gamma\mu\lambda' + \delta\mu\mu' = 0$$

where

(3a) $$\alpha = \begin{vmatrix} v_1 & v_2 & v_3 \\ a_1 & a_2 & a_3 \\ a'_1 & a'_2 & a'_3 \end{vmatrix}$$

$$= v_1(a_2 a'_3 - a_3 a'_2) + v_2(a_3 a'_1 - a_1 a'_3) + v_3(a_1 a'_2 - a_2 a'_1)$$

(3b) $$\beta = \begin{vmatrix} v_1 & v_2 & v_3 \\ a_1 & a_2 & a_3 \\ b'_1 & b'_2 & b'_3 \end{vmatrix}$$

$$= v_1(a_2 b'_3 - a_3 b'_2) + v_2(a_3 b'_1 - a_1 b'_3) + v_3(a_1 b'_2 - a_2 b'_1)$$

(3c) $$\gamma = \begin{vmatrix} v_1 & v_2 & v_3 \\ b_1 & b_2 & b_3 \\ a'_1 & a'_2 & a'_3 \end{vmatrix}$$

$$= v_1(b_2 a'_3 - b_3 a'_2) + v_2(b_3 a'_1 - b_1 a'_3) + v_3(b_1 a'_2 - b_2 a'_1)$$

(3d) $$\delta = \begin{vmatrix} v_1 & v_2 & v_3 \\ b_1 & b_2 & b_3 \\ b'_1 & b'_2 & b'_3 \end{vmatrix}$$

$$= v_1(b_2 b'_3 - b_3 b'_2) + v_2(b_3 b'_1 - b_1 b'_3) + v_3(b_1 b'_2 - b_2 b'_1)$$

It is important to note that unless they are all zero, the coefficients of v_1, v_2, v_3 in the respective equations (3a), (3b), (3c), and (3d) are the coordinates of the lines AA', AB', BA', and BB'. Moreover, since the points in at most one of the pairs (A,A'), (A,B'), (B,A'), (B,B') can be coincident, the coefficients of v_1, v_2, and v_3 can all be zero in at most one of the equations (3a), (3b), (3c), (3d). Furthermore, it is clear that not all of the lines containing the pairs of points (A,A'), (A,B'), (B,A'), (B,B') can be concurrent. Hence, by Theorem 2, Sec. 4.1, there is at least one set of three equations from the system (3) in which the determinant of the coefficients of the v's is different from zero.

The coefficients α, β, γ, δ in Eq. (2) are always such that $\alpha\delta - \beta\gamma \neq 0$; for if $\alpha\delta - \beta\gamma = 0$, we may take $\alpha = k\beta$, $\gamma = k\delta$ and write Eq. (2) in the form

$$k\beta\lambda\lambda' + \beta\lambda\mu' + k\delta\mu\lambda' + \delta\mu\mu' = (\beta\lambda + \delta\mu)(k\lambda' + \mu') = 0$$

which clearly cannot be satisfied by the parameters (λ,μ) and (λ',μ') of *all* pairs of corresponding points in a perspectivity. In fact, this equation asserts

[1] This is most conveniently done by means of the addition theorem for determinants, Theorem 7, Appendix 1.

that any point on l, except the point whose parameters are $(\delta, -\beta)$ corresponds to the point $(1, -k)$ on l', that is, is collinear with V and the point $(1, -k)$ on l', which is impossible.

Equations (3a) to (3d) constitute a system of four homogeneous linear equations satisfied by the quantities $-1, v_1, v_2, v_3$. Since these quantities are obviously not all zero, it follows, from Theorem 5, Sec. 3.6, that the determinant of the coefficients in the system must vanish. Hence

$$(4) \qquad \begin{vmatrix} \alpha & a_2a_3' - a_3a_2' & a_3a_1' - a_1a_3' & a_1a_2' - a_2a_1' \\ \beta & a_2b_3' - a_3b_2' & a_3b_1' - a_1b_3' & a_1b_2' - a_2b_1' \\ \gamma & b_2a_3' - b_3a_2' & b_3a_1' - b_1a_3' & b_1a_2' - b_2a_1' \\ \delta & b_2b_3' - b_3b_2' & b_3b_1' - b_1b_3' & b_1b_2' - b_2b_1' \end{vmatrix} = 0$$

or
$$\rho_1\alpha + \rho_2\beta + \rho_3\gamma + \rho_4\delta = 0$$

where ρ_i is the cofactor of the element in the ith row and first column in this determinant. Thus, *if there is a perspectivity between the line l and the line l' in which the point with parameters (λ',μ') on l' corresponds to the point with parameters (λ,μ) on l, then there exist four numbers α, β, γ, δ such that*

$$(5a) \qquad \alpha\delta - \beta\gamma \neq 0$$

$$(5b) \qquad \alpha\lambda\lambda' + \beta\lambda\mu' + \gamma\mu\lambda' + \delta\mu\mu' = 0$$

$$(5c) \qquad \rho_1\alpha + \rho_2\beta + \rho_3\gamma + \rho_4\delta = 0$$

where the ρ's are known functions of the coordinates of the base points on l and l'.

Conversely, given two parametrized lines, l and l', and four numbers α, β, γ, δ satisfying*(5a), (5b), and (5c), where ρ_1, ρ_2, ρ_3, and ρ_4 are the appropriate functions of the coordinates of the base points on l and l', there exists a perspectivity between l and l' in which the point with parameters (λ',μ') on l' corresponds to the point with parameters (λ,μ) on l. To see this, we first observe that if α, β, γ, and δ satisfy Eq. (5c), then the system (3) is consistent, and hence it is possible to solve for the coordinates of $V:(v_1,v_2,v_3)$ from these equations. It then follows from Eq. (5b) and the definition of α, β, γ, and δ in terms of the coordinates of V, that the collinearity condition (1) is satisfied. In other words, given any point $X:(\lambda,\mu)$ on l, the corresponding point, $X':(\lambda',\mu')$ on l' whose parameters are determined by Eq. (5b) is such that X, X', and the center V are collinear, as required by the definition of a perspectivity. Finally, since the system (3) always contains a subset of three equations in which the determinant of the coefficients of the v's is different from zero, it follows that when Eq. (5c) is satisfied and the system is consistent, the solution for V is unique.

Example 1

Let l be the line $x_1 - 2x_2 = 0$, parametrized in terms of the base points $A:(2,1,1)$ and $B:(0,0,1)$ with parameters (λ,μ), and let l' be the line $x_2 + x_3 = 0$, parametrized in terms of the base points $A':(0,1,-1)$ and $B':(1,0,0)$

with parameters (λ',μ'). Show that the equation $\alpha\lambda\lambda' + \beta\lambda\mu' + \gamma\mu\lambda' + \delta\mu\mu' = 0$ defines a perspectivity between l and l' if and only if $\alpha + 2\beta - 2\gamma - 4\delta = 0$. Find the center of the perspectivity corresponding to the particular set of values $\alpha = -2$, $\beta = 2$, $\gamma = 3$, $\delta = -1$.

With A, B, A', B' given, it is an easy matter to construct the determinant (4), and we find without difficulty

$$\begin{vmatrix} \alpha & -2 & 2 & 2 \\ \beta & 0 & 1 & -1 \\ \gamma & -1 & 0 & 0 \\ \delta & 0 & 1 & 0 \end{vmatrix} = 0$$

Expanding this in terms of the elements of the first column, we obtain $\alpha + 2\beta - 2\gamma - 4\delta = 0$ as the condition which α, β, γ, δ must satisfy in order that $\alpha\lambda\lambda' + \beta\lambda\mu' + \gamma\mu\lambda' + \delta\mu\mu' = 0$ should define a perspectivity. The values $\alpha = -2$, $\beta = 2$, $\gamma = 3$, $\delta = -1$ clearly satisfy this condition, and hence the equation $-2\lambda\lambda' + 2\lambda\mu' + 3\mu\lambda' - \mu\mu' = 0$ defines a perspectivity between l and l'. To find the center of this perspectivity, we return to the system of Eqs. (3) and solve it for the coordinates of V. In the present case the system becomes

$$\begin{aligned} -2 &= -2v_1 + 2v_2 + 2v_3 \\ 2 &= v_2 - v_3 \\ 3 &= - v_1 \\ -1 &= v_2 \end{aligned}$$

from which we conclude at once that $v_1 = -3$, $v_2 = -1$, $v_3 = -3$; that is, the center of the given perspectivity is the point $(3,1,3)$.

Exercises

1. If l: $x_1 + x_2 = 0$ is parametrized in terms of the base points A:$(1,-1,0)$ and B:$(-2,2,1)$ with parameters (λ,μ), and if l': $x_1 - x_2 + 2x_3 = 0$ is parametrized in terms of the base points A':$(1,1,0)$ and B':$(1,3,1)$ with parameters (λ',μ'), find the equation of the perspectivity between l and l' which is defined by the center V:$(0,1,0)$.

2. Work Exercise 1 if the base points on l are A:$(1,-1,2)$ and B:$(1,-1,0)$ and if the base points on l' are A':$(-1,1,1)$ and B':$(2,0,-1)$. Verify that the equation obtained in this case assigns the same image to an arbitrary point of l as does the equation obtained in Exercise 1.

3. In Example 1, find the center of the perspectivity corresponding to the following values:

(a) $\alpha = 2$, $\beta = 3$, $\gamma = -2$, $\delta = 3$ (b) $\alpha = 4$, $\beta = 3$, $\gamma = 1$, $\delta = 2$

4. Let l be the line $x_1 + x_2 + x_3 = 0$, parametrized in terms of the base points $A:(1,1,-2)$ and $B:(1,2,-3)$ with parameters (λ,μ), and let l' be the line $x_1 - x_2 + x_3 = 0$, parametrized in terms of the base points $A':(1,1,0)$ and $B':(0,1,1)$ with parameters (λ',μ'). Show that the equation $\alpha\lambda\lambda' + \beta\lambda\mu' + \gamma\mu\lambda' + \delta\mu\mu' = 0$ defines a perspectivity between l and l' if and only if $4\alpha - 2\beta - \gamma + \delta = 0$. Find the center of the perspectivity corresponding to the following values:

(a) $\alpha = 1$, $\beta = 3$, $\gamma = 2$, $\delta = 4$ $\qquad$ (b) $\alpha = 2$, $\beta = 1$, $\gamma = 3$, $\delta = -3$

5. Let l be the line $x_1 - x_2 + x_3 = 0$, parametrized in terms of the base points $A:(1,2,1)$ and $B:(1,1,0)$ with parameters (λ,μ), and let l' be the line $x_1 + 2x_2 = 0$, parametrized in terms of the base points $A':(-2,1,3)$ and $B':(-2,1,0)$. Determine the condition which α, β, γ, δ must satisfy if the equation $\alpha\lambda\lambda' + \beta\lambda\mu' + \gamma\mu\lambda' + \delta\mu\mu' = 0$ is to describe a perspectivity between l and l'.

6. Let l be the line $x_1 - 2x_2 = 0$, parametrized in terms of the base points $A:(2,1,1)$ and $B:(0,0,1)$ with parameters (λ,μ), and let l' be the line $x_2 + x_3 = 0$, parametrized in terms of the base points $A':(0,1,-1)$ and $B':(t,1,-1)$. Given an arbitrary set of numbers α, β, γ, δ subject only to the condition that $\alpha\delta - \beta\gamma \neq 0$, is it possible to determine t so that the equation $\alpha\lambda\lambda' + \beta\lambda\mu' + \gamma\mu\lambda' + \delta\mu\mu' = 0$ will describe a perspectivity between l and l'?

7. (a) If B is between A and C on a line l in the euclidean plane, and if A', B', C' are, respectively, the images of A, B, C under a perspectivity between l and l', is B' necessarily between A' and C'?
(b) Is an arbitrary perspectivity in the euclidean plane always, sometimes, or never a one-to-one correspondence between the two lines?

8. Given the lines $l_1: x_1 = 0$, $l_2: x_2 = 0$, $l_3: x_3 = 0$, and the points $V_1: (1,-1,1)$, $V_2:(-1,1,1)$, $V_3:(1,1,1)$. Show in two different ways that under the three perspectivities
$$l_1 \overset{V_3}{\wedge} l_2 \qquad l_2 \overset{V_1}{\wedge} l_3 \qquad l_3 \overset{V_2}{\wedge} l_1$$
the final image of an arbitrary point P_1 on l_1 is P_1 itself.

9. Work Exercise 8 if V_1, V_2, V_3 are, respectively, the points $(1,4,1)$, $(1,4,2)$, $(1,2,1)$.

10. (a) Given the lines $l_1: x_2 - x_3 = 0$, $l_2: x_1 - x_3 = 0$, $l_3: x_1 + x_2 = 0$, and the points $U:(2,0,3)$ and $V:(2,4,1)$. Show that the composition of the perspectivities $l_1 \overset{U}{\wedge} l_2$ and $l_2 \overset{V}{\wedge} l_3$ is a perspectivity, and find its center. Is this the case if V is the point $(1,2,3)$?
(b) Work part (a) if l_3 is the line $x_1 - x_2 = 0$.

4.5 Projectivities

One of the properties of a perspectivity is that it establishes a one-to-one correspondence between the points of two distinct lines. As a consequence, it follows that a perspectivity $l_1(A_1,B_1,C_1,...) \overset{V_1}{\wedge} l_2(A_2,B_2,C_2,...)$ mapping l_1 onto l_2 and a perspectivity $l_2(A_2,B_2,C_2,...) \overset{V_2}{\wedge} l_3(A_3,B_3,C_3,...)$ mapping l_2 onto l_3 together induce a one-to-one correspondence between the points of l_1 and the points of l_3 in which $A_1 \sim A_3,\ B_1 \sim B_3,\ C_1 \sim C_3,\ ...$ (Fig. 4.7a). It is then natural to ask whether this correspondence is also a perspectivity, i.e., whether the lines $A_1A_3,\ B_1B_3,\ C_1C_3,\ ...$ are all concurrent. By considering particular examples, either algebraically or geometrically, it is easy to see that in general this is not the case. The mapping between l_1 and l_3 is, nonetheless, a very important one, known as a **projectivity,** which we shall study in this section.

Definition 1

A one-to-one correspondence between points $(A_1,B_1,C_1,...)$ of a line l_1 and points $(A_n,B_n,C_n,...)$ of a line l_n which is the result of a finite sequence of perspectivities,

$$l_1(A_1,B_1,C_1,...) \overset{V_1}{\wedge} l_2(A_2,B_2,C_2,...)),\qquad ...,$$

$$l_{n-1}(A_{n-1},B_{n-1},C_{n-1},...) \overset{V_{n-1}}{\wedge} l_n(A_n,B_n,C_n,...)$$

relating l_1 to l_n is called a projectivity between l_1 and l_n.

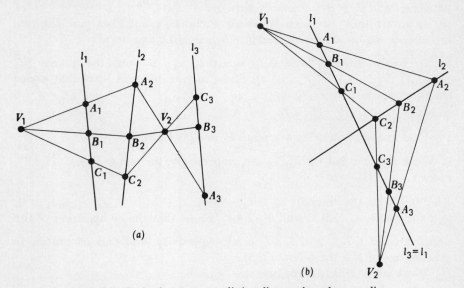

(a)

(b)

Fig. 4.7   Projective ranges on distinct lines and on the same line.

The fact that a set of points $(A_1, B_1, C_1, \ldots)$ on a line l_1 is projective with a set of points $(A_n, B_n, C_n, \ldots)$ on a line l_n is indicated by writing

$$l_1(A_1, B_1, C_1, \ldots) \sim l_n(A_n, B_n, C_n, \ldots)$$

or, when it is unnecessary to describe the correspondence explicitly, simply

$$l_1 \sim l_n$$

It is important to note that, unlike the definition of a perspectivity, the definition of a projectivity does not require that l_1 and l_n be distinct lines (Fig. 4.7b). When l_1 and l_n are the same line, the totality of points on l_1 is, of course, identical with the totality of points on l_n. However, it is convenient to think of the image set and the object, or preimage, set as being different ranges on $l_1 (= l_n)$. The two ranges identified in this fashion by a projective mapping of a line onto itself are usually referred to as **cobasal ranges.**

Dually, we have the concept of a projectivity between two pencils of lines (Fig. 4.8a):

Definition 2

A one-to-one correspondence between lines $(a_1, b_1, c_1, \ldots)$ on a point L_1 and lines $(a_n, b_n, c_n, \ldots)$ on a point L_n which is the result of a finite sequence

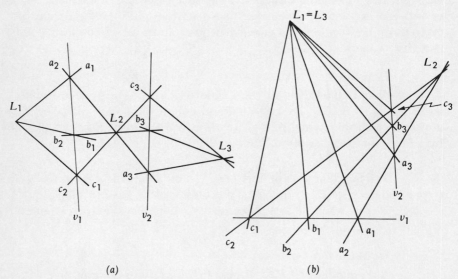

(a)   (b)

Fig. 4.8   Projective pencils on distinct points and on the same point.

of perspectivities

$$L_1(a_1,b_1,c_1,\ldots) \overset{v_1}{\overline{\wedge}} L_2(a_2,b_2,c_2,\ldots), \qquad \ldots,$$

$$L_{n-1}(a_{n-1},b_{n-1},c_{n-1},\ldots) \overset{v_{n-1}}{\overline{\wedge}} L_n(a_n,b_n,c_n,\ldots)$$

relating the lines of the pencil on L_1 to the lines of the pencil on L_n is called a projectivity between L_1 and L_n.

The fact that a set of lines $(a_1,b_1,c_1,\ldots)$ on a point L_1 is projective with a set of lines $(a_n,b_n,c_n,\ldots)$ on a point L_n is indicated by writing

$$L_1(a_1,b_1,c_1,\ldots) \sim L_n(a_n,b_n,c_n,\ldots)$$

or, when it is unnecessary to describe the correspondence explicitly, simply

$$L_1 \sim L_n$$

It is important to note that the definition of a projectivity between a point L_1 and a point L_2 does not require that these points be distinct (Fig. 4.8b). When $L_1 = L_n$, the totality of lines on L_1 is, of course, identical with the set of lines on L_n. It is customary, however, to think of the image set and the object, or preimage, set as being different pencils on $L_1(=L_n)$. The two pencils identified in this fashion by a projective mapping of the set of lines on a point, L, onto itself are usually referred to as **cobasal pencils.**

The concept of a projectivity between two sets of collinear points or between two sets of concurrent lines leads naturally to the notion of a projectivity between a set of concurrent lines and a set of collinear points, and vice versa:

Definition 3

A set of lines $(a,b,c,\ldots)$ on a point V is said to be projective with a set of points $(A,B,C,\ldots)$ on a line v if and only if the points $(A,B,C,\ldots)$ are projective with the points $(A',B',C',\ldots)$ in which the lines $(a,b,c,\ldots)$ are met by a line v' not passing through V.

The fact that a set of lines $(a,b,c,\ldots)$ on a point V is projective with a set of points $(A,B,C,\ldots)$ on a line v is indicated by writing $V(a,b,c,\ldots) \sim v(A,B,C,\ldots)$ or, when it is unnecessary to describe the correspondence explicitly, simply $V \sim v$.

Definition 4

A set of points $(A,B,C,\ldots)$ on a line v is said to be projective with a set of lines $(a,b,c,\ldots)$ on a point V if and only if the lines $(a,b,c,\ldots)$ are projective with the lines $(a',b',c',...)$ determined by the points $(A,B,C,...)$ and a point V not lying on v.

The fact that a set of points $(A,B,C,...)$ on a line v is projective with a set of lines $(a,b,c,...)$ on a point V is indicated by writing $v(A,B,C,...) \sim V(a,b,c,...)$ or, when it is unnecessary to describe the correspondence explicitly, simply $v \sim V$.

The following theorem, which asserts the symmetric character of the relations described in Definitions 3 and 4, is almost obvious, and trivial to prove.

Theorem 1

If $V(a,b,c,...) \sim v(A,B,C,...)$, then $v(A,B,C,...) \sim V(a,b,c,...)$.

To obtain the algebraic description of a general projectivity between two ranges, it is convenient to express the component perspectivities in matric form. To do this, we observe first that the equation

$$(1) \qquad \alpha\lambda\lambda' + \beta\lambda\mu' + \gamma\mu\lambda' + \delta\mu\mu' = 0$$

can be written in the form $\lambda'(\alpha\lambda + \gamma\mu) + \mu'(\beta\lambda + \delta\mu) = 0$, which implies that $\lambda' = k(\beta\lambda + \delta\mu)$ and $\mu' = -k(\alpha\lambda + \gamma\mu)$, $k \neq 0$. Since only ratios are significant, it is no restriction to take $k = 1$. When this is done, Eq. (1) can be written in matric form as (see Exercise 5, Sec. 3.4)

$$\Lambda' = M\Lambda$$

where

$$\Lambda' = \left\| \begin{matrix} \lambda' \\ \mu' \end{matrix} \right\| \qquad M = \left\| \begin{matrix} \beta & \delta \\ -\alpha & -\gamma \end{matrix} \right\| \qquad \Lambda = \left\| \begin{matrix} \lambda \\ \mu \end{matrix} \right\|$$

$$|M| = \alpha\delta - \beta\gamma \neq 0$$

Let us now suppose that we have a projectivity between a line l_1 and a line l_n defined by a sequence of perspectivities $T_1, T_2, \ldots, T_{n-1}$, where

$$T_i: \Lambda_{i+1} = M_i\Lambda_i \qquad \Lambda_{i+1} = \left\| \begin{matrix} \lambda_{i+1} \\ \mu_{i+1} \end{matrix} \right\|$$

$$M_i = \left\| \begin{matrix} \beta_i & \delta_i \\ -\alpha_i & -\gamma_i \end{matrix} \right\| \qquad \Lambda_i = \left\| \begin{matrix} \lambda_i \\ \mu_i \end{matrix} \right\|$$

and for each perspectivity the coefficients α_i, β_i, γ_i, δ_i satisfy the inequality $\alpha_i\delta_i - \beta_i\gamma_i \neq 0$ and the appropriate condition $\rho_{1i}\alpha_i + \rho_{2i}\beta_i + \rho_{3i}\gamma_i + \rho_{4i}\delta_i = 0$, corresponding to Eq. (5c), Sec. 4.4. As we showed in Sec. 3.4, the equation of the composition of these transformations, obtained by eliminating $\Lambda_2, \Lambda_3, \ldots, \Lambda_{n-1}$, is simply

$$\Lambda_n = M\Lambda_1 \qquad \text{where} \qquad M = M_{n-1}M_{n-2} \cdots M_2M_1$$

Moreover, since $|M| = |M_{n-1}| \cdot |M_{n-2}| \cdots |M_2| \cdot |M_1|$, and since $|M_i| = \alpha_i \delta_i - \beta_i \gamma_i \neq 0$, it follows that the matrix $M = M_{n-1}M_{n-2} \cdots M_2 M_1$ is nonsingular. Thus we have shown that *if a projectivity exists between a line l_1 and a line l_n, the parameters (λ_1, μ_1) of a general point on l_1 and the parameters (λ_n, μ_n) of its image on l_n satisfy an equation of the form $\Lambda_n = M\Lambda_1$, $|M| \neq 0$, or in scalar form*

$$\alpha \lambda_1 \lambda_n + \beta \lambda_1 \mu_n + \gamma \mu_1 \lambda_n + \delta \mu_1 \mu_n = 0 \qquad \alpha \delta - \beta \gamma \neq 0$$

The converse of this is also true; i.e., *the condition $\Lambda_n = M\Lambda_1$ or $\alpha \lambda_1 \lambda_n + \beta \lambda_1 \mu_n + \gamma \mu_1 \lambda_n + \delta \mu_1 \mu_n = 0$, $|M| = \alpha \delta - \beta \gamma \neq 0$, is a sufficient condition that a projectivity should exist between l_1 and l_n in which the image of a general point (λ_1, μ_1) on l_1 is the point (λ_n, μ_n) on l_n.* To prove this, let us suppose first that l_1 and l_n are distinct, parametrized lines and that a one-to-one correspondence effected by the equation

$$(2) \qquad T: \left\| \begin{matrix} \lambda_n \\ \mu_n \end{matrix} \right\| = \left\| \begin{matrix} \beta & \delta \\ -\alpha & -\gamma \end{matrix} \right\| \cdot \left\| \begin{matrix} \lambda_1 \\ \mu_1 \end{matrix} \right\| \qquad \alpha \delta - \beta \gamma \neq 0$$

exists between the points of l_1 and the points of l_n. We shall show that T is a projectivity by exhibiting two perspectivities whose composition is T. There are infinitely many ways of constructing these perspectivities, but since we need only to prove the existence of at least one pair, we shall proceed in a somewhat special way. Of the two base points, A_1 and B_1, on l_1, at least one, say A_1, must be distinct from the intersection of l_1 and l_n. Since the parameters of A_1 are $(1,0)$, the parameters of its image, as determined by the equation of T, are $(\beta, -\alpha)$. Now let l' be any line on A_1 which is distinct from l_1 and from the line determined by A_1 and its image on l_n. Then the parameters of the intersection of l' and l_n, say (s,t), are such that

$$(3) \qquad \alpha s + \beta t \neq 0$$

On l', let us take the base points A' and B' to be, respectively, A_1 and the intersection, $sA_n + tB_n$, of l' and l_n (Fig. 4.9).

Consider now the general perspectivity

$$(4a) \qquad \left\| \begin{matrix} \lambda' \\ \mu' \end{matrix} \right\| = \left\| \begin{matrix} b & d \\ -a & -c \end{matrix} \right\| \cdot \left\| \begin{matrix} \lambda_1 \\ \mu_1 \end{matrix} \right\| \qquad ad - bc \neq 0$$

$$(4b) \qquad \rho_1 a + \rho_2 b + \rho_3 c + \rho_4 d = 0$$

between l_1 and l'. Since $A_1 = A'$, it follows that the last three elements in the first row of the determinant (4) in Sec. 4.4 are zero. Hence the cofactors ρ_2, ρ_3, ρ_4 are each zero, and Eq. (4b) reduces to $a = 0$.

Next let us consider the general perspectivity

$$(5a) \qquad \left\| \begin{matrix} \lambda_n \\ \mu_n \end{matrix} \right\| = \left\| \begin{matrix} f & h \\ -e & -g \end{matrix} \right\| \cdot \left\| \begin{matrix} \lambda' \\ \mu' \end{matrix} \right\| \qquad eh - fg \neq 0$$

$$(5b) \qquad \sigma_1 e + \sigma_2 f + \sigma_3 g + \sigma_4 h = 0$$

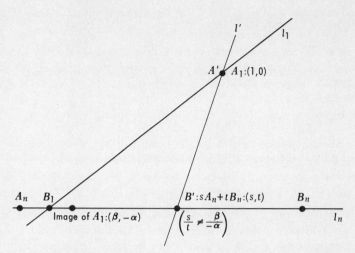

Fig. 4.9   Figure used in the decomposition of a projectivity into two perspectivities.

between l' and l_n. In this case, since B' is a point on l_n, and therefore collinear with A_n and B_n, it follows that the last three elements in the third row of the determinant (4) in Sec. 4.4 are proportional to the last three elements in the fourth row of this determinant, since each triple is a set of coordinates for l_n. In fact, since $B' = sA_n + tB_n$, the last three elements in the fourth row are just $-s/t$ times the corresponding elements in the third row. Thus in Eq. (5b), σ_1 and σ_2 are both zero since each is a third-order determinant whose last two rows are proportional. Moreover, from the proportionality we have just noted, it follows that $t\sigma_3 = s\sigma_4$. Hence Eq. (5b) reduces to

$$sg + th = 0$$

The matrix of the composition of the two perspectivities (4a) and (5a) is, of course,

$$\left\Vert \begin{matrix} f & h \\ -e & -g \end{matrix} \right\Vert \cdot \left\Vert \begin{matrix} b & d \\ -a & -c \end{matrix} \right\Vert = \left\Vert \begin{matrix} bf - ah & df - ch \\ -be + ag & -de + cg \end{matrix} \right\Vert$$

and this will be equal to the matrix of the given mapping of l_1 onto l_n if and only if

$$(6) \qquad \begin{aligned} \alpha &= be - ag \\ \beta &= bf - ah \\ \gamma &= de - gc \\ \delta &= df - ch \end{aligned}$$

Our proof that the mapping T is a projectivity will be complete if we can solve these equations for $a, b, \ldots, g, h$ subject to the restrictions

$$a = 0 \qquad sg + th = 0 \qquad ad - bc \neq 0 \qquad eh - fg \neq 0$$

From the first and third of Eqs. (6) we find, since $a = 0$,

(7) $$bcg = d\alpha - b\gamma$$

and from the second and fourth of Eqs. (6) we find

(8) $$bch = d\beta - b\delta$$

Substituting from these equations into the equation $sg + th = 0$, we obtain easily

(9) $$b(s\gamma + t\delta) - d(s\alpha + t\beta) = 0$$

Hence $b = k(s\alpha + t\beta)$, $d = k(s\gamma + t\delta)$, and, from (3), $b \neq 0$. Now, after c is assigned an arbitrary nonzero value, g and h can be determined from (7) and (8), respectively. Finally, e and f can be found from the first two of Eqs. (6). Since $a = 0$, while neither b nor c is zero, it follows that $ad - bc = -bc \neq 0$. Then from the fact that Eqs. (6) have been satisfied, so that $(\alpha\delta - \beta\gamma) = (ad - bc)(eh - fg)$, it follows that $eh - fg \neq 0$, as required. The determination of the two perspectivities can now be completed by finding their centers, using Eqs. (3), Sec. 4.4. The fact that we have found two perspectivities whose composition is the given transformation T proves that T is a projectivity, as asserted.

If $l_1 = l_n$, the intermediate line l' coincides with l_1 and l_n and the preceding argument cannot be applied, since the only perspectivity which can exist between a line and itself is the identity, which maps every point onto itself. However, our conclusion can be established in this case also by first projecting l_1 onto any convenient line, l_1', distinct from l_1, by an arbitrary perspectivity $l_1 \overset{V_1}{\wedge} l_1'$. The given relation mapping l_1 onto $l_n (= l_1)$ will then lead to a relation of the same form mapping l_1' onto l_n, and this, as we showed above, can be reduced to two perspectivities $l_1' \overset{V_2}{\wedge} l'$ and $l' \overset{V_3}{\wedge} l_n$. Thus the given mapping between l_1 and $l_n (= l_1)$ is the composition of the three perspectivities

$$l_1 \overset{V_1}{\wedge} l_1' \qquad l_1' \overset{V_2}{\wedge} l' \qquad l' \overset{V_3}{\wedge} l_n$$

and hence is a projectivity.

The foregoing observations are summarized in the following important theorem.

Theorem 2

The necessary and sufficient condition that there should exist a projectivity between a line l_1 and a line l_n in which an arbitrary point

(λ_1,μ_1) on l_1 is mapped into the point (λ_n,μ_n) on l_n is that the parameters (λ_1,μ_1) and (λ_n,μ_n) should satisfy an equation of the form

$$\alpha\lambda_1\lambda_n + \beta\lambda_1\mu_n + \gamma\mu_1\lambda_n + \delta\mu_1\mu_n = 0$$

or $\qquad \left\|\begin{matrix} \lambda_n \\ \mu_n \end{matrix}\right\| = \left\|\begin{matrix} \beta & \delta \\ -\alpha & -\gamma \end{matrix}\right\| \cdot \left\|\begin{matrix} \lambda_1 \\ \mu_1 \end{matrix}\right\| \qquad \alpha\delta - \beta\gamma \neq 0$

If l_1 and l_n are distinct lines, any projectivity between l_1 and l_n can be reduced to at most two perspectivities. If l_1 and l_n coincide, any projectivity between l_1 and l_n $(= l_1)$ can be reduced to at most three perspectivities.

Dually, of course, we have the companion theorem:

Theorem 3

The necessary and sufficient condition that there should exist a projectivity between a point L_1 and a point L_n in which an arbitrary line (λ_1,μ_1) on L_1 is mapped into the line (λ_n,μ_n) on L_n is that the parameters (λ_1,μ_1) and (λ_n,μ_n) should satisfy an equation of the form

$$\alpha\lambda_1\lambda_n + \beta\lambda_1\mu_n + \gamma\mu_1\lambda_n + \delta\mu_1\mu_n = 0$$

or $\qquad \left\|\begin{matrix} \lambda_n \\ \mu_n \end{matrix}\right\| = \left\|\begin{matrix} \beta & \delta \\ -\alpha & -\gamma \end{matrix}\right\| \cdot \left\|\begin{matrix} \lambda_1 \\ \mu_1 \end{matrix}\right\| \qquad \alpha\delta - \beta\gamma \neq 0$

If L_1 and L_n are distinct points, any projectivity between L_1 and L_n can be reduced to at most two perspectivities. If L_1 and L_n coincide, any projectivity between L_1 and $L_n(= L_1)$ can be reduced to at most three perspectivities.

Example 1

Find a chain of at most three perspectivities whose composition will be the projectivity between the line $x_3 = 0$ and itself defined by the equation

$$\lambda_1\lambda_n + 3\lambda_1\mu_n + \mu_1\lambda_n + 2\mu_1\mu_n = 0 \qquad \text{or} \qquad \left\|\begin{matrix} \lambda_n \\ \mu_n \end{matrix}\right\| = \left\|\begin{matrix} 3 & 2 \\ -1 & -1 \end{matrix}\right\| \cdot \left\|\begin{matrix} \lambda_1 \\ \mu_1 \end{matrix}\right\|$$

if the parameters (λ_1,μ_1) refer to the base points $A_1:(1,-1,0)$ and $B_1:(1,0,0)$ and the parameters (λ_n,μ_n) refer to the base points $A_n:(1,1,0)$ and $B_n:(0,1,0)$.

Since we are given a projectivity between coincident lines, $l_1 = l_n: x_3 = 0$, it is first necessary that we project l_1 onto another line, say l_1': $x_2 = 0$. Specifically, let us employ the perspectivity defined by the center $V_1:(1,1,1)$, and let us parametrize l_1' in terms of the base points $A_1':(0,0,1)$ and $B_1' = B_1:(1,0,0)$ with parameters (λ_1',μ_1') (Fig. 4.10). Now a point (λ_1,μ_1) on l_1 and a point (λ_1',μ_1') on l', will be collinear with the center $V_1:(1,1,1)$ if and only if

$$\begin{vmatrix} 1 & 1 & 1 \\ \lambda_1 + \mu_1 & -\lambda_1 & 0 \\ \mu_1' & 0 & \lambda_1' \end{vmatrix} = 0$$

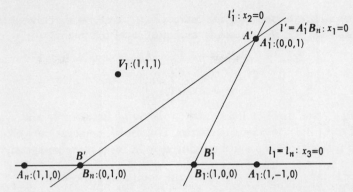

Fig. 4.10   Figure used in the decomposition of the projectivity of Example 1.

Expanding this, we obtain

$$2\lambda_1\lambda_1' - \lambda_1\mu_1' + \mu_1\lambda_1' = 0 \qquad \text{or} \qquad \left\|\begin{matrix}\lambda_1 \\ \mu_1\end{matrix}\right\| = \left\|\begin{matrix}1 & 0 \\ -2 & 1\end{matrix}\right\| \cdot \left\|\begin{matrix}\lambda_1' \\ \mu_1'\end{matrix}\right\|$$

as the equation of the perspectivity between l_1 and l_1'. Combining this equation with the equation of the given projectivity between l_1 and l_n $(=l_1)$, we have

(10)
$$
\begin{aligned}
\left\|\begin{matrix}\lambda_n \\ \mu_n\end{matrix}\right\| &= \left\|\begin{matrix}3 & 2 \\ -1 & -1\end{matrix}\right\| \cdot \left\|\begin{matrix}\lambda_1 \\ \mu_1\end{matrix}\right\| \\
&= \left\|\begin{matrix}3 & 2 \\ -1 & -1\end{matrix}\right\| \cdot \left\|\begin{matrix}1 & 0 \\ -2 & 1\end{matrix}\right\| \cdot \left\|\begin{matrix}\lambda_1' \\ \mu_1'\end{matrix}\right\| \\
&= \left\|\begin{matrix}-1 & 2 \\ 1 & -1\end{matrix}\right\| \cdot \left\|\begin{matrix}\lambda_1' \\ \mu_1'\end{matrix}\right\|
\end{aligned}
$$

as the equation of the mapping of l_1' onto l_n $(=l_1)$. Clearly, it is also the equation of a projectivity.

To determine two perspectivities whose composition will be the projectivity (10), it is sufficient, as we observed above, to choose as our intermediate line, l', the line $A_1'B_n$, provided that its intersections with l_1' and l_n are not corresponding points in the projectivity (10). Now from Eqs. (10) we find that the image of the point $A_1':(1,0)$ on l_1' is the point $(-1,1)$ on l_n, and clearly this point is different from B_n, whose parameters are $(0,1)$. Hence we can take $A_1'B_n: x_1 = 0$ to be the line l'. For convenience, we shall parametrize l' in terms of the base points $A' = A_1':(0,0,1)$ and $B' = B_n:(0,1,0)$, with (λ',μ') as parameters.

Now if

$$\left\|\begin{matrix}\lambda' \\ \mu'\end{matrix}\right\| = \left\|\begin{matrix}b & d \\ -a & -c\end{matrix}\right\| \cdot \left\|\begin{matrix}\lambda_1' \\ \mu_1'\end{matrix}\right\| \qquad \text{or} \qquad a\lambda_1'\lambda' + b\lambda_1'\mu' + c\mu_1'\lambda' + d\mu_1'\mu' = 0$$

is the equation of the general perspectivity mapping l_1' onto l', it follows from Eq. (4), Sec. 4.4, that a, b, c, d must satisfy the equation

$$\begin{vmatrix} a & 0 & 0 & 0 \\ b & -1 & 0 & 0 \\ c & 0 & -1 & 0 \\ d & 0 & 0 & -1 \end{vmatrix} = 0 \quad \text{or} \quad a = 0$$

Similarly, if

$$\left\| \begin{matrix} \lambda_n \\ \mu_n \end{matrix} \right\| = \left\| \begin{matrix} f & h \\ -e & -g \end{matrix} \right\| \cdot \left\| \begin{matrix} \lambda' \\ \mu' \end{matrix} \right\| \quad \text{or} \quad e\lambda'\lambda_n + f\lambda'\mu_n + g\mu'\lambda_n + h\mu'\mu_n = 0$$

is the equation of the general perspectivity mapping l' onto l_n, the coefficients e, f, g, h must satisfy the equation

$$\begin{vmatrix} e & -1 & 1 & 0 \\ f & -1 & 0 & 0 \\ g & 0 & 0 & -1 \\ h & 0 & 0 & 0 \end{vmatrix} = 0 \quad \text{or} \quad h = 0$$

Furthermore, if the composition of these two perspectivities is to be the projectivity (10) which maps l_1' onto l_n, it follows that

$$\left\| \begin{matrix} f & 0 \\ -e & -g \end{matrix} \right\| \cdot \left\| \begin{matrix} b & d \\ 0 & -c \end{matrix} \right\| = \left\| \begin{matrix} -1 & 2 \\ 1 & -1 \end{matrix} \right\|$$

or $fb = -1$, $fd = 2$, $-be = 1$, $-de + cg = -1$. Any solution of these four equations such that $ad - bc = -bc \neq 0$ and $eh - fg = -fg \neq 0$ will define two perspectivities whose composition is the projectivity (10) between l_1' and l_n. In particular, we may take $b = -1$, $c = -1$, $d = 2$, $e = 1$, $f = 1$, $g = -1$. Then, using Eqs. (3), Sec. 4.4, we find that the center of the first perspectivity is $V_2:(1,1,-2)$ and the center of the second perspectivity is $V_3:(-1,0,1)$. The given projectivity between l_1 and l_n $(=l_1)$ is thus the composition of the three perspectivities

$$l_1 \overset{V_1}{\wedge} l_1' \qquad l_1' \overset{V_2}{\wedge} l' \qquad l' \overset{V_3}{\wedge} l_n$$

where V_1 is the point $(1,1,1)$, V_2 is the point $(1,1,-2)$, V_3 is the point $(-1,0,1)$, l_1' is the line $x_2 = 0$, and l' is the line $x_1 = 0$.

Exercises

1. Given $l_1: x_1 = 0$, $l_2: x_2 = 0$, $l_3: x_3 = 0$. If l_2 is perspective with l_1 from $U:(1,-1,0)$, and if l_3 is perspective with l_2 from $V:(1,1,1)$, determine whether the correspondence thus established between l_1 and l_3 is a perspectivity.

2. Given the lines $l_1: x_1 + x_2 + x_3 = 0$ with base points $A_1:(1,-1,0)$ and $B_1:(1,0,-1)$, $l_2: x_1 - x_2 = 0$ with base points $A_2:(1,1,0)$ and $B_2:(0,0,1)$, $l_3: x_2 + x_3 = 0$ with base points $A_3:(1,0,0)$ and $B_3:(0,1,-1)$, and the points $U:(0,1,0)$ and $V:(1,0,1)$. If $l_1 \overset{U}{\wedge} l_2$ and $l_2 \overset{V}{\wedge} l_3$, find the equation of the projectivity thus established between l_1 and l_3.

3. Let $l_1: x_1 = 0$ be parametrized in terms of $A_1:(0,1,0)$ and $B_1:(0,0,1)$ as base points, with parameters (λ_1,μ_1), let $l_2: x_1 + x_2 = 0$ be parametrized in terms of $A_2:(0,0,1)$ and $B_2:(1,-1,0)$ as base points, with parameters (λ_2,μ_2), and let $l': x_2 - x_3 = 0$ be parametrized in terms of $A':(1,0,0)$ and $B':(0,1,1)$ as base points, with parameters (λ',μ'). Find points U and V such that the composition of the two perspectivities $l_1 \overset{U}{\wedge} l'$ and $l' \overset{V}{\wedge} l_2$ will be the projectivity between l_1 and l_2 defined by the equation $\lambda_1\lambda_2 - \lambda_1\mu_2 + 2\mu_1\lambda_2 + 3\mu_1\mu_2 = 0$.

4. (*a*) Work Exercise 3 if the projectivity between l_1 and l_2 is defined by the equation $2\lambda_1\lambda_2 + 3\lambda_1\mu_2 + \mu_1\lambda_2 + 2\mu_1\mu_2 = 0$.
(*b*) Work Exercise 3 if the projectivity between l_1 and l_2 is defined by the equation $\lambda_1\lambda_2 + \lambda_1\mu_2 + \mu_1\lambda_2 - 2\mu_1\mu_2 = 0$.
(*c*) Work Exercise 3 if l' is the line $x_1 + x_2 + x_3 = 0$ with base points $A':(0,1,-1)$ and $B':(1,0,-1)$.

5. Let $l_1: x_1 = 0$ be parametrized in terms of $A_1:(0,1,1)$ and $B_1:(0,1,-1)$ as base points, with parameters (λ_1,μ_1), let $l_2: x_2 = 0$ be parametrized in terms of $A_2:(1,0,0)$ and $B_2:(0,0,1)$ as base points, with parameters (λ_2,μ_2), and let $T: \lambda_1\lambda_2 + \lambda_1\mu_2 + \mu_1\lambda_2 + 2\mu_1\mu_2 = 0$ be a projectivity between l_1 and l_2. If l' is the line $x_1 + x_2 - x_3 = 0$, parametrized in terms of $A':(0,1,1)$ and $B':(1,0,1)$ as base points, find the loci of the centers U and V of the perspectivities $l_1 \overset{U}{\wedge} l'$ and $l' \overset{V}{\wedge} l_2$ whose composition is T.

6. (*a*) Work Exercise 5 if B' is the point $(1,1,2)$.
(*b*) Work Exercise 5 if l' is the line $2x_1 + x_2 - x_3 = 0$.
(*c*) Work Exercise 5 if T is the projectivity $\lambda_1\lambda_2 - \lambda_1\mu_2 - \mu_1\lambda_2 + 2\mu_1\mu_2 = 0$.

7. In Exercise 5, find the locus of U as l' varies in the pencil on A_1 if B' is the point of intersection of l' and l_2 and if V is the point: (*a*) $(1,1,0)$ (*b*) $(1,2,1)$ (*c*) $(-1,1,2)$.

8. In Exercise 5, show that the locus of V is independent of the particular line, l', of the pencil on A_1.

9. Let $T: a\lambda_1\lambda_2 + b\lambda_1\mu_2 + c\mu_1\lambda_2 + d\mu_1\mu_2 = 0$ be a projectivity between the line l_1 and the line l_2, let P_1 be a point on l_1 distinct from the intersection

of l_1 and l_2, and let l' be an arbitrary line of the pencil on P_1. If T is the composition of the perspectivities $l_1 \overset{U}{\wedge} l'$ and $l' \overset{V}{\wedge} l_2$, show that the locus of V is independent of l'.

10. Prove Theorem 1.

11. Let l_1 be the line $x_1 = 0$ with base points $A_1:(0,1,0)$ and $B_1:(0,0,1)$, let l_2 be the line $x_2 - x_3 = 0$ with base points $A_2:(1,0,0)$ and $B_2:(0,1,1)$, and let $T: \left\| \begin{matrix} \lambda_2 \\ \mu_2 \end{matrix} \right\| = \left\| \begin{matrix} 1 & -1 \\ 1 & 3 \end{matrix} \right\| \cdot \left\| \begin{matrix} \lambda_1 \\ \mu_1 \end{matrix} \right\|$ be the equation of a projectivity between l_1 and l_2. Determine what points on l_1, if any, have the property that they and their images under T are collinear with the point $U:(1,0,1)$.

12. (a) Work Exercise 11 if the base points on l_2 are $A_2:(-3,1,1)$ and $B_2:(1,2,2)$.

(b) Work Exercise 11 if U is the point $(1,1,3)$.

(c) Work Exercise 11 if T is the projectivity $\left\| \begin{matrix} \lambda_2 \\ \mu_2 \end{matrix} \right\| = \left\| \begin{matrix} 3 & -4 \\ 1 & 2 \end{matrix} \right\| \cdot \left\| \begin{matrix} \lambda_1 \\ \mu_1 \end{matrix} \right\|$.

13. Given $l_1: x_1 = 0$, $l_2: x_2 = 0$, $l: x_1 + x_2 + x_3 = 0$, $U:(0,1,0)$, and $V:(1,0,0)$. Let P_1 be an arbitrary point on l_1, let W be the intersection of P_1V and l, and let P_2 be the intersection of UW and l_2. Show that the correspondence between l_1 and l_2 in which P_2 is the image of P_1 is a projectivity, and find its equation if the base points on l_1 are $A_1:(0,0,1)$ and $B_1:(0,1,0)$ and the base points on l_2 are:
(a) $A_2:(0,0,1)$ and $B_2:(1,0,0)$ (b) $A_2:(1,0,1)$ and $B_2:(1,0,-1)$

14. Given $l_1: x_1 - x_2 = 0$, $l_2: x_1 + x_2 + x_3 = 0$, the conic $\Gamma: x_1x_2 - x_3{}^2 = 0$, and the points $U:(1,0,0)$ and $V:(0,1,0)$ on Γ. Let P_1 be an arbitrary point on l_1, let W be the second intersection of P_1U and Γ, and let P_2 be the intersection of VW and l_2. Show that the correspondence between l_1 and l_2 in which P_2 is the image of P_1 is a projectivity, and find its equation if the base points on l_1 are $A_1:(1,1,0)$ and $B_1:(0,0,1)$ and if the base points on l_2 are:
(a) $A_2:(1,0,-1)$ and $B_2:(1,-1,0)$ (b) $A_2:(1,-2,1)$ and $B_2:(1,2,-3)$

15. (a) Given $l_1: x_2 - x_3 = 0$, $l_2: x_2 + x_3 = 0$, $W:(0,1,0)$, the conic $\Gamma: x_2{}^2 - x_1x_3 = 0$, and the point $U:(0,0,1)$ on Γ. Let P_1 be an arbitrary point on l_1, let W_1 be the second intersection of P_1U and Γ, let W_2 be the second intersection of W_1W and Γ, and let P_2 be the intersection of UW_2 and l_2. If the base points on l_1 are $A_1:(1,0,0)$ and $B_1:(0,1,1)$, and if the base points on l_2 are $A_2:(1,0,0)$ and $B_2:(0,1,-1)$, find the equation of the correspondence between l_1 and l_2 in which P_2 is the image of P_1, and show that this correspondence is a projectivity.

(b) Work part (a) given $W:(1,1,0)$.

16. Given $l_1: x_1 - x_2 = 0$, $l_2: x_1 + x_2 + x_3 = 0$, the family of conics $ax_2x_3 + x_1x_3 + x_1x_2 = 0$, the particular conic $\Gamma: x_2x_3 + x_1x_3 + 2x_1x_2 = 0$, and the point $U:(1,1,-1)$ on Γ. Let P_1 be an arbitrary point on l_1, let Φ be the conic of the given family which passes through P_1, let V be the intersection of Φ and Γ which is distinct from $(1,0,0)$, $(0,1,0)$, $(0,0,1)$, and let P_2 be the intersection of UV and l_2. Find the equation of the correspondence between l_1 and l_2 in which P_2 is the image of P_1 if the base points on l_1 are $A_1:(1,1,0)$ and $B_1:(1,1,1)$ and if the base points on l_2 are:
(*a*) $A_2:(1,-1,0)$ and $B_2:(0,1,-1)$ (*b*) $A_2:(1,0,-1)$ and $B_2:(1,-2,1)$

17. (*a*) Given $l_1: x_1 - x_2 = 0$, $l_2: x_1 + x_2 = 0$, and the family of conics $ax_2x_3 + x_1x_3 + x_1x_2 = 0$. Let P_1 be an arbitrary point on l_1, let Γ be the conic of the given family which passes through P_1, and let P_2 be the intersection of Γ and l_2 which is distinct from $(0,0,1)$. If the base points on l_1 are $A_1:(1,1,0)$ and $B_1:(0,0,1)$, and if the base points on l_2 are $A_2:(1,-1,0)$ and $B_2:(0,0,1)$, find the equation of the correspondence between l_1 and l_2 in which P_2 is the image of P_1. Is this correspondence a projectivity?
(*b*) Work part (*a*), given $A_2:(1,-1,1)$ and $B_2:(1,-1,0)$.

18. Work Exercise 17 using the family of cubic curves $x_1^3 - ax_1^2x_3 + x_2^2x_3 = 0$ instead of the given family of conics.

19. Work Exercise 17 using the family of quartic curves $x_1^4 - ax_2^3x_3 = 0$ instead of the given family of conics.

20. Given $l_1: x_1 = 0$, $l_2: x_2 + x_3 = 0$, the conic $\Gamma: x_2x_3 = x_1^2$, and the points $U:(1,1,1)$ and $V:(1,-1,-1)$ on Γ. Let P_1 be an arbitrary point on l_1, let U' be the second intersection of P_1U and Γ, let V' be the second intersection of P_1V and Γ, and let P_2 be the intersection of $U'V'$ and l_2. Is the correspondence between l_1 and l_2 in which P_2 is the image of P_1 a projectivity?

4.6 The Determination of Projectivities

In this section we shall investigate the question of how many points on one line may be assigned arbitrary images on another line before a perspectivity or a projectivity between the lines is determined. Our first result concerns perspectivities.

Theorem 1

A perspectivity between two lines is uniquely determined when two points on the first line are assigned distinct images on the second line, provided none of these points is the point of intersection of the two lines.

Proof   Let l and l' be two lines intersecting in a point P, let Q and R be two points on l distinct from P, let Q' and R' be two points on l' distinct from P, and, if possible, let Q' and R' be, respectively, the images of Q and R in a perspectivity between l and l'. From the definition of a perspectivity, it follows that the center, V, of the required perspectivity must lie on the line QQ' and also on the line RR'. Moreover, it is clear from what we are given that these lines are distinct and their intersection does not lie on either l or l'. Thus V, and hence the perspectivity itself, is uniquely determined by the given data, as asserted.

Dually, of course, we have the following theorem.

Theorem 2

A perspectivity between two points is uniquely determined when two lines on the first point are assigned distinct images on the second point, provided none of these lines is the line determined by the two points.

The requirement that two points on one line have arbitrary, distinct images on another line, while sufficient to determine a unique perspectivity between the two lines, is not enough to determine a unique projectivity between the lines. In fact, given *three* points on one line, it is always possible to find a projectivity, i.e., a sequence of perspectivities, which will map these points onto arbitrary, distinct images on a second line. To verify this, let L_1, L_2, L_3 be three points on a line l, let M_1, M_2, M_3 be three points on a second line m, and, if possible, let M_i be the image of L_i in a projectivity between l and m. Now if l' is an arbitrary line on M_1 distinct from m and from L_1M_1, and if V_1 is an arbitrary point on L_1M_1 distinct from L_1 and M_1, we have

$$l(L_1,L_2,L_3) \overset{V_1}{\wedge} l'(L_1',L_2',L_3')$$

where L_1' is, of course, just the point M_1 (Fig. 4.11). Following this, if V_2 is the intersection of the lines $L_2'M_2$ and $L_3'M_3$, we have

$$l'(L_1',L_2',L_3') \equiv l'(M_1,L_2',L_3') \overset{V_2}{\wedge} m(M_1,M_2,M_3)$$

Hence the composition of the two perspectivities we have constructed is a projectivity which maps L_1 onto M_1, L_2 onto M_2, and L_3 onto M_3, as required.

The preceding observations make it clear that there is always at least one projectivity which will map three collinear points onto three other collinear points. However, because of the freedom we have in choosing l' and V_1, it is not clear that there is just one such projectivity. Conceivably,

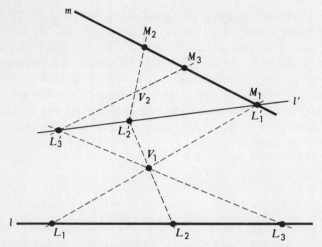

Fig. 4.11 The projection of three collinear points into three other collinear points.

using different choices for l' and V_1, we might obtain different projectivities each of which would map L_1 onto M_1, L_2 onto M_2, and L_3 onto M_3 but which at the same time would map a fourth point on l into points on m which were not the same. Actually, this cannot happen, as the next theorem assures us.

Theorem 3

A projectivity between two lines is uniquely determined when three distinct points on the second line are assigned as the respective images of three arbitrary, distinct points on the first line.

Proof Let A_1, B_1, C_1 be three points on a line l_1, let A_2, B_2, C_2 be three points on a line l_2, and let l_1 and l_2 be parametrized with parameters (λ_1, λ_1') and (λ_2, λ_2'), respectively, so that, in particular, $A_1 : (\alpha_1, \alpha_1')$, $B_1 : (\beta_1, \beta_1')$, $C_1 : (\gamma_1, \gamma_1')$ and $A_2 : (\alpha_2, \alpha_2')$, $B_2 : (\beta_2, \beta_2')$, $C_2 : (\gamma_2, \gamma_2')$. Then if

$$l_1(A_1, B_1, C_1) \sim l_2(A_2, B_2, C_2)$$

a relation of the form

$$a\lambda_1\lambda_2 + b\lambda_1\lambda_2' + c\lambda_1'\lambda_2 + d\lambda_1'\lambda_2' = 0$$

must be satisfied by the parameters of the pairs of corresponding points (A_1, A_2), (B_1, B_2), (C_1, C_2). Thus a, b, c, d must satisfy the equations

$$
\begin{aligned}
a\alpha_1\alpha_2 + b\alpha_1\alpha_2' + c\alpha_1'\alpha_2 + d\alpha_1'\alpha_2' &= 0 \\
a\beta_1\beta_2 + b\beta_1\beta_2' + c\beta_1'\beta_2 + d\beta_1'\beta_2' &= 0 \\
a\gamma_1\gamma_2 + b\gamma_1\gamma_2' + c\gamma_1'\gamma_2 + d\gamma_1'\gamma_2' &= 0
\end{aligned}
$$

(1)

In general, a system of equations of this form has, to within an irrelevant factor of proportionality, a unique solution. It is conceivable, however, that this system may have two linearly independent solutions. According to Theorem 3, Sec. 3.6, this will be the case if and only if the rank of the coefficient matrix

$$
(2) \qquad \begin{Vmatrix} \alpha_1\alpha_2 & \alpha_1\alpha_2' & \alpha_1'\alpha_2 & \alpha_1'\alpha_2' \\ \beta_1\beta_2 & \beta_1\beta_2' & \beta_1'\beta_2 & \beta_1'\beta_2' \\ \gamma_1\gamma_2 & \gamma_1\gamma_2' & \gamma_1'\gamma_2 & \gamma_1'\gamma_2' \end{Vmatrix}
$$

is at most 2; that is, if and only if the determinant of every (3,3) submatrix in the coefficient matrix is zero.

To see that this can never happen, let us consider the possibility that the determinants of the matrices obtained by deleting first the third column and then the fourth column are zero. This gives us the conditions

$$
\begin{vmatrix} \alpha_1\alpha_2 & \alpha_1\alpha_2' & \alpha_1'\alpha_2' \\ \beta_1\beta_2 & \beta_1\beta_2' & \beta_1'\beta_2' \\ \gamma_1\gamma_2 & \gamma_1\gamma_2' & \gamma_1'\gamma_2' \end{vmatrix} = 0 \qquad \begin{vmatrix} \alpha_1\alpha_2 & \alpha_1\alpha_2' & \alpha_1'\alpha_2 \\ \beta_1\beta_2 & \beta_1\beta_2' & \beta_1'\beta_2 \\ \gamma_1\gamma_2 & \gamma_1\gamma_2' & \gamma_1'\gamma_2 \end{vmatrix} = 0
$$

or, expanding each determinant in terms of the elements of its last column,

$$
(3) \quad \alpha_2'[\alpha_1'\beta_1\gamma_1(\beta_2\gamma_2' - \beta_2'\gamma_2)] + \beta_2'[\alpha_1\beta_1'\gamma_1(\gamma_2\alpha_2' - \gamma_2'\alpha_2)]
$$
$$
+ \gamma_2'[\alpha_1\beta_1\gamma_1'(\alpha_2\beta_2' - \alpha_2'\beta_2)] = 0
$$

$$
(4) \quad \alpha_2[\alpha_1'\beta_1\gamma_1(\beta_2\gamma_2' - \beta_2'\gamma_2)] + \beta_2[\alpha_1\beta_1'\gamma_1(\gamma_2\alpha_2' - \gamma_2'\alpha_2)]
$$
$$
+ \gamma_2[\alpha_1\beta_1\gamma_1'(\alpha_2\beta_2' - \alpha_2'\beta_2)] = 0
$$

Now since A_2, B_2, C_2 are distinct points, their parameters cannot be proportional, and hence

$$
\beta_2\gamma_2' - \beta_2'\gamma_2 \neq 0 \qquad \gamma_2\alpha_2' - \gamma_2'\alpha_2 \neq 0 \qquad \alpha_2\beta_2' - \alpha_2'\beta_2 \neq 0
$$

Therefore, solving Eqs. (3) and (4) for the quantities in brackets, using Corollary 1, Theorem 6, Sec. 3.6, we have

$$
\frac{\alpha_1'\beta_1\gamma_1(\beta_2\gamma_2' - \beta_2'\gamma_2)}{-(\beta_2\gamma_2' - \beta_2'\gamma_2)} = \frac{\alpha_1\beta_1'\gamma_1(\gamma_2\alpha_2' - \gamma_2'\alpha_2)}{-(\gamma_2\alpha_2' - \gamma_2'\alpha_2)} = \frac{\alpha_1\beta_1\gamma_1'(\alpha_2\beta_2' - \alpha_2'\beta_2)}{-(\alpha_2\beta_2' - \alpha_2'\beta_2)}
$$

or simply

$$
(5) \qquad \alpha_1'\beta_1\gamma_1 = \alpha_1\beta_1'\gamma_1 = \alpha_1\beta_1\gamma_1'
$$

If the unprimed parameters are all different from zero, it follows that $\alpha_1'/\alpha_1 = \beta_1'/\beta_1 = \gamma_1'/\gamma_1$, which is impossible since $A_1:(\alpha_1,\alpha_1')$, $B_1:(\beta_1,\beta_1')$, and $C_1:(\gamma_1,\gamma_1')$ are distinct points. On the other hand, if even one of the unprimed coordinates, say α_1, is equal to zero, we also have an immediate contradiction. For if $\alpha_1 = 0$, it follows from (5) that $\alpha_1'\beta_1\gamma_1 = 0$. But α_1' cannot be zero since α_1 is known to be zero and no point has parameters $(0,0)$. More-over, β_1 cannot be zero, because if it were, then both A_1 and B_1 would have parameters of the form $(0,1)$, which is impossible since A_1 and B_1 are distinct. Similarly, since A_1 and C_1 are distinct points, it follows that γ_1 is not zero. Thus in all cases our assumption that the rank of the coefficient matrix (2) is less than 3 has led us to a contradiction. Hence, to within a factor of proportionality, the solution of the system (1) for a, b, c, d is unique.

It remains now to show that the mapping we have obtained is independent of the particular parametrizations of l_1 and l_2 used to describe it. To do this, let us consider a second parametrization of l_1, with parameters $(\bar{\lambda}_1, \bar{\lambda}_1')$, and a second parametrization of l_2, with parameters $(\bar{\lambda}_2, \bar{\lambda}_2')$. Then, as we showed in Example 2, Sec. 4.2, the two sets of parameters on the respective lines are related by equations of the form

$$
(6) \qquad \left\| \begin{matrix} \lambda_1 \\ \lambda_1' \end{matrix} \right\| = M_1 \left\| \begin{matrix} \bar{\lambda}_1 \\ \bar{\lambda}_1' \end{matrix} \right\| \qquad \text{and} \qquad \left\| \begin{matrix} \lambda_2 \\ \lambda_2' \end{matrix} \right\| = M_2 \left\| \begin{matrix} \bar{\lambda}_2 \\ \bar{\lambda}_2' \end{matrix} \right\|
$$

where M_1 and M_2 are nonsingular $(2,2)$ matrices. Let us suppose, now, that a, b, c, d have been determined so that the matric form of the equation of the projectivity in terms of the original parametrizations of l_1 and l_2, namely,

$$
\left\| \begin{matrix} \lambda_2 \\ \lambda_2' \end{matrix} \right\| = \left\| \begin{matrix} b & d \\ -a & -c \end{matrix} \right\| \cdot \left\| \begin{matrix} \lambda_1 \\ \lambda_1' \end{matrix} \right\|, \text{ is satisfied by the pairs of matrices}
$$

$$
\left(\left\| \begin{matrix} \alpha_1 \\ \alpha_1' \end{matrix} \right\|, \ \left\| \begin{matrix} \alpha_2 \\ \alpha_2' \end{matrix} \right\| \right) \qquad \left(\left\| \begin{matrix} \beta_1 \\ \beta_1' \end{matrix} \right\|, \ \left\| \begin{matrix} \beta_2 \\ \beta_2' \end{matrix} \right\| \right) \qquad \left(\left\| \begin{matrix} \gamma_1 \\ \gamma_1' \end{matrix} \right\|, \ \left\| \begin{matrix} \gamma_2 \\ \gamma_2' \end{matrix} \right\| \right)
$$

Then, using Eqs. (6) we can write

$$
M_2 \left\| \begin{matrix} \bar{\lambda}_2 \\ \bar{\lambda}_2' \end{matrix} \right\| = \left\| \begin{matrix} b & d \\ -a & -c \end{matrix} \right\| \left(M_1 \left\| \begin{matrix} \bar{\lambda}_1 \\ \bar{\lambda}_1' \end{matrix} \right\| \right)
$$

or

$$
\left\| \begin{matrix} \bar{\lambda}_2 \\ \bar{\lambda}_2' \end{matrix} \right\| = \left(M_2^{-1} \left\| \begin{matrix} b & d \\ -a & -c \end{matrix} \right\| M_1 \right) \left\| \begin{matrix} \bar{\lambda}_1 \\ \bar{\lambda}_1' \end{matrix} \right\|
$$

Clearly, this equation has the form of the equation of a projectivity, and is satisfied by the pairs of matrices

$$
\left(\left\| \begin{matrix} \bar{\alpha}_1 \\ \bar{\alpha}_1' \end{matrix} \right\|, \ \left\| \begin{matrix} \bar{\alpha}_2 \\ \bar{\alpha}_2' \end{matrix} \right\| \right) \qquad \left(\left\| \begin{matrix} \bar{\beta}_1 \\ \bar{\beta}_1' \end{matrix} \right\|, \ \left\| \begin{matrix} \bar{\beta}_2 \\ \bar{\beta}_2' \end{matrix} \right\| \right) \qquad \left(\left\| \begin{matrix} \bar{\gamma}_1 \\ \bar{\gamma}_1' \end{matrix} \right\|, \ \left\| \begin{matrix} \bar{\gamma}_2 \\ \bar{\gamma}_2' \end{matrix} \right\| \right)
$$

determined by the parameters of the pairs of points (A_1, A_2), (B_1, B_2), (C_1, C_2) in the second parametrization of l_1 and l_2. But by the first part of our proof there is, to within a factor of proportionality, only one such equation. Hence if we were to consider the equation

$$
\bar{a}\bar{\lambda}_1\bar{\lambda}_2 + \bar{b}\bar{\lambda}_1\bar{\lambda}_2' + \bar{c}\bar{\lambda}_1'\bar{\lambda}_2 + \bar{d}\bar{\lambda}_1'\bar{\lambda}_2' = 0 \qquad \text{or} \qquad \left\| \begin{matrix} \bar{\lambda}_2 \\ \bar{\lambda}_2' \end{matrix} \right\| = \left\| \begin{matrix} \bar{b} & \bar{d} \\ -\bar{a} & -\bar{c} \end{matrix} \right\| \cdot \left\| \begin{matrix} \bar{\lambda}_1 \\ \bar{\lambda}_1' \end{matrix} \right\|
$$

and determine $\bar{a}$, $\bar{b}$, $\bar{c}$, $\bar{d}$ by working directly in the second parametrizations on l_1 and l_2, we would find that the matrix $\left\| \begin{matrix} \bar{b} & \bar{d} \\ -\bar{a} & -\bar{c} \end{matrix} \right\|$ was proportional to the matrix $M_2^{-1} \cdot \left\| \begin{matrix} b & d \\ -a & -c \end{matrix} \right\| \cdot M_1$. Thus both the mapping $(\lambda_1, \lambda_1') \rightarrow (\lambda_2, \lambda_2')$ defined by the equation constructed using the first parametrizations and the mapping $(\bar{\lambda}_1, \bar{\lambda}_1') \rightarrow (\bar{\lambda}_2, \bar{\lambda}_2')$ defined by the equation constructed using the second parametrizations assign the same image on l_2 to a general point on l_1 and are therefore identical. This completes our proof.

Corollary 1

A projectivity between distinct lines in which the point of intersection of the lines is self-corresponding is a perspectivity.

Example 1

Find the equation of the projectivity between $l_1: x_1 = 0$ and $l_2: x_2 = 0$ which maps the points $A_1:(0,1,1)$, $B_1:(0,2,1)$, $C_1:(0,1,3)$ onto the points $A_2:(1,0,2)$, $B_2:(1,0,1)$, $C_2:(2,0,1)$, respectively, if:

1. l_1 is parametrized in terms of the base points $(0,0,1)$ and $(0,1,0)$ and l_2 is parametrized in terms of the base points $(1,0,0)$ and $(1,0,1)$.

2. l_1 is parametrized in terms of the base points $(0,1,1)$ and $(0,1,-1)$ and l_2 is parametrized in terms of the base points $(1,0,-1)$ and $(0,0,1)$.

In part 1 it is easy to verify that the parameters of the three given points on l_1 are $A_1:(1,1)$, $B_1:(1,2)$, $C_1:(3,1)$ and that the parameters of the corresponding points on l_2 are $A_2:(-1,2)$, $B_2:(0,1)$, $C_2:(1,1)$. Hence the coefficients in the equation of the required projectivity, $a\lambda_1\lambda_2 + b\lambda_1\lambda_2' + c\lambda_1'\lambda_2 + d\lambda_1'\lambda_2' = 0$, must satisfy the equations

$$-a + 2b - c + 2d = 0$$
$$b \qquad + 2d = 0$$
$$3a + 3b + c + d = 0$$

Therefore, solving, we have $a = -7$, $b = 4$, $c = 11$, $d = -2$, and the equation of the projectivity is

$$-7\lambda_1\lambda_2 + 4\lambda_1\lambda_2' + 11\lambda_1'\lambda_2 - 2\lambda_1'\lambda_2' = 0$$

or

$$\left\| \begin{matrix} \lambda_2 \\ \lambda_2' \end{matrix} \right\| = \left\| \begin{matrix} 4 & -2 \\ 7 & -11 \end{matrix} \right\| \cdot \left\| \begin{matrix} \lambda_1 \\ \lambda_1' \end{matrix} \right\|$$

Using the base points identified in part 2, the parameters of the given points on l_1 are $A_1:(1,0)$, $B_1:(3,1)$, $C_1:(2,-1)$, and the parameters of the corresponding points on l_2 are $A_2:(1,3)$, $B_2:(1,2)$, $C_2:(2,3)$. Hence the coefficients in the equation of the required projectivity, $\bar{a}\bar{\lambda}_1\bar{\lambda}_2 + \bar{b}\bar{\lambda}_1\bar{\lambda}'_2 + \bar{c}\bar{\lambda}'_1\bar{\lambda}_2 + \bar{d}\bar{\lambda}'_1\bar{\lambda}'_2 = 0$, must satisfy the equations

$$
\begin{array}{rcl}
\bar{a} + 3\bar{b} & = 0 \\
3\bar{a} + 6\bar{b} + \bar{c} + 2\bar{d} & = 0 \\
4\bar{a} + 6\bar{b} - 2\bar{c} - 3\bar{d} & = 0
\end{array}
$$

From these we find at once that $\bar{a} = 3$, $\bar{b} = -1$, $\bar{c} = 21$, $\bar{d} = -12$, and the equation of the projectivity is

$$
3\bar{\lambda}_1\bar{\lambda}_2 - \bar{\lambda}_1\bar{\lambda}'_2 + 21\bar{\lambda}'_1\bar{\lambda}_2 - 12\bar{\lambda}'_1\bar{\lambda}'_2 = 0 \quad \text{or} \quad \left\|\begin{matrix}\bar{\lambda}_2 \\ \bar{\lambda}'_2\end{matrix}\right\| = \left\|\begin{matrix}1 & 12 \\ 3 & 21\end{matrix}\right\| \cdot \left\|\begin{matrix}\bar{\lambda}_1 \\ \bar{\lambda}'_1\end{matrix}\right\|
$$

Since the equations of the given projectivity are different in the two sets of parametrizations, it is interesting to verify that they do indeed define the same mapping of l_1 onto l_2. Using the first parametrization, a general point $P:(0,p_2,p_3)$ on l_1 has parameters (p_3,p_2), and its image on l_2 has parameters defined by the equation

$$
\left\|\begin{matrix}\lambda_2 \\ \lambda'_2\end{matrix}\right\| = \left\|\begin{matrix}4 & -2 \\ 7 & -11\end{matrix}\right\| \cdot \left\|\begin{matrix}p_3 \\ p_2\end{matrix}\right\| \quad \text{or} \quad (4p_3 - 2p_2, 7p_3 - 11p_2)
$$

The coordinate-vector of its image is therefore

$$
(4p_3 - 2p_2)\left\|\begin{matrix}1 \\ 0 \\ 0\end{matrix}\right\| + (7p_3 - 11p_2)\left\|\begin{matrix}1 \\ 0 \\ 1\end{matrix}\right\| = \left\|\begin{matrix}-13p_2 + 11p_3 \\ 0 \\ -11p_2 + 7p_3\end{matrix}\right\|
$$

Using the second parametrization, the point $P:(0,p_2,p_3)$ on l_1 has parameters $(p_2 + p_3, p_2 - p_3)$, and its image on l_2 has parameters defined by the equation

$$
\left\|\begin{matrix}\bar{\lambda}_2 \\ \bar{\lambda}'_2\end{matrix}\right\| = \left\|\begin{matrix}1 & 12 \\ 3 & 21\end{matrix}\right\| \cdot \left\|\begin{matrix}p_2 + p_3 \\ p_2 - p_3\end{matrix}\right\| \quad \text{or} \quad (13p_2 - 11p_3, 24p_2 - 18p_3)
$$

The coordinate-vector of its image is therefore

$$
(13p_2 - 11p_3)\left\|\begin{matrix}1 \\ 0 \\ -1\end{matrix}\right\| + (24p_2 - 18p_3)\left\|\begin{matrix}0 \\ 0 \\ 1\end{matrix}\right\| = \left\|\begin{matrix}13p_2 - 11p_3 \\ 0 \\ 11p_2 - 7p_3\end{matrix}\right\|
$$

Since the two coordinate-vectors of the image of the general point $P:(0,p_2,p_3)$ differ only by the factor -1, they represent the same point, and the two equations do, indeed, define the same projectivity.

Exercises

1. Determine the equation of the projectivity defined by each of the following sets of corresponding points:

(a) $(1,0) \to (1,1)$, $(0,1) \to (2,3)$, $(1,1) \to (3,4)$
(b) $(1,1) \to (0,1)$, $(1,-2) \to (1,0)$, $(1,-1) \to (2,1)$
(c) $(1,-1) \to (1,1)$, $(2,-1) \to (0,1)$, $(0,1) \to (1,2)$
(d) $(1,-1) \to (1,-1)$, $(2,3) \to (2,3)$, $(-2,1) \to (0,1)$
(e) $(1,2) \to (2,1)$, $(2,1) \to (1,2)$, $(1,1) \to (1,1)$

2. Find the equation of the projectivity between $l_1: x_1 - x_2 = 0$ and $l_2: x_1 + x_2 = 0$ which maps $P_1:(1,1,0)$, $Q_1:(1,1,1)$, $R_1:(2,2,1)$ onto the points $P_2:(1,-1,1)$, $Q_2:(1,-1,0)$, $R_2:(0,0,1)$, respectively, if:

(a) The base points on l_1 are $A_1:(0,0,1)$ and $B_1:(1,1,0)$ and the base points on l_2 are $A_2:(0,0,1)$ and $B_2:(2,-2,1)$.
(b) The base points on l_1 are $A_1:(1,1,1)$ and $B_1:(1,1,0)$ and the base points on l_2 are $A_2:(1,-1,0)$ and $B_2:(1,-1,1)$.

3. Find an equation for the projectivity which maps the points $P_1:(1,0,1)$, $Q_1:(0,1,1)$, $R_1:(1,-1,0)$ on the line $l_1: x_1 + x_2 - x_3 = 0$ onto the respective image points:

(a) $P_2:(1,1,-1)$, $Q_2:(1,0,-1)$, $R_2:(0,1,0)$ on the line $l_2: x_1 + x_3 = 0$
(b) $P_2:(1,0,0)$, $Q_2:(0,1,-2)$, $R_2:(2,-1,2)$ on the line $l_2: 2x_2 + x_3 = 0$
(c) $P_2:(0,1,-1)$, $Q_2:(1,0,-1)$, $R_2:(3,2,-5)$ on the line $l_2: x_1 + x_2 + x_3 = 0$

4. (a) Complete the proof of Theorem 3 by showing that the values of a, b, c, d are such that $ad - bc \neq 0$.
(b) Since a system of homogeneous equations of the form (1) *always* has a nontrivial solution, why cannot a projectivity be found which will assign the same image on l_2 to two distinct points on l_1 or assign distinct images on l_2 to one point on l_1?

5. Give a proof of Theorem 3 based on the assumption that the determinants obtained by deleting first the first column and then the second column in the matrix (2) are zero.

6. If l_1, l_2, l_3 are three lines on a point L, and if m_1, m_2, m_3 are three lines on a point M, show how to construct the image of an arbitrary line on L in the projectivity between L and M in which $l_1 \to m_1$, $l_2 \to m_2$, $l_3 \to m_3$.

4.7 Cross Ratio

In both E_2^+ and Π_2 neither distance nor angle measures are defined. Hence properties such as congruence, similarity, betweenness, perpendicularity, and conic types, which depend upon metric ideas, are meaningless. In their stead, relations which do not depend upon measurement, such as those described by Desargues' theorem and the theorem of Pappus, are the subject of investigation. One property of this sort, of fundamental importance in projective geometry, is the so-called **cross ratio** of four collinear points or of four concurrent lines.

Definition 1

If (P_1,P_2,P_3,P_4) are four collinear points whose parameters with respect to a given pair of base points on the line they determine are (λ_1,λ_1'), (λ_2,λ_2'), (λ_3,λ_3'), (λ_4,λ_4'), the quantity

$$\frac{(\lambda_1\lambda_3' - \lambda_3\lambda_1')(\lambda_2\lambda_4' - \lambda_4\lambda_2')}{(\lambda_1\lambda_4' - \lambda_4\lambda_1')(\lambda_2\lambda_3' - \lambda_3\lambda_2')} = \frac{\begin{vmatrix} \lambda_1 & \lambda_3 \\ \lambda_1' & \lambda_3' \end{vmatrix} \cdot \begin{vmatrix} \lambda_2 & \lambda_4 \\ \lambda_2' & \lambda_4' \end{vmatrix}}{\begin{vmatrix} \lambda_1 & \lambda_4 \\ \lambda_1' & \lambda_4' \end{vmatrix} \cdot \begin{vmatrix} \lambda_2 & \lambda_3 \\ \lambda_2' & \lambda_3' \end{vmatrix}}$$

is called the cross ratio of the set (P_1,P_2,P_3,P_4).

Definition 2

If (p_1,p_2,p_3,p_4) are four concurrent lines whose parameters with respect to a given pair of base lines on the point they determine are $[\lambda_1,\lambda_1']$, $[\lambda_2,\lambda_2']$, $[\lambda_3,\lambda_3']$, $[\lambda_4,\lambda_4']$, the quantity

$$\frac{(\lambda_1\lambda_3' - \lambda_3\lambda_1')(\lambda_2\lambda_4' - \lambda_4\lambda_2')}{(\lambda_1\lambda_4' - \lambda_4\lambda_1')(\lambda_2\lambda_3' - \lambda_3\lambda_2')} = \frac{\begin{vmatrix} \lambda_1 & \lambda_3 \\ \lambda_1' & \lambda_3' \end{vmatrix} \cdot \begin{vmatrix} \lambda_2 & \lambda_4 \\ \lambda_2' & \lambda_4' \end{vmatrix}}{\begin{vmatrix} \lambda_1 & \lambda_4 \\ \lambda_1' & \lambda_4' \end{vmatrix} \cdot \begin{vmatrix} \lambda_2 & \lambda_3 \\ \lambda_2' & \lambda_3' \end{vmatrix}}$$

is called the cross ratio of the set (p_1,p_2,p_3,p_4).

We shall denote the cross ratio of four collinear points (P_1,P_2,P_3,P_4) by the symbol $R\,(P_1P_2,P_3P_4)$ and the cross ratio of four concurrent lines (p_1,p_2,p_3,p_4) by the symbol $R\,(p_1p_2,p_3p_4)$.

From Definition 1 it appears that the cross ratio of four points depends on the two base points used to determine the parameters of the points, but it is easy to show that this is actually not the case. To do this, we observe first that the typical binomial $(\lambda_i\lambda_j' - \lambda_j\lambda_i')$ appearing in the cross-ratio formula can be written in the form

$$\| \lambda_i \quad \lambda_i' \| \cdot \left\| \begin{matrix} 0 & 1 \\ -1 & 0 \end{matrix} \right\| \cdot \left\| \begin{matrix} \lambda_j \\ \lambda_j' \end{matrix} \right\|$$

Next, we note the following lemma, whose proof, by direct calculation, we leave as an exercise.

Lemma 1

If M is any (2,2) matrix, then

$$M^T \cdot \left\| \begin{matrix} 0 & 1 \\ -1 & 0 \end{matrix} \right\| \cdot M = |M| \left\| \begin{matrix} 0 & 1 \\ -1 & 0 \end{matrix} \right\|$$

Now let us suppose that the line l which contains the points P_1, P_2, P_3, P_4 is parametrized in two different ways, one with parameters (λ, λ'), the other with parameters $(\bar{\lambda}, \bar{\lambda}')$. Then, as we observed in Example 2, Sec. 4.2, the two sets of parameters are connected by a relation of the form $\left\| \begin{matrix} \lambda \\ \lambda' \end{matrix} \right\| = M \left\| \begin{matrix} \bar{\lambda} \\ \bar{\lambda}' \end{matrix} \right\|$, where M is a nonsingular (2,2) matrix. Hence, making this substitution in the typical binomial in the formula for the cross ratio, and remembering that the transpose of the product of two matrices is the product of the transposes in the reverse order, we have

$$(\lambda_i \lambda'_j - \lambda_j \lambda'_i) = \left\| \lambda_i \quad \lambda'_i \right\| \cdot \left\| \begin{matrix} 0 & 1 \\ -1 & 0 \end{matrix} \right\| \cdot \left\| \begin{matrix} \lambda_j \\ \lambda'_j \end{matrix} \right\|$$

$$= \left\| \bar{\lambda}_i \quad \bar{\lambda}'_i \right\| M^T \cdot \left\| \begin{matrix} 0 & 1 \\ -1 & 0 \end{matrix} \right\| \cdot M \left\| \begin{matrix} \bar{\lambda}_j \\ \bar{\lambda}'_j \end{matrix} \right\|$$

$$= |M| \cdot \left\| \bar{\lambda}_i \quad \bar{\lambda}'_i \right\| \cdot \left\| \begin{matrix} 0 & 1 \\ -1 & 0 \end{matrix} \right\| \cdot \left\| \begin{matrix} \bar{\lambda}_j \\ \bar{\lambda}'_j \end{matrix} \right\|$$

$$= |M| \cdot (\bar{\lambda}_i \bar{\lambda}'_j - \bar{\lambda}_j \bar{\lambda}'_i)$$

When this result is used to express the cross ratio itself in terms of the new parameters, the factor $|M|^2$ cancels from the numerator and denominator, and we have

$$R(P_1 P_2, P_3 P_4) = \frac{(\lambda_1 \lambda'_3 - \lambda_3 \lambda'_1)(\lambda_2 \lambda'_4 - \lambda_4 \lambda'_2)}{(\lambda_1 \lambda'_4 - \lambda_4 \lambda'_1)(\lambda_2 \lambda'_3 - \lambda_3 \lambda'_2)} = \frac{(\bar{\lambda}_1 \bar{\lambda}'_3 - \bar{\lambda}_3 \bar{\lambda}'_1)(\bar{\lambda}_2 \bar{\lambda}'_4 - \bar{\lambda}_4 \bar{\lambda}'_2)}{(\bar{\lambda}_1 \bar{\lambda}'_4 - \bar{\lambda}_4 \bar{\lambda}'_1)(\bar{\lambda}_2 \bar{\lambda}'_3 - \bar{\lambda}_3 \bar{\lambda}'_2)}$$

which shows that the cross ratio is independent of the base points in terms of which the parameters are defined. Dually, of course, the cross ratio of four concurrent lines is independent of the base lines in terms of which the parameters of the lines are defined.

Since four points (or four lines) can be arranged in $4! = 24$ different ways, it would seem that there are 24 possible values for the cross ratio of four collinear points (or four concurrent lines). This is not the case, however, and by using the following theorems, whose proofs we leave as exercises, it is easy to show that the 24 possible arrangements form six sets of four, the arrangements in each set yielding the same cross ratio.

Theorem 1

The cross ratio of four collinear points (or four concurrent lines) is left unchanged if any two of the points (lines) are interchanged, provided the other two are also interchanged.

Theorem 2

If in a set of four collinear points (or four concurrent lines), either the points (lines) in the first pair or the points (lines) in the second pair are interchanged, the cross ratio of the resulting arrangement is the reciprocal of the cross ratio of the original arrangement.

Theorem 3

If in a set of four collinear points (or four concurrent lines), the first and last points (lines) or the second and third points (lines) are interchanged, the sum of the cross ratios of the original arrangement and the new arrangement is equal to 1.

For some purposes it is convenient to extend the notion of the cross ratio to **singular tetrads,** i.e., sets of fewer than four points. This we do, when possible, by taking the formal content of the cross-ratio formula when two or more of the points P_1, P_2, P_3, P_4 coincide. Specifically,

If $P_1 = P_2$ or if $P_3 = P_4$, then $R(P_1P_2, P_3P_4) = 1$, and conversely.
If $P_1 = P_3$ or if $P_2 = P_4$, then $R(P_1P_2, P_3P_4) = 0$, and conversely.
If $P_1 = P_4$ or if $P_2 = P_3$, then $R(P_1P_2, P_3P_4) = \infty$, and conversely.

If more than two of the points (P_1, P_2, P_3, P_4) coincide, the value of the cross ratio is an indeterminate of the form $\frac{0}{0}$ and we leave it undefined.

One of the most important properties of the cross ratio is described in the next theorem.

Theorem 4

The cross ratio of four concurrent lines is equal to the cross ratio of the four points in which the lines are intersected by an arbitrary line not passing through the point of concurrence of the lines.

Proof Let l_i: $a_ix_1 + b_ix_2 + c_ix_3 = 0$, $i = 1, 2, 3, 4$, be four concurrent lines, and let l: $ax_1 + bx_2 + cx_3 = 0$ be an arbitrary line which does not contain the point of concurrence of l_1, l_2, l_3, l_4. Let the pencil containing l_1, l_2, l_3, l_4 be parametrized in terms of l_1 and l_2, so that the parameters of the four lines are l_1:$[1,0]$, l_2:$[0,1]$, l_3:$[\lambda_3, \lambda_3']$, l_4:$[\lambda_4, \lambda_4']$. This implies, of course, that the equations of l_3 and l_4 can be written in the form

$$l_3: \lambda_3(a_1x_1 + b_1x_2 + c_1x_3) + \lambda_3'(a_2x_1 + b_2x_2 + c_2x_3) = 0$$
$$l_4: \lambda_4(a_1x_1 + b_1x_2 + c_1x_3) + \lambda_4'(a_2x_1 + b_2x_2 + c_2x_3) = 0$$

Furthermore, let P_i be the intersection of l_i and l, and let the range on l be parametrized in terms of P_1 and P_2, so that the parameters of the four points of intersection are $P_1:(1,0)$, $P_2:(0,1)$, $P_3:(\bar{\lambda}_3,\bar{\lambda}_3')$, $P_4:(\bar{\lambda}_4,\bar{\lambda}_4')$ (Fig. 4.12). Now the coordinates of the intersection of l_3 and l can be read from the matrix of coefficients

$$\left\| \begin{matrix} a_3 & b_3 & c_3 \\ a & b & c \end{matrix} \right\| = \left\| \begin{matrix} \lambda_3 a_1 + \lambda_3' a_2 & \lambda_3 b_1 + \lambda_3' b_2 & \lambda_3 c_1 + \lambda_3' c_2 \\ a & b & c \end{matrix} \right\|$$

and are, in fact,

$$\left(\lambda_3 \begin{vmatrix} b_1 & c_1 \\ b & c \end{vmatrix} + \lambda_3' \begin{vmatrix} b_2 & c_2 \\ b & c \end{vmatrix}, \quad -\lambda_3 \begin{vmatrix} a_1 & c_1 \\ a & c \end{vmatrix} - \lambda_3' \begin{vmatrix} a_2 & c_2 \\ a & c \end{vmatrix}, \right.$$
$$\left. \lambda_3 \begin{vmatrix} a_1 & b_1 \\ a & b \end{vmatrix} + \lambda_3' \begin{vmatrix} a_2 & b_2 \\ a & b \end{vmatrix} \right)$$

But these are just linear combinations of the coordinates of P_1 and P_2, the constants of combination being the parameters (λ_3,λ_3') of l_3. Thus the parameters $(\bar{\lambda}_3,\bar{\lambda}_3')$ of P_3 are actually (λ_3,λ_3'). Similarly, the parameters $(\bar{\lambda}_4,\bar{\lambda}_4')$ of the intersection of l_4 and l are actually the parameters (λ_4,λ_4') of l_4. Therefore l_1, l_2, l_3, l_4 and P_1, P_2, P_3, P_4 have, respectively, the same parameters and hence the same cross ratio, as asserted.

   Dually, *the cross ratio of four collinear points is the same as the cross ratio of the four lines which join these points to any point not on the line which contains them.*

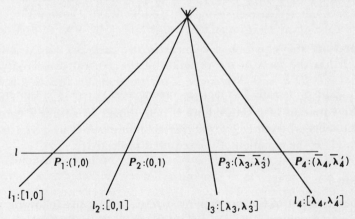

Fig. 4.12   The relation between the parameters of four lines and the parameters of their intersections with an arbitrary line.

From the last theorem, it is clear that if P_1', P_2', P_3', P_4' are, respectively, the images of four collinear points P_1, P_2, P_3, P_4 under any perspectivity, then

$$R(P_1'P_2',P_3'P_4') = R(P_1P_2,P_3P_4)$$

Therefore, since any projectivity is the composition of a finite number of perspectivities, we have the following important theorem.

Theorem 5

The cross ratio of four collinear points and the cross ratio of their images under any projectivity are equal.

Corollary 1

Four points (P_1,P_2,P_3,P_4) on a line l are projective with four points (P_1',P_2',P_3',P_4') on a line l' if and only if

$$R(P_1P_2,P_3P_4) = R(P_1'P_2',P_3'P_4')$$

Using the fact that the cross ratio of four points is invariant under any projectivity, we can obtain the equation of the projectivity determined by three pairs of corresponding points in a way that is often simpler than the method we developed in Sec. 4.6. Suppose that we are required to find the equation of the projectivity which assigns to the points $P_1:(\lambda_1,\lambda_1')$, $P_2:(\lambda_2,\lambda_2')$, $P_3:(\lambda_3,\lambda_3')$ on a line l the respective images $\bar{P}_1:(\bar{\lambda}_1,\bar{\lambda}_1')$, $\bar{P}_2:(\bar{\lambda}_2,\bar{\lambda}_2')$, $\bar{P}_3:(\bar{\lambda}_3,\bar{\lambda}_3')$ on a line $\bar{l}$. Then if $P:(\lambda,\lambda')$ is an arbitrary point on l, and if $\bar{P}:(\bar{\lambda},\bar{\lambda}')$ is its image on $\bar{l}$, it follows from Theorem 5 that $R(P_1P_2,P_3P) = R(\bar{P}_1\bar{P}_2,\bar{P}_3\bar{P})$; that is,

$$(1) \qquad \frac{(\lambda_1\lambda_3 - \lambda_3\lambda_1')(\lambda_2\lambda' - \lambda\lambda_2')}{(\lambda_1\lambda' - \lambda\lambda_1')(\lambda_2\lambda_3 - \lambda_3\lambda_2')} = \frac{(\bar{\lambda}_1\bar{\lambda}_3 - \bar{\lambda}_3\bar{\lambda}_1')(\bar{\lambda}_2\bar{\lambda}' - \bar{\lambda}\bar{\lambda}_2')}{(\bar{\lambda}_1\bar{\lambda}' - \bar{\lambda}\bar{\lambda}_1')(\bar{\lambda}_2\bar{\lambda}_3 - \bar{\lambda}_3\bar{\lambda}_2')}$$

By inspection, this equation is bilinear in the variables (λ,λ') and $(\bar{\lambda},\bar{\lambda}')$; that is, it has the form of the equation of the general projectivity $a\lambda\bar{\lambda} + b\lambda\bar{\lambda}' + c\lambda'\bar{\lambda} + d\lambda'\bar{\lambda}' = 0$. Moreover, it is clear that this equation is satisfied by the pairs $[(\lambda_1,\lambda_1'), (\bar{\lambda}_1,\bar{\lambda}_1')]$, $[(\lambda_2,\lambda_2'), (\bar{\lambda}_2,\bar{\lambda}_2')]$, $[(\lambda_3,\lambda_3'), (\bar{\lambda}_3,\bar{\lambda}_3')]$. Therefore, since the equation of a projectivity is determined to within a constant of proportionality when three pairs of corresponding points are given, it follows that Eq. (1) is the equation of the required projectivity.

Example 1

Find the equation of the projectivity determined by the following pairs of corresponding points: $P_1:(1,1) \to \bar{P}_1:(1,0)$, $P_2:(1,2) \to \bar{P}_2:(0,1)$, $P_3:(2,1) \to \bar{P}_3:(1,3)$.

Letting $\bar{P}:(\bar{\lambda},\bar{\lambda}')$ be the image of an arbitrary point $P:(\lambda,\lambda')$, and setting up the equation $\mathrm{R}(P_1P_2,P_3P) = \mathrm{R}(\bar{P}_1\bar{P}_2,\bar{P}_3\bar{P})$, we have

$$\frac{(-1)(\lambda' - 2\lambda)}{(\lambda' - \lambda)(-3)} = \frac{(3)(-\bar{\lambda})}{(\bar{\lambda}')(-1)} \qquad \text{or} \qquad 9\lambda\bar{\lambda} - 2\lambda\bar{\lambda}' - 9\lambda'\bar{\lambda} + \lambda'\bar{\lambda}' = 0$$

Of course, until we are given the parametrized lines on which the points $P_1, P_2, \ldots, \bar{P}_3$ are located, we really do not know what projectivity it is that we have found!

If (P_1,P_2,P_3,P_4) are four points such that $\mathrm{R}(P_1P_2,P_3P_4) = -1$, they are said to form a **harmonic tetrad.** From Theorem 2, it is clear that if $\mathrm{R}(P_1P_2,P_3P_4) = -1$, then $\mathrm{R}(P_1P_2,P_4P_3) = -1$; that is, if (P_1P_2,P_3P_4) is a harmonic tetrad, so is (P_1P_2,P_4P_3). Thus in any harmonic tetrad, the last two points play symmetric roles with respect to the first two points, and for this reason either is said to be the **harmonic conjugate** of the other with respect to the first two points. Furthermore, from Theorem 1 it follows that if $\mathrm{R}(P_1P_2,P_3P_4) = -1$, then $\mathrm{R}(P_4P_3,P_2P_1) = -1$. Hence if (P_1P_2,P_3P_4) is a harmonic tetrad, so is (P_4P_3,P_2P_1); and therefore in any harmonic tetrad the first two points are also harmonic conjugates with respect to the last two points.

Harmonic tetrads are of considerable importance in projective geometry, and the notion of harmonic conjugates is fundamental in the definition of many important transformations. For this reason it is convenient to be able to identify quickly the harmonic conjugate of a given point with respect to two other points collinear with it. Such a result is contained in the following lemma.

Lemma 2

If P_1 and P_2 are chosen as base points on the line l which they determine, then the harmonic conjugate of a general point $P_3:(\lambda_3,\lambda_3')$ on l with respect to P_1 and P_2 is the point P_4 whose parameters are $(\lambda_3,-\lambda_3')$.

Proof When P_1 and P_2 are chosen as base points on l, their parameters are, respectively, $(1,0)$ and $(0,1)$. Hence, if the parameters of P_3 and P_4 are (λ_3,λ_3') and (λ_4,λ_4'), then

$$\mathrm{R}(P_1P_2,P_3P_4) = \frac{(\lambda_3')(-\lambda_4)}{(\lambda_4')(-\lambda_3)}$$

This will have the value -1, and P_4 will be the harmonic conjugate of P_3 with respect to P_1 and P_2, if and only if $\lambda_4/\lambda_4' = -\lambda_3/\lambda_3'$, as asserted.

Example 2

Let C be the point (c_1, c_2, c_3), let l be the line $x_2 = 0$, let $Y:(y_1, y_2, y_3)$ be a general point, and let C' be the intersection of l and the line CY. Find the harmonic conjugate of Y with respect to C and C'.

The equation of the line CY is

$$\begin{vmatrix} x_1 & x_2 & x_3 \\ c_1 & c_2 & c_3 \\ y_1 & y_2 & y_3 \end{vmatrix} = 0 \quad \text{or} \quad \begin{aligned} (c_2 y_3 - c_3 y_2)x_1 - (c_1 y_3 - c_3 y_1)x_2 \\ + (c_1 y_2 - c_2 y_1)x_3 = 0 \end{aligned}$$

Hence the intersection of CY and the line $l: x_2 = 0$ is the point $C':(-c_1 y_2 + c_2 y_1, 0, c_2 y_3 - c_3 y_2)$. Now in order that we may use Lemma 2, we choose C and C' as base points for the parametrization of the line CY. The parameters of Y are then numbers (λ, λ') such that

$$\begin{Vmatrix} y_1 \\ y_2 \\ y_3 \end{Vmatrix} = \lambda \begin{Vmatrix} c_1 \\ c_2 \\ c_3 \end{Vmatrix} + \lambda' \begin{Vmatrix} c_2 y_1 - c_1 y_2 \\ 0 \\ -c_3 y_2 + c_2 y_3 \end{Vmatrix}$$

or $(y_2, 1)$. By Lemma 2, the parameters of the harmonic conjugate of Y with respect to C and C' are therefore $(y_2, -1)$, and the coordinate-vector of this point, Y', is

$$\begin{Vmatrix} y_1' \\ y_2' \\ y_3' \end{Vmatrix} = y_2 \begin{Vmatrix} c_1 \\ c_2 \\ c_3 \end{Vmatrix} - \begin{Vmatrix} c_2 y_1 - c_1 y_2 \\ 0 \\ -c_3 y_2 + c_2 y_3 \end{Vmatrix} = \begin{Vmatrix} -c_2 y_1 + 2c_1 y_2 \\ c_2 y_2 \\ 2c_3 y_2 - c_2 y_3 \end{Vmatrix}$$

The transformation

$$T: \begin{aligned} y_1' &= -c_2 y_1 + 2c_1 y_2 \\ y_2' &= c_2 y_2 \\ y_3' &= 2c_3 y_2 - c_2 y_3 \end{aligned}$$

obtained in Example 2 is an example of an important class of linear transformations known as **harmonic homologies:**

Definition 3

Let C be an arbitrary fixed point in either E_2^+ or Π_2, let l be an arbitrary fixed line not passing through C, let Y be a general point, and let C' be the intersection of l and the line CY. Then the transformation which maps the general point Y into its harmonic conjugate with respect to C and C' is known as a harmonic homology.

The point C and the line l, which together define a harmonic homology, are known, respectively, as the **center** and **axis** of the homology.

The transformation T has an interesting interpretation in the euclidean plane when $c_1 = c_3 = 0$ and the center C becomes the ideal point on the line $x_1 = 0$, that is, the y axis. In this case, after dividing out the nonzero factor $-c_2$, the equations of the transformation become

$$y_1' = y_1$$
$$y_2' = -y_2$$
$$y_3' = y_3$$

or, reverting to nonhomogeneous coordinates by the relations

$$x = \frac{y_1}{y_3} \qquad y = \frac{y_2}{y_3} \qquad x' = \frac{y_1'}{y_3'} \qquad y' = \frac{y_2'}{y_3'}$$

simply $x' = x$ and $y' = -y$. These we recognize as the equations of the transformation of the euclidean plane defined by reflection in the x axis.

Exercises

1. Find $R(P_1P_2,P_3P_4)$, $R(P_1P_2,P_4P_3)$, and $R(P_1P_3,P_2P_4)$ for the following sets of points:

(a) $P_1:(1,2)$, $P_2:(2,3)$, $P_3:(-1,0)$, $P_4:(1,-1)$
(b) $P_1:(-1,2)$, $P_2:(2,-1)$, $P_3:(1,1)$, $P_4:(3,1)$
(c) $P_1:(2,1)$, $P_2:(0,1)$, $P_3:(1,3)$, $P_4:(-2,3)$
(d) $P_1:(1,-1)$, $P_2:(2,3)$, $P_3:(4,1)$, $P_4:(2,1)$

2. Verify that the points in each of the following sets are collinear, and find $R(P_1P_2,P_3P_4)$, $R(P_4P_3,P_2P_1)$, and $R(P_1P_3,P_4P_2)$ in each case:

(a) $P_1:(0,1,1)$, $P_2:(1,0,1)$, $P_3:(1,-1,0)$, $P_4:(1,1,2)$
(b) $P_1:(1,2,0)$, $P_2:(-1,1,1)$, $P_3:(0,3,1)$, $P_4:(3,0,-2)$
(c) $P_1:(2,0,1)$, $P_2:(1,-1,1)$, $P_3:(1,1,0)$, $P_4:(0,2,-1)$

3. Let O, U, Ω, and P be, respectively, the origin, the unit point, the ideal point, and a general point on the x axis in E_2^+. Show that $R(PU,O\Omega) = x$, where x is the euclidean x coordinate of P.

4. (a) Prove Theorem 1   (b) Prove Theorem 2   (c) Prove Theorem 3

5. If $R(P_1P_2,P_3P_4) = r$, verify the following:

$$R(P_1P_2,P_3P_4) = R(P_2P_1,P_4P_3) = R(P_4P_3,P_2P_1) = R(P_3P_4,P_1P_2) = r$$

$$R(P_1P_2,P_4P_3) = R(P_2P_1,P_3P_4) = R(P_4P_3,P_1P_2) = R(P_3P_4,P_2P_1) = \frac{1}{r}$$

$$R(P_1P_3,P_2P_4) = R(P_3P_1,P_4P_2) = R(P_4P_2,P_3P_1) = R(P_2P_4,P_1P_3) = 1 - r$$

$$R(P_1P_3,P_4P_2) = R(P_3P_1,P_2P_4) = R(P_4P_2,P_1P_3) = R(P_2P_4,P_3P_1) = \frac{1}{1-r}$$

$$R(P_1P_4,P_3P_2) = R(P_4P_1,P_2P_3) = R(P_3P_2,P_1P_4) = R(P_2P_3,P_4P_1) = \frac{r}{r-1}$$

$$R(P_1P_4,P_2P_3) = R(P_4P_1,P_3P_2) = R(P_3P_2,P_4P_1) = R(P_2P_3,P_1P_4) = \frac{r-1}{r}$$

6. Find $R(P_1P_2,P_3P_4)$ if P_1, P_2, P_3, P_4 are, respectively:

(a) The points in which the line joining $(1,1,0)$ and $(0,1,1)$ intersects $x_1 = 0$, $x_2 = 0$, $x_3 = 0$, $2x_1 + x_2 + x_3 = 0$.

(b) The points in which the line joining $(1,1,1)$ and $(1,1,-1)$ intersects the lines $x_1 + 2x_2 + x_3 = 0$, $x_1 - x_3 = 0$, $2x_1 - x_2 + x_3 = 0$, $x_1 - 3x_2 + x_3 = 0$.

(c) The points in which the line $x_1 + x_2 - 2x_3 = 0$ intersects the lines $2x_1 + x_2 = 0$, $x_1 - 3x_2 = 0$, $x_1 + x_2 + x_3 = 0$, $x_1 - 2x_2 + 3x_3 = 0$.

7. What is the cross ratio of the four lines which join the point $(1,1,1)$ to the points $P_1:(1,0,0)$, $P_2:(0,1,0)$, $P_3:(0,0,1)$, $P_4:(1,2,3)$?

8. If $P_1:(x_1,y_1,z_1)$, $P_2:(x_2,y_2,z_2)$, $P_3:(x_3,y_3,z_3)$, $P_4:(x_4,y_4,z_4)$ are four points on a line $ax + by + cz = 0$, prove that

$$R(P_1P_2,P_3P_4) = \frac{(x_1z_3 - x_3z_1)(x_2z_4 - x_4z_2)}{(x_1z_4 - x_4z_1)(x_2z_3 - x_3z_2)}$$

$$= \frac{(y_1z_3 - y_3z_1)(y_2z_4 - y_4z_2)}{(y_1z_4 - y_4z_1)(y_2z_3 - y_3z_2)}$$

$$= \frac{(x_1y_3 - x_3y_1)(x_2y_4 - x_4y_2)}{(x_1y_4 - x_4y_1)(x_2y_3 - x_3y_2)}$$

whenever these expressions are meaningful. Under what conditions, if any, will one or more of these expressions fail to be meaningful?

9. Using the results of Exercise 8, prove that if P, Q, R, S are four collinear points in the euclidean plane, then

$$R(PQ,RS) = \frac{(PR)(QS)}{(PS)(QR)}$$

where (AB) denotes the length of the segment $\overline{AB}$.

10. If A and B are distinct finite points in the extended euclidean plane, show that the midpoint of the segment $\overline{AB}$ is the harmonic conjugate of the ideal point on the line AB with respect to A and B.

11. If P, Q, R are three collinear points in the euclidean plane, and if Ω is the ideal point on the line which they determine, prove that $R(PQ,R\Omega) = -(PR)/(RQ)$, where (PR) and (RQ) denote, respectively, the lengths of the segments $\overline{PR}$ and $\overline{RQ}$.

12. In the euclidean plane prove that four collinear points P_1, P_2, P_3, P_4 form a harmonic tetrad if and only if P_3 and P_4 divide the segment $\overline{P_1P_2}$ internally and externally in the same ratio.

13. If P_1, P_2, Q_1, Q_2, Q_3 are five collinear points, prove that

$$\text{R}(P_1P_2,Q_1Q_2) \cdot \text{R}(P_1P_2,Q_2Q_3) = \text{R}(P_1P_2,Q_1Q_3)$$

(This useful result is known as the **multiplication theorem** for cross ratios.)

14. What condition(s) must be satisfied by the coefficients in the equation of the line l: $a_1x_1 + a_2x_2 + a_3x_3 = 0$ in order for the intersections of l with the lines $x_1 = 0$ and $x_2 = 0$ to be harmonic conjugates with respect to the intersections of l with the lines $x_3 = 0$ and $x_1 + x_2 + x_3 = 0$?

15. Give an elementary trigonometric proof of the fact that in the euclidean plane if four collinear points P, Q, R, S are perspective with four other collinear points P', Q', R', S', then $\text{R}(PQ,RS) = \text{R}(P'Q',R'S')$.

16. Give a purely algebraic proof of the fact that the cross ratio of four collinear points is the same as the cross ratio of their images under an arbitrary projectivity.

17. (a) If P_1 and P_2 are the points whose parameters are the roots of the equation $2\lambda^2 - 7\lambda\mu + 6\mu^2 = 0$, and if P_3 and P_4 are the points whose parameters are the roots of the equation $2\lambda^2 - 5\lambda\mu - 3\mu^2 = 0$, what is the value of $\text{R}(P_1P_2,P_3P_4)$?
(b) What condition(s) must be satisfied by the coefficients of the equations $a_0\lambda^2 + 2a_1\lambda\mu + a_2\mu^2 = 0$ and $b_0\lambda^2 + 2b_1\lambda\mu + b_2\mu^2 = 0$ in order for the points whose parameters are the roots of the first equation to be harmonic conjugates with respect to the points whose parameters are the roots of the second equation?

18. (a) Can a meaningful answer be given to the question: What is the harmonic conjugate of P with respect to P and Q?
(b) Can a meaningful answer be given to the question: What is the harmonic conjugate of P with respect to Q and Q?

19. Given P_1:$(1,0,0)$, P_2:$(0,1,0)$, P_3:$(0,0,1)$, and U:(u_1,u_2,u_3), with u_1, u_2, $u_3 \neq 0$. Let l_1 be the harmonic conjugate of the line P_1U with respect to P_1P_2 and P_1P_3, and let Q_1 be the intersection of l_1 and P_2P_3. Similarly, let Q_2 be the intersection of P_1P_3 and the harmonic conjugate of P_2U with respect to P_2P_1 and P_2P_3, and let Q_3 be the intersection of P_1P_2 and the harmonic conjugate of P_3U with respect to P_3P_1 and P_3P_2. Show that Q_1, Q_2, Q_3 are collinear. Is this result still true if U lies on one of the sides of $\triangle P_1P_2P_3$?

20. Let $\triangle P_1P_2P_3$ be an arbitrary triangle, and let U be an arbitrary point. If Q_i is the intersection of the line P_jP_k and the harmonic conjugate of P_iU with respect to P_iP_j and P_iP_k, show that Q_1, Q_2, Q_3 are always collinear. (The line $Q_1Q_2Q_3$ is called the **polar** of the point U with respect to $\triangle P_1P_2P_3$.)

21. What is the dual of the result of Exercise 20?

22. Use the results of Exercises 10 and 21 to give another proof of the fact that the medians of a triangle are concurrent.

23. Prove Corollary 1, Theorem 5.

24. Given $C_1:(1,0,1)$, $C_2:(1,-1,0)$, $l_1: x_1 + x_3 = 0$, $l_2: x_2 + x_3 = 0$. Find the equations of the harmonic homology whose center and axis are, respectively:

(a) C_1 and l_1 (b) C_1 and l_2 (c) C_2 and l_1 (d) C_2 and l_2

25. In the euclidean plane, let (A,B,C,D) be four points which are collinear in the order A, C, B, D. Show that $R(AB,CD)$ is equal to $- \tan^2 \theta$, where 2θ is the angle of intersection of the circles on $\overline{AB}$ and $\overline{CD}$ as diameters, and show, further, that the other values of the cross ratio of the four points are $\cos^2 \theta$, $\sin^2 \theta$, $\sec^2 \theta$, $\csc^2 \theta$, $- \cot^2 \theta$.

26. Show that the lines $l_1x_1 + l_2x_2 + l_3x_3 = 0$ and $m_1x_1 + m_2x_2 + m_3x_3 = 0$ are perpendicular if and only if their intersections with the ideal line, $x_3 = 0$, are harmonic conjugates with respect to the circular points at infinity, $I:(1,i,0)$ and $J:(1,-i,0)$.

4.8 Projectivities Between Cobasal Ranges

When a projectivity maps a line onto itself, it is natural to ask if there are any points of the line which coincide with their images. Given a projectivity

$$T: \left\| \begin{matrix} \lambda_2 \\ \lambda_2' \end{matrix} \right\| = \left\| \begin{matrix} a_{11} & a_{12} \\ a_{21} & a_{22} \end{matrix} \right\| \cdot \left\| \begin{matrix} \lambda_1 \\ \lambda_1' \end{matrix} \right\| \qquad |A| = \left| \begin{matrix} a_{11} & a_{12} \\ a_{21} & a_{22} \end{matrix} \right| \neq 0$$

between cobasal ranges assumed to be parametrized in terms of the same base points, a point will be invariant, i.e., will map into itself, if and only if its parameters and those of its image are proportional. Thus the problem of determining the invariant points of the transformation T is equivalent to the problem of finding pairs of numbers (λ,λ'), not both of which are zero, such that

$$k \left\| \begin{matrix} \lambda \\ \lambda' \end{matrix} \right\| = \left\| \begin{matrix} k & 0 \\ 0 & k \end{matrix} \right\| \cdot \left\| \begin{matrix} \lambda \\ \lambda' \end{matrix} \right\| = \left\| \begin{matrix} a_{11} & a_{12} \\ a_{21} & a_{22} \end{matrix} \right\| \cdot \left\| \begin{matrix} \lambda \\ \lambda' \end{matrix} \right\|$$

or, combining the matrices,

(1)
$$\left\| \begin{matrix} a_{11} - k & a_{12} \\ a_{21} & a_{22} - k \end{matrix} \right\| \cdot \left\| \begin{matrix} \lambda \\ \lambda' \end{matrix} \right\| = O$$

This matric equation is equivalent to the pair of homogeneous linear equations

(2)
$$(a_{11} - k)\lambda + a_{12}\lambda' = 0$$
$$a_{21}\lambda + (a_{22} - k)\lambda' = 0$$

in the parameters (λ, λ'); and, according to Theorem 5, Sec. 3.7, this system will have a solution other than the trivial solution $\lambda = \lambda' = 0$ if and only if the determinant of the coefficient matrix is equal to zero. In other words, there will be a nontrivial solution for λ and λ' if and only if

(3)
$$\begin{vmatrix} a_{11} - k & a_{12} \\ a_{21} & a_{22} - k \end{vmatrix} = k^2 - (a_{11} + a_{22})k + a_{11}a_{22} - a_{12}a_{21} = 0$$

This quadratic equation, which is known as the **characteristic equation** of both the matrix A and the transformation T, always has at least one root, and in general will have two roots, $k = k_1, k_2$, which may be either real or complex. For these values of k, which are known as the **characteristic values** or **characteristic roots** of the matrix A, and for no others, Eq. (1) is satisfied by a nontrivial vector, and the transformation T has a fixed point. For either characteristic value, Eqs. (2) become dependent, and the solution for (λ, λ') can be read from either equation, provided, of course, that the left-hand member of that equation is not identically zero. Thus, using the first equation, we find the parameters of the fixed points corresponding to the values $k = k_1$ and $k = k_2$ to be $(-a_{12}, a_{11} - k_1)$ and $(-a_{12}, a_{11} - k_2)$, respectively. From this it is clear that except possibly when $a_{12} = 0$, the fixed points arising from different characteristic values are always distinct. If $a_{12} = 0$, the characteristic values are $k_1 = a_{11}$, $k_2 = a_{22}$, and the corresponding solutions, read now from the second of Eqs. (2), are $(a_{22} - a_{11}, -a_{21})$ and $(0, - a_{21})$, which are distinct if $a_{11} \neq a_{22}$, that is, if $k_1 \neq k_2$. Thus a projectivity between cobasal ranges will in general have either one or two fixed points, according as the discriminant of Eq. (3), namely, $(a_{11} - a_{22})^2 + 4a_{12}a_{21}$, is equal to or different from zero. The only exception occurs when $a_{12} = a_{21} = 0$ and $a_{11} = a_{22}$. In this case, the left member of each of Eqs. (2) vanishes identically, and every point is a fixed point.

Example 1

Determine the parameters of the fixed points of the projectivity

$$\begin{Vmatrix} \lambda_2 \\ \lambda_2' \end{Vmatrix} = \begin{Vmatrix} 1 & 2 \\ 4 & 3 \end{Vmatrix} \cdot \begin{Vmatrix} \lambda_1 \\ \lambda_1' \end{Vmatrix}$$

The characteristic equation of the given transformation is

$$\begin{vmatrix} 1 - k & 2 \\ 4 & 3 - k \end{vmatrix} = 0 \qquad \text{or} \qquad k^2 - 4k - 5 = 0$$

and its roots are $k = -1, 5$. When $k = -1$, Eqs. (2) become

$$2\lambda + 2\lambda' = 0$$
$$4\lambda + 4\lambda' = 0$$

Hence the solution corresponding to $k = -1$ is $\lambda = 1$, $\lambda' = -1$. When $k = 5$, Eqs. (2) become

$$-4\lambda + 2\lambda' = 0$$
$$4\lambda - 2\lambda' = 0$$

from which we read the second solution $\lambda = 1$, $\lambda' = 2$. The actual co-ordinates of the invariant points of the given transformation, of course, cannot be found until the base points of the parametrization are known.

One interesting property of the fixed points of a projectivity is contained in the following theorem.

Theorem 1

If a projectivity has distinct fixed points, F and G, and if P_2 is the image of a general point P_1 under the projectivity, then $R(FG, P_1 P_2)$ is a constant independent of P_1.

Proof Let T: $\begin{Vmatrix} \lambda_2 \\ \lambda_2' \end{Vmatrix} = \begin{Vmatrix} a_{11} & a_{12} \\ a_{21} & a_{22} \end{Vmatrix} \cdot \begin{Vmatrix} \lambda_1 \\ \lambda_1' \end{Vmatrix}$ be a projectivity with distinct fixed points, F and G, and let us take F and G as base points of the parametrization in terms of which the projectivity T is described, so that $F:(1,0)$ and $G:(0,1)$. Then both the vector $\begin{Vmatrix} 1 \\ 0 \end{Vmatrix}$ and the vector $\begin{Vmatrix} 0 \\ 1 \end{Vmatrix}$ must correspond to itself in the equation of T. From the first of these conditions it follows by direct substitution that $a_{21} = 0$, and from the second it follows that $a_{12} = 0$. Hence the equation of T, referred to its fixed points as base points, is

$$\begin{Vmatrix} \lambda_2 \\ \lambda_2' \end{Vmatrix} = \begin{Vmatrix} a_{11} & 0 \\ 0 & a_{22} \end{Vmatrix} \cdot \begin{Vmatrix} \lambda_1 \\ \lambda_1' \end{Vmatrix} \qquad a_{11} \neq a_{22}$$

From this it follows that the image of a general point $P_1:(\lambda_1, \lambda_1')$ is the point $P_2:(a_{11}\lambda_1, a_{22}\lambda_1')$. Hence

$$R(FG, P_1 P_2) = \frac{(\lambda_1')(-a_{11}\lambda_1)}{(a_{22}\lambda_1')(-\lambda_1)} = \frac{a_{11}}{a_{22}}$$

Since this is independent of the parameters (λ_1, λ_1') of the particular point P_1, the theorem is established.

In Π_2, where the distinction between real and imaginary elements is irrelevant, there are just two types of nonidentical projectivities: those with distinct fixed points and those with a single fixed point. On the other hand, in E_2^+, where all elements are real, the distinction between projectivities with real fixed points and projectivities with complex fixed points is significant, and thus in E_2^+ there are not two but three types of nonidentical projectivities. In E_2^+, a projectivity between cobasal ranges whose fixed points are real and distinct is said to be **hyperbolic,** one whose fixed points are coincident is said to be **parabolic,** and one whose fixed points are strictly complex is said to be **elliptic.**

Example 2

Show that the equation of any parabolic projectivity can be reduced to the form

$$\begin{Vmatrix} \lambda_2 \\ \lambda_2' \end{Vmatrix} = \begin{Vmatrix} a_{11} & a_{12} \\ 0 & a_{11} \end{Vmatrix} \cdot \begin{Vmatrix} \lambda_1 \\ \lambda_1' \end{Vmatrix} \qquad a_{11}, a_{12} \neq 0$$

Let T: $\begin{Vmatrix} \lambda_2 \\ \lambda_2' \end{Vmatrix} = \begin{Vmatrix} a_{11} & a_{12} \\ a_{21} & a_{22} \end{Vmatrix} \cdot \begin{Vmatrix} \lambda_1 \\ \lambda_1' \end{Vmatrix}$ be a projectivity with a single fixed point F. Then if F is taken as the base point $(1,0)$ in the parametrization used to describe T, it follows that the vector $\begin{Vmatrix} 1 \\ 0 \end{Vmatrix}$ must correspond to itself in the equation of T, which is possible if and only if $a_{21} = 0$. Now for all we presently know, T may well have a second fixed point G. To investigate this possibility, we proceed, as in Example 1, to determine the fixed points of T by finding the roots of the characteristic equation of the matrix of T, with $a_{21} = 0$:

$$\begin{vmatrix} a_{11} - k & a_{12} \\ 0 & a_{22} - k \end{vmatrix} = 0 \qquad \text{or} \qquad (k - a_{11})(k - a_{22}) = 0$$

Clearly, unless $a_{11} = a_{22}$, this equation will have two distinct roots, and hence T will have two distinct fixed points, contrary to our requirement that T be parabolic. Furthermore, it is necessary that a_{12} be different from zero; for if $a_{12} = 0$, with $a_{11} = a_{22}$, then T reduces to the identity projectivity, which leaves every point invariant and hence is not parabolic.

It is interesting to note that certain simple transformations on the euclidean line are in fact projectivities. For instance, consider on the x axis the translation defined by the equation $x' = x + h$. If we parametrize the x axis in terms of the ideal point $(1,0,0)$ and the origin $(0,0,1)$, the parameters (λ_1, λ_1') of a general point $P_1:(x,0,1)$ are clearly $\lambda_1 = x$ and $\lambda_1' = 1$. Similarly, the parameters (λ_2, λ_2') of the image point $P_2:(x',0,1)$ are $\lambda_2 = x'$ and $\lambda_2' = 1$. Hence we may write

$$x = \frac{\lambda_1}{\lambda_1'} \qquad x' = \frac{\lambda_2}{\lambda_2'}$$

and the equation of the given translation becomes $\lambda_2/\lambda_2' = \lambda_1/\lambda_1' + h$, which is equivalent to the pair of equations $\lambda_2 = \lambda_1 + \lambda_1'h$, $\lambda_2' = \lambda_1'$ or to the matric equation

$$\left\|\begin{matrix} \lambda_2 \\ \lambda_2' \end{matrix}\right\| = \left\|\begin{matrix} 1 & h \\ 0 & 1 \end{matrix}\right\| \cdot \left\|\begin{matrix} \lambda_1 \\ \lambda_1' \end{matrix}\right\|$$

It is easy to see that if $h \neq 0$, this is the equation of a parabolic projectivity whose fixed point has parameters $(1,0)$ and is therefore the ideal point $(1,0,0)$.

Given a projectivity

$$T: \quad \Lambda_2 = A\Lambda_1 \qquad A = \left\|\begin{matrix} a_{11} & a_{12} \\ a_{21} & a_{22} \end{matrix}\right\| \qquad \Lambda_i = \left\|\begin{matrix} \lambda_i \\ \lambda_i' \end{matrix}\right\|$$

on a line l, it is easy to see by specific examples that if the image of a general point P_1 under T is P_2, then the image of P_2 under T is not P_1 but rather some new point P_3. Thus, repeated applications of T to a general point P_1 yield a succession of images, $P_2, P_3, \ldots$, known as **iterates** of P_1. It is interesting to consider the properties of this sequence of images and in particular to determine under what conditions, if any, the nth iterate of P_1 is P_1 itself.

At the outset, we observe that if T is the projectivity $\Lambda_2 = A\Lambda_1$, then, as in Sec. 3.4, the composition of T with itself is the projectivity $T^2: \Lambda_3 = A^2\Lambda_1$, and in general, the composition of T with itself $n - 1$ times is the projectivity $T^n: \Lambda_{n+1} = A^n\Lambda_1$. For this reason, the nth iterate, P_{n+1}, of a point P_1 is often denoted by the symbol T^nP_1. In particular, if F is a fixed point of the projectivity T, then applying T a second time will also leave F fixed. Hence F is also a fixed point of the projectivity T^2. Similarly, of course, F is a fixed point of each of the projectivities $T^3, T^4, \ldots$. Thus we have established the following theorem.

Theorem 2

Any fixed point of a projectivity $T: \Lambda_2 = A\Lambda_1$ is also a fixed point of the projectivity $T^n: \Lambda_{n+1} = A^n\Lambda_1$ for every positive integer n.

Let us now consider any four consecutive iterates of an arbitrary point P_1, say $P_{n+1}, P_{n+2}, P_{n+3}, P_{n+4}$. Then if we let J denote the matrix $\left\|\begin{matrix} 0 & 1 \\ -1 & 0 \end{matrix}\right\|$, we have, as we observed in Sec. 4.7,

$$\mathbf{R}(P_{n+1}P_{n+2}, P_{n+3}P_{n+4}) = \frac{(\Lambda_{n+1}^T J \Lambda_{n+3})(\Lambda_{n+2}^T J \Lambda_{n+4})}{(\Lambda_{n+1}^T J \Lambda_{n+4})(\Lambda_{n+2}^T J \Lambda_{n+3})}$$

$$= \frac{[\Lambda_1^T (A^n)^T J A^{n+2} \Lambda_1][\Lambda_1^T (A^{n+1})^T J A^{n+3} \Lambda_1]}{[\Lambda_1^T (A^n)^T J A^{n+3} \Lambda_1][\Lambda_1^T (A^{n+1})^T J A^{n+2} \Lambda_1]}$$

Now by repeated applications of Lemma 1, Sec. 4.7, $|A|^n$ and $|A|^{n+1}$ can be removed from the respective factors in both the numerator and the denominator of the last fraction. Hence

$$R(P_{n+1}P_{n+2},P_{n+3}P_{n+4}) = \frac{(\Lambda_1^T JA^2\Lambda_1)(\Lambda_1^T JA^2\Lambda_1)}{(\Lambda_1^T JA^3\Lambda_1)(\Lambda_1^T JA\Lambda_1)}$$

Now by a completely straightforward calculation, we find

$$\Lambda_1^T JA\Lambda_1 = a_{21}\lambda_1^2 + (a_{22} - a_{11})\lambda_1\lambda_1' - a_{12}(\lambda_1')^2$$
$$\Lambda_1^T JA^2\Lambda_1 = (a_{11} + a_{22})[a_{21}\lambda_1^2 + (a_{22} - a_{11})\lambda_1\lambda_1' - a_{21}(\lambda_1')^2]$$
$$\Lambda_1^T JA^3\Lambda_1 = (a_{11}^2 + a_{11}a_{22} + a_{12}a_{21} + a_{22}^2)$$
$$\times [a_{21}\lambda_1^2 + (a_{22} - a_{11})\lambda_1\lambda_1' - a_{21}(\lambda_1')^2]$$

Hence, substituting,

$$R(P_{n+1}P_{n+2},P_{n+3}P_{n+4}) = \frac{(a_{11} + a_{22})^2}{a_{11}^2 + a_{11}a_{22} + a_{12}a_{21} + a_{22}^2}$$

Since this expression is independent of both n and P, we have thus established the following interesting result.

Theorem 3

If T is a projectivity on a line l, the cross ratio of any four consecutive iterates of a point P_1 under T is the same for all points of l.

As we observed above, if T is the projectivity $\Lambda_2 = A\Lambda_1$, then the composition of T with itself $n - 1$ times is the projectivity $\Lambda_{n+1} = A^n\Lambda_1$. Hence if the nth iterate of a general point P_1 is to be P_1 itself, A^n must be a scalar multiple of the identity matrix, and in fact it is no specialization to assume the scalar to be 1, so that $A^n = I$. A projectivity whose matrix A has the property that $A^n = I$ is said to be **periodic,** and if n is the smallest integer for which $A^n = I$, n is said to be the **period** of T.

One important property of periodic projectivities is contained in the following theorem.

Theorem 4

If T is a periodic projectivity on a line l, then the fixed points of T are always distinct.

Proof To prove this theorem, let us assume the contrary and suppose that $T: \Lambda_2 = A\Lambda_1$ is a parabolic projectivity, i.e., a projectivity with a single fixed point. Then, as we showed in Example 2, the matrix of T must be of the form

$$A = \left\| \begin{array}{cc} a_{11} & a_{12} \\ 0 & a_{11} \end{array} \right\| \qquad a_{11}, a_{12} \neq 0$$

Now an easy induction shows that

$$\left\| \begin{matrix} a_{11} & a_{12} \\ 0 & a_{11} \end{matrix} \right\|^n = \left\| \begin{matrix} a_{11}{}^n & na_{11}^{n-1}a_{12} \\ 0 & a_{11}{}^n \end{matrix} \right\|$$

and for this to be the identity matrix, or even proportional to the identity matrix, it is necessary that either $a_{11} = 0$ or $a_{12} = 0$, each of which is impossible. Thus no parabolic transformation can be periodic, and our theorem is established.

To see that there actually are projectivities which are periodic, let us consider a general projectivity T with distinct fixed points. Then if the fixed points are taken as the base points of the parametrization which describes T, it follows that the equation of T can be written in the form

$$\left\| \begin{matrix} \lambda_2 \\ \lambda_2' \end{matrix} \right\| = \left\| \begin{matrix} a_{11} & 0 \\ 0 & a_{22} \end{matrix} \right\| \cdot \left\| \begin{matrix} \lambda_1 \\ \lambda_1' \end{matrix} \right\| \qquad a_{11} \neq a_{22}$$

Now an obvious induction shows that

$$\left\| \begin{matrix} a_{11} & 0 \\ 0 & a_{22} \end{matrix} \right\|^n = \left\| \begin{matrix} a_{11}{}^n & 0 \\ 0 & a_{22}{}^n \end{matrix} \right\|$$

Hence T will be periodic provided $a_{11}{}^n = a_{22}{}^n$. Moreover, if n is the smallest integer for which $a_{11}{}^n = a_{22}{}^n$, then the period of T is n. In particular, if a_{11} is the nth root of unity of minimum nonzero amplitude, and if $a_{22} = 1$, then the period of T is actually n. Thus it is clear that corresponding to any positive integer n there are projectivities of period n.

Since a projectivity is usually defined in terms of a parametrization whose base points are not the fixed points of the projectivity, it is a matter of some importance to be able to tell from the general equation

$$T: \quad \left\| \begin{matrix} \lambda_2 \\ \lambda_2' \end{matrix} \right\| = \left\| \begin{matrix} a_{11} & a_{12} \\ a_{21} & a_{22} \end{matrix} \right\| \cdot \left\| \begin{matrix} \lambda_1 \\ \lambda_1' \end{matrix} \right\|$$

whether or not T is of period 2, 3, ... Several such results are contained in the following theorems.

Theorem 5

The projectivity $\left\| \begin{matrix} \lambda_2 \\ \lambda_2' \end{matrix} \right\| = \left\| \begin{matrix} a_{11} & a_{12} \\ a_{21} & a_{22} \end{matrix} \right\| \cdot \left\| \begin{matrix} \lambda_1 \\ \lambda_1' \end{matrix} \right\|$ is of period 2 if and only if $a_{11} + a_{22} = 0$.

Proof For a projectivity to be of period 2 it is necessary and sufficient that the square of its matrix, A, be a scalar multiple of the identity matrix.

Hence we have the condition

$$A^2 = \left\| \begin{matrix} a_{11} & a_{12} \\ a_{21} & a_{22} \end{matrix} \right\|^2 = \left\| \begin{matrix} a_{11}{}^2 + a_{12}a_{21} & a_{12}(a_{11} + a_{22}) \\ a_{21}(a_{11} + a_{22}) & a_{12}a_{21} + a_{22}{}^2 \end{matrix} \right\| = \left\| \begin{matrix} k & 0 \\ 0 & k \end{matrix} \right\|$$

which implies that $a_{12}(a_{11} + a_{22}) = 0$ and $a_{21}(a_{11} + a_{22}) = 0$. These will be satisfied if $a_{11} + a_{22} = 0$ or if $a_{12} = a_{21} = 0$. If $a_{11} + a_{22} = 0$, then, clearly,

A^2 becomes $\left\| \begin{matrix} a_{11}{}^2 + a_{12}a_{21} & 0 \\ 0 & a_{11}{}^2 + a_{12}a_{21} \end{matrix} \right\|$, which is proportional to the

identity matrix. Moreover, if $a_{11} + a_{22} = 0$, it is obvious that A itself is not the identity. Hence $a_{11} + a_{22} = 0$ is a sufficient condition that A be of period

2. On the other hand, if $a_{12} = a_{21} = 0$, then A^2 becomes $\left\| \begin{matrix} a_{11}{}^2 & 0 \\ 0 & a_{22}{}^2 \end{matrix} \right\|$, and

this will be proportional to the identity matrix if and only if $a_{22} = \pm a_{11}$. The case $a_{11} = a_{22}$ must be rejected since it implies that A is proportional to I and therefore is of period 1. Hence, if A is to be of period 2, we must have $a_{22} = -a_{11}$. Thus even when $a_{12} = a_{21} = 0$ it follows that $a_{11} + a_{22} = 0$. Therefore this condition is both necessary and sufficient in order that A be of period 2, and our proof is complete.

Corollary 1

The projectivity $a\lambda_1\lambda_2 + b\lambda_1\lambda_2' + c\lambda_1'\lambda_2 + d\lambda_1'\lambda_2' = 0$ is of period 2 if and only if $b = c$.

The condition that a projectivity be of period 3 is given in the next theorem, whose proof we leave as an exercise.

Theorem 6

The projectivity $\left\| \begin{matrix} \lambda_2 \\ \lambda_2' \end{matrix} \right\| = \left\| \begin{matrix} a_{11} & a_{12} \\ a_{21} & a_{22} \end{matrix} \right\| \cdot \left\| \begin{matrix} \lambda_1 \\ \lambda_1' \end{matrix} \right\|$ is of period 3 if and only if $a_{11}{}^2 + a_{11}a_{22} + a_{12}a_{21} + a_{22}{}^2 = 0$.

Corollary 1

The projectivity $a\lambda_1\lambda_2 + b\lambda_1\lambda_2' + c\lambda_1'\lambda_2 + d\lambda_1'\lambda_2' = 0$ is of period 3 if and only if $b^2 - bc - ad + c^2 = 0$.

Projectivities of period 2 are called **involutions,** and are of fundamental importance in projective geometry. Some of their properties are given in the following theorems.

Theorem 7

Any pair of corresponding points in an involution are harmonic conjugates with respect to the fixed points of the involution.

Proof Let F and G be the fixed points of an involution T, and let P_2 be the image of an arbitrary point P_1 under T. Then from the proof of Theorem 1, we know that $\text{R}(FG,P_1P_2) = a_{11}/a_{22}$. But since T is an involution, it follows that $a_{11} = -a_{22}$. Hence $\text{R}(FG,P_1P_2) = -1$, which means that (FG,P_1P_2) is a harmonic tetrad and P_1 and P_2 are harmonic conjugates with respect to F and G, as asserted.

Theorem 8

A necessary and sufficient condition that a projectivity T be an involution is that there should exist two distinct points each of which is the image of the other under T.

Proof Suppose first that the projectivity T: $\Lambda_2 = A\Lambda_1$ is an involution, and let P_2 be the image of the point P_1 under T. If we let P_1 and P_2 denote not only the points themselves but also their parameter-vectors, this means that $P_2 = AP_1$. But if this is the case, then multiplying on the left by A and remembering that $A^2 = I$, we have $AP_2 = A^2P_1 = IP_1 = P_1$, which shows that not only is P_2 the image of P_1 but P_1 is also the image of P_2. Thus if T is an involution, then there exists at least one pair of points each of which is the image of the other under T.

On the other hand, suppose that T is a projectivity such that there are at least two distinct points, P_1 and P_2, each of which is the image of the other under T. Then $P_2 = AP_1$ and $P_1 = AP_2$. Hence, substituting, $P_2 = A(AP_2) = A^2P_2$ and, equally well, $P_1 = A(AP_1) = A^2P_1$. Thus both P_1 and P_2 are fixed points of the projectivity T^2: $\Lambda_3 = A^2\Lambda_1$. Moreover, since T has at least one fixed point, F, distinct from P_1 and P_2, it follows from Theorem 2 that F is also a fixed point of T^2. Thus T^2 has at least three distinct fixed points and is therefore the identity. In other words, $A^2 = I$, which proves that T is an involution, as asserted.

Corollary 1

If P_2 is the image of an arbitrary point P_1 under an involution T, then P_1 is also the image of P_2 under T.

Theorem 9

Any projectivity which is not an involution can be expressed as the composition of two involutions.

Proof Let T: $\Lambda_2 = A\Lambda_1$ be any projectivity which is not an involution, let P be a point which is not a fixed point of T, let $AP = Q$, and let $A^{-1}P = R$. Then $P \neq Q$, for otherwise P would be a fixed point of T, contrary to hypothesis. Similarly, $P \neq R$, for if this were the case, we would have

$A^{-1}P = P$ or $P = AP$, which again implies that P is a fixed point. Finally, $Q \neq R$, for otherwise we would have $A^{-1}P = Q$ and hence $P = AQ$, and this, together with the relation $Q = AP$, would imply that T was an involution, contrary to hypothesis. Thus, since P, Q, and R are all distinct, there is, by Theorem 3, Sec. 4.6, and the last theorem, a unique involution $T_1 \colon \Lambda_2 = A_1\Lambda_1$ having P as a fixed point and Q and R as a pair of mutually corresponding points. Now consider the composition of T and T_1, say $T_2 = T_1T \colon \Lambda_2 = A_2\Lambda_1$, where $A_2 = A_1A$. Under T_2 the image of R is $A_2R = A_1AR = A_1P = P$, and the image of P is $A_2P = A_1AP = A_1Q = R$. Hence P and R are images of each other under T_2, and therefore T_2 is an involution. Now since T_1 is an involution, it follows that $A_1{}^2 = I$. Therefore

$$A = IA = A_1{}^2A = A_1(A_1A) = A_1A_2$$

which shows that T is the composition of the two involutions T_2 and T_1, as asserted.

Other properties of involutions will be found in the exercises at the end of this section.

Exercises

1. Find the parameters of the fixed points of each of the following projectivities:

(a) $\left\| \begin{matrix} \lambda_2 \\ \lambda_2' \end{matrix} \right\| = \left\| \begin{matrix} 2 & 2 \\ 3 & 1 \end{matrix} \right\| \cdot \left\| \begin{matrix} \lambda_1 \\ \lambda_1' \end{matrix} \right\|$
(b) $\left\| \begin{matrix} \lambda_2 \\ \lambda_2' \end{matrix} \right\| = \left\| \begin{matrix} 0 & 3 \\ 2 & 5 \end{matrix} \right\| \cdot \left\| \begin{matrix} \lambda_1 \\ \lambda_1' \end{matrix} \right\|$

(c) $\left\| \begin{matrix} \lambda_2 \\ \lambda_2' \end{matrix} \right\| = \left\| \begin{matrix} 3 & 2 \\ 1 & 4 \end{matrix} \right\| \cdot \left\| \begin{matrix} \lambda_1 \\ \lambda_1' \end{matrix} \right\|$
(d) $\lambda_1\lambda_2 + \lambda_1\lambda_2' + \lambda_1'\lambda_2 - 3\lambda_1'\lambda_2' = 0$

(e) $\lambda_1\lambda_2 + 2\lambda_1\lambda_2' - 3\lambda_1'\lambda_2 - 6\lambda_1'\lambda_2' = 0$

(f) $2\lambda_1\lambda_2 + \lambda_1\lambda_2' - 5\lambda_1'\lambda_2 + 2\lambda_1'\lambda_2' = 0$

2. Express each of the projectivities in Exercise 1 as the composition of two involutions.

3. Give a purely algebraic proof of the fact that every projectivity can be obtained as the composition of two involutions.

4. In the proof of Theorem 2, discuss the possibility that either $a_{11} + a_{22}$ or $a_{11}{}^2 + a_{11}a_{22} + a_{12}a_{21} + a_{22}{}^2$ is zero.

5. Under what conditions, if any, is it possible for four consecutive iterates of a point under a projectivity T to form a harmonic tetrad?

6. Prove that the mapping of the euclidean x axis onto itself defined by the equation $x' = ax$ is a projectivity, and find its fixed points.

7. Show that the projectivity $X' = AX$ is periodic, and find its period if:

(a) $A = \begin{Vmatrix} 0 & 1 \\ 2 & 0 \end{Vmatrix}$

(b) $A = \begin{Vmatrix} 1 & -5 \\ 1 & 3 \end{Vmatrix}$

(c) $A = \begin{Vmatrix} \sqrt{5} & 11 \\ -1 & 2 - \sqrt{5} \end{Vmatrix}$

(d) $A = \begin{Vmatrix} 1 & -1 \\ 3 & 1 \end{Vmatrix}$

(e) $A = \begin{Vmatrix} 1 & 7 \\ -1 & 2 \end{Vmatrix}$

(f) $A = \begin{Vmatrix} 5 & 7 \\ -1 & 1 \end{Vmatrix}$

(g) $A = \begin{Vmatrix} 2 & 3 \\ -1 & -1 \end{Vmatrix}$

(h) $A = \begin{Vmatrix} 2 & 5 \\ -2 & 4 \end{Vmatrix}$

8. If M is a nonsingular (2,2) matrix, and if A is the matrix of a projectivity of period n, show that $M^{-1}AM$ is also the matrix of a projectivity of period n.

9. Prove Theorem 6 and its corollary.

10. If $A = \begin{Vmatrix} a & b \\ c & d \end{Vmatrix}$, determine the condition on a, b, c, d which will ensure that the transformation $X' = AX$ is of period 4, of period 5, of period 6.

11. Given two projectivities $T_1: X' = A_1X$ and $T_2: X' = A_2X$ on the same line. Show that there is always at least one point, P, for which T_1 and T_2 commute; i.e., show that there is always at least one point, P, such that $T_2T_1P = T_1T_2P$.

12. Find the point(s) for which the following projectivities commute:

(a) $X' = \begin{Vmatrix} 2 & 1 \\ 3 & 4 \end{Vmatrix} X, \quad X' = \begin{Vmatrix} 1 & 0 \\ 1 & 2 \end{Vmatrix} X$

(b) $X' = \begin{Vmatrix} -1 & -3 \\ 2 & 4 \end{Vmatrix} X, \quad X' = \begin{Vmatrix} -2 & -2 \\ 4 & 5 \end{Vmatrix} X$

(c) $X' = \begin{Vmatrix} 4 & 6 \\ -1 & -1 \end{Vmatrix} X, \quad X' = \begin{Vmatrix} 0 & -3 \\ -1 & -2 \end{Vmatrix} X$

(d) $X' = \begin{Vmatrix} 1 & 2 \\ 3 & 5 \end{Vmatrix} X, \quad X' = \begin{Vmatrix} -5 & 4 \\ 6 & 3 \end{Vmatrix} X$

(e) $X' = \begin{Vmatrix} 5 & 4 \\ -2 & -1 \end{Vmatrix} X, \quad X' = \begin{Vmatrix} -2 & -5 \\ 3 & 6 \end{Vmatrix} X$

(f) $X' = \begin{Vmatrix} -1 & -3 \\ 2 & 4 \end{Vmatrix} X, \quad X' = \begin{Vmatrix} -1 & -3 \\ 4 & 7 \end{Vmatrix} X$

13. If (A,A') and (B,B') are two pairs of mates in a projectivity having distinct fixed points F_1 and F_2, show that there is also a projectivity having F_1 and F_2, as fixed points in which (A,B) and (A',B') are pairs of mates.

14. Prove that two involutions on the same line always have one and only one pair of mates in common.

15. Prove that two involutions

$$X' = \begin{Vmatrix} a_1 & b_1 \\ c_1 & -a_1 \end{Vmatrix} X \qquad \text{and} \qquad X' = \begin{Vmatrix} a_2 & b_2 \\ c_2 & -a_2 \end{Vmatrix} X$$

are commutative if and only if $2a_1a_2 + b_1c_2 + b_2c_1 = 0$. Prove, further, that if two involutions commute, the fixed points of either are mates in the other.

16. Given $l: x_1 - x_2 = 0$ and the family of conics $x_3{}^2 = ax_1x_2$. Let P_1 be an arbitrary point on l, let Γ be the conic of the given family which contains P_1, and let P_2 be the second intersection of Γ and l. Show that the mapping of l onto itself in which P_2 is the image of P_1 is an involution. What is the geometrical significance of the fixed points of this involution? (This exercise and the one which follows illustrate the important result known as **Desargues' conic theorem,** which is discussed in Sec. 8.4.)

17. (a) Work Exercise 16 given $l: x_1 + x_2 + x_3 = 0$ and the family of conics $x_1{}^2 = ax_2x_3$.
(b) Work Exercise 16 given $l: x_1 + x_3 = 0$ and the family of conics $x_1{}^2 + x_2{}^2 - ax_3{}^2 = 0$.

18. Given $l: x_1 - x_2 = 0$ and the family of cubic curves $x_1{}^3 = ax_1x_3{}^2 - x_2{}^3$. Let P_1 be an arbitrary point on l, let C be the cubic of the given family which contains P_1, and let P_2 be the intersection of C and l which is distinct from P_1 and from the point $(0,0,1)$. Show that the mapping of l onto itself in which P_2 is the image of P_1 is an involution.

19. Given $l: x_1 - x_3 = 0$, $\Gamma: x_1x_2 - x_3{}^2 = 0$, $U:(1,0,0)$, and $V:(v_1,v_2,v_3)$. Let P_1 be an arbitrary point on l, let W_1 be the second intersection of Γ and UP_1, let W_2 be the second intersection of Γ and VW_1, and let P_2 be the intersection of UW_2 and l. Show that the mapping of l onto itself in which P_2 is the image of P_1 is an involution.

20. Given $l_1: x_1 = 0$, $l_2: x_2 = 0$, $l_3: x_3 = 0$, $U:(1,1,1)$, $V:(1,-1,1)$, and $W:(w_1,w_2,w_3)$. Determine the locus of points W such that $l_1 \overset{U}{\wedge} l_2 \overset{V}{\wedge} l_3 \overset{W}{\wedge} l_1$ is an involution.

21. Given $l_1: x_1 = 0$, $l_2: x_2 = 0$, $l_3: x_3 = 0$, $U:(1,-1,1)$, $V:(1,2,1)$, and $W:(1,1,1)$. Determine the points on l_1 which coincide with their images under the chain of perspectivities $l_1 \overset{U}{\wedge} l_2 \overset{V}{\wedge} l_3 \overset{W}{\wedge} l_1$.

22. Given $l_1: x_1 = 0$, $l_2: x_2 = 0$, $l_3: x_3 = 0$, $U:(1,-1,1)$, $V:(1,1,2)$, $W:(w_1,w_2,w_3)$. Determine the locus of W if under the chain of perspectivities

$l_1 \overset{U}{\wedge} l_2 \overset{V}{\wedge} l_3 \overset{W}{\wedge} l_1$ there is exactly one point on l_1 which coincides with its image.

23. (a) Work Exercise 22 if U is the point $(1,1,1)$ and V is the point $(1,-1,1)$.
(b) Work Exercise 22 given $l_1: x_2 = 0$, $l_2: x_1 + x_3 = 0$, $l_3: x_2 + x_3 = 0$.

24. Work Exercises 22 and 23 if it is required to determine the locus of W so that the mapping of l_1 onto itself is an involution.

25. Given $l: x_1 - x_2 = 0$, $U:(1,0,0)$, $V:(0,1,0)$, and $l': a_1 x_1 + a_2 x_2 + a_3 x_3 = 0$. Let P_1 be an arbitrary point on l, let W be the intersection of $P_1 U$ and l', and let P_2 be the intersection of VW and l. Show that the mapping of l onto itself in which P_2 is the image of P_1 is an involution if and only if l' is a line of the pencil whose vertex is the harmonic conjugate of the intersection of l and UV with respect to U and V.

4.9 Conics

By a conic in E_2^+ or Π_2 we mean the locus of points whose coordinates satisfy a homogeneous quadratic equation

$$a_{11}x_1{}^2 + 2a_{12}x_1x_2 + 2a_{13}x_1x_3 + a_{22}x_2{}^2 + 2a_{23}x_2x_3 + a_{33}x_3{}^2 = 0$$

As we observed in Sec. 3.7, this equation can be written in the matric form

$$X^T A X = 0 \qquad \text{where} \qquad X = \begin{Vmatrix} x_1 \\ x_2 \\ x_3 \end{Vmatrix}$$

and
$$A = \begin{Vmatrix} a_{11} & a_{12} & a_{13} \\ a_{21} & a_{22} & a_{23} \\ a_{31} & a_{32} & a_{33} \end{Vmatrix} \qquad a_{ij} = a_{ji}$$

One of the fundamental problems involving conics is to determine the points in which a given line intersects a given conic. Suppose, specifically, that we have a line l determined by the points

$$P = \begin{Vmatrix} p_1 \\ p_2 \\ p_3 \end{Vmatrix} \qquad \text{and} \qquad Q = \begin{Vmatrix} q_1 \\ q_2 \\ q_3 \end{Vmatrix}$$

Taking P and Q as base points on the line l, the coordinate vector of any point X on l can be written

(1)
$$X = \lambda P + \mu Q$$

Hence the intersections of the line $l: PQ$ and the conic $\Gamma: X^T A X = 0$ are determined by the values of the parameters (λ, μ) which satisfy the equation

$(\lambda P + \mu Q)^T A(\lambda P + \mu Q) = 0$ obtained by solving simultaneously the equation of l and the equation of Γ. As we observed in Sec. 3.7, this equation can be simplified to the form

$$(2) \qquad \lambda^2 P^T A P + 2\lambda\mu P^T A Q + \mu^2 Q^T A Q = 0$$

Moreover, since A is a known matrix and since P and Q are the coordinate-vectors of known points, this is a quadratic equation in the ratio λ/μ with known coefficients, whose solution presents no difficulty. With the solutions (λ_1, μ_1) and (λ_2, μ_2) determined, the coordinates of the intersections of l and Γ can be found at once from Eq. (1).

Example 1

Find the coordinates of the points in which the line joining the points $P = \left\|\begin{matrix} 1 \\ -1 \\ 2 \end{matrix}\right\|$ and $Q = \left\|\begin{matrix} 1 \\ 5 \\ 2 \end{matrix}\right\|$ meets the conic

$$x_1{}^2 - 2x_1x_2 - 4x_1x_3 + x_2{}^2 + 2x_2x_3 + x_3{}^2 = 0$$

When the given conic is written in the matric form $X^T A X = 0$, the matrix A is $\left\|\begin{matrix} 1 & -1 & -2 \\ -1 & 1 & 1 \\ -2 & 1 & 1 \end{matrix}\right\|$. Hence, evaluating, we have $P^T A P = -4$, $P^T A Q = -4$, $Q^T A Q = 32$. Therefore the equation determining the parameters of the intersections is

$$-4\lambda^2 - 8\lambda\mu + 32\mu^2 = 0 \qquad \text{or} \qquad -4(\lambda + 4\mu)(\lambda - 2\mu) = 0$$

The parameters of the intersections are thus $(4, -1)$ and $(2, 1)$, and the coordinate-vectors of the intersections are

$$4\left\|\begin{matrix} 1 \\ -1 \\ 2 \end{matrix}\right\| - \left\|\begin{matrix} 1 \\ 5 \\ 2 \end{matrix}\right\| = \left\|\begin{matrix} 3 \\ -9 \\ 6 \end{matrix}\right\| \sim \left\|\begin{matrix} 1 \\ -3 \\ 2 \end{matrix}\right\|$$

and

$$2\left\|\begin{matrix} 1 \\ -1 \\ 2 \end{matrix}\right\| + \left\|\begin{matrix} 1 \\ 5 \\ 2 \end{matrix}\right\| = \left\|\begin{matrix} 3 \\ 3 \\ 6 \end{matrix}\right\| \sim \left\|\begin{matrix} 1 \\ 1 \\ 2 \end{matrix}\right\|$$

If in the preceding discussion, P is a fixed point but Q is considered variable, say $Q = X$, then the equation

$$P^T A Q \equiv P^T A X = 0$$

defines a line, known as the **polar** of the point P with respect to the conic $\Gamma: X^T A X = 0$. The polar line is of fundamental importance in the study of conics, and the following theorems describe some of its more significant properties.

In the first place, it is important to know if a point always has a polar line with respect to an arbitrary conic or if it is possible for the polar of a point with respect to a given conic to be indeterminate. Clearly, the polar of a point P with respect to a conic $\Gamma: X^T A X = 0$ will be indeterminate if and only if the expression $P^T A X$ is identically zero; i.e., if and only if the coefficients of x_1, x_2, and x_3 in the equation of the polar of P are all zero. Now if P is the point (p_1, p_2, p_3), the equation of the polar of P when written at length in scalar form becomes

$$(a_{11}p_1 + a_{21}p_2 + a_{31}p_3)x_1 + (a_{12}p_1 + a_{22}p_2 + a_{32}p_3)x_2$$
$$+ (a_{13}p_1 + a_{23}p_2 + a_{33}p_3)x_3 = 0$$

Hence, if there is a point P whose polar with respect to Γ is indeterminate, its coordinates (p_1, p_2, p_3) must satisfy the equations

$$a_{11}p_1 + a_{21}p_2 + a_{31}p_3 = 0$$
$$a_{12}p_1 + a_{22}p_2 + a_{32}p_3 = 0$$
$$a_{13}p_1 + a_{23}p_2 + a_{33}p_3 = 0$$

or, in matric form, $A^T P = O$. This system of homogeneous linear equations will have a nontrivial solution if and only if the determinant of the coefficients, $|A^T| = |A|$, is equal to zero. Hence we have the following theorem.

Theorem 1

If $\Gamma: X^T A X = 0$ is a conic for which $|A| \neq 0$, then without exception every point has a well-defined polar with respect to Γ. If $|A| = 0$, then there is always at least one point whose polar with respect to Γ is indeterminate.

Corollary 1

It is possible for two distinct points to have the same polar with respect to a conic $\Gamma: X^T A X = 0$ if and only if $|A| = 0$.

Proof Let $P^T A X = 0$ and $Q^T A X = 0$ be the polars of two distinct points, P and Q, with respect to a conic $\Gamma: X^T A X = 0$. Clearly, the polar lines of P and Q will be the same if and only if the equations of these lines are proportional; i.e., if and only if there is a constant k such that

$$P^T A X - k Q^T A X = (P^T - k Q^T)A X = (P - k Q)^T A X$$

vanishes identically. But from the proof of Theorem 1, it is clear that this can happen if and only if $|A| = 0$, as asserted.

Definition 1

Any point whose polar line with respect to a conic Γ is indeterminate is said to be a vertex of Γ.

Theorem 2

If a conic has a vertex, V, then V lies on the conic.

Proof   If the conic $\Gamma: X^T A X = 0$ has a vertex, V, then V satisfies the equation $A^T V = O$. From this, by taking the transpose, we conclude that $V^T A = O$. Finally, by multiplying this equation on the right by V we obtain $V^T A V = 0$, which proves that V is a point of Γ, as asserted.

Theorem 3

If a conic $\Gamma: X^T A X = 0$ has a vertex, V, then the polar of every point with respect to Γ passes through V.

Proof   Suppose that the conic $\Gamma: X^T A X = 0$ has a vertex, so that $A^T V = AV = O$. Now the polar of an arbitrary point, P, with respect to Γ is $P^T A X = 0$, and V will lie on this line if and only if $P^T A V = 0$. But this is true, since, by hypothesis, $AV = O$.

Theorem 4

If a conic has a vertex, then every point on the line joining any point of the conic to the vertex lies on the conic.

Proof   Let V be a vertex of the conic $\Gamma: X^T A X = 0$, and let Q be any other point on the conic. Then, as we saw above, the parameters of the intersections of the line VQ and Γ are determined by the equation

$$(3) \qquad \lambda^2 V^T A V + 2\lambda\mu V^T A Q + \mu^2 Q^T A Q = 0$$

However, $V^T A V$ and $Q^T A Q$ are both zero, since both V and Q lie on Γ. Moreover, $V^T A Q = 0$, since, by hypothesis, V is a vertex of Γ and therefore has a polar with respect to Γ which vanishes identically. Thus every term in Eq. (3) is zero, which means that this equation is satisfied by all values of λ and μ. Finally, this implies that every point on VQ is a point of the conic, Γ, since the coordinates of every such point satisfy the equation of Γ.

From Theorem 4, it is clear that if a conic has a vertex, the conic consists of one or more lines passing through the vertex. Because of their exceptional nature, such conics are said to be **singular:**

Definition 2

A conic which contains all the points of one or more lines is said to be singular.

Definition 3

A conic is said to be nonsingular if there is no line all of whose points belong to the conic.

Definition 4

The determinant of the coefficients, $|A|$, in the equation of a conic $\Gamma\colon X^TAX = 0$ is called the discriminant of the conic.

Theorem 5

A conic is singular if and only if its discriminant is equal to zero.

Proof Let us suppose first that $\Gamma\colon X^TAX = 0$ is a conic whose discriminant is zero. Then from Theorem 1 it is clear that Γ has at least one vertex, and from Theorem 4 it follows that Γ is singular. Thus if the discriminant of a conic is zero, the conic is singular. Conversely, let Γ be a conic which is singular, and let P and Q be distinct points on one of the lines which Γ therefore contains. Then the equation $\lambda^2 P^TAP + 2\lambda\mu P^TAQ + \mu^2 Q^TAQ = 0$ must be satisfied by all values of λ and μ, which implies that $P^TAP = P^TAQ = Q^TAQ = 0$. Now the polar of P with respect to Γ is the line $P^TAX = 0$, and this contains not only the point P, since $P^TAP = 0$, but also the point Q, since $P^TAQ = 0$. Thus the polar of P is the line PQ. Similarly, of course, the polar of Q with respect to Γ is also the line PQ. Therefore the distinct points P and Q have the same polar with respect to Γ. However, by Corollary 1, Theorem 1, this is possible only if $|A| = 0$. Hence, if Γ is singular, its discriminant is zero, and our proof is complete.

Theorem 6

If V_1 and V_2 are distinct vertices of a conic Γ, then any point on the line V_1V_2 is also a vertex of Γ.

Proof Let V_1 and V_2 be distinct vertices of a conic $\Gamma\colon X^TAX = 0$, and consider the polar of any point $P\colon \lambda V_1 + \mu V_2$ on the line V_1V_2. Its equation, of course, is

$$P^TAX = (\lambda V_1 + \mu V_2)^TAX = 0 \qquad \text{or} \qquad \lambda V_1{}^TAX + \mu V_2{}^TAX = 0$$

and this vanishes identically, since $V_1{}^TAX$ and $V_2{}^TAX$ are identically zero because V_1 and V_2 are vertices of Γ. Hence the polar of every point of V_1V_2 is indeterminate, which proves that every point of V_1V_2 is a vertex of Γ, as asserted.

The condition that a conic Γ: $X^T A X = 0$ have two vertices, and hence a line of vertices, is contained in the following theorem, whose proof follows immediately from Theorem 3, Sec. 3.6.

Theorem 7

A conic Γ: $X^T A X = 0$ has a line of vertices if and only if the rank of A is equal to 1.

Corollary 1

A conic Γ: $X^T A X = 0$ is nonsingular if A is nonsingular. It consists of two distinct lines if A is of rank 2 and consists of one line if A is of rank 1. If A is of rank 0, the equation of Γ vanishes identically.

Example 2

Show that the conic Γ: $8x_1{}^2 + 2x_1x_2 - 14x_1x_3 - 3x_2{}^2 + 2x_2x_3 + 5x_3{}^2 = 0$ is singular, and find its vertex or vertices and the lines of which it is composed.

The matrix of Γ is $A = \begin{Vmatrix} 8 & 1 & -7 \\ 1 & -3 & 1 \\ -7 & 1 & 5 \end{Vmatrix}$ and it is easy to check that $|A| = 0$. Hence, by Theorem 5, Γ is singular. Moreover, since the rank of A is clearly 2, it follows from Theorem 1 and its corollary that Γ has a single vertex and therefore consists of two distinct lines. Finally, as we observed in the proof of Theorem 1, the coordinates of the vertex of Γ satisfy the (dependent) equations

$$\begin{aligned} 8x_1 + x_2 - 7x_3 &= 0 \\ x_1 - 3x_2 + x_3 &= 0 \\ -7x_1 + x_2 + 5x_3 &= 0 \end{aligned}$$

Using the first two of these, we find easily that the coordinates of the vertex, V, are $(4,3,5)$.

The fact that Γ is singular assures us that it must contain at least one linear component, and hence that its equation must be factorable. It would not be difficult in the present problem to factor the equation of Γ by inspection, and thus find the equations of the lines which comprise Γ. However, it is more straightforward to proceed as follows. By Theorem 4, since Γ is singular, the line joining the vertex, V, to any point of Γ which is distinct from V is a component of Γ. In particular, if we solve the equation of Γ simultaneously with the equation of any line which does not pass through the vertex, V, we shall obtain one or more points of Γ which are distinct from V and which will therefore determine with V a component of Γ. Choosing the line $x_3 = 0$, for convenience, we have thus $8x_1{}^2 + 2x_1x_2 - 3x_2{}^2 = 0$ or

$(4x_1 + 3x_2)(2x_1 - x_2) = 0$. Hence the points $P_1:(-3,4,0)$ and $P_2:(1,2,0)$ are two points on Γ and distinct from V. The lines VP_1 and VP_2, whose equations are easily found to be $VP_1: 4x_1 + 3x_2 - 5x_3 = 0$ and $VP_2: 2x_1 - x_2 - x_3 = 0$, are then the components of Γ.

The next two theorems make clear the geometric relation between a nonsingular conic and the polar line of a general point with respect to that conic.

Theorem 8

If $\Gamma: X^TAX = 0$ is a nonsingular conic, and if P is a point which does not lie on Γ, then the intersections of Γ and the polar of P with respect to Γ are the points of contact of the tangents to Γ from P.

Proof Let Q be an intersection of the conic $\Gamma: X^TAX = 0$ and the polar line of a point P with respect to Γ. Then since Q lies on both Γ and the polar of P with respect to Γ, it follows that $Q^TAQ = 0$ and $P^TAQ = 0$. Hence the equation which defines the parameters of the intersections of Γ and the line PQ [Eq. (2)] reduces to $\lambda^2P^TAP = 0$. Moreover, since our hypothesis is that P is not a point of Γ, it follows that $P^TAP \neq 0$. Hence $\lambda^2 = 0$, and the equation has only one solution, namely, $(0,1)$. Thus the line PQ intersects Γ in a single point; and since Γ is nonsingular, PQ is therefore a tangent to Γ. Furthermore, since the parameters of the intersection of Γ and PQ are $(0,1)$, the intersection, i.e., the point of tangency, is Q itself, as asserted.

Theorem 9

If P is a point on a nonsingular conic $\Gamma: X^TAX = 0$, its polar with respect to Γ is the tangent to Γ at P.

Proof Let P be a point on a nonsingular conic $\Gamma: X^TAX = 0$, and let Q be an arbitrary point on the polar of P with respect to Γ. Then $P^TAP = 0$ and $P^TAQ = 0$. Hence the equation whose roots determine the parameters of the intersections of Γ and the polar of P becomes simply $\mu^2Q^TAQ = 0$. Since this equation has the single root $\mu = 0$, the line PQ intersects Γ in a single point, i.e., is tangent to Γ. Furthermore, since the parameters of the intersection are $(1,0)$, the point of tangency is P, as asserted.

Corollary 1

The tangent at a general point of a nonsingular conic is unique.

Figure 4.13 shows the relations between a nonsingular conic and the polar of an arbitrary point with respect to that conic.

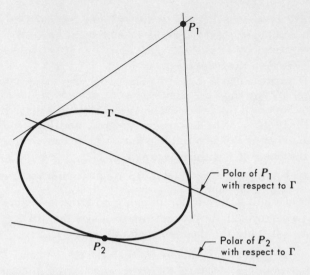

Fig. 4.13   Polars of points with respect to a nonsingular conic.

Theorem 10

If Q lies on the polar of P with respect to a conic Γ, then P lies on the polar of Q with respect to Γ.

Proof   If Γ is the conic $X^T A X = 0$, the polars of P and Q with respect to Γ are, respectively, $P^T A X = 0$ and $Q^T A X = 0$. By hypothesis, Q lies on the polar of P; hence $P^T A Q = 0$. Moreover, since $P^T A Q$ is a scalar, i.e., a $(1,1)$ matrix, it is equal to its transpose; and therefore, since A is symmetric, $P^T A Q = 0 \Rightarrow (P^T A Q)^T = 0 \Rightarrow Q^T A P = 0$. But the condition $Q^T A P = 0$ is precisely the condition that P lies on the polar of Q with respect to Γ, as asserted.

Theorem 11

Let l be any line through a point P_1 which is not on a conic Γ, and let P_2 be the intersection of l and the polar of P_1 with respect to Γ. Then the intersections of l and Γ are harmonic conjugates with respect to P_1 and P_2.

Proof   Let $\Gamma\colon X^T A X = 0$ be an arbitrary conic, let P_1 be any point which is not on Γ, let l be any line through P_1, and let P_2 be the intersection of l and the polar of P_1 with respect to Γ. Then $P_1{}^T A P_2 = 0$, and hence if l is parametrized in terms of P_1 and P_2, the equation defining the parameters of the intersections of l and Γ becomes $\lambda^2 P_1{}^T A P_1 + \mu^2 P_2{}^T A P_2 = 0$. Therefore the parameters of the intersections are $(\sqrt{P_2{}^T A P_2},\ \sqrt{-P_1{}^T A P_1})$ and $(\sqrt{P_2{}^T A P_2},$

$-\sqrt{-P_1{}^TAP_1}$). Hence, by Lemma 2, Sec. 4.7, the intersections of l and Γ are harmonic conjugates with respect to P_1 and P_2, as asserted.

Theorem 12

If P varies along a line, its polar with respect to a given conic passes through a fixed point.

Proof Let $\Gamma: X^TAX = 0$ be an arbitrary conic, and let $l = P_1P_2$ be an arbitrary line. Then if P is a variable point on l, we can write $P = \lambda P_1 + \mu P_2$. Hence the polar of P with respect to Γ is

$$P^TAX = (\lambda P_1 + \mu P_2)^TAX = \lambda P_1{}^TAX + \mu P_2{}^TAX = 0$$

For all values of λ and μ, that is, for all points on l, the last equation defines a line which passes through the point of intersection of the two lines $P_1{}^TAX = 0$ and $P_2{}^TAX = 0$. Thus the theorem is established.

Example 3

Let C be the point (c_1, c_2, c_3), let Γ be the conic $x_1{}^2 + x_2{}^2 - x_3{}^2 = 0$, let $Y:(y_1, y_2, y_3)$ be a general point, and let P and Q be the intersections of Γ and the line CY. Find the harmonic conjugate of Y with respect to P and Q.

By Theorem 11, the harmonic conjugate of Y with respect to P and Q is the intersection of CY and the polar of Y with respect to Γ. Now the equation of CY is

$$\begin{vmatrix} x_1 & x_2 & x_3 \\ c_1 & c_2 & c_3 \\ y_1 & y_2 & y_3 \end{vmatrix} = 0 \quad \text{or} \quad (c_2 y_3 - c_3 y_2)x_1 - (c_1 y_3 - c_3 y_1)x_2 \\ + (c_1 y_2 - c_2 y_1)x_3 = 0$$

and the polar of Y with respect to Γ is

$$\|y_1 \ \ y_2 \ \ y_3\| \cdot \begin{Vmatrix} 1 & 0 & 0 \\ 0 & 1 & 0 \\ 0 & 0 & -1 \end{Vmatrix} \cdot \begin{Vmatrix} x_1 \\ x_2 \\ x_3 \end{Vmatrix} = 0 \quad \text{or} \quad y_1 x_1 + y_2 x_2 - y_3 x_3 = 0$$

Hence the coordinates of the required harmonic conjugate, $Y':(y_1', y_2', y_3')$ can be read from the matrix

$$\begin{Vmatrix} y_1 & y_2 & -y_3 \\ c_2 y_3 - c_3 y_2 & c_3 y_1 - c_1 y_3 & c_1 y_2 - c_2 y_1 \end{Vmatrix}$$

and are

$$\begin{aligned} y_1' &= -c_2 y_1 y_2 + c_3 y_1 y_3 + c_1 y_2{}^2 - c_1 y_3{}^2 \\ y_2' &= c_2 y_1{}^2 - c_1 y_1 y_2 + c_3 y_2 y_3 - c_2 y_3{}^2 \\ y_3' &= c_3 y_1{}^2 - c_1 y_1 y_3 + c_3 y_2{}^2 - c_2 y_2 y_3 \end{aligned}$$

The transformation T defined by the last set of equations in Example 3 is an example of a class of nonlinear transformations known as **quadratic inversions:**

Definition 5

Let C be an arbitrary fixed point, let Γ be an arbitrary nonsingular conic, let Y be a general point, and let P and Q be the intersections of Γ and the line CY. Then the transformation which maps Y into its harmonic conjugate with respect to P and Q is known as a quadratic inversion.

The transformation T has an interesting interpretation in the euclidean plane when $c_1 = c_2 = 0$, and the point C becomes the origin and Γ becomes the unit circle, $x^2 + y^2 = 1$. In this case, after dividing by the nonzero constant c_3, the equations of the transformation become

$$y_1' = y_1 y_3$$
$$y_2' = y_2 y_3$$
$$y_3' = y_1{}^2 + y_2{}^2$$

Now, reverting to nonhomogeneous coordinates by means of the relations $y_1/y_3 = x$, $y_2/y_3 = y$ and $y_1'/y_3' = x'$, $y_2'/y_3' = y'$, we obtain

$$x' = \frac{x}{x^2 + y^2} \qquad y' = \frac{y}{x^2 + y^2}$$

These are the equations of the important euclidean transformation known as **inversion in the unit circle.** Some of the properties of this transformation will be found among the exercises at the end of this section.

Up to this point in our discussion of conics, we have considered them exclusively as loci, i.e., as sets of points whose coordinates satisfy an equation of the form $X^T A X = 0$. However, the principle of duality suggests that it should be possible to consider a conic not only as the set of points which lie on it but also as the set of lines which are tangent to it. It is not convenient to explore this question in detail in this chapter, but we shall conclude this section with a theorem giving us the condition satisfied by any line which is tangent to a given conic.

Theorem 13

A line Λ: $\begin{Vmatrix} l_1 \\ l_2 \\ l_3 \end{Vmatrix}$ is tangent to a nonsingular conic Γ: $X^T A X = 0$

if and only if $\Lambda^T A^{-1} \Lambda = 0$.

Proof   Let P be an arbitrary point on a nonsingular conic Γ: $X^T A X = 0$. Then, by Theorem 9, $P^T A X = 0$ is the equation of the tangent to Γ at P.

If we compare this with the general equation of a line, namely, $l_1x_1 + l_2x_2 + l_3x_3 = 0$ or $\Lambda^T X = 0$, it follows that the coordinate vector, Λ, of the tangent line is defined by the relation $\Lambda^T = P^T A$. This in turn implies that

$$(P^T A)A^{-1} = \Lambda^T A^{-1} \qquad \text{or} \qquad P^T = \Lambda^T A^{-1}$$

Finally, since P necessarily lies on the tangent to Γ at P, it follows that $P^T \Lambda = (\Lambda^T A^{-1})\Lambda = 0$, which proves the necessity of the condition of the theorem.

Conversely, let l be any line whose coordinate vector Λ satisfies the equation $\Lambda^T A^{-1}\Lambda = 0$, and consider the point P defined by the relation $P = A^{-1}\Lambda$. Since $\Lambda^T A^{-1}\Lambda = 0$ can be written in the form

$$(\Lambda^T A^{-1})A(A^{-1}\Lambda) = (A^{-1}\Lambda)^T A(A^{-1}\Lambda) = 0$$

it follows that the coordinates of the point $P = A^{-1}\Lambda$ satisfy the equation $X^T A X = 0$. Hence P is a point of Γ. Moreover, the tangent to Γ at P is the line $P^T A X = 0$, or $(\Lambda^T A^{-1})AX = 0$, or finally $\Lambda^T X = 0$. Hence Λ is the coordinate vector of the tangent to Γ at the point $P = A^{-1}\Lambda$, and the sufficiency of the condition of the theorem is established.

Exercises

1. Find the intersections of the conic Γ and the line PQ if:

(a) Γ: $x_1^2 - 2x_1x_2 + 4x_1x_3 + 3x_2^2 - x_2x_3 - 2x_3^2 = 0$, P:(0,0,1), Q:(1,1,1)

(b) Γ: $3x_1^2 + 10x_1x_2 - 3x_1x_3 - 8x_2^2 - 18x_3^2 = 0$, P:(2,$-$3,1), Q:(4,3,$-$4)

(c) Γ: $x_1^2 + 4x_1x_2 + 2x_1x_3 - 6x_2x_3 - 4x_3^2 = 0$, P:(2,2,$-$1), Q:(1,1,$-$2)

(d) Γ: $x_1^2 + 4x_1x_2 - 2x_1x_3 + x_2^2 - 2x_2x_3 = 0$, P:(1,0,1), Q:(3,$-$1,2)

2. Given P_1:(1,2,$-$3), P_2:(2,0,$-$1), P_3:(2,$-$2,3), Γ_1: $3x_1^2 + 2x_2x_3 = 0$, Γ_2: $2x_1x_2 - 3x_2^2 - 6x_2x_3 + 4x_3^2 = 0$, Γ_3: $x_1^2 - 2x_1x_3 + 3x_2^2 + 4x_2x_3 - x_3^2 = 0$. Find the polar of each of these points with respect to each of these conics.

3. (a) What is the equation of the most general conic which contains the points $(1,0,0)$, $(0,1,0)$, $(0,0,1)$?

(b) What is the equation of the most general conic which is tangent to $x_2 = 0$ at the point $(0,0,1)$ and tangent to $x_3 = 0$ at the point $(0,1,0)$?

(c) Given P_1:(1,0,0), P_2:(0,1,0), P_3:(0,0,1). What is the equation of the most general conic with respect to which the polar of the point P_i is the line P_jP_k?

4. (a) A triangle is said to be **self-polar** with respect to a conic Γ if each side of the triangle is the polar of the opposite vertex with respect to Γ. Given P_1:(1,$-$1,0), P_2:(0,1,0), P_3:(1,1,2); find the equation of the most general conic with respect to which $\triangle P_1P_2P_3$ is self-polar.

(b) Work part (a), given P_1:(1,1,1), P_2:($-$1,2,1), P_3:(0,1,1).

5. (a) Given $\Gamma: x_1{}^2 - 2x_1x_2 + 4x_1x_3 - x_3{}^2 = 0$ and $l: 2x_1 - x_2 + 4x_3 = 0$. Find the point whose polar with respect to Γ is l. (The point whose polar with respect to a given conic Γ is a given line l is called the **pole** of l with respect to Γ.)

(b) Work part (a), given $\Gamma: 2x_1{}^2 - x_2{}^2 + x_2x_3 + x_3{}^2 = 0$ and $l: x_1 - 2x_2 + 5x_3 = 0$.

(c) Work part (a), given $\Gamma: x_1{}^2 + 2x_1x_2 + 2x_2{}^2 - 2x_2x_3 + 5x_3{}^2 = 0$ and $l: 3x_1 + 2x_2 + 13x_3 = 0$.

6. Show that each of the following conics is singular, find its vertex or vertices, and determine the lines of which it is composed:

(a) $2x_1{}^2 - 9x_1x_2 + 10x_1x_3 + 9x_2{}^2 - 24x_2x_3 + 12x_3{}^2 = 0$
(b) $4x_1{}^2 - 12x_1x_2 + 16x_1x_3 + 9x_2{}^2 - 24x_2x_3 + 16x_3{}^2 = 0$
(c) $6x_1{}^2 - x_1x_2 + 6x_1x_3 - 12x_2{}^2 + 25x_2x_3 - 12x_3{}^2 = 0$
(d) $2x_1{}^2 - 2x_1x_2 + 2x_1x_3 + x_2{}^2 + x_3{}^2 = 0$

7. Determine the points of intersection of the following pairs of conics:

(a) $x_1{}^2 - x_2x_3 = 0$, $3x_1{}^2 - x_1x_2 + 2x_1x_3 - 4x_2x_3 = 0$
(b) $x_1{}^2 - x_2x_3 = 0$, $x_1{}^2 + x_1x_3 - 2x_2x_3 = 0$

8. Determine what lines, if any, are tangent to each of the conics in the following pairs:

(a) $-x_1{}^2 + 4x_2x_3 = 0$, $4x_1{}^2 + 4x_1x_2 + x_2{}^2 - 8x_2x_3 = 0$
(b) $x_1{}^2 + 4x_2x_3 + 4x_3{}^2 = 0$, $x_1{}^2 + x_2{}^2 - x_3{}^2 = 0$

9. (a) Given $A_1:(1,0,0)$, $A_2:(0,1,0)$, $A_3:(0,0,1)$, $A_4:(1,1,1)$. What is the locus of a point P with the property that the cross ratio of the lines PA_1, PA_2, PA_3, PA_4 is a constant?

(b) Work part (a), given $A_1:(1,-1,0)$, $A_2:(0,1,1)$, $A_3:(1,0,0)$, $A_4:(1,1,-1)$.

10. (a) Let the pencil of lines on the point $(1,0,0)$ be parametrized in terms of the base lines $x_2 = 0$ and $x_3 = 0$ with parameters (λ_1,μ_1), let the pencil of lines on $(0,1,0)$ be parametrized in terms of the base lines $x_1 = 0$ and $x_3 = 0$ with parameters (λ_2,μ_2), and let the pencils be related by the projectivity $\left\|\begin{matrix}\lambda_2\\\mu_2\end{matrix}\right\| = \left\|\begin{matrix}1 & 0\\2 & 1\end{matrix}\right\| \cdot \left\|\begin{matrix}\lambda_1\\\mu_1\end{matrix}\right\|$. What is the locus of the intersections of corresponding lines of the two pencils?

(b) Work part (a), given the pencil on $(1,-1,0)$ parametrized in terms of the lines $x_3 = 0$ and $x_1 + x_2 = 0$ and the pencil on $(1,0,1)$ parametrized in terms of the lines $x_2 = 0$ and $x_1 - x_3 = 0$.

11. Show that there is a unique triangle which is self-polar with respect to the two conics in each of the following pairs, and find the triangle:

(a) $x_1x_2 + 2x_1x_3 - 2x_2x_3 - 2x_3{}^2 = 0$, $x_1x_2 - x_1x_3 + x_2x_3 + x_3{}^2 = 0$
(b) $2x_1{}^2 + 2x_1x_2 + x_1x_3 - 5x_3{}^2 = 0$, $3x_1{}^2 + 3x_1x_2 + x_2x_3 - 7x_3{}^2 = 0$
(c) $x_1{}^2 + x_1x_2 - 7x_1x_3 + 5x_2x_3 = 0$, $x_1{}^2 + x_1x_2 - 4x_1x_3 + 3x_2x_3 - x_3{}^2 = 0$

12. Given $l_1: x_1 = 0$, $l_2: x_2 = 0$, and $\Gamma: x_3^2 = 2x_1x_2$. What condition must be satisfied by the coefficients in the equation of a general line

$$l: a_1x_1 + a_2x_2 + a_3x_3 = 0$$

in order for the intersections of l and Γ to be harmonic conjugates with respect to the intersections of l and l_1 and l_2? What is the envelope of the lines l which meet this condition?

13. Work Exercise 12, given $l_1: x_1 + x_2 = 0$, $l_2: x_1 - x_3 = 0$, and $\Gamma: x_1^2 - 2x_1x_2 + x_3^2 = 0$.

14. Given $l_1: x_1 = 0$, $l_2: x_2 = 0$, $l_3: x_1 - x_3 = 0$, $l_4: x_2 - x_3 = 0$, and $\Gamma: x_1^2 + x_2^2 - x_3^2 = 0$. Determine the points, if any, on Γ at which the tangent has the property that its intersections with l_1 and l_2 are harmonic conjugates with respect to its intersections with l_3 and l_4.

15. Given $\Gamma_1: x_1^2 + 2x_2x_3 = 0$, $\Gamma_2: x_2^2 + 2x_1x_3 = 0$, and $\Gamma_3: x_3^2 + 2x_1x_2 = 0$. If t is an arbitrary tangent to Γ_i, prove that the intersections of t and Γ_j are harmonic conjugates with respect to the intersections of t and Γ_k.

16. Given $\Gamma_1: X^T A_1 X = 0$ and $\Gamma_2: X^T A_2 X = 0$. What is the envelope of the polars of an arbitrary point, P, with respect to the conics of the family $\lambda_1 X^T A_1 X + \lambda_2 X^T A_2 X = 0$?

17. Given $C_1:(0,0,1)$, $C_2:(1,0,1)$, $\Gamma_1: x_3^2 - 2x_1x_2 = 0$, and $\Gamma_2: x_1^2 - 2x_1x_2 + x_3^2 = 0$. Find the equations of the quadratic inversions defined by:
(a) C_1 and Γ_1 (b) C_1 and Γ_2 (c) C_2 and Γ_1 (d) C_2 and Γ_2

18. Given $\Gamma: x_3^2 - x_1x_2 = 0$ and the point $C_1:(1,1,1)$ on Γ. Let $Y:(y_1,y_2,y_3)$ be an arbitrary point of Π_2, and let C_2 be the second intersection of Γ and the line C_1Y. Find the equation of the transformation which maps Y into its harmonic conjugate, $Y':(y_1',y_2',y_3')$, with respect to C_1 and C_2.

19. Prove the following properties of the transformation of inversion in the unit circle, Γ:
(a) If P is the image of Q, then Q is the image of P.
(b) If P is the image of Q, then one of the points (P,Q) is inside Γ and one is outside Γ.
(c) Every point of Γ is invariant.
(d) Every line through the origin is invariant.
(e) The image of a line l which does not pass through the origin is a circle which passes through the origin and whose tangent at the origin is parallel to l.
(f) The image of a circle which does not pass through the origin is another circle which does not pass through the origin.
(g) A circle is transformed into itself if and only if it is orthogonal to Γ.
(h) (P,Q) is a pair of images if and only if every circle containing P and Q is orthogonal to Γ.

(*i*) The measure of the angle at which two lines or two circles intersect is equal to the measure of the angle at which their images intersect.

20. (*a*) Describe a euclidean construction for the inverse of a point in the unit circle.
(*b*) Generalize inversion in the unit circle to inversion in an arbitrary circle.

21. Let Γ be a nonsingular conic, let l be a line meeting Γ in two points, P_1 and P_2, let l' be a line meeting Γ in two points, P_1' and P_2', and let t_1, t_2, t_1', t_2' be the tangents to Γ at P_1, P_2, P_1', P_2', respectively. Show that there is a conic Φ to which l, l', t_1, t_2, t_1', t_2' are all tangent. What is the dual of this result? *Hint:* Verify that without loss of generality Γ, l, and l' can be chosen so that $P_1:(1,0,0)$, $P_2:(0,1,0)$, $P_1':(0,0,1)$, $P_2':(1,1,1)$.

4.10 Conclusion

In this chapter we continued our investigation of the extended euclidean plane and the algebraic representation by first introducing the useful notion of the parametrization of a line. Then, using this technique, we proved the famous theorems of Desargues and Pappus. We discussed perspectivities and projectivities and showed that any projectivity could be obtained as the composition of at most three perspectivities. Then we established the important result that a projectivity is uniquely determined when three points on one line and their images on a second line are given, and as a corollary we observed that a projectivity between distinct lines in which the point of intersection of the two lines is self-corresponding is necessarily a perspectivity. We then introduced the notion of the cross ratio of four collinear points, or four concurrent lines, and showed that it was invariant under any projectivity. Then we considered the special case of projectivities between ranges on the same line and introduced the important idea of an involution; i.e., a projectivity of period 2. Finally, we studied some of the properties of conics in E_2^+ and Π_2, in particular their so-called polar properties, and we developed a necessary and sufficient condition for a conic to consist of one or two lines.

Of the many important ideas developed in this chapter, the theorems of Desargues and Pappus, together with the theorem on the determination of a projectivity by the assignment of three points and their images and its corollary, have a significance greater than might be anticipated. Although we proved each of these results, we did so only in the context of the extended euclidean plane and the algebraic representation, and there are geometrical systems in which they are not valid. In fact, in the axiomatic development of projective geometry which we will begin in Chap. 6, it will be necessary to assume at least one of these theorems as an axiom if our work is to lead us to the classical projective plane.

five

LINEAR TRANSFORM= ATIONS IN Π_2

5.1 Introduction

When we investigated projectivities between cobasal ranges in Sec. 4.8, we were, in fact, studying linear mappings of a line onto itself, and our work provided us with information about the existence of invariant points, interesting particular cases, and euclidean specializations of such mappings. In this chapter, after first discussing the possibility of different coordinate systems in Π_2, we shall undertake a similar investigation of linear mappings of the entire projective plane onto itself. As before, we shall determine the invariant elements of such transformations and classify the transformations according to the possible invariant configurations. For each type we shall obtain a standard form for its equations, and we shall note a number of specializations of interest in euclidean geometry. Then, after we have introduced the concept of a *group*, we shall show that the set of all nonsingular projective transformations is a group under the operation of composition, and we shall identify four subgroups of this group defining, respectively, affine geometry, similarity geometry, equiareal geometry, and finally euclidean geometry.

5.2 Coordinate Transformations

In Sec. 2.5 we investigated the possibility of obtaining euclidean special-izations of E_2^+ by identifying $x_1 = 0$ or $x_2 = 0$, as well as $x_3 = 0$, as the exceptional line in the extended plane. Clearly, this is equivalent to the introduction of new coordinates (x_1',x_2',x_3') according to one or the other of the following schemes:

$$
\begin{array}{cccc}
x_1' = x_2 & x_1' = x_3 & x_1' = x_1 & x_1' = x_3 \\
x_2' = x_3 & x_2' = x_2 & x_2' = x_3 & x_2' = x_1 \\
x_3' = x_1 & x_3' = x_1 & x_3' = x_2 & x_3' = x_2
\end{array}
$$

The triangle of reference for each of these new coordinate systems consists, of course, of the same three lines, namely, the axes of the original coordinate system, and in each case only the roles played by the axes have been changed. This is not the only way in which new coordinates can be introduced, how-ever, and we shall now show how to establish a coordinate system in Π_2, or in E_2^+ if we restrict ourselves to real quantities, whose triangle of reference consists of any three nonconcurrent lines.

To do this, let

$$
\begin{aligned}
\lambda_1 &: l_{11}x_1 + l_{12}x_2 + l_{13}x_3 = 0 \\
\lambda_2 &: l_{21}x_1 + l_{22}x_2 + l_{23}x_3 = 0 \\
\lambda_3 &: l_{31}x_1 + l_{32}x_2 + l_{33}x_3 = 0
\end{aligned}
$$

be three nonconcurrent lines which we wish to be, respectively, the lines $x_1' = 0$, $x_2' = 0$, $x_3' = 0$ of a new coordinate system. In other words, in our new coordinate system, points for which $l_{11}x_1 + l_{12}x_2 + l_{13}x_3 = 0$, and only these points, are to have new coordinates such that $x_1' = 0$; points for which $l_{21}x_1 + l_{22}x_2 + l_{23}x_3 = 0$, and only these points, are to have new coordinates such that $x_2' = 0$; and points for which $l_{31}x_1 + l_{32}x_2 + l_{33}x_3 = 0$, and only these points, are to have new coordinates such that $x_3' = 0$. These require-ments will be met if we define the new coordinates (x_1',x_2',x_3') by expressions of the form

$$
\begin{aligned}
x_1' &= h_1(l_{11}x_1 + l_{12}x_2 + l_{13}x_3) \\
x_2' &= h_2(l_{21}x_1 + l_{22}x_2 + l_{23}x_3) \\
x_3' &= h_3(l_{31}x_1 + l_{32}x_2 + l_{33}x_3)
\end{aligned}
$$

(1)

However, x_1', x_2', x_3' are not determined by these equations until specific values are assigned to h_1, h_2, h_3. This we do by selecting an arbitrary point $U:(u_1,u_2,u_3)$, not on λ_1, λ_2, or λ_3, to have coordinates $(h,h,h) \sim (1,1,1)$ in the new coordinate system. This fixes h_1, h_2, h_3 to within an arbitrary factor of proportionality and completes the determination of the required transformation of coordinates. Once this has been done, we shall, for convenience, put $h_i l_{ij} = a_{ij}$ and write Eqs. (1) in the standard form

(2)
$$\begin{aligned}
x'_1 &= a_{11}x_1 + a_{12}x_2 + a_{13}x_3 \\
x'_2 &= a_{21}x_1 + a_{22}x_2 + a_{23}x_3 \\
x'_3 &= a_{31}x_1 + a_{32}x_2 + a_{33}x_3
\end{aligned} \qquad \text{or} \qquad X' = AX$$

where $\quad X' = \left\| \begin{matrix} x'_1 \\ x'_2 \\ x'_3 \end{matrix} \right\| \quad A = \left\| \begin{matrix} a_{11} & a_{12} & a_{13} \\ a_{21} & a_{22} & a_{23} \\ a_{31} & a_{32} & a_{33} \end{matrix} \right\| \quad X = \left\| \begin{matrix} x_1 \\ x_2 \\ x_3 \end{matrix} \right\|$

and, of course, a_{ij} is not necessarily equal to a_{ji}.

From Eqs. (2), it is clear that to within an arbitrary proportionality constant, a point (x_1,x_2,x_3) has a unique set of new coordinates (x'_1,x'_2,x'_3). Moreover, since λ_1, λ_2, λ_3 are nonconcurrent lines, it follows that the matrix $L = \|l_{ij}\|$ is nonsingular. Hence $|A| = h_1 h_2 h_3 |L|$ is different from zero, and therefore Eqs. (2) can be solved for the original coordinates in terms of the new ones. Doing this, we obtain $X = A^{-1}X'$, which, in turn, shows that to each triple of new coordinates (x'_1,x'_2,x'_3) there corresponds a unique point $P:(x_1,x_2,x_3)$. Thus (2) establishes a one-to-one correspondence between the points of Π_2 and sets of triples of the form (hx'_1,hx'_2,hx'_3); that is, we have indeed established a new coordinate system in Π_2 whose axes are λ_1, λ_2, and λ_3, as required.

Dually, we can establish a new coordinate system for the lines of Π_2 by choosing any three noncollinear points, say P_1, P_2, P_3, whose equations are

$$\begin{aligned}
P_1 &: p_{11}l_1 + p_{12}l_2 + p_{13}l_3 = 0 \\
P_2 &: p_{21}l_1 + p_{22}l_2 + p_{23}l_3 = 0 \\
P_3 &: p_{31}l_1 + p_{32}l_2 + p_{33}l_3 = 0
\end{aligned}$$

and then writing

$$\begin{aligned}
l'_1 &= k_1(p_{11}l_1 + p_{12}l_2 + p_{13}l_3) \\
l'_2 &= k_2(p_{21}l_1 + p_{22}l_2 + p_{23}l_3) \\
l'_3 &= k_3(p_{31}l_1 + p_{32}l_2 + p_{33}l_3)
\end{aligned}$$

The determination of the transformation will be complete when k_1, k_2, k_3 are determined, and this we do by selecting an arbitrary line $v:[v_1,v_2,v_3]$, not passing through P_1, P_2, or P_3, and requiring that its coordinates in the new system be $[k,k,k] \sim [1,1,1]$. Once the values of k_1, k_2, k_3 are fixed, we put $k_i p_{ij} = b_{ij}$ and write the equations of the transformation in the standard form

(3)
$$\Lambda' = B\Lambda$$

where $\quad \Lambda' = \left\| \begin{matrix} l'_1 \\ l'_2 \\ l'_3 \end{matrix} \right\| \quad B = \left\| \begin{matrix} b_{11} & b_{12} & b_{13} \\ b_{21} & b_{22} & b_{23} \\ b_{31} & b_{32} & b_{33} \end{matrix} \right\| \quad \Lambda = \left\| \begin{matrix} l_1 \\ l_2 \\ l_3 \end{matrix} \right\|$

and, of course, b_{ij} is not necessarily equal to b_{ji}. Since the points P_1, P_2, P_3 are noncollinear, it follows that the matrix B is nonsingular. Hence Eq. (3) can be solved for Λ, and we have the inverse transformation

$$(4) \qquad\qquad \Lambda = B^{-1}\Lambda'$$

Equations (3) and (4), together, show that the correspondence we have established between the lines of Π_2 and the sets of triples $[kl_1', kl_2', kl_3']$ is one to one, and that we have, indeed, a new coordinate system for the lines of Π_2.

Example 1

New point coordinates are established in Π_2 in such a way that the lines $x_1 + x_3 = 0$, $x_1 + x_2 - x_3 = 0$, and $x_2 - x_3 = 0$ become, respectively, the lines $x_1' = 0$, $x_2' = 0$, and $x_3' = 0$, and the point $(0,2,1)$ becomes the new **unit point**, i.e., the point whose new coordinates are $(1,1,1)$. At the same time, new line coordinates are established in such a way that the points $l_1 - l_2 - l_3 = 0$, $l_1 - 2l_2 - l_3 = 0$, and $l_2 + l_3 = 0$ become, respectively, the points $l_1' = 0$, $l_2' = 0$, and $l_3' = 0$, and the line $[2,2,-1]$ becomes the new **unit line,** i.e., the line whose new coordinates are $[1,1,1]$. Find the equations of these coordinate transformations, and determine the condition that the point whose new coordinates are (x_1', x_2', x_3') should lie on the line whose new coordinates are $[l_1', l_2', l_3']$.

The equations expressing the new point-coordinates in terms of the old will be

$$\begin{aligned} x_1' &= h_1(x_1 \qquad\ + x_3) \\ x_2' &= h_2(x_1 + x_2 - x_3) \\ x_3' &= h_3(\qquad x_2 - x_3) \end{aligned}$$

provided h_1, h_2, h_3 are determined so that the new coordinates of the point $(0,2,1)$ are $(1,1,1)$. Substituting these two sets of values, we find immediately that $h_1 = h_2 = h_3 = 1$. Hence the equation of the point-coordinate transformation is

$$(5) \qquad X' = AX \quad \text{where} \quad A = \begin{Vmatrix} 1 & 0 & 1 \\ 1 & 1 & -1 \\ 0 & 1 & -1 \end{Vmatrix}$$

From this, by a straightforward calculation, we find that the equation of the inverse transformation is

$$(6) \qquad X = A^{-1}X' \quad \text{where} \quad A^{-1} = \begin{Vmatrix} 0 & 1 & -1 \\ 1 & -1 & 2 \\ 1 & -1 & 1 \end{Vmatrix}$$

Similarly, the equations expressing the new line-coordinates in terms of the old will be

$$l'_1 = k_1(l_1 - l_2 - l_3)$$
$$l'_2 = k_2(l_1 - 2l_2 - l_3)$$
$$l'_3 = k_3(\quad\; l_2 + l_3)$$

provided k_1, k_2, k_3 are determined so that the new coordinates of the line $[2,2,-1]$ are $[1,1,1]$. Substituting these two sets of values, we find that $k_1 = 1$, $k_2 = -1$, and $k_3 = 1$. Hence the equation of the line-coordinate transformation is

(7) $\Lambda' = B\Lambda$ where $B = \begin{Vmatrix} 1 & -1 & -1 \\ -1 & 2 & 1 \\ 0 & 1 & 1 \end{Vmatrix}$

An easy calculation then shows that the inverse transformation is

(8) $\Lambda = B^{-1}\Lambda'$ where $B^{-1} = \begin{Vmatrix} 1 & 0 & 1 \\ 1 & 1 & 0 \\ -1 & -1 & 1 \end{Vmatrix}$

To determine the condition that the point with coordinate-vector X' should lie on the line whose coordinate-vector is Λ', it is convenient to observe that the incidence condition derived in Sec. 2.3, namely, $l_1x_1 + l_2x_2 + l_3x_3 = 0$, can be written in the matric form $\Lambda^T X = 0$. Hence in terms of the new point- and line-coordinates defined by Eqs. (5) and (7), or more conveniently by Eqs. (6) and (8), the incidence condition becomes

$$(B^{-1}\Lambda')^T(A^{-1}X') = 0 \quad \text{or} \quad (\Lambda')^T(B^{-1})^T A^{-1}X' = 0$$

Now

$$(B^{-1})^T A^{-1} = \begin{Vmatrix} 1 & 1 & -1 \\ 0 & 1 & -1 \\ 1 & 0 & 1 \end{Vmatrix} \cdot \begin{Vmatrix} 0 & 1 & -1 \\ 1 & -1 & 2 \\ 1 & -1 & 1 \end{Vmatrix} = \begin{Vmatrix} 0 & 1 & 0 \\ 0 & 0 & 1 \\ 1 & 0 & 0 \end{Vmatrix}$$

Hence the required condition is

$$\|l'_1 \; l'_2 \; l'_3\| \cdot \begin{Vmatrix} 0 & 1 & 0 \\ 0 & 0 & 1 \\ 1 & 0 & 0 \end{Vmatrix} \cdot \begin{Vmatrix} x'_1 \\ x'_2 \\ x'_3 \end{Vmatrix} = 0 \quad \text{or} \quad l'_3x'_1 + l'_1x'_2 + l'_2x'_3 = 0$$

Clearly, this is *not* the same condition on the new coordinates that we had on the original coordinates.

The results of the last example are somewhat surprising and perhaps even a little disturbing, since they show that the form of the incidence condition for a point and a line is not preserved when point- and line-coordinates are independently transformed. Clearly, if there is any possibility of the symmetric incidence condition $l_1x_1 + l_2x_2 + l_3x_3 = 0$ being invariant under simultaneous point- and line-coordinate transformations, some special relation must exist between the structures of the two coordinate systems. Since we naturally want the incidence condition to be preserved, it is important that we determine whether such special relations exist and, if so, just what they are.

Suppose, then, that we have the point-coordinate transformation $X' = AX$ and its inverse $X = A^{-1}X'$ and the line-coordinate transformation $\Lambda' = B\Lambda$ and its inverse $\Lambda = B^{-1}\Lambda'$. Under these transformations the symmetric incidence condition $\Lambda^T X = 0$ becomes $(B^{-1}\Lambda')^T(A^{-1}X') = 0$ or $(\Lambda')^T(B^{-1})^TA^{-1}X' = 0$. Now in order that this should be identically $(\Lambda')^TX' = 0$, it is necessary that $(B^{-1})^TA^{-1} = I$, or, taking the transpose of both sides and then multiplying on the right by B,

$$(A^{-1})^T = B$$

Therefore $b_{ij} = A_{ij}/|A|$, and hence, neglecting the irrelevant proportionality constant $1/|A|$, the equations of the line-coordinate transformation must be

$$l'_1 = A_{11}l_1 + A_{12}l_2 + A_{13}l_3$$
$$l'_2 = A_{21}l_1 + A_{22}l_2 + A_{23}l_3$$
$$l'_3 = A_{31}l_1 + A_{32}l_2 + A_{33}l_3$$

Thus if the form of the incidence condition is to be preserved when both point- and line-coordinates are transformed, the vertices of the triangle used to define the new line-coordinates must be the points whose equations are

$$P_1: A_{11}l_1 + A_{12}l_2 + A_{13}l_3 = 0$$
$$P_2: A_{21}l_1 + A_{22}l_2 + A_{23}l_3 = 0$$
$$P_3: A_{31}l_1 + A_{32}l_2 + A_{33}l_3 = 0$$

To identify these points, we observe first that their coordinates in the original coordinate system are, respectively,

$$(A_{11}, A_{12}, A_{13}) \qquad (A_{21}, A_{22}, A_{23}) \qquad (A_{31}, A_{32}, A_{33})$$

Then, since the sides of the triangle of reference used to define the new point-coordinates are the lines

$$\lambda_1: a_{11}x_1 + a_{12}x_2 + a_{13}x_3 = 0$$
$$\lambda_2: a_{21}x_1 + a_{22}x_2 + a_{23}x_3 = 0$$
$$\lambda_3: a_{31}x_1 + a_{32}x_2 + a_{33}x_3 = 0$$

it follows from Theorem 9, Appendix 1, that P_1 lies on both λ_2 and λ_3, that P_2 lies on both λ_1 and λ_3, and that P_3 lies on both λ_1 and λ_2. In other words, in the new point-coordinate system, P_1 is the point $(1,0,0)$, P_2 is the point $(0,1,0)$, and P_3 is the point $(0,0,1)$. Finally, the invariance of the incidence condition requires that the equation of the new unit line in terms of the new point-coordinates be $x_1' + x_2' + x_3' = 0$. To identify this line, we note that if U is the new unit point, then the equation of the line $P_1 U$ is $x_2' - x_3' = 0$; and, by Lemma 2, Sec. 4.7, its harmonic conjugate with respect to $x_2' = 0$ and $x_3' = 0$ is $x_2' + x_3' = 0$, which intersects $x_1' = 0$ in the point $(0,1,-1)$. Similarly, the intersection of $x_2' = 0$ and the harmonic conjugate of $P_2 U$ with respect to $x_1' = 0$ and $x_3' = 0$ is $(-1,0,1)$; and the intersection of $x_3' = 0$ and the harmonic conjugate of $P_3 U$ with respect to $x_1' = 0$ and $x_2' = 0$ is the point $(1,-1,0)$. It is easy to verify that each of these points lies on the new unit line $x_1' + x_2' + x_3' = 0$, which is thus the polar[1] of the new unit point with respect to the common triangle of reference of the two new coordinate systems. Thus, summarizing, we have the following theorem.

Theorem 1

In order that the incidence condition $l_1 x_1 + l_2 x_2 + l_3 x_3 = 0$ be invariant under a simultaneous transformation of point- and line-coordinates, the vertex P_i of the triangle of reference for the new line-coordinates must be the intersection of the sides λ_j and λ_k of the triangle of reference for the new point-coordinates, and the new unit line must be the polar of the new unit point with respect to the triangle whose sides are λ_1, λ_2, λ_3.

Exercises

1. Verify that under a transformation of point-coordinates the equation of a line always goes into a linear equation in the new coordinates. Is the corresponding result true for the equation of a point?

2. Verify that the form of the condition that three points be collinear is preserved under any transformation of point-coordinates.

3. Verify that the form of the condition that three lines be concurrent is preserved under any transformation of line-coordinates.

4. If $\lambda_1: x_1 + x_2 + x_3 = 0$, $\lambda_2: x_2 + x_3 = 0$, and $\lambda_3: x_1 - x_3 = 0$ are chosen as the x_1', x_2', and x_3' axes, respectively, and if $U:(1,-1,2)$ is chosen as the new unit point, find the equations expressing the new point-coordinates in terms of the old, and vice versa.

5. Given the euclidean circles $C_1: x^2 + y^2 = \frac{1}{2}$, $C_2: x^2 + y^2 = 1$, and $C_3: x^2 + y^2 = 4$. Find the equations of each of these circles if the coordinate transformation of Exercise 4 is specialized so that the ideal line is:
(a) λ_1 (b) λ_2 (c) λ_3
What type of conic does each of the new equations represent? Why?

[1] See Exercise 20, Sec. 4.7.

6. Given $\lambda_1: x_3 = 0$, $\lambda_2: x_1 - x_2 = 0$, $\lambda_3: x_1 - x_3 = 0$, $U:(0,1,1)$, and $v:[2,-1,-2]$.

(a) Find the equations of the point-coordinate transformation for which λ_1, λ_2, λ_3 are, respectively, the x_1', x_2', x_3' axes and for which U is the new unit point. Find the equations of the line-coordinate transformation for which $\lambda_2\lambda_3$, $\lambda_1\lambda_3$, $\lambda_1\lambda_2$ are, respectively, the points $l_1' = 0$, $l_2' = 0$, $l_3' = 0$ and for which v is the new unit line. What is the incidence condition in the new coordinates?

(b) Work part (a) if $\lambda_2\lambda_3$, $\lambda_1\lambda_2$, $\lambda_1\lambda_3$ are, respectively, the points $l_1' = 0$, $l_2' = 0$, $l_3' = 0$.

(c) Work part (a) if $\lambda_1\lambda_3$, $\lambda_1\lambda_2$, $\lambda_2\lambda_3$ are, respectively, the points $l_1' = 0$, $l_2' = 0$, $l_3' = 0$.

7. (a) Find the equations of the coordinate transformation which assigns to the points $P_1:(1,1,-1)$, $P_2:(1,0,2)$, $P_3:(1,-1,0)$, $U:(2,-1,1)$ the new coordinates $(1,0,0)$, $(0,1,0)$, $(0,0,1)$, $(1,1,1)$, respectively.

(b) Work part (a), given $P_1:(1,0,-1)$, $P_2:(0,1,1)$, $P_3:(1,1,1)$, $U:(1,2,0)$.

8. Given $\lambda_1: x_2 + x_3 = 0$, $\lambda_2: x_1 + x_3 = 0$, $\lambda_3: x_1 + x_2 = 0$, and $U:(1,0,1)$. Find the equations of the point-coordinate transformation which assigns to λ_1, λ_2, λ_3 the new equations $x_1' = 0$, $x_2' = 0$, $x_3' = 0$, respectively, and makes U the new unit point. Find the equations of the line-coordinate transformation based on the triangle whose vertices are the points $\lambda_2\lambda_3$, $\lambda_1\lambda_3$, $\lambda_1\lambda_2$ and for which $x_1 + x_2 - 2x_3 = 0$ is the new unit line. What is the incidence condition for the new coordinate systems? What line must be chosen for the new unit line if the form of the incidence condition is to be preserved?

9. (a) It is desired to convert the equation of the parabola $y = x^2$ into the equation of a circle by suitably specializing a new coordinate system in which $x_1 = 0$, $x_1 + x_2 = 0$, and $a_1x_1 + a_2x_2 + a_3x_3 = 0$ are, respectively, the lines $x_1' = 0$, $x_2' = 0$, and $x_3' = 0$. Determine what relations a_1, a_2, a_3 must satisfy and what the new unit point must be if this is to be accomplished.

(b) Work part (a) if the equation of the hyperbola $xy = 1$ is to be converted into the equation of a circle.

10. Discuss the possibility of introducing new homogeneous coordinates in E_2 in the following way. Let $\triangle A_1A_2A_3$ be an arbitrary triangle, let P be an arbitrary point, and associate with P the ordered triple (kx_1', kx_2', kx_3'), where x_i' is the area of $\triangle PA_jA_k$. Show that this procedure is a special case of the general process of introducing new homogeneous coordinates which we developed in this section. Can this process be extended to E_2^+?

5.3 Collineations

In the last section we considered the equation $X' = AX$ to be a device which introduced a new triangle of reference and thereby assigned new coordinates

(x_1',x_2',x_3') to every point (x_1,x_2,x_3) in Π_2. In the remainder of this chapter we shall adopt a different point of view and consider $X' = AX$ to be a linear transformation which maps each point $P:(x_1,x_2,x_3)$ onto another point, P', whose coordinates *with respect to the same triangle of reference* are (x_1',x_2',x_3'). When we obtained $X' = AX$ as the equation of a transformation of point-coordinates, the matrix A was necessarily nonsingular since it was the matrix of the coefficients in the equations of three nonconcurrent lines. However, when $X' = AX$ is interpreted as a mapping of Π_2, there is no a priori restriction on the matrix A, and we must consider the possibility that it is singular.

If $|A| \neq 0$, then, by Theorem 5, Sec. 3.6, the equation $AX = O$ is satisfied only by the trivial vector $X = O$. Hence there is no point, P, in Π_2 to which the transformation $T: X' = AX$ fails to assign a unique image, P'. Moreover, if $|A| \neq 0$, then A^{-1} exists and T has an inverse, $T^{-1}: X = A^{-1}X'$, which, by the same reasoning, assigns a unique preimage, $P = A^{-1}P'$, to each point P' in Π_2. Thus if A is nonsingular, i.e., if T is a **nonsingular linear transformation,** the mapping effected by T is one to one over all of Π_2. Furthermore, this is true *only* if $|A| \neq 0$; for if $|A| = 0$, the equation $AX = O$ is satisfied by at least one nontrivial vector $X = P$, which implies that there is at least one point P which does not have a determinate image under T. Since we want the transformation $T: X' = AX$ to be a one-to-one mapping of Π_2 onto itself, we shall henceforth consider only nonsingular linear transformations, or **collineations,** as we shall call them.

Since a collineation is, by definition, a nonsingular linear transformation, it is clear that every collineation has an inverse which is also a linear transformation. Hence it follows that the locus of the images of the points of an arbitrary line, $l_1x_1 + l_2x_2 + l_3x_3 = \Lambda^T X = 0$, under a collineation $X' = AX$ is defined by the equation $\Lambda^T A^{-1} X' = 0$, which is also the equation of a straight line. Thus any collineation transforms a line into a line, and, moreover, does this in such a way that the points of the line are in one-to-one correspondence with the points of its image line. It is thus natural to ask if this induced transformation is a projectivity between the line and its image. That the answer to this question is Yes, is guaranteed by the next theorem.

Theorem 1

If l' is the image of a line l under a collineation, T, then the mapping of l onto l' induced by T is a projectivity.

Proof Let l' be the image of an arbitrary line l under a collineation $T: X' = AX$, let l be parametrized in terms of the base points P_1 and P_2, let $P_1' = AP_1$ and $P_2' = AP_2$ be the points on l' which are the images of P_1 and P_2 under T, and let l' be parametrized in terms of P_1' and P_2'. Then if $P = \lambda P_1 + \mu P_2$

is an arbitrary point on l, its image on l' is the point

$$P' = AP = A(\lambda P_1 + \mu P_2) = \lambda AP_1 + \mu AP_2 = \lambda P'_1 + \mu P'_2$$

Thus in terms of parameters (λ,μ) and (λ',μ') referred to the respective pairs of base points (P_1,P_2) and (P'_1,P'_2), the mapping of l onto l' is defined by the equations $\lambda' = \lambda$ and $\mu' = \mu$, or in matric form

$$\left\| \begin{matrix} \lambda' \\ \mu' \end{matrix} \right\| = \left\| \begin{matrix} 1 & 0 \\ 0 & 1 \end{matrix} \right\| \cdot \left\| \begin{matrix} \lambda \\ \mu \end{matrix} \right\|$$

which is the equation of a projectivity, as asserted.

Corollary 1

The cross ratio of four collinear points and the cross ratio of their images under any collineation are the same.

It is now a matter of some interest to determine the coordinates of the line l' onto which a collineation $T: X' = AX$ maps an arbitrary line l. To do this, let $l_1x_1 + l_2x_2 + l_3x_3 = 0$, that is, $\Lambda^T X = 0$, be the equation of an arbitrary line, l. Then since $X = A^{-1}X'$, it follows that the equation of the image of l is $\Lambda^T A^{-1}X' = 0$. Hence the coordinate vector, Λ', of the image of l is given by the equation $(\Lambda')^T = \Lambda^T A^{-1}$, from which, by taking the transpose of each side, we have $\Lambda' = (A^{-1})^T\Lambda$. Thus we have established the following theorem.

Theorem 2

The pointwise mapping of Π_2 onto itself effected by the collineation $X' = AX$ induces a mapping of the lines of Π_2 onto the lines of Π_2 defined by the equation $\Lambda' = (A^{-1})^T\Lambda$.

We now propose to determine what points, if any, coincide with their images, i.e., are invariant, under a collineation $T: X' = AX$. Since X and X', in the equation of T, refer to the same coordinate system, it is clear that a point P will be invariant if and only if its coordinate-vector and the coordinate-vector of its image, P', are proportional, i.e., if and only if $P' = kP$. Hence we must investigate the existence and nature of solutions of the equation $kX = AX$, or, writing $X = IX$ and then using the distributive law for matric multiplication,

$$(A - kI)X = O$$

This matric equation is equivalent to a system of three homogeneous linear equations and hence, according to Theorem 5, Sec. 3.6, a nontrivial solution for (x_1,x_2,x_3) will exist if and only if the determinant of the coefficients of the system is equal to zero, i.e., if and only if

$$|A - kI| = \begin{vmatrix} a_{11} - k & a_{12} & a_{13} \\ a_{21} & a_{22} - k & a_{23} \\ a_{31} & a_{32} & a_{33} - k \end{vmatrix} = 0$$

This is a cubic equation in the proportionality constant k, and for those values of k which satisfy this equation, and for no others, the equation $kX = AX$ will be satisfied by a nontrivial coordinate-vector X, and we shall have an invariant point. The equation $|A - kI| = 0$ is called the **characteristic equation** of the matrix A, the values of k which satisfy this equation are called the **characteristic values** or **characteristic roots** of A, and the solution vectors of the equation $(A - kI)X = O$ for these values of k are called **characteristic vectors**[1] of A.

Example 1

Find the invariant points of the collineation

$$T: X' = AX \qquad \text{where} \qquad A = \begin{Vmatrix} 4 & -4 & -6 \\ 3 & -4 & -9 \\ -1 & 2 & 5 \end{Vmatrix}$$

To find the invariant points of T we must first solve the characteristic equation

$$|A - kI| = \begin{vmatrix} 4 - k & -4 & -6 \\ 3 & -4 - k & -9 \\ -1 & 2 & 5 - k \end{vmatrix} = -(k^3 - 5k^2 + 8k - 4) = 0$$

to find the values of k for which there are nontrivial solutions of the equation $(A - kI)X = O$. Then we must find these nontrivial solutions themselves by solving the system of equations

$$
\begin{aligned}
(1) \qquad (4 - k)x_1 &\quad - 4x_2 &\quad - 6x_3 &= 0 \\
3x_1 &- (4 + k)x_2 &\quad - 9x_3 &= 0 \\
-x_1 &\quad + 2x_2 &+ (5 - k)x_3 &= 0
\end{aligned}
$$

for each of these values of k.

In this problem it is easy to solve the characteristic equation by factoring, since

$$k^3 - 5k^2 + 8k - 4 = (k - 1)(k - 2)^2$$

Hence the possible values of k are 1 and 2. When $k = 1$, the system of equations (1) becomes

$$
\begin{aligned}
3x_1 - 4x_2 - 6x_3 &= 0 \\
3x_1 - 5x_2 - 9x_3 &= 0 \\
-x_1 + 2x_2 + 4x_3 &= 0
\end{aligned}
$$

---

[1] Some writers graft the German word *eigen* meaning *own, peculiar,* or *proper* onto the words *values* and *vectors* and use the terms *eigenvalues* and *eigenvectors*.

Since the rank of the coefficient matrix in this case is 2, it follows from Theorem 3, Sec. 3.6, that there is a single independent solution for the x's, and from Theorem 6, Sec. 3.6, this solution can be read from the matrix of coefficients in the first two equations. The result is $x_1 = 2, x_2 = 3, x_3 = -1$; that is, the transformation leaves invariant the point $F_1:(2,3,-1)$.

When $k = 2$, the system (1) becomes

$$2x_1 - 4x_2 - 6x_3 = 0$$
$$3x_1 - 6x_2 - 9x_3 = 0$$
$$-x_1 + 2x_2 + 3x_3 = 0$$

In this case, the rank of the coefficient matrix is 1, the equations are all proportional, and there are two linearly independent solutions, which can be read from any one of the equations. Using the last equation, and letting first x_2 and then x_3 be zero, we find the independent solutions $x_1 = 3$, $x_2 = 0$, $x_3 = 1$ and $x_1 = 2$, $x_2 = 1$, $x_3 = 0$. Moreover, by Theorem 1, Sec. 3.6, any vector linearly dependent upon the two solution vectors $\begin{Vmatrix} 3 \\ 0 \\ 1 \end{Vmatrix}$ and $\begin{Vmatrix} 2 \\ 1 \\ 0 \end{Vmatrix}$ is also a solution. Hence any point on the line determined by the points $F_2:(3,0,1)$ and $F_3:(2,1,0)$, that is, the line $x_1 - 2x_2 - 3x_3 = 0$, is invariant. Clearly, the point F_1 does not lie on the line F_2F_3. Hence it follows that not only is the line F_2F_3 invariant, but so too is any line of the pencil on F_1, since every such line contains two invariant points, namely, F_1 and the intersection of that line with the line F_2F_3.

It should be noted, however, that the lines of the pencil on F_1 are invariant in a different sense than the line F_2F_3, since the former are transformed into themselves in such a way that only two of their points are invariant while F_2F_3 is transformed into itself in such a way that each of its points is invariant. In other words, the lines of the pencil on F_1 are mapped onto themselves by a nonidentical projectivity, while F_2F_3 is mapped onto itself by the identity projectivity.

In the last example we found that the invariant elements consisted of a fixed point with the property that every line through it was transformed into itself and a fixed line with the property that every point of it was transformed into itself. In other words, the configuration of invariant lines was the dual of the configuration of invariant points, and vice versa. This interesting property is not peculiar to the transformation of Example 1, but is actually true of any collineation. To verify this, we note first that if a transformation T sends a point P into a point P', then the inverse transformation sends P' back into P. Hence the points which a transformation maps into themselves are also left invariant by the inverse transformation. Equally well, of course, the lines which map into themselves under the line transformation induced by a collineation T are left invariant by the inverse of that transformation. Now by Theorem 2, the equation of the line transformation induced by a collineation T: $X' = AX$ is $\Lambda' = (A^{-1})^T \Lambda$, and by the preceding discussion, the invariant lines of this transformation are the same as the invariant lines of the inverse transformation $\Lambda = A^T \Lambda'$. Moreover, although $A^T \neq A$, it is true that

$$A^T - kI = A^T - kI^T = (A - kI)^T$$

Hence, since a determinant is unaltered by an interchange of its rows and columns, it follows that

$$|A^T - kI| = |A - kI|$$

that is, A and A^T have the same characteristic equation and therefore the same characteristic roots. Furthermore, if k_i is any characteristic root of A and A^T, every subdeterminant of the matrix $A - k_i I$ appears somewhere in $A^T - k_i I$, and vice versa; hence $A - k_i I$ and $A^T - k_i I$ have the same rank. Thus the structure of the configuration of the invariant lines of T is identical to the structure of the configuration of the invariant points of T; that is, either configuration is the dual of the other, as asserted.

We shall now attempt to classify collineations on the basis of the structure of their invariant configurations. However, before we can do this we must first prove several preliminary theorems.

Theorem 3

If $|A - kI| = -k^3 + \beta_1 k^2 - \beta_2 k + \beta_3 = 0$ is the characteristic equation of the matrix A, then β_i is equal to the sum of all the principal minors of order i in A.

Proof Let

$$|A - kI| = \begin{vmatrix} a_{11} - k & a_{12} & a_{13} \\ a_{21} & a_{22} - k & a_{23} \\ a_{31} & a_{32} & a_{33} - k \end{vmatrix} = -k^3 + \beta_1 k^2 - \beta_2 k + \beta_3 = 0$$

be the characteristic equation of the matrix A. Then to obtain β_3, that is, the constant term in the equation $|A - kI| = 0$, we need only set $k = 0$, getting $\beta_3 = |A|$. To obtain β_2, we observe that the terms containing the first power of k in the expansion of $|A - kI|$ are obtained by multiplying the term $-k$ in each diagonal element by the k-free part of the cofactor of that element. Thus the coefficient of k in the equation $|A - kI| = 0$ is $-\beta_2 = -A_{11} - A_{22} - A_{33}$; hence $\beta_2 = A_{11} + A_{22} + A_{33}$. Finally, to obtain β_1, we note that the terms containing k^2 in the expansion of $|A - kI|$ arise only by multiplying the terms $-k$ in every pair of diagonal elements by the k-free part of the remaining diagonal element. Hence the coefficient of k^2 is $\beta_1 = a_{11} + a_{22} + a_{33}$, and our proof is complete.

Theorem 4

If k_i is a characteristic root of multiplicity r of a $(3,3)$ matrix A, the rank of $A - k_i I$ is equal to or greater than $3 - r$.

Proof If k_i is a repeated root of multiplicity r of the characteristic equation $|A - kI| = 0$ of a matrix A, then if we put $k = k_i + h$, it is clear that $h = 0$ is a repeated root of multiplicity r of the equation $|A - (k_i + h)I| = |(A - k_i I) - hI| = 0$. Hence the expanded form of the last equation, say $-h^3 + \sigma_1 h^2 - \sigma_2 h + \sigma_3 = 0$, must contain h^r, and no higher power of h, as a factor. Therefore in this equation $\sigma_3 = \cdots = \sigma_{3-(r-1)} = 0$ and $\sigma_{3-r} \neq 0$. But by Theorem 3, σ_{3-r} is equal to the sum of the principal minors of order $3 - r$ of the matrix $A - k_i I$, and if $\sigma_{3-r} \neq 0$, then there is at least one minor of this order which is different from zero. In other words, the rank of $A - k_i I$ is at least as great as $3 - r$, as asserted.

Theorem 5

A characteristic vector of a matrix A cannot correspond to two distinct characteristic values.

Proof Let k_1 and k_2 be distinct characteristic values of a matrix A, and let X_1 be a characteristic vector of A corresponding, if possible, to both k_1 and k_2. Then, simultaneously,

$$(A - k_1 I)X_1 = O \quad \text{and} \quad (A - k_2 I)X_1 = O$$

Therefore, subtracting these two equations, it follows that

$$(2) \qquad (k_2 - k_1)IX_1 = (k_2 - k_1)X_1 = O$$

Now by hypothesis $k_2 \neq k_1$. Furthermore, a characteristic vector is, by definition, a *nontrivial* solution vector of $(A - kI)X = O$. Hence $X_1 \neq O$, and therefore Eq. (2) cannot hold. Thus the assumption that a characteristic vector can correspond to two distinct characteristic values must be abandoned, and the theorem is established.

Theorem 6

If X_1, X_2, X_3 are characteristic vectors corresponding to distinct characteristic values of a matrix A, then X_1, X_2, X_3 are linearly independent.

Proof Let X_1, X_2, X_3 be characteristic vectors corresponding to distinct characteristic values, k_1, k_2, k_3, of a matrix A, and let us assume, contrary to the theorem, that X_1, X_2, X_3 are linearly dependent. In other words, let us assume that the X's satisfy an equation of the form

$$(3) \qquad\qquad c_1 X_1 + c_2 X_2 + c_3 X_3 = O$$

in which at least one of the c's, say c_3, is different from zero. Then multiplying Eq. (3) on the left by A, we obtain

$$c_1 A X_1 + c_2 A X_2 + c_3 A X_3 = O$$

However, for each value of i, we have $A X_i = k_i X_i$. Hence the last equation becomes

$$(4) \qquad\qquad c_1 k_1 X_1 + c_2 k_2 X_2 + c_3 k_3 X_3 = O$$

Finally, if we subtract k_3 times Eq. (3) from Eq. (4), we obtain

$$(5) \qquad\qquad c_1(k_1 - k_3)X_1 + c_2(k_2 - k_3)X_2 = O$$

in which, by hypothesis, both $(k_1 - k_3)$ and $(k_2 - k_3)$ are different from zero. Now if $c_1 = c_2 = 0$, it follows from Eq. (3) that $X_3 = O$, which is impossible since X_3 is not a null vector. Similarly, if just one of the c's, say c_1, is equal to zero, it follows from Eq. (5) that $X_2 = O$, which is also impossible. Finally, if neither c_1 nor c_2 is equal to zero, it follows from Eq. (5) that X_1 and X_2 are proportional, which contradicts Theorem 5. Thus in every case the assumption that the X's are linearly dependent leads to a contradiction, and the theorem is established.

We are now in a position to begin the classification of collineations by enumerating the algebraic characteristics of the possible invariant configurations of a collineation, T. In the first place, we observe that the characteristic equation of the matrix of T, being a cubic equation, must have either

1. Three simple roots
2. One simple root and one double root
3. One triple root

Furthermore, since $|A - kI| = 0$ for any characteristic root, it follows from Theorem 4 that the rank of the matrix $A - kI$ for any simple root is necessarily 2, the rank of $A - kI$ for any double root is either 1 or 2, and the rank of $A - kI$ for any triple root is either 0, 1, or 2. Hence we have only the six possibilities listed in Table 5.1.

In the next section we shall show that these possibilities indeed provide a basis for classifying collineations by showing that the algebraic characteristics of each case are invariant under any coordinate transformation. Then in succeeding sections we shall examine each case in detail,

Table 5.1

Collineation type	Characteristic roots	Multiplicity of roots	Rank of $A - k_i I$
I	k_1 k_2 k_3	1 1 1	2 2 2
II	k_1 k_2	1 2	2 2
III	k_1 k_2	1 2	2 1
IV	k_1	3	2
V	k_1	3	1
VI	k_1	3	0

determine the configuration of its invariant elements, reduce its equations to a canonical form, and, where possible, note specializations of interest in euclidean geometry.

Exercises

1. Given $O:(1,1,2)$ and $l: 3x_1 - x_2 + x_3 = 0$. Find the equations of the transformation which assigns to a general point P the intersection of OP and l as its image. Is this transformation singular or nonsingular? What points, if any, are invariant under this transformation? What points, if any, do not have a unique image under this transformation?

2. (a) Given $\Gamma: x_1 x_2 = x_3^2$ and $\lambda: x_1 + x_2 - x_3 = 0$. Find the equations of the transformation which maps a general point P onto the intersection of λ and the polar of P with respect to Γ. Is this transformation singular or nonsingular? What points, if any, are invariant under this transformation? What points, if any, do not have a unique image under this transformation? (b) Work part (a), given $\Gamma: x_1^2 + x_1 x_2 - x_2 x_3 + x_3^2 = 0$ and $\lambda: -x_1 + x_2 + 7x_3 = 0$.

3. (a) Given $l: x_1 + x_2 - 3x_3 = 0$ and $O:(0,0,1)$. Find the equations of the transformation which maps a general point P onto the intersection of l and the harmonic conjugate of OP with respect to the lines $x_1 = 0$ and $x_2 = 0$. Is this transformation singular or nonsingular? What points, if any, are

invariant under this transformation? What points, if any, do not have a unique image under this transformation?

(b) Work part (a), given $l: x_1 - 2x_2 = 0$.

4. Is there a linear transformation which maps every point of Π_2 onto the point $(2,1,-3)$?

5. Prove that the images of three noncollinear points under a linear transformation $X' = AX$ will be collinear if and only if A is a singular matrix.

6. (a) Given $\Gamma_1: x_1{}^2 + 2x_1x_2 + x_2{}^2 - x_3{}^2 = 0$ and $\Gamma_2: x_1{}^2 + x_1x_3 - x_2x_3 + x_3{}^2 = 0$. Find the equations of the transformation which maps a point P onto the point P' whose polar with respect to Γ_2 is the same as the polar of P with respect to Γ_1. Is this transformation singular or nonsingular?

(b) Find the equations of the transformation which maps P onto the intersection of the polars of P with respect to Γ_1 and Γ_2. Is this transformation a collineation?

7. Given $\Gamma_1: X^T A_1 X = 0$ and $\Gamma_2: X^T A_2 X = 0$. Find the equation of the transformation which maps a point P onto the point P' whose polar with respect to Γ_2 is the same as the polar of P with respect to Γ_1, and show that this transformation is a collineation if and only if Γ_1 and Γ_2 are nonsingular conics.

8. What is the equation of the line transformation induced by the point transformation $X' = AX$ if:

(a) $A = \begin{Vmatrix} 1 & 2 & -1 \\ 0 & 1 & 1 \\ 2 & -1 & 0 \end{Vmatrix}$ (b) $A = \begin{Vmatrix} 1 & 2 & 2 \\ 2 & 0 & 1 \\ 1 & 1 & 1 \end{Vmatrix}$

(c) $A = \begin{Vmatrix} 1 & 0 & 3 \\ 2 & -1 & 0 \\ 1 & 1 & 7 \end{Vmatrix}$

9. What is the relation between the characteristic roots of a nonsingular matrix, A, and the characteristic roots of its inverse, A^{-1}?

10. Find the fixed points of the collineation $X' = AX$ if:

(a) $A = \begin{Vmatrix} 1 & 1 & -2 \\ -2 & -2 & 2 \\ 2 & 1 & -3 \end{Vmatrix}$ (b) $A = \begin{Vmatrix} 3 & 0 & -1 \\ 10 & 4 & -8 \\ 6 & 1 & -3 \end{Vmatrix}$

(c) $A = \begin{Vmatrix} 2 & 1 & -1 \\ 1 & -1 & 2 \\ 6 & 1 & 0 \end{Vmatrix}$

11. If A is a nonsingular matrix, the transformation $\Lambda = AX$ is called a **correlation,** because it associates, or *correlates,* a line Λ with every point P,

and conversely. Show that under any correlation the lines associated with the points of a range form a pencil.

12. If $\Lambda = AX$ is a correlation, show that the set of points which lie on their associated lines is a conic locus, and find its equation. Show also that the set of lines which contain their associated points is a conic envelope, and find its equation. Show that in general the conic envelope is not the set of tangents to the conic locus.

13. If $\Gamma: X^T A X = 0$ is an arbitrary nonsingular conic, show that the transformation which associates with an arbitrary point its polar with respect to Γ is a correlation, and find its equation. What is the locus of points which lie on their image lines in this correlation? What is the envelope of the lines which contain their image points in this correlation?

14. Show that a correlation $\Lambda = AX$ is the polarity defined by some non-singular conic if and only if A is a symmetric matrix.

15. (*a*) Is the transformation which associates with each point its polar with respect to a given triangle a polarity?
(*b*) Is the following transformation a correlation? Let l_1 and l_2 be two lines intersecting in a point O, let A_1 be a point on l_1 distinct from O, and let A_2 be a point on l_2 distinct from O. If P is an arbitrary point, let Q_1 be the intersection of PA_2 and l_1, let Q_2 be the intersection of PA_1 and l_2, and, finally, let the line Q_1Q_2 be the image of the point P.

16. Prove Theorem 2 by computing the coordinates of the image of a general line P_1P_2 from the coordinates of the images of P_1 and P_2.

5.4 *The Invariance of the Classification*

In the last section we discovered what appeared to be a basis for classifying collineations into six different types. However, before we can consider such a classification significant, we must be sure that it is based on intrinsic characteristics of the collineations and not on superficial properties associated with the particular coordinate system used to describe them.

Let us suppose, then, that in a collineation $T: X' = AX$, both X and X' are referred to a new coordinate system related to the original one by the equation $\bar{X} = BX$, where, of course, $|B| \neq 0$. Then $X = B^{-1}\bar{X}$ and $X' = B^{-1}\bar{X}'$; and in terms of the new coordinate system the equation of the collineation becomes $B^{-1}\bar{X}' = A(B^{-1}\bar{X})$ or, multiplying through on the left by B,

$$\bar{X}' = (BAB^{-1})\bar{X}$$

We shall now show that the transformation T has exactly the same configuration of invariant points whether it is described in the first coordinate system by the equation $X' = AX$ or in the second coordinate system by the equation $\bar{X}' = (BAB^{-1})\bar{X}$.

Theorem 1

The matrix A and the matrix BAB^{-1} have the same characteristic roots.

Proof The characteristic roots of the matrix BAB^{-1} are the roots of the characteristic equation $|BAB^{-1} - kI| = 0$. Now by obvious steps we have

$$BAB^{-1} - kI = BAB^{-1} - kBIB^{-1} = B(A - kI)B^{-1}$$

Furthermore, since the determinant of a product of square matrices is equal to the product of the determinants of the individual matrices, and since $|B^{-1}| = 1/|B|$, we have

$$|BAB^{-1} - kI| = |B(A - kI)B^{-1}| = |B| \cdot |A - kI| \cdot |B^{-1}| = |A - kI|$$

Thus BAB^{-1} and A have the same characteristic equation and hence the same characteristic roots, as asserted.[1]

Lemma 1

If X_1 and X_2 are linearly dependent vectors, then for any conformable matrix, B, the matrices BX_1 and BX_2 are also linearly dependent.

Proof If X_1 and X_2 are linearly dependent vectors, they must satisfy an equation of the form $c_1X_1 + c_2X_2 = O$ in which at least one of the c's is different from zero. Hence, multiplying through on the left by B, we have $c_1BX_1 + c_2BX_2 = O$, which proves that BX_1 and BX_2 are linearly dependent, as asserted.

Lemma 2

If X_1 is a characteristic vector of the matrix BAB^{-1} corresponding to the characteristic value $k = k_1$, then $B^{-1}X_1$ is a characteristic vector of the matrix A corresponding to the characteristic value $k = k_1$.

Proof Let k_1 be a characteristic value of the matrix BAB^{-1}, and hence of the matrix A, and let X_1 be a characteristic vector of BAB^{-1} corresponding to k_1.

---

[1] For any nonsingular matrix, B, the matrix $C = BAB^{-1}$ is said to be **similar** to the matrix A. From this, since $C = BAB^{-1}$ implies that $A = B^{-1}CB$, it follows that if C is similar to A, then A is similar to C. Thus Theorem 1 can be restated in the concise form: *similar matrices have the same characteristic roots.*

Then $(BAB^{-1} - k_1 I)X_1 = O$, and, by obvious steps,

$$(BAB^{-1} - k_1 I)X_1 = (BAB^{-1} - k_1 BIB^{-1})X_1 = B(A - k_1 I)(B^{-1}X_1) = O$$

Finally, multiplying through on the left by B^{-1}, the last equation becomes $(A - k_1 I)(B^{-1}X_1) = O$, which proves that $B^{-1}X_1$ is a characteristic vector of the matrix A, as asserted.

Lemma 3

If k_1 is a characteristic value of both C and DCD^{-1}, then the rank of the matrix $C - k_1 I$ is equal to or less than the rank of the matrix $DCD^{-1} - k_1 I$.

Proof Let k_1 be a characteristic root of the matrix DCD^{-1}, and let the rank of the matrix $DCD^{-1} - k_1 I$ be r. It follows, then, that the equation $(DCD^{-1} - k_1 I)X = O$ has exactly $n - r$ linearly independent solution vectors, $X_1, X_2, \ldots, X_{n-r}$. Now by Lemma 2, the vectors

$$D^{-1}X_1, \qquad D^{-1}X_2, \qquad \ldots, \qquad D^{-1}X_{n-r}$$

are solutions of the equation $(C - k_1 I)X = O$. Moreover, these vectors are linearly independent, for if they were dependent, it would follow by Lemma 1 that the vectors

$$D(D^{-1}X_1) = X_1, \qquad D(D^{-1}X_2) = X_2, \qquad \ldots, \qquad D(D^{-1}X_{n-r}) = X_{n-r}$$

were also linearly dependent, contrary to hypothesis. Hence the equation $(C - k_1 I)X = O$ has at least $n - r$ linearly independent solution vectors, which implies that the rank of the matrix $C - k_1 I$ is equal to or less than r. In other words, the rank of $C - k_1 I$ is equal to or less than the rank of $DCD^{-1} - k_1 I$, as asserted.

Theorem 2

If k_1 is a characteristic value of both A and BAB^{-1}, then the matrix $A - k_1 I$ and the matrix $BAB^{-1} - k_1 I$ have the same rank.

Proof By Lemma 3, with $C = A$ and $D = B$, it follows that the rank of $A - k_1 I$ is equal to or less than the rank of $BAB^{-1} - k_1 I$. Similarly, with $C = BAB^{-1}$ and $D = B^{-1}$, it follows that the rank of $BAB^{-1} - k_1 I$ is equal to or less than the rank of $B^{-1}(BAB^{-1})B - k_1 I = A - k_1 I$. Clearly, the only way in which these two inequalities can hold is for the rank of $A - k_1 I$ and the rank of $BAB^{-1} - k_1 I$ to be equal, as asserted.

From Theorems 1 and 2, it is now clear that not only do the matrices A and BAB^{-1} have the same characteristic roots but, moreover, for each of these roots, the matrices $A - kI$ and $BAB^{-1} - kI$ have the same rank. Thus the configuration of invariant points of $X' = AX$ and the configuration of invariant points of $X' = BAB^{-1}X$ are the same. In other words, the structure of the configuration of invariant points of a collineation, T, is independent of the particular coordinate system in terms of which the equation of T is expressed. The classification of collineations which we proposed at the end of the last section is therefore based on intrinsic rather than accidental properties, and the six types of collineations we listed are fundamentally different. In the following sections we shall analyze each of these in some detail.

Exercises

1. If X_1 and X_2 are linearly independent vectors, is it necessarily true that BX_1 and BX_2 are linearly independent?

2. Do A and $B^{-1}AB$ always have the same characteristic values?

3. Given $A = \begin{Vmatrix} 1 & 0 & 0 \\ 2 & 2 & -1 \\ 2 & 1 & 0 \end{Vmatrix}$ and $B = \begin{Vmatrix} -1 & 0 & 1 \\ 4 & 1 & -3 \\ 2 & 1 & -2 \end{Vmatrix}$. Verify that A and BAB^{-1} have the same characteristic values. Verify that $A - kI$ and $BAB^{-1} - kI$ have the same rank for each characteristic value. Find the characteristic vectors of BAB^{-1}, and verify that multiplying each on the left by B^{-1} yields a characteristic vector for A.

4. Work Exercise 3, given

$$A = \begin{Vmatrix} -3 & -2 & 2 \\ 14 & 7 & -5 \\ 10 & 4 & -2 \end{Vmatrix} \quad \text{and} \quad B = \begin{Vmatrix} 1 & 2 & 0 \\ -1 & 1 & 1 \\ -2 & 1 & 2 \end{Vmatrix}$$

5. Work Exercise 3, given

$$A = \begin{Vmatrix} 2 & 1 & -2 \\ -2 & 5 & -7 \\ -2 & 2 & -3 \end{Vmatrix} \quad \text{and} \quad B = \begin{Vmatrix} 1 & 0 & 1 \\ 0 & 2 & -1 \\ 1 & 1 & 0 \end{Vmatrix}$$

5.5 Collineations of Type I

By definition, the characteristic equation, $|A - kI| = 0$, of a collineation of type I has three unrepeated roots, k_1, k_2, k_3, and by Theorem 4, Sec. 5.3, for each of these roots the rank of the matrix $A - kI$ is equal to 2. Hence each of these roots gives rise to a single invariant point, and by Theorem 6,

Sec. 5.3, these three points, F_1, F_2, F_3, are noncollinear. Since the configuration of invariant lines is always the dual of the configuration of invariant points, it follows that there must also be three nonconcurrent invariant lines, and these are obviously the lines, $f_1 = F_2F_3$, $f_2 = F_3F_1$, $f_3 = F_1F_2$, determined by the three invariant points taken two at a time. Let us now try to find the equations of a collineation of type I, with characteristic roots k_1, k_2, k_3, in terms of a coordinate system whose triangle of reference is the triangle determined by the three invariant points.

If the invariant points of a collineation, T, of type I are $F_1:(1,0,0)$, $F_2:(0,1,0)$, and $F_3:(0,0,1)$, then under T

$$(1,0,0) \rightarrow (k_1,0,0) \qquad (0,1,0) \rightarrow (0,k_2,0) \qquad (0,0,1) \rightarrow (0,0,k_3)$$

Hence, imposing these requirements on the equations of the general collineation, namely,

$$x_1' = a_{11}x_1 + a_{12}x_2 + a_{13}x_3$$
$$x_2' = a_{21}x_1 + a_{22}x_2 + a_{23}x_3$$
$$x_3' = a_{31}x_1 + a_{32}x_2 + a_{33}x_3$$

we find that $a_{11} = k_1$, $a_{22} = k_2$, $a_{33} = k_3$ and that $a_{12} = a_{13} = a_{21} = a_{23} = a_{31} = a_{32} = 0$. Therefore, in terms of a coordinate system whose triangle of reference is the triangle whose vertices are the invariant points, the equations of a collineation of type I are

(1)
$$\begin{aligned} x_1' &= k_1x_1 \\ x_2' &= k_2x_2 \\ x_3' &= k_3x_3 \end{aligned} \qquad \begin{array}{l} k_1,\ k_2,\ k_3 \text{ all different and all} \\ \text{different from zero} \end{array}$$

The characteristic roots, k_1, k_2, k_3, have an interesting geometrical interpretation. To discover it, let us take the equation of the transformation in the form (1), and let us consider an arbitrary point $P:(p_1,p_2,p_3)$, distinct from each of the invariant points, and its image $P':(p_1',p_2',p_3') = (k_1p_1,k_2p_2, k_3p_3)$. The equation of the line joining P to the invariant point $F_3:(0,0,1)$ is $p_2x_1 - p_1x_2 = 0$, and the intersection of this line and the opposite side of the triangle of reference, $f_3: x_3 = 0$, is the point $Q_3:(p_1,p_2,0)$. Similarly, the equation of the line joining P' to $(0,0,1)$ is $p_2'x_1 - p_1'x_2 = 0$ or $k_2p_2x_1 - k_1p_1x_2 = 0$, and its intersection with the invariant line $x_3 = 0$ is the point $Q_3':(k_1p_1,k_2p_2,0)$ (Fig. 5.1). Now if the line $F_1F_2: x_3 = 0$ is parametrized in terms of F_1 and F_2 as base points, the parameters of Q_3 and Q_3' are, respectively, (p_1,p_2) and (k_1p_1,k_2p_2). Hence

$$R(F_1F_2,Q_3Q_3') = \frac{\begin{vmatrix} 1 & p_1 \\ 0 & p_2 \end{vmatrix} \cdot \begin{vmatrix} 0 & k_1p_1 \\ 1 & k_2p_2 \end{vmatrix}}{\begin{vmatrix} 1 & k_1p_1 \\ 0 & k_2p_2 \end{vmatrix} \cdot \begin{vmatrix} 0 & p_1 \\ 1 & p_2 \end{vmatrix}} = \frac{k_1}{k_2}$$

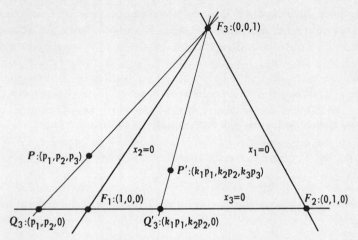

Fig. 5.1   The significance of the characteristic values of a collineation of type I.

In the same way, if Q_1 and Q_1' are the intersections of PF_1 and $P'F_1$ with the line F_2F_3: $x_1 = 0$, we have

$$\text{R}\,(F_2F_3,Q_1Q_1') = \frac{k_2}{k_3}$$

and if Q_2 and Q_2' are the intersections of PF_2 and $P'F_2$ with the line F_3F_1: $x_2 = 0$, we have

$$\text{R}\,(F_3F_1,Q_2Q_2') = \frac{k_3}{k_1}$$

Thus we have established the following theorem.

Theorem 1

If T is a collineation of type I with characteristic roots k_1, k_2, k_3 and corresponding invariant points F_1, F_2, F_3, if P' is the image under T of an arbitrary point P, distinct from F_1, F_2, F_3, and if Q_i and Q_i' are, respectively, the intersections of PF_i and $P'F_i$ with the invariant line F_jF_k, then

$$\text{R}\,(F_jF_k,Q_iQ_i') = k_j/k_k$$

A particularly interesting special case of a collineation of type I arises if the characteristic roots are assumed to be $e^{i\theta}$, $e^{-i\theta}$, and 1 $[\theta \neq n\pi]$, with corresponding invariant points $(1,i,0)$, $(1,-i,0)$, and $(0,0,1)$. If the equations of the general collineation are

$$x_1' = a_{11}x_1 + a_{12}x_2 + a_{13}x_3$$
$$x_2' = a_{21}x_1 + a_{22}x_2 + a_{23}x_3$$
$$x_3' = a_{31}x_1 + a_{32}x_2 + a_{33}x_3$$

the requirement that $(0,0,1) \rightarrow (0,0,1)$ implies that $a_{13} = a_{23} = 0$ and $a_{33} = 1$. Similarly, the requirements that $(1,i,0) \rightarrow (e^{i\theta}, ie^{i\theta}, 0)$ and $(1,-i,0) \rightarrow (e^{-i\theta}, -ie^{-i\theta}, 0)$ imply, respectively, that

$$\begin{aligned}
e^{i\theta} &= a_{11} + ia_{12} & e^{-i\theta} &= a_{11} - ia_{12} \\
ie^{i\theta} &= a_{21} + ia_{22} & -ie^{-i\theta} &= a_{21} - ia_{22} \\
0 &= a_{31} + ia_{32} & 0 &= a_{31} - ia_{32}
\end{aligned}$$

If we solve simultaneously the first equations in each of these sets, we find

$$a_{11} = \frac{e^{i\theta} + e^{-i\theta}}{2} = \cos\theta \quad \text{and} \quad a_{12} = \frac{e^{i\theta} - e^{-i\theta}}{2i} = \sin\theta$$

Likewise, by solving simultaneously the second equations in each set, we find

$$a_{21} = \frac{-e^{i\theta} + e^{-i\theta}}{2i} = -\sin\theta \quad \text{and} \quad a_{22} = \frac{e^{i\theta} + e^{-i\theta}}{2} = \cos\theta$$

Finally, solving simultaneously the last equations in each set, we have $a_{31} = a_{32} = 0$. Hence the equations of the collineation become

$$(2) \qquad \begin{aligned}
x_1' &= x_1 \cos\theta + x_2 \sin\theta \\
x_2' &= -x_1 \sin\theta + x_2 \cos\theta \\
x_3' &= x_3
\end{aligned}$$

Reverting to nonhomogeneous coordinates by putting

$$\frac{x_1}{x_3} = x \qquad \frac{x_2}{x_3} = y \quad \text{and} \quad \frac{x_1'}{x_3'} = x' \qquad \frac{x_2'}{x_3'} = y'$$

we obtain

$$\begin{aligned}
x' &= x \cos\theta + y \sin\theta \\
y' &= -x \sin\theta + y \cos\theta
\end{aligned}$$

which we recognize as the equations of a clockwise rotation of the euclidean plane about the origin through an angle whose measure is θ. Conversely, it is easy to verify that the characteristic roots of the matrix

$$A = \begin{Vmatrix} \cos\theta & \sin\theta & 0 \\ -\sin\theta & \cos\theta & 0 \\ 0 & 0 & 1 \end{Vmatrix}$$

of the transformation (2), which describes the most general rotation of the euclidean plane about the origin, are $e^{i\theta}$, $e^{-i\theta}$, 1 and that the corresponding invariant points are $(1,i,0)$, $(1,-i,0)$, $(0,0,1)$. Hence we have established the following theorem.

Theorem 2

The euclidean specialization of any collineation of type I whose characteristic roots are $e^{i\theta}$, $e^{-i\theta}$, 1 and whose associated invariant points are

$(1,i,0)$, $(1,-i,0)$, $(0,0,1)$ is a clockwise rotation about the origin through an angle of measure θ, and conversely provided $\theta \neq n\pi$.

The points $(1,i,0)$ and $(1,-i,0)$ are often called the **circular points at infinity.** They play a fundamental role in the study of euclidean geometry as a specialization of the geometry of Π_2.

Exercises

1. Find the fixed points of the collineation $X' = AX$ if:

(a) $A = \begin{Vmatrix} 0 & 6 & 2 \\ 1 & -1 & -1 \\ 2 & 6 & 0 \end{Vmatrix}$

(b) $A = \begin{Vmatrix} 1 & -4 & -4 \\ -1 & 1 & -1 \\ 1 & -2 & 0 \end{Vmatrix}$

(c) $A = \begin{Vmatrix} 1 & 2 & 2 \\ -2 & 6 & 3 \\ 2 & -4 & -1 \end{Vmatrix}$

(d) $A = \begin{Vmatrix} 0 & -5 & -3 \\ 0 & 3 & 2 \\ -1 & -5 & -2 \end{Vmatrix}$

2. Prove that no collineation of type I can be of period 2.

3. Under what conditions, if any, can a collineation of type I be of period 3? of period 4? of period n?

4. Is a rotation of the euclidean plane about an arbitrary point the specialization of a collineation of type I?

5. Determine the nature of the euclidean specialization of a collineation of type I whose characteristic values are 1, $e^{i\theta}$, $e^{-i\theta}$ if the corresponding fixed points are $(0,0,1)$, $(1,-i,0)$, $(1,i,0)$.

6. Is the composition of two collineations of type I always, sometimes, or never a collineation of type I?

7. Find the equations of a collineation of type I whose characteristic values are k_1, k_2, k_3 if the corresponding fixed points are:
(a) $(1,0,0)$, $(0,1,0)$, $(1,1,1)$ (b) $(1,0,0)$, $(0,1,1)$, $(0,1,-1)$
(c) $(0,1,1)$, $(1,0,1)$, $(1,1,0)$

8. Show that there are no nonsingular conics which are transformed into themselves by a collineation of type I.

9. What is the locus of a point P with the property that it and its image under a collineation of type I are collinear with a given point O? What is the locus of the images of such points?

10. Let Γ_1 and Γ_2 be two conics having the triangle of reference as a common self-polar triangle, and let T_1 and T_2 be the polarities determined by Γ_1 and

Γ_2, respectively. Under what conditions, if any, is the transformation $T_2^{-1}T_1$ a collineation of type I?

11. Let T be a collineation of type I with characteristic values k_1, k_2, k_3 and corresponding fixed points $F_1:(1,0,0)$, $F_2:(0,1,0)$, $F_3:(0,0,1)$. Let Γ be the conic $ax_2x_3 + bx_1x_3 + cx_1x_2 = 0$, and let Γ_1 be the image of Γ under T. Clearly, Γ and Γ_1 intersect in the fixed points of T and in some fourth point, P. Find the coordinates of P. If this process be iterated indefinitely, show that the intersection of Γ_n and Γ_{n+1} which is distinct from F_1, F_2, F_3 is T^nP.

12. In the proof of Theorem 2, why is it necessary to assume that $\theta \neq n\pi$?

5.6 Collineations of Types II and III

Even though their properties are quite different, it is natural to consider collineations of types II and III together, since the characteristic equation of each has one simple root and one double root.

By definition, the characteristic equation, $|A - kI| = 0$, of a collineation of type II has one simple root and one double root, and for each root the rank of the matrix $A - kI$ is equal to 2. Hence each of these roots gives rise to a single invariant point, and by Theorem 6, Sec. 5.3, these points, F_1 and F_2, are distinct. Clearly, the line, f_1, determined by the fixed points F_1 and F_2 is invariant. Moreover, since the configuration of invariant lines must be the dual of the configuration of invariant points, the collineation must have a second invariant line, f_2, necessarily passing through either F_1 or F_2, for otherwise its intersection with the line F_1F_2 would be a third invariant point on F_1F_2, which is impossible. We shall soon see that in every case this second invariant line must pass through the invariant point which arises from the repeated root of the characteristic equation.

To obtain a standard form for the equations of a collineation of type II, let us suppose the coordinate system chosen so that the fixed point corresponding to the unrepeated characteristic root, k_1, is the point $F_1:(0,0,1)$ and the fixed point corresponding to the repeated characteristic root, k_2, is the point $F_2:(1,0,0)$. Then since $(0,0,1) \rightarrow (0,0,k_1)$, it follows that in the equations of the general collineation, namely,

$$
\begin{aligned}
x_1' &= a_{11}x_1 + a_{12}x_2 + a_{13}x_3 \\
x_2' &= a_{21}x_1 + a_{22}x_2 + a_{23}x_3 \\
x_3' &= a_{31}x_1 + a_{32}x_2 + a_{33}x_3
\end{aligned}
\tag{1}
$$

we must have $a_{13} = a_{23} = 0$, $a_{33} = k_1$. Similarly, since $(1,0,0) \rightarrow (k_2,0,0)$, we must have $a_{11} = k_2$, $a_{21} = a_{31} = 0$. The equations of the collineation have thus been reduced to the form

$$
\begin{aligned}
x_1' &= k_2x_1 + a_{12}x_2 \\
x_2' &= a_{22}x_2 \\
x_3' &= a_{32}x_2 + k_1x_3
\end{aligned}
$$

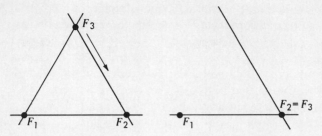

Fig. 5.2   A collineation of type II as a limiting case of a collineation of type I.

Moreover, since $k = k_2$ is to be a double root of the characteristic equation

$$|A - kI| = \begin{vmatrix} k_2 - k & a_{12} & 0 \\ 0 & a_{22} - k & 0 \\ 0 & a_{32} & k_1 - k \end{vmatrix} = 0$$

it follows that $a_{22} = k_2$. Also, since the rank of $A - k_2I$ is to be 2, it is clear that $a_{12} \neq 0$. Let us now suppose that the coordinate system is specialized further so that the point $(0,1,0)$ is a point on the second invariant line. As we observed above, this point and its image, namely, the point (a_{12},k_2,a_{32}), must be collinear with one or the other of the fixed points $F_1:(0,0,1)$ and $F_2:(1,0,0)$. Since

$$\begin{vmatrix} 0 & 1 & 0 \\ a_{12} & k_2 & a_{32} \\ 0 & 0 & 1 \end{vmatrix} = -a_{12} \neq 0$$

it follows that the second invariant line does not pass through $F_1:(0,0,1)$ and hence must pass through $F_2:(1,0,0)$. This requires that

$$\begin{vmatrix} 0 & 1 & 0 \\ a_{12} & k_2 & a_{32} \\ 1 & 0 & 0 \end{vmatrix} = a_{32} = 0$$

Thus the second invariant line, f_2, is the line $x_3 = 0$, and the equations of the collineation reduce to the standard form

$$\begin{aligned} x'_1 &= k_2x_1 + a_{12}x_2 \\ x'_2 &= \qquad\quad k_2x_2 \\ x'_3 &= \qquad\qquad\qquad k_1x_3 \end{aligned} \qquad \begin{aligned} k_1, k_2, a_{12} &\neq 0 \\ k_1 &\neq k_2 \end{aligned}$$

For descriptive purposes, it may be helpful to consider a collineation of type II as a limiting case of a collineation of type I in which two of the three fixed points of the latter approach coincidence, the line along which they approach each other being the second invariant line of the collineation of type II. Figure 5.2 suggests this behavior.

By definition, the characteristic equation of a collineation of type III has an unrepeated root, k_1, for which the rank of $A - kI$ is 2, and a double root, k_2, for which the rank of $A - kI$ is 1. There is thus a single invariant point, F_1, corresponding to the simple root, k_1, and a line of invariant points, f_1, arising from the two independent fixed points which correspond to the double root k_2. Moreover, by Theorem 6, Sec. 5.3, the invariant point F_1 does not lie on the line of invariant points, f_1. Dually, the configuration of invariant lines consists of a fixed line, f_1, and a point of fixed lines, which clearly is the point F_1 since every line on F_1 contains two fixed points, namely, F_1 and its intersection with f_1. The fixed point F_1 is often called the **center** of the collineation, and the line of fixed points, f_1, is often called the **axis** of the collineation.

If we now choose our coordinate system so that the fixed point, F_1, is the point $(1,0,0)$ and the line of fixed points, f_1, is the line $x_1 = 0$, we must have $(1,0,0) \rightarrow (k_1,0,0)$. Moreover, since every point on f_1 must be an invariant point corresponding to the characteristic root $k = k_2$, we must have for all values of p_2 and p_3, $(0,p_2,p_3) \rightarrow (0,k_2p_2,k_2p_3)$. When the first of these requirements is imposed on the equations of the general collineation (1), we find that $a_{11} = k_1$, $a_{21} = a_{31} = 0$. Then from the second requirement we find that $a_{22} = a_{33} = k_2$ and $a_{12} = a_{13} = a_{23} = a_{32} = 0$. Thus for our choice of coordinate system, the equations of a collineation of type III assume the standard form

$$(2) \qquad \begin{aligned} x_1' &= k_1x_1 \\ x_2' &= k_2x_2 \\ x_3' &= k_2x_3 \end{aligned} \qquad \begin{aligned} k_1, k_2 &\neq 0 \\ k_1 &\neq k_2 \end{aligned}$$

As we found in the case of a collineation of type I, so also for a collineation of type III the ratio of the characteristic roots has an important geometrical significance. Specifically, we have the following theorem.

Theorem 1

Let F_1 be the center and f_1 be the axis of a collineation of type III. Let P be an arbitrary point, distinct from F_1 and not lying on f_1, and let F_2 be the intersection of the line PF_1 and the axis f_1. Finally, let P' be the image of P under the collineation. Then $\text{R}(F_1F_2,PP') = k_1/k_2$, where k_1 is the simple root and k_2 is the double root of the characteristic equation of the matrix of the collineation.

Proof Let F_1 be the center and f_1 be the axis of a collineation of type III. Then since the lines which pass through F_1 are all invariant, though not in a point-by-point fashion, it follows that a general point P and its image P' are always collinear with F_1 and the cross ratio appearing in the statement of the theorem is meaningful. Now let $P:(p_1,p_2,p_3)$ be any point distinct from F_1 and not lying on f_1. Then using the standard form (2) of the equations of the collineation, so that $F_1:(1,0,0)$ and $f_1: x_1 = 0$, we have for the line PF_1 the equation $p_3x_2 - p_2x_3 = 0$. The intersection of this line and the invariant line f_1 is therefore the point $F_2:(0,p_2,p_3)$. Furthermore, the image of P is the point $P':(k_1p_1,k_2p_2,k_2p_3)$. Hence, if the line PF_1 is parametrized in terms of F_1 and F_2 as base points, the parameters of P are $(p_1,1)$ and the parameters of P' are (k_1p_1,k_2). Therefore

$$ \text{R}(F_1F_2,PP') = \frac{\begin{vmatrix} 1 & p_1 \\ 0 & 1 \end{vmatrix} \cdot \begin{vmatrix} 0 & k_1p_1 \\ 1 & k_2 \end{vmatrix}}{\begin{vmatrix} 1 & k_1p_1 \\ 0 & k_2 \end{vmatrix} \cdot \begin{vmatrix} 0 & p_1 \\ 1 & 1 \end{vmatrix}} = \frac{k_1}{k_2} $$

as asserted.

Recalling the definition of a harmonic homology from Sec. 4.7, we have the following corollary of Theorem 1.

Corollary 1

A collineation of type III whose characteristic roots are the negatives of each other is a harmonic homology, and conversely.

There are several euclidean specializations of harmonic homologies which are worthy of note. In the first place, if in Eqs. (2) we take $k_1 = -1$ and $k_2 = 1$ and then revert to nonhomogeneous coordinates in the usual fashion, we obtain the euclidean transformation

$$ \begin{aligned} x' &= -x \\ y' &= y \end{aligned} $$

which is the equation of a reflection in the y axis. On the other hand, if instead of taking the center to be the point $(1,0,0)$ and the axis to be the line $x_1 = 0$, as we did in deriving the standard form (2), we take the center to be the point $(0,1,0)$ and the axis to be the line $x_2 = 0$, we obtain the alternate standard form

$$ \begin{aligned} x_1' &= k_2x_1 \\ x_2' &= k_1x_2 \\ x_3' &= k_2x_3 \end{aligned} $$

Now if we take $k_1 = -1$ and $k_2 = 1$ and revert to nonhomogeneous coordinates, we obtain the euclidean transformation

$$ \begin{aligned} x' &= x \\ y' &= -y \end{aligned} $$

which describes a reflection in the x axis. Finally, if we take the center to be the point $(0,0,1)$ and the axis to be the (ideal) line $x_3 = 0$, we obtain the alternate standard form

$$
\begin{aligned}
x_1' &= k_2 x_1 \\
x_2' &= k_2 x_2 \\
x_3' &= k_1 x_3
\end{aligned}
$$

and if we now take $k_1 = 1$, $k_2 = -1$ and revert to nonhomogeneous coordinates, we obtain the euclidean transformation

$$
\begin{aligned}
x' &= -x \\
y' &= -y
\end{aligned}
$$

which is a reflection in the origin. These results are summarized in the following theorem.

Theorem 2

The euclidean specialization of a harmonic homology whose axis is the y axis and whose center is the ideal point in the direction perpendicular to the y axis is a reflection in the y axis, and conversely. The euclidean specialization of the harmonic homology whose axis is the x axis and whose center is the ideal point in the direction perpendicular to the x axis is a reflection in the x axis, and conversely. The euclidean specialization of the harmonic homology whose center is the origin and whose axis is the ideal line is a reflection in the origin, and conversely.

Exercises

1. Show that no collineation of type II can be periodic.

2. Under what conditions, if any, can a collineation of type III be of period 2? Of period 3? Of period 4? Of period n?

3. Show that the plane perspective transformation described in Chap. 1 is a collineation of type III.

4. If P is a point on the line $F_1 F_2$, and if P' is its image under a collineation of type II, what is $R(F_1 F_2, PP')$?

5. Find the fixed points of the transformation $X' = AX$ if:

(a) $A = \left\| \begin{array}{rrr} 3 & 1 & -1 \\ -6 & -2 & 3 \\ -4 & -2 & 3 \end{array} \right\|$
  (b) $A = \left\| \begin{array}{rrr} -4 & 3 & -3 \\ -4 & 4 & -5 \\ 2 & -1 & 0 \end{array} \right\|$

(c) $A = \left\| \begin{array}{rrr} 9 & 3 & -7 \\ 4 & 1 & -2 \\ 12 & 4 & -9 \end{array} \right\|$
  (d) $A = \left\| \begin{array}{rrr} -3 & -2 & 2 \\ 12 & 7 & -6 \\ 8 & 4 & -3 \end{array} \right\|$

6. What is the locus of a point with the property that it and its image under a collineation of type II are collinear with a given point O? What is the locus of the images of such points?

7. What are the equations of a collineation of type II with characteristic values k_1, k_2, k_2 if:
(a) The corresponding fixed points are $F_1:(0,1,0)$ and $F_2:(1,1,0)$, and the second fixed line is $f_2: x_1 - x_2 = 0$?
(b) The corresponding fixed points are $F_1:(1,0,0)$ and $F_2:(0,-1,1)$, and the second fixed line is $f_2: x_1 = 0$?

8. (a) Show that any collineation of type I can be obtained as the composition of two collineations of type III.
(b) Can every collineation of type III be obtained as the composition of two collineations of type II?

9. If T_1 is the harmonic homology whose center is the point $C_1:(1,-1,0)$ and whose axis is the line $l: x_1 - x_2 = 0$, and if T_2 is the quadratic inversion defined by the point $C_2:(0,0,1)$ and the conic $\Gamma: x_3{}^2 - x_1 x_2 = 0$, find the equations of T_1 and T_2 and show that T_1 and T_2 commute. Show also that $T = T_1 T_2$ is of period 2.

10. In Exercise 9, what is the image of each side of the triangle of reference under the transformation $T = T_1 T_2$? What can be said about the curve into which T transforms a conic which is tangent to each side of the triangle of reference? What can be said of the curve into which T transforms a conic which intersects each side of the triangle of reference in two real points? What are the images under T of the vertices of the triangle of reference? What is the image under T of a conic which passes through the vertices of the triangle of reference?

5.7 Collineations of Types IV, V, and VI

Collineations of types IV, V, and VI are alike in that their characteristic equations have a triple root, $k = k_1$; they differ in that the rank of the matrix $A - k_1 I$ is 2, 1, and 0 for the respective types.

By definition, the characteristic equation of a collineation of type IV has a triple root, $k = k_1$, for which the rank of the matrix $A - kI$ is 2. The collineation therefore has a single invariant point, F_1, and dually, a single invariant line, f_1. In this case, the fixed line must pass through the fixed point, for if it did not, then the projectivity induced on the fixed line by the collineation would have at least one fixed point, necessarily distinct from F_1, which is impossible.

If we choose a coordinate system such that the invariant point, F_1, is the point $(0,0,1)$ and the invariant line, f_1, is the line $x_1 = 0$, then $(0,0,1) \rightarrow (0,0,k_1)$, and any point whose first coordinate is zero must be mapped into another point whose first coordinate is zero. The first of these conditions requires that in the equations of a general collineation, namely,

$$
(1) \qquad
\begin{aligned}
x_1' &= a_{11}x_1 + a_{12}x_2 + a_{13}x_3 \\
x_2' &= a_{21}x_1 + a_{22}x_2 + a_{23}x_3 \\
x_3' &= a_{31}x_1 + a_{32}x_2 + a_{33}x_3
\end{aligned}
$$

we must have $a_{13} = a_{23} = 0$ and $a_{33} = k_1$. The second condition requires further that $a_{12} = 0$. Therefore, since $k = k_1$ must be a triple root of the characteristic equation

$$
\begin{vmatrix}
a_{11} - k & 0 & 0 \\
a_{21} & a_{22} - k & 0 \\
a_{31} & a_{32} & k_1 - k
\end{vmatrix} = 0
$$

it follows that $a_{11} = a_{22} = k_1$. Moreover, since the rank of $A - k_1 I$ must be 2, neither a_{21} nor a_{32} can be zero. The equations of a collineation of type IV can therefore be reduced to the standard form

$$
\begin{aligned}
x_1' &= k_1 x_1 \\
x_2' &= a_{21}x_1 + k_1 x_2 \qquad\qquad a_{21},\, a_{32},\, k_1 \neq 0 \\
x_3' &= a_{31}x_1 + a_{32}x_2 + k_1 x_3
\end{aligned}
$$

By definition, the characteristic equations of a collineation of type V have a triple root, $k = k_1$, for which the rank of the matrix $A - kI$ is 1. The collineation therefore has a line, f_1, of invariant points and a point, F_1, of invariant lines.[1] Since the vertex, F_1, of the pencil of invariant lines is obviously an invariant point, it follows that F_1 must be a point of f_1. If we now choose a coordinate system such that f_1 is the line $x_3 = 0$, then for all values of p_1 and p_2 we have $(p_1,p_2,0) \rightarrow (k_1 p_1, k_1 p_2, 0)$. This implies that in the general equations (1) we must have $a_{12} = a_{21} = a_{31} = a_{32} = 0$ and $a_{11} = a_{22} = k_1$. Moreover, since $k = k_1$ is to be a triple root of the characteristic equation

$$
\begin{vmatrix}
k_1 - k & 0 & a_{13} \\
0 & k_1 - k & a_{23} \\
0 & 0 & a_{33} - k
\end{vmatrix} = 0
$$

it follows that $a_{33} = k_1$.

From the equations of the collineation as we have thus far specialized them, the image of a general point $P:(p_1,p_2,p_3)$ is the point $P':(k_1 p_1 + a_{13}p_3,\ k_1 p_2 + a_{23}p_3,\ k_1 p_3)$. Furthermore, any point P and its image P' must

---

[1] As in the case of a collineation of type III, so here the line of fixed points, f_1, and the point of fixed lines, F_1, are often called, respectively, the **axis** and the **center** of the collineation.

be collinear with F_1. Hence, letting F_1 be the point $(u,v,0)$, we must have

$$\begin{vmatrix} p_1 & p_2 & p_3 \\ k_1p_1 + a_{13}p_3 & k_1p_2 + a_{23}p_3 & k_1p_3 \\ u & v & 0 \end{vmatrix} = 0$$

or, expanding and simplifying, $a_{13}v - a_{23}u = 0$, which implies that $a_{13} = \alpha u$ and $a_{23} = \alpha v$. Thus the equations of a collineation of type V can be reduced to the standard form

$$\begin{aligned} x_1' &= k_1x_1 && + \alpha ux_3 \\ x_2' &= && k_1x_2 + \alpha vx_3 \\ x_3' &= && k_1x_3 \end{aligned}$$

or, dividing out the proportionality constant k_1 and then setting $a = \alpha u/k_1$ and $b = \alpha v/k_1$,

$$\begin{aligned} x_1' &= x_1 && + ax_3 \\ x_2' &= && x_2 + bx_3 \\ x_3' &= && x_3 \end{aligned}$$

From this, by reverting to nonhomogeneous coordinates, we obtain the equations

$$\begin{aligned} x' &= x + a \\ y' &= y + b \end{aligned}$$

which are the equations of a translation in the euclidean plane. Hence we have established the following theorem.

Theorem 1

The euclidean specialization of a collineation of type V whose line of invariant points is the ideal line is a translation in the direction of the ideal point which is the vertex of the pencil of invariant lines of the collineation.

Finally, we recall that by definition the characteristic equations of a collineation of type VI have a triple root $k = k_1$ for which the rank of the matrix $A - kI$ is equal to zero. This means that every element in the matrix $A - k_1I$ must be zero, which implies that $a_{12} = a_{21} = a_{13} = a_{31} = a_{23} = a_{32} = 0$ and $a_{11} = a_{22} = a_{33} = k_1$. Thus the equations of any collineation of type VI must be

$$\begin{aligned} x_1' &= k_1x_1 \\ x_2' &= k_1x_2 \\ x_3' &= k_1x_3 \end{aligned}$$

or simply

$$\begin{aligned} x_1' &= x_1 \\ x_2' &= x_2 \\ x_3' &= x_3 \end{aligned}$$

which are the equations of the **identity collineation** which leaves every point in the plane invariant.

Exercises

1. Show that no collineation of type V can be periodic.

2. Under what conditions, if any, can a collineation of type IV be periodic?

3. Show that the composition of two collineations of type V with the same axis is always a collineation of type V.

4. Show that the composition of two collineations of type V with distinct axes will also be a collineation of type V if and only if the center of each transformation is the point of intersection of the two axes.

5. Find the equations of the most general collineation of type IV having the fixed elements:
(a) $F_1:(0,1,0), f_1: x_3 = 0$ (b) $F_1:(1,0,1), f_1: x_2 = 0$
(c) $F_1:(1,0,0), f_1: x_2 - x_3 = 0$

6. Find the equations of the most general collineation of type V having the fixed elements:
(a) $F_1:(s,0,t), f_1: x_2 = 0$ (b) $F_1:(1,0,1), f_1: x_1 - x_3 = 0$
(c) $F_1:(1,1,1), f_1: x_1 - x_2 = 0$

7. Under what conditions, if any, will two collineations of type V commute with each other?

8. Show that in general the composition of two collineations of type V is a collineation of type I. Under what conditions, if any, will the composition be a collineation of type II? Of type III? Of type IV?

9. What is the locus of a point with the property that it and its image under a collineation of type IV are collinear with a given point O? What is the locus of the images of such points?

10. Find the fixed points of the transformation $X' = AX$ if:

(a) $A = \begin{Vmatrix} 3 & -1 & 1 \\ 6 & -2 & 3 \\ 2 & -1 & 2 \end{Vmatrix}$ (b) $A = \begin{Vmatrix} -2 & -4 & -1 \\ 3 & 5 & 1 \\ -1 & -2 & 0 \end{Vmatrix}$

(c) $A = \begin{Vmatrix} 3 & 2 & -4 \\ 8 & 3 & -8 \\ 8 & 4 & -9 \end{Vmatrix}$ (d) $A = \begin{Vmatrix} -2 & -1 & 3 \\ 5 & 3 & -4 \\ -4 & -1 & 5 \end{Vmatrix}$

5.8 The Determination of a Collineation

In the last three sections we have investigated in some detail the six types of nonsingular linear transformations, or collineations, which exist in Π_2.

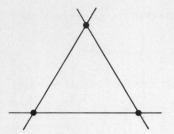

Type I: Three noncollinear
fixed points and three
nonconcurrent fixed lines

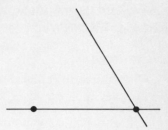

Type II: Two fixed points
and two fixed lines

Type III: A line of fixed points
and a pencil of fixed lines, the
vertex of the pencil not lying
on the line of fixed points

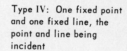

Type IV: One fixed point
and one fixed line, the
point and line being
incident

Type V: A line of fixed points
and a pencil of fixed lines, the
vertex of the pencil lying on
the line of fixed points

Type VI: Every point fixed
and every line fixed

Fig. 5.3   The possible configurations of the invariant elements of a collineation.

In particular, we determined the fixed points and fixed lines for each case and obtained the invariant configurations shown in Fig. 5.3.

Suppose now that P_1, P_2, P_3, and P_4 are four points no three of which are collinear, and suppose that T is a collineation which leaves these four points invariant. By comparing this set of fixed points with the invariant configurations shown in Fig. 5.3, it is clear that T cannot be of type I, II, III, IV, or V; hence if it exists, it must be of type VI. Moreover, the identity collineation does indeed leave P_1, P_2, P_3, and P_4 invariant since in fact it leaves every point invariant. Thus we have established the following simple but important theorem.

Theorem 1

If a collineation T leaves invariant each of four points no three of which are collinear, then T is the identity.

Using Theorem 1, we can now prove the following fundamental result on the determination of collineations.

Theorem 2

If (P_1,P_2,P_3,P_4) and (Q_1,Q_2,Q_3,Q_4) are two sets of four points such that no three points in either set are collinear, then there is at most one collineation which simultaneously maps P_1 into Q_1, P_2 into Q_2, P_3 into Q_3, and P_4 into Q_4.

Proof   Let (P_1,P_2,P_3,P_4) be four points no three of which are collinear, let (Q_1,Q_2,Q_3,Q_4) be four points no three of which are collinear, and let us suppose that both $T_1: X' = A_1 X$ and $T_2: X' = A_2 X$ are collineations which map P_1 into Q_1, P_2 into Q_2, P_3 into Q_3, and P_4 into Q_4. Then T_2^{-1} maps Q_1 into P_1, Q_2 into P_2, Q_3 into P_3, and Q_4 into P_4; and therefore, by Theorem 1, the transformation $T_2^{-1}T_1$ is the identity. Now the matrix of the transformation $T_2^{-1}T_1$ is $A_2^{-1}A_1$, and since $T_2^{-1}T_1$ is the identity, it is clear that $A_2^{-1}A_1 = I$. Finally, multiplying this equation through on the left by A_2, we conclude that $A_1 = A_2$, which proves that T_1 and T_2 are the same transformation, as asserted.

Theorem 2 tells us that *if* there is a collineation which maps four given points no three of which are collinear into four given points no three of which are collinear, *then* that collineation is unique, but it does not tell us that such a transformation actually exists. That this is the case, however, is guaranteed by the next theorem.

Theorem 3

If (P_1,P_2,P_3,P_4) and (Q_1,Q_2,Q_3,Q_4) are two sets of four points such that no three points in either set are collinear, then there exists at least one collineation which maps P_1 into Q_1, P_2 into Q_2, P_3 into Q_3, and P_4 into Q_4.

Proof   Let (P_1,P_2,P_3,P_4) be four points no three of which are collinear, and let $a_{11}x_1 + a_{12}x_2 + a_{13}x_3 = 0$ be the equation of the line P_2P_3, let $a_{21}x_1 + a_{22}x_2 + a_{23}x_3 = 0$ be the equation of the line P_3P_1, and let $a_{31}x_1 + a_{32}x_2 +$

$a_{33}x_3 = 0$ be the equation of the line P_1P_2. Then the transformation

$$x_1' = k_1(a_{11}x_1 + a_{12}x_2 + a_{13}x_3)$$
$$x_2' = k_2(a_{21}x_1 + a_{22}x_2 + a_{23}x_3)$$
$$x_3' = k_3(a_{31}x_1 + a_{32}x_2 + a_{33}x_3)$$

clearly has the property that it maps P_1 into $(1,0,0)$, P_2 into $(0,1,0)$, and P_3 into $(0,0,1)$. Furthermore, since P_4 does not lie on any of the lines P_2P_3, P_3P_1, P_1P_2, it follows that nonzero values of k_1, k_2, k_3 can be found such that P_4 will map into $(1,1,1)$. Let T_1: $X' = A_1X$ be the transformation which thus maps P_1 into $(1,0,0)$, P_2 into $(0,1,0)$, P_3 into $(0,0,1)$, and P_4 into $(1,1,1)$. Since P_2P_3, P_3P_1, P_1P_2 are nonconcurrent lines, the matrix A_1 is nonsingular, and T_1 is a collineation. Similarly, let T_2: $X' = A_2X$ be the collineation which maps Q_1, Q_2, Q_3, Q_4 into $(1,0,0)$, $(0,1,0)$, $(0,0,1)$, $(1,1,1)$, respectively. Now, clearly, the transformation T_2^{-1}: $X' = A_2^{-1}X$ maps $(1,0,0)$ into Q_1, $(0,1,0)$ into Q_2, $(0,0,1)$ into Q_3, and $(1,1,1)$ into Q_4. Therefore the composition of T_1 and T_2^{-1}, namely, T: $X' = A_2^{-1}A_1X$, is a collineation which maps P_1 into Q_1, P_2 into Q_2, P_3 into Q_3, and P_4 into Q_4, as required.

Combining and restating the last two theorems, we have the following result, which is often called the **fundamental theorem of projective geometry.**

Theorem 4

If (P_1,P_2,P_3,P_4) and (Q_1,Q_2,Q_3,Q_4) are two sets of four points such that no three points in either set are collinear, then there exists one and only one collineation which maps P_1 into Q_1, P_2 into Q_2, P_3 into Q_3, and P_4 into Q_4.

Example 1

Given P_1:$(1,-1,0)$, P_2:$(1,1,2)$, P_3:$(0,1,3)$, P_4:$(1,2,6)$ and Q_1:$(1,0,2)$, Q_2:$(1,-1,1)$, Q_3:$(1,2,3)$, Q_4:$(1,3,3)$, find the collineation which maps P_1 into Q_1, P_2 into Q_2, P_3 into Q_3, and P_4 into Q_4.

By steps which are now completely familiar to us, we find without difficulty that the equations of P_2P_3, P_3P_1, and P_1P_2 are, respectively, $x_1 - 3x_2 + x_3 = 0$, $3x_1 + 3x_2 - x_3 = 0$, and $x_1 + x_2 - x_3 = 0$. Hence any collineation of the family

$$x_1' = k_1(\ x_1 - 3x_2 + x_3)$$
$$x_2' = k_2(3x_1 + 3x_2 - x_3)$$
$$x_3' = k_3(\ x_1 + \ x_2 - x_3)$$

will map P_1 into $(1,0,0)$, P_2 into $(0,1,0)$, and P_3 into $(0,0,1)$. Furthermore, substituting $(x_1,x_2,x_3) = (1,2,6)$ and $(x_1',x_2',x_3') = (1,1,1)$, we find that these points will also correspond, as required, if $k_1 = 1$, $k_2 = \frac{1}{3}$, and $k_3 = -\frac{1}{3}$. Hence, inserting these values and then multiplying through by the proportionality constant 3, we find that the collineation

$$T_1: \quad \begin{aligned} x_1' &= 3x_1 - 9x_2 + 3x_3 \\ x_2' &= 3x_1 + 3x_2 - x_3 \\ x_3' &= -x_1 - x_2 + x_3 \end{aligned} \quad \text{or} \quad X' = A_1X$$

accomplishes the mapping $P_1 \rightarrow (1,0,0)$, $P_2 \rightarrow (0,1,0)$, $P_3 \rightarrow (0,0,1)$, $P_4 \rightarrow (1,1,1)$. Similarly we find that the collineation

$$T_2: \quad \begin{aligned} x_1' &= 5x_1 + 2x_2 - 3x_3 \\ x_2' &= 8x_1 + 2x_2 - 4x_3 \\ x_3' &= 2x_1 + x_2 - x_3 \end{aligned} \quad \text{or} \quad X' = A_2X$$

accomplishes the mapping $Q_1 \rightarrow (1,0,0)$, $Q_2 \rightarrow (0,1,0)$, $Q_3 \rightarrow (0,0,1)$, $Q_4 \rightarrow (1,1,1)$. Hence the inverse of T_2, namely, $T_2^{-1}: X' = A_2^{-1}X$, will map $(1,0,0)$ into Q_1, $(0,1,0)$ into Q_2, $(0,0,1)$ into Q_3, and $(1,1,1)$ into Q_4; and the composition of T_1 and T_2^{-1}, namely, $T: X' = A_2^{-1}A_1X$, will accomplish the mapping asked for in the problem. Specifically, we find

$$A_2^{-1} = -\frac{1}{2} \begin{Vmatrix} 2 & -1 & -2 \\ 0 & 1 & -4 \\ 4 & -1 & -6 \end{Vmatrix} \quad \text{and} \quad A_2^{-1}A_1 = -\frac{1}{2} \begin{Vmatrix} 5 & -19 & 5 \\ 7 & 7 & -5 \\ 15 & -33 & 7 \end{Vmatrix}$$

Hence, neglecting the irrelevant proportionality constant $-\frac{1}{2}$, the collineation required by the problem is

$$T: \quad \begin{aligned} x_1' &= 5x_1 - 19x_2 + 5x_3 \\ x_2' &= 7x_1 + 7x_2 - 5x_3 \\ x_3' &= 15x_1 - 33x_2 + 7x_3 \end{aligned}$$

Exercises

1. Prove Theorem 4 by showing that the determination of the required collineation is equivalent to solving a system of 12 homogeneous linear equations in 13 variables when the rank of the matrix of the coefficients is 12.

2. Find the transformation which effects each of the following mappings:

(a)
$(0,0,1) \rightarrow (1,-1,2)$
$(1,-1,2) \rightarrow (3,0,5)$
$(1,1,1) \rightarrow (0,3,1)$
$(1,0,0) \rightarrow (1,-1,0)$

(b)
$(1,0,0) \rightarrow (0,1,1)$
$(1,-1,0) \rightarrow (1,-1,0)$
$(2,0,1) \rightarrow (1,0,2)$
$(1,-1,1) \rightarrow (0,1,0)$

(c)
$(1,0,0) \rightarrow (1,1,1)$
$(0,1,0) \rightarrow (1,0,1)$
$(-1,1,1) \rightarrow (0,0,1)$
$(1,1,1) \rightarrow (2,2,1)$

(d)
$(0,0,1) \rightarrow (0,1,3)$
$(1,1,0) \rightarrow (3,0,1)$
$(0,3,1) \rightarrow (3,1,0)$
$(1,1,1) \rightarrow (-3,1,2)$

3. Show that there is more than one transformation T which will accomplish the following mappings, find the most general form of T, and explain why T is not unique:

(a) $(2,0,1) \rightarrow (1,1,0)$
$(2,2,3) \rightarrow (0,0,1)$
$(1,1,1) \rightarrow (1,0,0)$
$(0,1,1) \rightarrow (2,2,1)$

(b) $(1,0,0) \rightarrow (1,0,0)$
$(1,1,0) \rightarrow (1,0,-1)$
$(3,0,-1) \rightarrow (1,-1,1)$
$(-1,2,1) \rightarrow (0,1,-2)$

4. Verify that there is no collineation which will accomplish either of the following mappings, and explain why:

(a) $(1,1,-1) \rightarrow (2,1,1)$
$(3,0,-1) \rightarrow (0,0,1)$
$(1,-1,0) \rightarrow (1,1,0)$
$(1,-2,1) \rightarrow (1,1,1)$

(b) $(3,2,0) \rightarrow (1,-1,0)$
$(2,2,1) \rightarrow (1,1,2)$
$(1,1,1) \rightarrow (1,0,1)$
$(4,2,-1) \rightarrow (1,1,0)$

5. (a) Let T be the collineation which leaves the points $(1,0,0)$, $(0,1,0)$, $(0,0,1)$ invariant and maps the point $P:(y_1,y_2,y_3)$ onto the point $(1,1,1)$. Determine the possible types for T, and find the locus of P corresponding to each type.

(b) Work part (a) if T effects the mapping $(1,0,0) \rightarrow (0,1,1)$, $(0,1,0) \rightarrow (1,0,1)$, $(0,0,1) \rightarrow (1,1,0)$, $(y_1,y_2,y_3) \rightarrow (1,1,1)$.

5.9 Groups

It is a fact, familiar to mathematicians and nonmathematicians alike, that geometry is an axiomatic science, unfolding deductively from a set of unproved assumptions, or *axioms*, by the application of the laws of logic. It was the Greek mathematician Euclid (365?–275? B.C.) who first made this clear by showing how the geometry of his day, now known in his honor as *euclidean* geometry, could be developed as a logical consequence of a relatively few initial assumptions. Since then, other sets of axioms differing in essential ways from Euclid's have been studied, and other, *noneuclidean*, geometries have been developed. A geometry can thus be thought of as the totality of properties which arise from a given set of axioms, much as an unfolding flower is the inevitable product of a particular kind of seed.

The axiomatic approach to geometry is not the only one which can be or has been taken, however. In fact, up to this point in our work in projective geometry we have had little or nothing to do with the axiomatic foundations of the subject, though we shall begin a consideration of these in the next chapter. In particular, in the inaugural lecture which he gave in 1872 when he was appointed professor of mathematics at the University of Erlangen, the German mathematician Felix Klein (1849–1925) suggested that a geometry might be considered to be the study of all properties of a given **space,** or collection of elements, which are invariant under a given set of transformations of that space onto itself.

In the proposal of Klein, which has since become known as the **Erlangen program,** the sets of transformations to which he referred were of the particular type known as *groups*. Since groups are among the most important algebraic structures studied in mathematics, and since they are of great interest in geometry, as the Erlangen program suggests, we shall devote this section to a brief discussion of the concept of a group, preparatory to considering certain particular groups of collineations in the remaining sections of this chapter.

Definition 1

A system G consisting of a collection of elements e_1, e_2, e_3, . . . and an operation ∘ by means of which two elements can be combined is said to be a group if it has the following properties:

1. If e_i and e_j are any elements of G, then $e_i \circ e_j$ is also an element of G (closure).

2. If e_i, e_j, and e_k are any elements of G, then $(e_i \circ e_j) \circ e_k = e_i \circ (e_j \circ e_k)$ (associativity).

3. There exists an element I of G such that $e_i \circ I = I \circ e_i = e_i$ for each element e_i in G (existence of an identity).

4. For each element e_i of G there exists an element e_i^{-1} of G such that $e_i \circ e_i^{-1} = e_i^{-1} \circ e_i = I$ (existence of inverses).

Definition 2

Any subset of the elements of a group G which is also a group with respect to the operation ∘ of G is said to be a subgroup of G.

Definition 3

A group which contains only a finite number n of elements is said to be a group of order n.

As we shall soon see, the nature of the binary operation ∘ by means of which two elements of G are combined to give a third element of G may differ greatly from group to group. It is customary, however, to refer to the operation ∘ as "multiplication" and to speak of the element $e_i \circ e_j$ as the "product" of e_i and e_j. Furthermore, in most cases the usual notation for products is adopted, and $e_i \circ e_j$ is written simply as $e_i e_j$.

Our first example of a group is important enough to be stated as a theorem.

Theorem 1

The set of all nonsingular linear transformations in Π_2 is a group under the operation of composition.

Proof   Let G be the set of all nonsingular linear transformations T: $X' = AX$ in Π_2. To prove the theorem we must show that each of the requirements of Definition 1 is fulfilled when the operation $\circ$ is interpreted to be composition, i.e., when $T_j \circ T_i = T_j T_i$ is interpreted to be the transformation resulting when T_i is followed by T_j.

To verify condition 1, let T_i: $X' = A_i X$ and T_j: $X' = A_j X$ be any nonsingular linear transformations. Then, as usual, $T_j T_i$ is the transformation $X' = A_j A_i X$, which surely has the form of a linear transformation. Moreover, since $|A_j A_i| = |A_j| \cdot |A_i|$ and since A_i and A_j are nonsingular, it follows that the matrix $A_j A_i$ is also nonsingular. Hence $T_j T_i$ is a nonsingular linear transformation, and condition 1 is verified.

To verify condition 2, let T_i: $X' = A_i X$, T_j: $X' = A_j X$, and T_k: $X' = A_k X$ be any nonsingular linear transformations. Then $T_k(T_j T_i)$ is the transformation whose matrix is $A_k(A_j A_i)$, and $(T_k T_j) T_i$ is the transformation whose matrix is $(A_k A_j) A_i$. However, matric multiplication is associative, and hence $(A_k A_j) A_i = A_k(A_j A_i)$. Therefore $(T_k T_j) T_i$ and $T_k(T_j T_i)$ are the same transformation, and condition 2 is verified.

Condition 3 is obviously fulfilled, since the transformation $X' = IX$, where I is the identity matrix, acts as the identity element of G.

Condition 4 is also obviously fulfilled, for if T: $X' = AX$ is any transformation in G, then A is nonsingular. Hence T^{-1}: $X' = A^{-1}X$ exists and has the property that $TT^{-1} = T^{-1}T$ is the identity transformation, since $AA^{-1} = A^{-1}A = I$ is the identity matrix. Thus the last of the conditions of Definition 1 is verified, and our proof is complete.

The fact that the set of all collineations is a group under the operation of composition illustrates an important characteristic of groups, namely, that in general the multiplication of elements in a group is not commutative. Specifically, if T_i: $X' = A_i X$ and T_j: $X' = A_j X$ are two collineations, then $T_j T_i$ is the transformation $X' = A_j A_i X$, while $T_i T_j$ is the transformation $X' = A_i A_j X$; and since matric multiplication is not commutative, it follows that in general $T_j T_i$ and $T_i T_j$ are not the same transformation. If it should happen, as it may, that for all elements of a group $e_j e_i = e_i e_j$, then the group is said to be **commutative** or **abelian.**[1]

We shall now consider several groups of finite order which illustrate interesting aspects of group theory. As a first example, let us take the four

[1] Named for the Norwegian mathematician Niels Abel (1802–1829), who was one of the founders of group theory.

numbers $(1,i,-1,-i)$ and determine how they combine under the process of ordinary multiplication. The results, which of course are completely familiar to us, are conveniently displayed in a multiplication table:

	1	i	-1	$-i$
1	1	i	-1	$-i$
i	i	-1	$-i$	1
-1	-1	$-i$	1	i
$-i$	$-i$	1	i	-1

in which the product of any two elements is entered in the cell at the intersection of the row corresponding to the first factor and the column corresponding to the second factor. Clearly the set $G:(1,i,-1,-i)$ is closed under multiplication; i.e., the product of any two elements of G is also an element of G. Moreover, since multiplication is associative for *all* complex numbers, it is associative for the four numbers in G. Also the number 1 serves as the identity element in G. Finally, for each element e in G there is an element e^{-1} in G such that $ee^{-1} = e^{-1}e = 1$; specifically, both 1 and -1 are their own inverses, while i and $-i$ are inverses of each other. Thus each of the requirements of Definition 1 is met, and G is a group under the operation of multiplication. Either by an inspection of the multiplication table of G or simply by observing that multiplication is commutative for all complex numbers, it is clear that G is in fact a commutative, or abelian, group.

As a seemingly quite different example, let us consider the four projectivities $T_1: X' = A_1X,$ $T_2: X' = A_2X,$ $T_3: X' = A_3X,$ $T_4: X' = A_4X$, where

$$A_1 = \begin{Vmatrix} 1 & 0 \\ 0 & 1 \end{Vmatrix} \qquad A_2 = \begin{Vmatrix} -1 & 0 \\ 0 & -1 \end{Vmatrix} \qquad A_3 = \begin{Vmatrix} 0 & -1 \\ 1 & 0 \end{Vmatrix}$$

$$A_4 = \begin{Vmatrix} 0 & 1 \\ -1 & 0 \end{Vmatrix}$$

and let us determine how they combine under the operation of composition. Since the composition of two projectivities $T_j T_i$ is another projectivity whose matrix is the product $A_j A_i$, it follows that the multiplication table for the given transformations can be constructed simply by multiplying the corresponding matrices. The results are easily found to be those given in the following table:

	T_1	T_2	T_3	T_4
T_1	T_1	T_2	T_3	T_4
T_2	T_2	T_1	T_4	T_3
T_3	T_3	T_4	T_2	T_1
T_4	T_4	T_3	T_1	T_2

Here, again, it is obvious that the system is closed. Moreover, since composition of linear transformations is equivalent to the multiplication of matrices, and since matric multiplication is associative, it is clear that composition of the given transformations is also associative. Finally, it is evident from the multiplication table that the projectivity T_1 acts as the identity element and that every element has an inverse such that the product of the two is the identity element T_1. Thus the system consisting of the four given projectivities and the operation of composition is also a group. Incidentally, although in general, matric multiplication is not commutative, it is for the matrices of the transformations of this example. As a consequence, this is also a commutative group, as its multiplication table makes clear.

Although the two examples which we have just given of groups of order 4 differ significantly in their definitions, an inspection of their multiplication tables reveals a certain similarity in structure. Specifically, in each case there are two elements which are their own inverses: 1 and -1 in the first example, T_1 and T_2 in the second, and two elements each of which is the inverse of the other: i and $-i$ in the first example, T_3 and T_4 in the second. If we rename the elements in each group according to the following scheme

	$e_1 = I$	e_2	e_3	e_4
First group	1	i	-1	$-i$
Second group	T_1	T_3	T_2	T_4

the multiplication table for each becomes

	e_1	e_2	e_3	e_4
e_1	e_1	e_2	e_3	e_4
e_2	e_2	e_3	e_4	e_1
e_3	e_3	e_4	e_1	e_2
e_4	e_4	e_1	e_2	e_3

In other words, when the elements in the two groups are properly matched, the multiplication tables of the groups are identical! Two groups, such as these, whose multiplication tables have the same structure are said to be **abstractly identical** or **isomorphic,** and the one-to-one correspondence between their elements which maps the product of any two elements in one group onto the product of the corresponding elements in the other group is called an **isomorphism.**

As a final example of a group of order 4, let us consider a regular tetrahedron $ABCD$ and the three lines P_1P_1', P_2P_2', P_3P_3' determined by the midpoints of opposite sides of the tetrahedron (Fig. 5.4a). Clearly if the tetrahedron is rotated through an angle of 180° about any one of the axes P_1P_1', P_2P_2', P_3P_3', it will return to its original position in space, although its vertices will be permuted in one way or another. The three rotations

$$R_1, \text{ about } P_1P_1', \qquad R_2, \text{ about } P_2P_2', \qquad R_3, \text{ about } P_3P_3'$$

where the axes are considered fixed in space and not fixed with respect to the tetrahedron itself, together with the identity transformation, which leaves each vertex unmoved, constitute the elements of our example; and the product of any two elements will mean, as usual, the composition of the corresponding rotations. To aid us in constructing the multiplication table for this system, we note that:

R_1 interchanges the vertices A and B and the vertices C and D.
R_2 interchanges the vertices A and C and the vertices B and D.
R_3 interchanges the vertices A and D and the vertices B and C.

With this information it is not difficult to determine how the vertices are interchanged if one rotation is followed by another. For example, to determine the product R_2R_1, we note that after R_1 has been performed, the tetrahedron occupies the position shown in Fig. 5.4b. Then performing R_2, remembering, as Fig. 5.4 reminds us, that the axes of rotation are fixed in

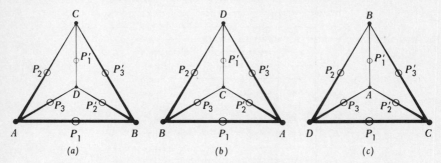

Fig. 5.4   Rotations of a tetrahedron into itself.

space, the tetrahedron moves into the position shown in Fig. 5.4c, which is the position it would occupy if R_3 were performed on its original position. Thus $R_2R_1 = R_3$. In this fashion all products of the given rotations can be worked out, and the multiplication table for the system can be constructed:

	I	R_1	R_2	R_3
I	I	R_1	R_2	R_3
R_1	R_1	I	R_3	R_2
R_2	R_2	R_3	I	R_1
R_3	R_3	R_2	R_1	I

Conditions 1, 3, and 4 of Definition 1 are obviously satisfied in this example, and it is not difficult to verify from the multiplication table that condition 2 is also fulfilled. Hence the system is a group, and in fact a commutative group. It is not, however, isomorphic to either of the other groups of order 4 which we have discussed. Specifically, it is evident from its multiplication table that in the last group each element is its own inverse, while in each of the other groups of order 4 only two elements have this property.

As a final example of a finite group, let us consider the six possible values of the cross ratio of four points, namely,

$$e_1 = r \qquad e_2 = \frac{1}{r} \qquad e_3 = 1 - r$$

$$e_4 = \frac{1}{1 - r} \qquad e_5 = \frac{r - 1}{r} \qquad e_6 = \frac{r}{r - 1}$$

and let us define e_ie_j to mean the result of substituting the expression e_j for the variable r in e_i. The algebra involved in computing the various products is easy, and we find that the multiplication table for the system is

	e_1	e_2	e_3	e_4	e_5	e_6
e_1	e_1	e_2	e_3	e_4	e_5	e_6
e_2	e_2	e_1	e_4	e_3	e_6	e_5
e_3	e_3	e_5	e_1	e_6	e_2	e_4
e_4	e_4	e_6	e_2	e_5	e_1	e_3
e_5	e_5	e_3	e_6	e_1	e_4	e_2
e_6	e_6	e_4	e_5	e_2	e_3	e_1

Clearly this system satisfies conditions 1, 3, and 4 of Definition 1. It is not obvious that condition 2 is also satisfied, but it is easy, though tedious, to verify this by considering each triple of elements (e_i, e_j, e_k) and checking that $(e_i e_j) e_k = e_i (e_j e_k)$. For example,

$$(e_2 e_5) e_3 = e_6 e_3 = e_5 \quad \text{and} \quad e_2 (e_5 e_3) = e_2 e_6 = e_5$$

Thus the six cross-ratio expressions form a group of order 6 under the operation of substitution. Incidentally, this group is not commutative; for instance, $e_4 e_6 = e_3$ while $e_6 e_4 = e_2$.

The multiplication table for the cross-ratio group contains the multiplication tables of several smaller groups. Specifically, we have

	e_1	e_2
e_1	e_1	e_2
e_2	e_2	e_1

	e_1	e_3
e_1	e_1	e_3
e_3	e_3	e_1

	e_1	e_6
e_1	e_1	e_6
e_6	e_6	e_1

	e_1	e_4	e_5
e_1	e_1	e_4	e_5
e_4	e_4	e_5	e_1
e_5	e_5	e_1	e_4

The fact that the orders of these groups, namely, 2 and 3, are divisors of the order, 6, of the group which contains them illustrates the following important theorem.

Theorem 2   Lagrange's theorem

In any finite group, the order of any subgroup is a factor of the order of the group.

Proof   Let $e_1, e_2, \ldots, e_k$ be the (distinct) elements of a subgroup, g, of order k of a group, G, of order n. Unless $k = n$, in which case the theorem is trivially true, there is at least one element of G which is not in the subgroup g. Let x be such an element, and consider the set, h_1, of the products of x with each element in g:

$$h_1 : (x e_1, x e_2, \ldots, x e_k)$$

Clearly, the elements defined by these products are all distinct, for if $x e_i = x e_j$, say, then multiplying on the left by x^{-1} we would have $e_i = e_j$, which is impossible since e_i and e_j are distinct elements of g. Moreover, none of the elements in h_1 is an element of g, for if this were the case, i.e., if $xe_i = e_j$, then multiplying on the right by e_i^{-1} we would have $x = e_j e_i^{-1}$. Since both e_j and e_i^{-1} are elements of the subgroup g, this implies that x is also an element of g, contrary to hypothesis. Thus, multiplying the elements of g on the left by the element x has identified k distinct elements of G none of which is an element of g.

  If every element of G is contained either in g or in the derived set h_1,[†] then $n = 2k$ and the theorem is proved. If $n > 2k$, then there is at least one more element, y, which is neither in g nor in h_1. Using this element, we can now construct a second set of products

$$h_2:(ye_1, ye_2, \ldots, ye_k)$$

Exactly as before, we can show that the elements in the set h_2 are all distinct and that none is an element of g. Moreover, none is a member of the set h_1, for if this were the case, i.e., if $ye_i = xe_j$, then multiplying on the right by e_i^{-1} we would have $y = x(e_j e_i^{-1})$, which implies that y is an element of h_1, contrary to hypothesis. Thus another set, h_2, of k elements, having no element in common with either g or h_1, has been identified in G. We now continue this process until it terminates, as it must since G is a finite group. The final result is that the elements of G have been separated into a number of sets, $g, h_1, h_2, \ldots, h_p$, each of which contains k elements and no two of which have any element in common. Hence $n = (p + 1)k$, which proves that k is a factor of n, as asserted.

Example 1

Under what conditions, if any, is it possible for the identity collineation and three harmonic homologies whose axes are not concurrent to form a group?

  Let H_1, H_2, and H_3 be three harmonic homologies whose axes are not concurrent. The coordinate system can then be chosen so that the axes of H_1, H_2, and H_3 are, respectively, the lines $x_1 = 0$, $x_2 = 0$, and $x_3 = 0$. Since the center and axis of a homology cannot be incident, the centers of H_1, H_2, and H_3 can without loss of generality be taken to be $O_1:(1, b_1, c_1)$, $O_2:(a_2, 1, c_2)$, and $O_3:(a_3, b_3, 1)$. Then exactly as in Example 2, Sec. 4.7, we find the equations of H_1, H_2, and H_3 to be

$$H_1: X' = A_1 X \qquad H_2: X' = A_2 X \qquad H_3: X' = A_3 X$$

[†] The set of elements obtained from the elements of a subgroup g of a group G by multiplying them on the left by an element x of G is called a **left coset** of g. The set derived from g by multiplying its elements on the right by an element x of G is called a **right coset** of g.

where
$$A_1 = \begin{Vmatrix} 1 & 0 & 0 \\ 2b_1 & -1 & 0 \\ 2c_1 & 0 & -1 \end{Vmatrix} \qquad A_2 = \begin{Vmatrix} -1 & 2a_2 & 0 \\ 0 & 1 & 0 \\ 0 & 2c_2 & -1 \end{Vmatrix}$$

$$A_3 = \begin{Vmatrix} -1 & 0 & 2a_3 \\ 0 & -1 & 2b_3 \\ 0 & 0 & 1 \end{Vmatrix}$$

Clearly, since H_1, H_2, and H_3 are harmonic homologies, each is its own inverse. Hence H_1H_2 cannot be the identity and therefore must be H_3 if H_1, H_2, H_3 and the identity are to form a group. Now $H_1H_2 = H_3$ if and only if $A_1A_2 = \cdot A_3$, that is, if and only if

$$\begin{Vmatrix} -1 & 2a_2 & 0 \\ -2b_1 & 4b_1a_2 - 1 & 0 \\ -2c_1 & 4c_1a_2 - 2c_2 & 1 \end{Vmatrix} = \begin{Vmatrix} -1 & 0 & 2a_3 \\ 0 & -1 & 2b_3 \\ 0 & 0 & 1 \end{Vmatrix}$$

This in turn requires that $b_1 = c_1 = a_2 = c_2 = a_3 = b_3 = 0$. Furthermore, under these conditions it is easy to verify that we also have $H_2H_1 = H_3$, $H_1H_3 = H_3H_1 = H_2$, and $H_2H_3 = H_3H_2 = H_1$. Hence these conditions are not only necessary but also sufficient for H_1, H_2, H_3, and I to form a group. Thus when H_1, H_2, H_3, and I form a group, the centers of the respective homologies are O_1:(1,0,0), O_2:(0,1,0), and O_3:(0,0,1), which are the vertices of the triangle formed by the axes of the homologies. In other words, *three harmonic homologies whose axes are not concurrent, together with the identity, form a group if and only if the center of each homology is the intersection of the axes of the other two homologies.*

The group G consisting of H_1, H_2, H_3, and I has a number of interesting geometric properties. In the first place, since any homology leaves invariant its axis and any line through its center, it follows that H_1, H_2, H_3, and of course I, all leave the triangle of reference invariant. Moreover, every conic of the family $ax_1^2 + bx_2^2 + cx_3^2 = 0$ is left invariant by each transformation in G. It is also easy to verify that the axis of any one of the three homologies is the polar of the center of that homology with respect to each of the invariant conics. On each invariant conic the points are arranged in sets of four, consisting of a general point P and its images, P_1, P_2, P_3, under H_1, H_2, H_3, respectively. From the definition of a homology, it follows immediately that P and P_i are collinear with O_i. Furthermore, since $H_iH_j = H_k$, it follows that $H_iH_jP = H_kP$ or $H_iP_j = P_k$, which shows that P_j and P_k, being images under H_i, are also collinear with O_i. Thus each set of four points, P, P_1, P_2, P_3, on any one of the invariant conics, is the set of vertices of an inscribed quadrilateral whose opposite sides intersect in the centers of the three homologies (Fig. 5.5).

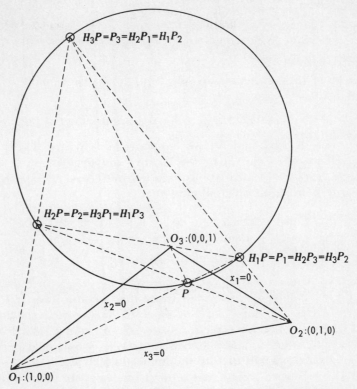

Fig. 5.5   Properties of the group of three homologies and the identity.

Exercises

1. Show that the set of all integers, positive, negative, and zero, is a group under the operation of addition.

2. Which, if any, of the following sets is a group under the operation of addition?
(*a*) The set of all nonnegative integers
(*b*) The set of all rational numbers
(*c*) The set of all real numbers
(*d*) The set of all complex numbers

3. Show that none of the following sets is a group under the operation of multiplication:
(*a*) The set of all integers   (*b*) The set of all rational numbers
(*c*) The set of all real numbers   (*d*) The set of all complex numbers

4. Are there any sets of numbers which form a group under the operation of multiplication?

5. Show that the set of all (2,2) matrices whose determinants are equal to 1 is a group under the operation of matric multiplication. Is it a group under the operation of matric addition?

6. Is the set of all (2,2) matrices whose determinants are equal to 2 a group under matric multiplication?

7. If a set of matrices M_1, M_2, . . . , M_k is a group under the operation of matric multiplication, show that the set

$$PM_1P^{-1}, \qquad PM_2P^{-1}, \qquad \ldots, \qquad PM_kP^{-1}$$

is also a group under matric multiplication.

8. Show that the matrices

$$\left\| \begin{matrix} 1 & 0 \\ 0 & 1 \end{matrix} \right\| \qquad \left\| \begin{matrix} -1 & 0 \\ 0 & -1 \end{matrix} \right\| \qquad \left\| \begin{matrix} 1 & 0 \\ 0 & -1 \end{matrix} \right\| \qquad \left\| \begin{matrix} -1 & 0 \\ 0 & 1 \end{matrix} \right\|$$

form a group under matric multiplication. To which groups that we have discussed, if any, is this group isomorphic?

9. Let S be the set of integers $(0,1,2,\ldots,n-1)$, and for any elements e_i and e_j in S let $e_i \circ e_j$ be the remainder when the sum $e_i + e_j$ is divided by n. Show that S is a group and construct its multiplication table if:
(a) $n = 3$ (b) $n = 4$ (c) $n = 5$

10. Let S be the set of integers $(1,2,\ldots,n-1)$ and for any elements e_i and e_j in S let $e_i \circ e_j$ be the remainder when the product $e_i e_j$ is divided by n. Show that S is a group and construct its multiplication table if:
(a) $n = 3$ (b) $n = 5$ (c) $n = 7$

11. Is the system described in Exercise 10 a group if:
(a) $n = 4$ (b) $n = 6$ (c) $n = 8$ (d) $n = 9$

12. Prove that any two groups of order 3 are isomorphic.

13. Prove that every group of order 4 is isomorphic to one or the other of the groups of order 4 which we discussed in this section.

14. Show that if a group of order 6 has three subgroups of order 2, it can have at most one subgroup of order 3. Must it always have at least one subgroup of order 3?

15. Prove that in any group, G, the identity element is unique.

16. Prove that in any group, G, the inverse of each element is unique.

17. Sometimes the third condition for a group, G, is weakened by requiring only that G contain a **right identity,** i.e., an element I_r such that $e_i \circ I_r = e_i$

for each e_i in G. Under this assumption, prove that I_r is also a **left identity,** i.e., prove that it is also true that $I_r \circ e_i = e_i$ for each e_i in G.

18. Sometimes the fourth condition for a group is weakened by requiring only that each element e_i in the group have a **right inverse,** i.e., an element $(e_i^{-1})_r$ such that $e_i \circ (e_i^{-1})_r = I_r (= I_l = I)$. Under this assumption, prove that $(e_i^{-1})_r$ is also a left inverse for e_i; that is, prove that $(e_i^{-1})_r \circ e_i = I_r (= I_l = I)$.

19. Prove that if two groups are isomorphic, their identity elements must correspond to each other.

20. Let 1, $\omega = (-1 + i\sqrt{3})/2$, and $\omega^2 = (-1 - i\sqrt{3})/2$ be the three cube roots of unity, and let $\{A\}$ be the set of all $(3,3)$ diagonal matrices whose diagonal elements are the three numbers 1, ω, ω^2 in any order. Show that the set of collineations $X' = AX$ (necessarily of type I) whose matrices are the members of the set $\{A\}$, together with the identity, is a group of order 3 under the operation of composition. Show also that the transformations of this group leave invariant every conic of each of the families $a_1 x_1^2 + b_1 x_2 x_3 = 0$, $a_2 x_2^2 + b_2 x_1 x_3 = 0$, $a_3 x_3^2 + b_3 x_1 x_2 = 0$.

21. Let $\{A\}$ be the set of all $(3,3)$ diagonal matrices whose diagonal elements, not necessarily distinct, are cube roots of unity. Show that the set of collineations $X' = AX$ (necessarily of types I, III, and VI) whose matrices are the members of the set $\{A\}$ is a group of order 9 under the operation of composition. Show also that this group contains four subgroups of order 3, and show further that for exactly three of these subgroups the images of an arbitrary point under the transformations of the subgroup are collinear.

22. Show that the nine images of an arbitrary point P under the transformations of the group described in Exercise 21 lie by three's on nine lines, three of which pass through each image point. Show, further, that the nine images of an arbitrary point P can be grouped to form three triangles such that no side of any one of the triangles contains more than two images of P. Finally, show that any two of these triangles are centrally perspective from each of the vertices of the remaining triangle.

23. Show that the triangle of reference and any one of the three triangles described in Exercise 22 are centrally perspective from each of the six images of P which is not a vertex of that triangle. What are the corresponding axes of perspectivity?

24. Show that at each vertex of each of the triangles described in Exercise 22 the two sides of the triangle passing through that vertex and the lines joining that vertex to any two of the points $(1,0,0)$, $(0,1,0)$, $(0,0,1)$ form a tetrad of lines whose cross ratio is independent of the point P.†

† Further properties of the configurations discussed in Exercises 20 to 24 and a different approach to them can be found in D. Hilbert and S. Cohn-Vossen, "Geometry and the Imagination," pp. 102–112. Chelsea Publishing Company, New York, 1952.

25. Let $\{A\}$ be the set of all (3,3) diagonal matrices whose diagonal elements are fourth roots of unity. Show that the set of collineations $X' = AX$ (necessarily of types I, III, and VI) whose matrices are the members of the set $\{A\}$ is a group of order 16 under the operation of composition, and determine its various subgroups. In particular, show that there are three subgroups of order 8, and show that each of these subgroups has the following property: if P is an arbitrary point, its images under the transformations of the subgroup form two harmonic tetrads, and the lines which contain these tetrads are harmonic conjugates with respect to two sides of the triangle of reference.

5.10 Affine Transformations

Although the general collineation leaves invariant such properties as collinearity, incidence, and cross ratio, it does not preserve such euclidean properties as length, angle measure, area, or even the relation of parallelism. There are some collineations, however, which leave one or more of these properties unchanged, and it is interesting to try to identify them. In this section we shall begin such an investigation by determining the set of all collineations which transform parallel lines into parallel lines.

The simplest way to identify the collineations which preserve the relation of parallelism is to recall that in any euclidean specialization, E_2, of Π_2, two lines are parallel if and only if:

1. They are real.

2. They intersect on the line of Π_2 which is chosen to be the ideal line of E_2.

Two lines which are parallel in E_2, that is, intersect in a point P on the ideal line in Π_2, will be transformed into lines which are also parallel if and only if the point P is transformed into some point P' which is also on the ideal line. In other words, a collineation will transform parallel lines into parallel lines if and only if it transforms into itself the line chosen as the ideal line. If we assume that the coordinate system is chosen so that the ideal line is the line whose equation is $x_3 = 0$, this requires that any point $(a_1, a_2, 0)$ whose third coordinate is zero be transformed into a point $(a_1', a_2', 0)$ whose third coordinate is also zero. Finally this will be the case for all values of a_1 and a_2 if and only if the equations of the general collineation, namely,

(1) $\qquad X' = AX \qquad$ or $\qquad \begin{aligned} x_1' &= a_{11}x_1 + a_{12}x_2 + a_{13}x_3 \\ x_2' &= a_{21}x_1 + a_{22}x_2 + a_{23}x_3 \\ x_3' &= a_{31}x_1 + a_{32}x_2 + a_{33}x_3 \end{aligned}$

satisfy the conditions $a_{31} = a_{32} = 0$.

It is of course unnecessary to give another proof of the last result, but it may help to clarify the work of the next three sections if we present a second argument emphasizing the preservation of the algebraic rather than the geometric condition for parallelism. Let us consider, then, two arbitrary parallel lines

$$l: l_1 x_1 + l_2 x_2 + l_3 x_3 = 0 \qquad \text{and} \qquad m: m_1 x_1 + m_2 x_2 + m_3 x_3 = 0$$

and their images

$$l': l_1' x_1' + l_2' x_2' + l_3' x_3' = 0 \qquad \text{and} \qquad m': m_1' x_1' + m_2' x_2' + m_3' x_3' = 0$$

under the general collineation (1). Since l and m are parallel, it follows that $l_1 m_2 - l_2 m_1 = 0$. Hence, since we want l' and m' to be parallel, it follows that we must determine conditions under which $l_1 m_2 - l_2 m_1 = 0$ will imply $l_1' m_2' - l_2' m_1' = 0$. Now from Sec. 5.3, we know that under the transformation of point-coordinates defined by (1), the coordinates of a line are transformed according to the equation

$$\Lambda' = (A^{-1})^T \Lambda \qquad \text{or} \qquad \Lambda = A^T \Lambda'$$

Therefore, under the transformation (1), the condition $l_1 m_2 - l_2 m_1 = 0$ becomes

$$(a_{11} l_1' + a_{21} l_2' + a_{31} l_3')(a_{12} m_1' + a_{22} m_2' + a_{32} m_3')$$
$$- (a_{12} l_1' + a_{22} l_2' + a_{32} l_3')(a_{11} m_1' + a_{21} m_2' + a_{31} m_3') = 0$$

or, simplifying,

$$(l_2' m_3' - l_3' m_2')(a_{21} a_{32} - a_{22} a_{31}) + (l_3' m_1' - l_1' m_3')(a_{12} a_{31} - a_{11} a_{32})$$
$$+ (l_1' m_2' - l_2' m_1')(a_{11} a_{22} - a_{12} a_{21}) = 0$$

From this it is clear that the condition $l_1 m_2 - l_2 m_1 = 0$ will imply the condition $l_1' m_2' - l_2' m_1' = 0$ for all lines in E_2 if and only if, simultaneously,

$$(2) \qquad \begin{aligned} -a_{22} a_{31} + a_{21} a_{32} &= 0 \\ a_{12} a_{31} - a_{11} a_{32} &= 0 \end{aligned}$$

and

$$(3) \qquad a_{11} a_{22} - a_{12} a_{21} \neq 0$$

However, since the left member of (3) is just the determinant of the coefficients in the system (2), regarded as a pair of equations in the quantities a_{31} and a_{32}, it follows from (3) that (2) has only the trivial solution $a_{31} = a_{32} = 0$, as we found before. Thus, summarizing, we have the following theorem.

Theorem 1

If the ideal line is $x_3 = 0$, the most general collineation which transforms parallel lines into parallel lines is defined by an equation of the form $X' = AX$, where

$$A = \begin{Vmatrix} a_{11} & a_{12} & a_{13} \\ a_{21} & a_{22} & a_{23} \\ 0 & 0 & a_{33} \end{Vmatrix}, \quad a_{ij} \text{ real.}$$

Any collineation whose coefficients are real and for which $a_{31} = a_{32} = 0$ is called an **affine transformation.** Since a_{31} and a_{32} are zero for any affine transformation, it is clear that a_{33} cannot be zero. Hence without loss of generality it is always possible to assume that $a_{33} = 1$, and it is often convenient to do this. From Fig. 5.3 it is evident that every collineation leaves at least one line invariant. Hence since the coordinate system can always be chosen so that this line is the line $x_3 = 0$, it follows that *there exist affine transformations which are collineations of each of the six possible types.*

One of the most fundamental properties of affine transformations is contained in the following theorem.

Theorem 2

The set of all affine transformations is a group under the operation of composition.

Proof   To prove that the set of all affine transformations is a group under the operation of composition, we must prove four things:

1. The composition of any two affine transformations is an affine transformation.

2. The composition of affine transformations is associative.

3. The set of affine transformations contains the identity.

4. The inverse of any affine transformation is also an affine transformation.

Obviously the identity collineation is an affine transformation. Furthermore, since composition is associative for *all* collineations, it is certainly associative for those which are affine transformations. Hence conditions 2 and 3 are fulfilled. To verify condition 3, it is sufficient to show that if A_1 and A_2 are matrices of affine transformations, then A_1A_2 is also the matrix of an affine transformation. This, in turn, requires that we show that the elements of A_1A_2 are real and that the first two elements in the third row of A_1A_2 are zero. The reality of the elements of A_1A_2 is obvious. The other requirement follows at once from the definition of matric multiplication and the fact that the first two elements in the third row of both A_1 and A_2 are zero. Finally, to verify condition 4, it is sufficient to show that if A is the matrix of an affine transformation, then so too is A^{-1}, which requires that we show that the elements of A^{-1} are real and that the first two elements in the third row of A^{-1} are zero. The reality of A^{-1} is obvious. The other requirement follows from the fact that the first two elements in the third row of A^{-1} are, respectively, $A_{13}/|A|$ and $A_{23}/|A|$ and each of the cofactors A_{13} and A_{23} is a (2,2) determinant whose second row consists entirely of zeros. Thus each of the conditions for a group is verified, and the theorem is established.

It is easy to verify by specific examples, or prove in general, that affine transformations neither preserve length nor multiply all lengths by a constant factor. They do, however, have a related property which is described in the following theorem.

Theorem 3

An affine transformation multiplies the lengths of all segments in a given direction by a factor which depends only on the transformation and the given direction.

Proof Let $U:(u_1,u_2,1)$ and $V:(v_1,v_2,1)$ be two points which are not on the ideal line, and let $U':(u_1',u_2',1)$ and $V':(v_1',v_2',1)$ be their images under an arbitrary affine transformation

$$\begin{aligned} x_1' &= a_{11}x_1 + a_{12}x_2 + a_{13}x_3 \\ x_2' &= a_{21}x_1 + a_{22}x_2 + a_{23}x_3 \\ x_3' &= \phantom{a_{21}x_1 + a_{22}x_2 + {}} x_3 \end{aligned}$$

Then by an easy calculation we find

$$\begin{aligned} v_1' - u_1' &= a_{11}(v_1 - u_1) + a_{12}(v_2 - u_2) \\ v_2' - u_2' &= a_{21}(v_1 - u_1) + a_{22}(v_2 - u_2) \end{aligned}$$

The square of the ratio of the distance $(U'V')$ to the distance (UV) is therefore

$$\frac{(a_{11}{}^2 + a_{21}{}^2)(v_1 - u_1)^2 + 2(a_{11}a_{12} + a_{21}a_{22})(v_1 - u_1)(v_2 - u_2) + (a_{12}{}^2 + a_{22}{}^2)(v_2 - u_2)^2}{(v_1 - u_1)^2 + (v_2 - u_2)^2}$$

If $(v_1 - u_1) = 0$, this expression reduces to $a_{12}{}^2 + a_{22}{}^2$, which depends only on the coefficients in the equation of the given transformation. On the other hand, if $(v_1 - u_1) \neq 0$, we can divide the numerator and the denominator of the last fraction by $(v_1 - u_1)^2$. Then, noting that $(v_2 - u_2)/(v_1 - u_1)$ is just the slope, m, of the line which contains the segment $\overline{UV}$, we have

$$\frac{(U'V')^2}{(UV)^2} = \frac{(a_{11}{}^2 + a_{21}{}^2) + 2(a_{11}a_{12} + a_{21}a_{22})m + (a_{12}{}^2 + a_{22}{}^2)m^2}{1 + m^2}$$

Since this depends only on the coefficients in the equations of the transformation and the slope of the given segment, our proof is complete.

Corollary 1

If P', Q', and R' are, respectively, the images of three collinear points, P, Q, and R, under an affine transformation, then the ratio in which Q divides the segment $\overline{PR}$ is equal to the ratio in which Q' divides the segment $\overline{P'R'}$.

Exercises

1. Prove Corollary 1, Theorem 3.

2. Prove that an affine transformation always transforms an ellipse into an ellipse, a parabola into a parabola, and a hyperbola into a hyperbola.

3. Is the group of affine transformations a commutative group?

4. Give an example of an affine transformation of each of the six possible types.

5. Give a proof of Theorem 3 based on the invariance of the cross ratio under an arbitrary collineation.

6. Prove that the images of all circles under a given affine transformation are similar ellipses whose major axes are all parallel.

7. Find the most general affine transformation of period 2.

8. Are there any directions such that the lengths of segments in these directions are unaltered by a given affine transformation?

9. Show that there is a maximum and a minimum to the value of the factor by which an affine transformation multiplies the length of an arbitrary segment.

10. Prove that the directions in which an affine transformation multiplies lengths by the greatest and least factors are perpendicular.

11. Let Γ': $a'(x_1')^2 + 2b'x_1'x_2' + c'(x_2')^2 + 2d'x_1'x_3' + 2e'x_2'x_3' + f'(x_3')^2 = 0$ be the image of the conic Γ: $ax_1^2 + 2bx_1x_2 + cx_2^2 + 2dx_1x_3 + 2ex_2x_3 + fx_3^2 = 0$ under an arbitrary affine transformation. What relation, if any, exists between the quantities $(b')^2 - a'c'$ and $b^2 - ac$? What relation, if any, exists between the discriminants of Γ and Γ'?

12. Let $\triangle PQR$ and $\triangle P'Q'R'$ be two finite triangles. Prove that there is a unique affine transformation which maps P into P', Q into Q', and R into R'.

13. Using the fact that the diagonals of a parallelogram bisect each other, give a geometric proof of the fact that if P', Q', R' are, respectively, the images of three collinear points P, Q, R under an affine transformation, and if Q is the midpoint of the segment $\overline{PR}$, then Q' is the midpoint of the segment $\overline{P'R'}$.

14. Using the appropriate property of a general parallelogram, prove that if P', Q', R' are, respectively, the images of three collinear points P, Q, R under an affine transformation, and if Q divides the segment $\overline{PR}$ in the ratio $1/3$, then Q' divides the segment $\overline{P'R'}$ in the ratio $1/3$.

15. Can Exercise 14 be generalized to provide a proof of the fact that if P', Q', R' are, respectively, the images of three collinear points P, Q, R, and if Q divides the segment $\overline{PR}$ in the ratio $1/n$, then Q' divides the segment $\overline{P'R'}$ in the ratio $1/n$?

5.11 Similarity Transformations

In this section we shall continue our investigation of the important subgroups of the projective group by determining the set of all collineations which preserve angle measures. Our main result is contained in the following theorem.

Theorem 1

The most general collineation which leaves angle measures invariant is defined by an equation of the form $X' = AX$, where

$$A = \begin{Vmatrix} a_{11} & a_{12} & a_{13} \\ -ka_{12} & ka_{11} & a_{23} \\ 0 & 0 & a_{33} \end{Vmatrix}$$

a_{ij} real, $k = \pm 1$.

Proof Let $l: l_1 x_1 + l_2 x_2 + l_3 x_3 = 0$ and $m: m_1 x_1 + m_2 x_2 + m_3 x_3 = 0$ be two real lines which form an angle whose measure, θ, is defined by the formula $\tan \theta = (l_1 m_2 - l_2 m_1)/(l_1 m_1 + l_2 m_2)$, and let $l' = l_1' x_1' + l_2' x_2' + l_3' x_3' = 0$ and $m': m_1' x_1' + m_2' x_2' + m_3' x_3' = 0$ be their images under an arbitrary collineation $X' = AX$. Since we require that real lines be transformed into real lines, it is clear that the elements of A must be real. Furthermore, since we want angle measures to be preserved, we must have

(1)
$$\frac{l_1 m_2 - l_2 m_1}{l_1 m_1 + l_2 m_2} = \pm \frac{l_1' m_2' - l_2' m_1'}{l_1' m_1' + l_2' m_2'}$$

To determine conditions which will ensure this, it is convenient to begin by expressing the fraction $(l_1 m_2 - l_2 m_1)/(l_1 m_1 + l_2 m_2)$ in terms of the coordinates of the image lines l' and m' by means of the equation $\Lambda = A^T \Lambda'$. The transformation of the numerator was carried out in detail in the proof of Theorem 1 in the last section, where we found

$$(2) \qquad l_1 m_2 - l_2 m_1 = (l_2' m_3' - l_3' m_2')(a_{21} a_{32} - a_{22} a_{31})$$
$$+ (l_3' m_1' - l_1' m_3')(a_{12} a_{31} - a_{11} a_{32})$$
$$+ (l_1' m_2' - l_2' m_1')(a_{11} a_{22} - a_{12} a_{21})$$

As a condition that (1) should hold, it is clearly necessary that the first two terms in the last expression should be zero. Hence we must have $-a_{22} a_{31} + a_{21} a_{32} = 0$ and $a_{12} a_{31} - a_{11} a_{32} = 0$, which imply, as we saw in the last section in the proof of Theorem 1, that $a_{31} = a_{32} = 0$. Using these values, we now have for the denominator of the left member of (1), the expression

$$(3) \quad (a_{11} l_1' + a_{21} l_2')(a_{11} m_1' + a_{21} m_2') + (a_{12} l_1' + a_{22} l_2')(a_{12} m_1' + a_{22} m_2')$$
$$= l_1' m_1'(a_{11}{}^2 + a_{12}{}^2) + (l_1' m_2' + l_2' m_1')(a_{11} a_{21} + a_{12} a_{22}) + l_2' m_2'(a_{21}{}^2 + a_{22}{}^2)$$

In order that the expression for $\tan \theta$ be invariant to within a plus or minus sign, it is further necessary that:

1. The second term in (3) be zero.
2. The coefficients of $l_1' m_1'$ and $l_2' m_2'$ in (3) be equal to each other and equal, numerically, to the coefficient of $l_1' m_2' - l_2' m_1'$ in (2).

Condition 1 will be met if $a_{11} a_{21} + a_{12} a_{22} = 0$, which implies that $a_{22} = k a_{11}$ and $a_{21} = -k a_{12}$. Condition 2 then requires that

$$a_{11}{}^2 + a_{12}{}^2 = a_{21}{}^2 + a_{22}{}^2 = \pm(a_{11} a_{22} - a_{12} a_{21})$$

or $\qquad a_{11}{}^2 + a_{12}{}^2 = k^2(a_{12}{}^2 + a_{11}{}^2) = \pm k(a_{11}{}^2 + a_{12}{}^2)$

Since this will hold if and only if $k^2 = 1$, our proof is complete.

Any collineation which preserves angle measures is called a **similarity transformation,** since it obviously maps any configuration into another which is similar to it.

Corollary 1

Every similarity transformation is an affine transformation.

Since the product of the matrices of two similarity transformations,

$$AB = \begin{Vmatrix} a_{11} & a_{12} & a_{13} \\ -h a_{12} & h a_{11} & a_{23} \\ 0 & 0 & a_{33} \end{Vmatrix} \cdot \begin{Vmatrix} b_{11} & b_{12} & b_{13} \\ -k b_{12} & k b_{11} & b_{23} \\ 0 & 0 & b_{33} \end{Vmatrix} \qquad h^2 = k^2 = 1$$

can be written in the form

$$
\left\|
\begin{array}{ccc}
(a_{11}b_{11} - ka_{12}b_{12}) & (a_{11}b_{12} + ka_{12}b_{11}) & (a_{11}b_{13} + a_{12}b_{23} + a_{13}b_{33}) \\
-hk(a_{11}b_{12} + ka_{12}b_{11})hk & (a_{11}b_{11} - ka_{12}b_{12}) & (-ha_{12}b_{13} + ha_{11}b_{23} + a_{23}b_{33}) \\
0 & 0 & a_{33}b_{33}
\end{array}
\right\|
$$

it is clear that the composition of two similarity transformations is also a similarity transformation. Furthermore, since the inverse of the matrix, A, of a similarity transformation can be written in the form

$$
A^{-1} = \frac{\left\|
\begin{array}{ccc}
(ha_{11}a_{33}) & (-a_{12}a_{33}) & (a_{12}a_{23} - ha_{11}a_{13}) \\
-h(-a_{12}a_{33}) & h(ha_{11}a_{33}) & -(ha_{12}a_{13} + a_{11}a_{23}) \\
0 & 0 & h(a_{11}{}^2 + a_{12}{}^2)
\end{array}
\right\|}{ha_{33}(a_{11}{}^2 + a_{12}{}^2)}
$$

it is clear that the inverse of a similarity transformation is also a similarity transformation. Finally, it is obvious that the composition of similarity transformations is associative and that the identity is included among the similarity transformations. Hence we have established the following theorem.

Theorem 2

The set of all similarity transformations is a group under the operation of composition.

Exercises

1. Are there similarity transformations of all six types? Give an example of each possible type.

2. Except in the case of triangles, equality of the measures of corresponding angles is not a sufficient condition for the similarity of two polygons. In view of this fact, what is the justification for calling the angle-preserving transformations of this section *similarity transformations?*

3. A similarity transformation for which $k = 1$ is said to be **direct;** one for which $k = -1$ is said to be **indirect.** Show that the set of all direct similarity transformations is a subgroup of the group of all similarity transformations. Do the indirect similarity transformations form a group?

4. Show that every direct similarity transformation leaves invariant the circular points at infinity, $I:(1,i,0)$ and $J:(1,-i,0)$. Show that every indirect similarity transformation interchanges I and J.

5. Show that the image of any circle under an arbitrary similarity transformation is a circle.

6. If A and A' are distinct points, and if B and B' are distinct points, show that there are exactly two similarity transformations, one direct and one indirect, which map A onto A' and B onto B'.

7. (*a*) Find a similarity transformation which will map the triangle whose vertices are $(0,0,1)$, $(2,0,1)$, $(0,2,1)$ onto the triangle whose vertices are $(-1,0,1)$, $(-1,1,1)$, $(0,0,1)$.
(*b*) Is there a similarity transformation which will map the triangle whose vertices are $(0,0,1)$, $(2,0,1)$, $(0,1,1)$ onto the triangle whose vertices are $(-1,0,1)$, $(-1,2,1)$, $(-4,0,1)$?

8. (*a*) Is there a similarity transformation which will map the square whose vertices are $(0,0,1)$, $(1,0,1)$, $(1,1,1)$, $(0,1,1)$ onto the square whose vertices are $(1,0,1)$, $(0,1,1)$, $(-1,0,1)$, $(0,-1,1)$? If so, find its equations.
(*b*) Is there a similarity transformation which will map the circle whose center is the origin and whose radius is 1 onto the circle whose center is the point $(4,3,1)$ and whose radius is 5? If so, find its equations.

9. By what factor is the length of a general segment altered by an arbitrary similarity transformation?

10. What conditions must be satisfied in order that a similarity transformation shall be of period 2? Give an example of such a transformation.

11. If $l_1 m_1 + l_2 m_2 = 0$, $\tan \theta$ is undefined and the proof of Theorem 1 appears to break down. Investigate this case and show that right angles are also preserved by the general similarity transformation.

5.12 Equiareal Transformations

Another interesting class of projective transformations is the set of those which leave the area of a general triangle invariant. These are identified in the following theorem.

Theorem 1

The most general collineation which leaves invariant the area of an arbitrary triangle is defined by an equation of the form $X' = AX$, where

$$A = \begin{Vmatrix} a_{11} & a_{12} & a_{13} \\ a_{21} & a_{22} & a_{23} \\ 0 & 0 & a_{33} \end{Vmatrix} \qquad a_{ij} \text{ real}$$

$$\text{abs} \left(\frac{a_{11}a_{22} - a_{12}a_{21}}{a_{33}^2} \right)^\dagger = 1$$

† By abs Q we mean the absolute value of the quantity Q.

Proof Let $U:(u_1,u_2,u_3)$, $V:(v_1,v_2,v_3)$, $W:(w_1,w_2,w_3)$ be the vertices of an arbitrary triangle and let $U':(u_1',u_2',u_3')$, $V':(v_1',v_2',v_3')$, $W':(w_1',w_2',w_3')$ be their images under an arbitrary collineation $X' = AX$. Then, assuming that the points, U, V, W, U', V', W' are all real and finite, the area of $\triangle UVW$ is

$$\tfrac{1}{2}\,\text{abs}\left(\begin{vmatrix} u_1 & u_2 & u_3 \\ v_1 & v_2 & v_3 \\ w_1 & w_2 & w_3 \end{vmatrix}\frac{1}{u_3 v_3 w_3}\right)$$

and, similarly, the area of the transformed triangle, $\triangle U'V'W'$, is

$$\tfrac{1}{2}\,\text{abs}\left(\begin{vmatrix} u_1' & u_2' & u_3' \\ v_1' & v_2' & v_3' \\ w_1' & w_2' & w_3' \end{vmatrix}\frac{1}{u_3' v_3' w_3'}\right)$$

or

$$\tfrac{1}{2}\,\text{abs}\left[\frac{\begin{vmatrix} u_1 & u_2 & u_3 \\ v_1 & v_2 & v_3 \\ w_1 & w_2 & w_3 \end{vmatrix} \cdot \begin{vmatrix} a_{11} & a_{21} & a_{31} \\ a_{12} & a_{22} & a_{32} \\ a_{13} & a_{23} & a_{33} \end{vmatrix}}{(a_{31}u_1 + a_{32}u_2 + a_{33}u_3)(a_{31}v_1 + a_{32}v_2 + a_{33}v_3) \atop \times\,(a_{31}w_1 + a_{32}w_2 + a_{33}w_3)}\right]$$

Now if $\triangle UVW$ and $\triangle U'V'W'$ are to have the same area for arbitrary U, V, and W, it is necessary that

$$a_{31} = a_{32} = 0 \qquad \text{and} \qquad \text{abs}\left(\begin{vmatrix} a_{11} & a_{21} & a_{31} \\ a_{12} & a_{22} & a_{32} \\ a_{13} & a_{23} & a_{33} \end{vmatrix}\frac{1}{a_{33}^{\ 3}}\right) = 1$$

which together imply that

$$\text{abs}\left(\frac{a_{11}a_{22} - a_{12}a_{21}}{a_{33}^{\ 2}}\right) = 1$$

as asserted.

If a collineation leaves invariant the area of an arbitrary triangle, it follows by a suitable limiting process that it also leaves invariant the area of any plane region. For this reason such transformations are called **equiareal transformations.**

Corollary 1
Every equiareal transformation is an affine transformation.

By an argument very much like that which we used to establish Theorem 2, Sec. 5.10, and Theorem 2, Sec. 5.11, it is now easy to prove the following theorem.

Theorem 2

The set of all equiareal transformations is a group under the operation of composition.

Exercises

1. Prove Theorem 2.

2. Are there equiareal transformations of all six types? Give an example of each possible type.

3. Determine which, if any, of the following mappings can be accomplished by an equiareal transformation, and when it exists, find the equations of the transformation:

(*a*) $(0,0,1) \rightarrow (0,0,1)$, $(1,0,1) \rightarrow (1,0,1)$, $(2,1,1) \rightarrow (-1,1,1)$
(*b*) $(0,0,1) \rightarrow (0,0,1)$, $(2,0,1) \rightarrow (0,3,1)$, $(2,1,1) \rightarrow (1,0,3)$
(*c*) $(0,0,1) \rightarrow (0,0,1)$, $(2,0,1) \rightarrow (3,1,1)$, $(1,1,1) \rightarrow (1,2,2)$

4. If A and B are distinct finite points, and if A' and B' are distinct finite points, show that there is an equiareal transformation which maps A into A' and B into B'. Is this transformation unique?

5. Show that for a given equiareal transformation there are in general two directions with the property that the lengths of segments in these directions are unaltered by the transformation.

5.13 Euclidean Transformations

The most familiar, and in many respects the most important, projective transformations are those which preserve distances, in other words the euclidean transformations:

Theorem 1

The most general collineation which preserves distances is defined by an equation of the form $X' = AX$, where

$$A = \begin{Vmatrix} a_{11} & a_{12} & a_{13} \\ -ka_{12} & ka_{11} & a_{23} \\ 0 & 0 & a_{33} \end{Vmatrix} \qquad \begin{matrix} a_{ij} \text{ real} \\ k^2 = 1 \\ a_{11}{}^2 + a_{12}{}^2 = a_{33}{}^2 \end{matrix}$$

Proof   Let $U:(u_1,u_2,u_3)$ and $V:(v_1,v_2,v_3)$ be the end points of an arbitrary segment, and let $U':(u_1',u_2',u_3')$ and $V':(v_1',v_2',v_3')$ be their images under an arbitrary collineation $X' = AX$. Then, as usual, the square of the length of the segment $\overline{UV}$ (if it exists) is

(1) $$(UV)^2 = \frac{(u_1v_3 - v_1u_3)^2 + (u_2v_3 - u_3v_2)^2}{u_3{}^2v_3{}^2}$$

and, similarly, the square of the length of the segment $\overline{U'V'}$ (if it exists) is

(2) $$(U'V')^2 = \frac{(u_1'v_3' - u_3'v_1')^2 + (u_2'v_3' - u_3'v_2')^2}{(u_3')^2(v_3')^2}$$

Clearly, in order that $(UV)^2 = (U'V')^2$ for arbitrary (finite) U and V, it is necessary that

$$u_3' = a_{31}u_1 + a_{32}u_2 + a_{33}u_3 \quad \text{and} \quad v_3' = a_{32}v_1 + a_{32}v_2 + a_{33}v_3$$

be proportional to u_3 and v_3, respectively. Hence, as a first requirement, we conclude that $a_{31} = a_{32} = 0$. Using this condition, the right-hand side of formula (2) reduces to

$$\frac{1}{a_{33}{}^2 u_3{}^2 v_3{}^2} \{[(a_{11}u_1 + a_{12}u_2 + a_{13}u_3)(a_{33}v_3) - (a_{11}v_1 + a_{12}v_2 + a_{13}v_3)(a_{33}u_3)]^2$$

$$+ [(a_{21}u_1 + a_{22}u_2 + a_{23}u_3)(a_{33}v_3) - (a_{21}v_1 + a_{22}v_2 + a_{23}v_3)(a_{33}u_3)]^2\}$$

$$= \frac{1}{a_{33}{}^2 u_3{}^2 v_3{}^2} [(a_{11}{}^2 + a_{21}{}^2)(u_1v_3 - u_3v_1)^2$$

$$+ 2(a_{11}a_{12} + a_{21}a_{22})(u_1v_3 - u_3v_1)(u_2v_3 - u_3v_2)$$

$$+ (a_{12}{}^2 + a_{22}{}^2)(u_2v_3 - u_3v_2)^2]$$

Comparing this with formula (1), it is clear that we require further that

(3) $$a_{11}a_{12} + a_{21}a_{22} = 0$$

and that

(4) $$\frac{a_{11}{}^2 + a_{21}{}^2}{a_{33}{}^2} = \frac{a_{12}{}^2 + a_{22}{}^2}{a_{33}{}^2} = 1$$

From (3) we infer that $a_{22} = ka_{11}$ and $a_{12} = -ka_{21}$; and then from (4) we infer that

$$\frac{a_{11}{}^2 + a_{21}{}^2}{a_{33}{}^2} = \frac{k^2(a_{11}{}^2 + a_{21}{}^2)}{a_{33}{}^2} = 1$$

which implies that $k^2 = 1$. Hence we have, equally well, $a_{21} = -ka_{12}$ and $a_{11}{}^2 + a_{12}{}^2 = a_{33}{}^2$, which completes our proof.

Corollary 1

Every distance-preserving transformation is an affine transformation.

Corollary 2

Every distance-preserving transformation is also angle-preserving.

Corollary 3

Every distance-preserving transformation is also area-preserving.

Because of the properties guaranteed by Corollaries 2 and 3, distance-preserving transformations are clearly **euclidean transformations.** More specifically, since a_{33} is necessarily different from zero, it is no specialization to take it to be 1. Then putting $a_{11} = \cos \theta$ and $a_{12} = \sin \theta$ and reverting to nonhomogeneous coordinates, we obtain from the general distance-preserving transformation the euclidean transformation

$$x' = x \cos \theta + y \sin \theta + a_{13}$$
$$y' = -kx \sin \theta + ky \cos \theta + a_{23} \qquad k = \pm 1$$

If $k = 1$, these are the equations of a rotation about the origin through an angle of measure $-\theta$ followed by a translation of a_{13} units in the x direction and a_{23} units in the y direction. If $a_{13} = a_{23} = 0$, we have a pure rotation, and if $\theta = 0$, we have a pure translation. If $k = -1$, these are the equations of a rotation about the origin followed by a translation and a reflection in the x axis. If $\theta = a_{13} = a_{23} = 0$, they define a pure reflection in the x axis.

On the other hand, if we put $a_{11} = -\cos \theta$ and $a_{12} = -\sin \theta$ and then revert to nonhomogeneous coordinates, we obtain the euclidean transformation

$$x' = -x \cos \theta - y \sin \theta + a_{13}$$
$$y' = kx \sin \theta - ky \cos \theta + a_{23}$$

If $k = 1$, these are the equations of a rotation about the origin followed by a translation and a reflection in the origin. If $\theta = a_{13} = a_{23} = 0$, we have a pure reflection in the origin. If $k = -1$, these are the equations of a rotation about the origin followed by a translation and a reflection in the y axis. If $\theta = a_{13} = a_{23} = 0$, they are the equations of a pure reflection in the y axis.

By an argument very much like that used to prove Theorem 2, Sec. 5.10, and Theorem 2, Sec. 5.11, it is now easy to prove the following theorem.

Theorem 2

The euclidean transformations form a group under the operation of composition.

Exercises

1. Prove Theorem 2.

2. Are there euclidean transformations of all six types? Give an example of each possible type.

3. Prove Theorem 1 using Theorem 1, Sec. 5.11, and the result of Exercise 9, Sec. 5.11.

4. Show that every euclidean transformation induces an involution on the ideal line which is the identity, or has $I:(1,i,0)$ and $J:(1,-i,0)$ as its fixed points, or has I and J as a pair of mates. Determine the geometric nature of euclidean transformations with each of these properties.

5. Show that the composition of two reflections in parallel lines is a translation in the direction perpendicular to the two lines.

6. Show that the composition of two reflections in lines which intersect in a finite point is a rotation about the point of intersection of the lines.

7. What is the composition of two reflections in distinct points?

8. Under what conditions will a euclidean transformation be of period 2?

9. Under what conditions will a euclidean transformation be of period 3?

10. Under what conditions will two euclidean transformations commute?

5.14 Conclusion

In this chapter we have studied linear transformations, first as a device for assigning new coordinates to the points of Π_2, second as a mapping which assigns to each point of Π_2 another point as image. In investigating linear transformations from the second point of view, we discovered that there was always at least one point which was its own image, and we classified the nonsingular linear transformations, or collineations, into six projectively different types on the basis of the configuration of their invariant elements. In each case we were able to obtain a convenient standard form for the equations of the transformation. Then after a brief discussion of the mathematical structures known as groups, we verified that the set of all nonsingular projective transformations formed a group under the operation of composition, and we studied four important subgroups of the projective group. These were the **affine group,** defined as the set of all collineations which preserved the relation of parallelism, the **similarity group,** defined as the set of all collineations which preserved angle measures, the **equiareal group,** defined as the set of all collineations which preserved areas, and the **euclidean group,** defined as the set of all collineations which preserved

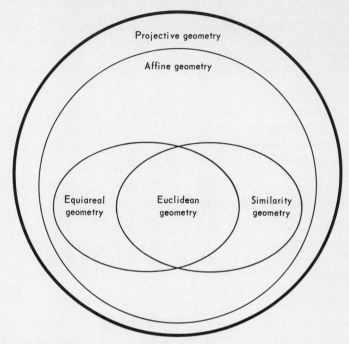

Fig. 5.6 The inclusion relations of projective geometry and its important subgeometries.

lengths and therefore areas and angle measures. The inclusion relations between the projective group and these four subgroups are shown in Fig. 5.6. In the spirit of Klein's Erlangen program, the study of the properties left invariant by these groups of transformations constitutes, respectively, projective geometry, affine geometry, similarity geometry, equiareal geometry, and euclidean geometry.

six

THE
AXIOMATIC
FOUNDATION

6.1 Introduction

In the preceding chapters we have developed
a number of important projective properties,
but we have done so only in the context of the
euclidean plane and the algebraic representation. As a consequence, we do not know
whether our results are valid in other systems
or not; and if they are, we have, at present,
no way of identifying these systems. Furthermore, our work up to this point has made
projective geometry seem (as indeed it was,
historically) only an outgrowth of euclidean
geometry, without any claim to an independent existence in the world of mathematics.

     In the rest of this book we shall
undertake an axiomatic development of
projective geometry which, hopefully, will
make clear its logical structure as an essentially self-contained deductive system. In
principle, the assumptions on which we shall
base our work are subject only to our
creative mathematical fancy and the universal

requirement of consistency. Actually, of course, they were suggested by our past mathematical experience and finally selected to provide the most effective basis for investigating deductively problems in which we were already interested.

Many of the theorems we shall prove from our axioms will already be familiar to us from the work of earlier chapters. This does not mean, however, that we will be merely repeating ourselves or belaboring the obvious, because now we shall be establishing our results under different assumptions and in a much broader context. Henceforth, at the end of a proof, we shall be able to say that our conclusions are valid not only in some specific system, such as E_2^+ or Π_2, but in *any* system in which our axioms can be verified.

In this chapter we introduce most though not all of the axioms we shall need in our work. Using these, we then prove the theorems of Desargues and Pappus and investigate the logical interrelations between these fundamental results. Also, to illustrate the variety of systems to which our conclusions might apply, we discuss several models, or specific representations, of our abstract axiomatic system, some finite, some infinite, but all quite different from the algebraic structure we studied in previous chapters.

6.2 The Axioms of Incidence and Connection

Any axiomatic development must begin with a set of undefined terms which acquire their meaning from the statements the axioms make about them. In plane projective, as in plane euclidean, geometry, three such terms suffice, namely, "point," "line," and a relation referred to variously as "lies on," "passes through," "contains," or "is incident with" which may or may not exist between a particular point and a particular line.

Inevitably, in drawing the figures which will summarize the information we are given and suggest the relations to be explored, we shall represent points by small round marks and lines by long thin marks. However, these marks, like the figures which contain them, are only reminders of more abstract concepts, and ultimately points and lines are, or can be thought of as, any objects whatsoever which have the properties the axioms attribute to points and lines. Just what these properties are are spelled out in the following axioms.

Axiom 1

For any two points, there is at least one line which contains both points.

Axiom 2

For any two points, there is at most one line which contains both points.

Axiom 3

For any two lines, there is at least one point which lies on both lines.

Axiom 4

Every line contains at least three points.

Axiom 5

All points do not lie on the same line.

Axiom 6

There exists at least one line.

Here, as always, the first question we face as we begin to work with our axioms is that of consistency. Our axioms may or may not be well motivated, the assertions they make may or may not seem obvious, but they must be consistent; for if it is possible to conclude from them that some statement is both true and false, the system is worthless. There is no absolute test for the consistency of a set of axioms, and relative consistency is usually all that can be demonstrated. To do this, we exhibit some particular system, usually drawn from elementary arithmetic or geometry, in which each of the axioms is verifiably true. Then any contradiction arising from the axioms would imply a contradiction within the structure of arithmetic or euclidean geometry, and 3,000 years of experience with these systems gives us convincing empirical evidence that no such contradiction exists. At any rate, once such a model is constructed, it is at least possible to say that the system under consideration is consistent *if* arithmetic and euclidean geometry are consistent.

In the present case, Π_2 provides us with a ready-made consistency model for the axioms we have thus far introduced. Defining a point as an equivalence class of triples

$$(kx_1, kx_2, kx_3) \qquad \begin{array}{l} k \neq 0 \\ x_1, x_2, x_3 \text{ not all zero} \end{array}$$

defining a line as an equivalence class of triples

$$[hl_1, hl_2, hl_3] \qquad \begin{array}{l} h \neq 0 \\ l_1, l_2, l_3 \text{ not all zero} \end{array}$$

Table 6.1

l_1	l_2	l_3	l_4	l_5	l_6	l_7
P_1	P_3	P_2	P_1	P_1	P_2	P_5
P_2	P_4	P_4	P_3	P_4	P_3	P_6
P_5	P_5	P_6	P_6	P_7	P_7	P_7

and defining the incidence relation between a point (x_1, x_2, x_3) and a line $[l_1, l_2, l_3]$ to mean that

$$l_1 x_1 + l_2 x_2 + l_3 x_3 = 0$$

it is easy to verify that each of our axioms holds.

Even more simply, the consistency of our axioms can be established by considering the system described by Table 6.1. Here we assign no specific interpretations to the terms "point" and "line" but merely use the names of

Table 6.2

l_1	l_2	l_3	l_4	l_5	l_6	l_7	l_8	l_9	l_{10}	l_{11}	l_{12}	l_{13}
P_1	P_3	P_2	P_3	P_2	P_1	P_1	P_2	P_3	P_1	P_4	P_7	P_{10}
P_6	P_5	P_4	P_4	P_6	P_5	P_4	P_5	P_6	P_2	P_5	P_8	P_{11}
P_7	P_8	P_9	P_7	P_8	P_9	P_8	P_7	P_9	P_3	P_6	P_9	P_{12}
P_{10}	P_{10}	P_{10}	P_{11}	P_{11}	P_{11}	P_{12}	P_{12}	P_{12}	P_{13}	P_{13}	P_{13}	P_{13}

seven points, $P_1, P_2, \ldots, P_7$, and the names of seven lines, $l_1, l_2, \ldots, l_7$, in a tabulation which identifies the points which lie on a particular line by listing the names of those points in a column under the name of that line. Axioms 4, 5, and 6 are obviously true in this model. To verify Axioms 1 and 2, we merely check that the names of the points in each of the $_7C_2 = 21$ possible pairs of points appear together in one and only one column. Finally, to verify Axiom 3, we note that without exception the columns in each of the $_7C_2 = 21$ pairs of columns have the name of at least one point in common.

Tables 6.2 and 6.3 describe other models, similar to the one described by Table 6.1, which can also be used to establish the consistency of Axioms 1 to 6. These three systems are examples of what are known as **finite projective geometries.** Infinitely many different finite projective

Table 6.3

l_1	l_2	l_3	l_4	l_5	l_6	l_7	l_8	l_9	l_{10}	l_{11}	l_{12}	l_{13}	l_{14}	l_{15}	l_{16}	l_{17}	l_{18}	l_{19}	l_{20}	l_{21}
P_3	P_4	P_1	P_2	P_2	P_1	P_4	P_3	P_4	P_3	P_2	P_1	P_1	P_2	P_3	P_4	P_1	P_5	P_9	P_{13}	P_{17}
P_6	P_5	P_8	P_7	P_8	P_7	P_6	P_5	P_6	P_5	P_8	P_7	P_6	P_5	P_8	P_7	P_6	P_6	P_{10}	P_{14}	P_{18}
P_{12}	P_{11}	P_{10}	P_9	P_{11}	P_{12}	P_9	P_{10}	P_{10}	P_9	P_{12}	P_{11}	P_9	P_{10}	P_{11}	P_{12}	P_3	P_7	P_{11}	P_{15}	P_{19}
P_{13}	P_{14}	P_{15}	P_{16}	P_{13}	P_{14}	P_{15}	P_{16}	P_{13}	P_{14}	P_{15}	P_{16}	P_{13}	P_{14}	P_{15}	P_{16}	P_4	P_8	P_{12}	P_{16}	P_{20}
P_{17}	P_{17}	P_{17}	P_{17}	P_{18}	P_{18}	P_{18}	P_{18}	P_{19}	P_{19}	P_{19}	P_{19}	P_{20}	P_{20}	P_{20}	P_{20}	P_{21}	P_{21}	P_{21}	P_{21}	P_{21}

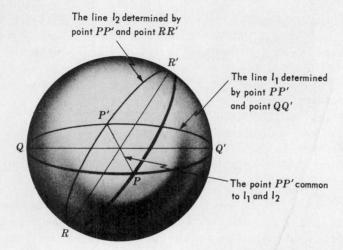

The line l_2 determined by
point PP' and point RR'

R'

The line l_1 determined
by point PP'
and point QQ'

P'

Q Q'

The point PP' common
to l_1 and l_2

P

R

Fig. 6.1   A spherical model of Axioms 1 to 6.

geometries are known to exist.[1] Some of the simpler properties of these
geometries will be found in the exercises at the end of this section.

Another model, this time containing infinitely many points and
infinitely many lines, is provided by the following system. Let S be an
arbitrary sphere in euclidean three-space, E_3, let the diameters of S be the
points of our model, let the great circles of S be the lines of our model, and
let a point be on a line if and only if the diametral segment which represents
the point is a diameter of the great circle which represents the line (Fig. 6.1).
It is now easy to verify that each of our axioms holds in this model. In fact
this model is isomorphic with the extended euclidean plane, E_2^+, although we
shall leave the proof of this as an exercise.

Clearly, in the last model a point can be described equally well as a
diameter, $\overline{PP'}$, of S or as a pair of diametrically opposite points, (P,P'), on S.
Carrying this one step further, we may consider either P or P' itself as
representative of a point of our system, since P determines P', and conversely.
This in turn means that we need retain only one hemisphere, H, of S,
provided we understand that the points of our system are the interior points
of the surface H together with all the *pairs* of diametrically opposite points
on the great circle which is the boundary of H. Thus we reach the topological
description of the classical projective plane: *the surface of a sphere with diametrically opposite points identified as the same point* or *a hemisphere with diametrically
opposite points on its edge "sewn" together.*

[1] See, for instance, O. Veblen and W. H. Bussey, Finite Projective Geometries, *Trans. Am.
Math. Soc.*, vol. 7, pp. 241–259, 1906, where it is shown that if $n - 1$ is a power of a prime,
then there exists a finite geometry in which every line contains exactly n points. Such a
geometry is usually denoted by the symbol PG_n.

There are a number of simple theorems which follow from Axioms 1 to 6. Though their assertions are obvious and their proofs trivial, it will be helpful to state them because of the light they shed on the important matter of duality:

Theorem 1

There exists at least one point.

Theorem 2

For any two lines, there is at most one point which lies on both lines.

Theorem 3

All lines do not pass through the same point.

Theorem 4

Every point lies on at least three lines.

If we now bring together Axioms 1 to 6 and Theorems 1 to 4, we see that they can be arranged in the following interesting way:

Axiom 1 ↔ Axiom 3
Axiom 2 ↔ Theorem 2
Axiom 3 ↔ Axiom 1
Axiom 4 ↔ Theorem 4
Axiom 5 ↔ Theorem 3
Axiom 6 ↔ Theorem 1

Each statement in the first column has exactly the same structure as the corresponding statement in the second column and differs from it only in the interchange of the words "point" and "line" and in the use of a different one of the equivalent phrases "lies on," "passes through," or "contains." In other words, the statement obtained from any axiom by interchanging the words "point" and "line" is either another axiom or else a theorem which we have already proved and therefore, in either case, true in our system.

Consider now any theorem which might be provable on the basis of our axioms. The theorem is, of course, some statement involving the terms "point," "line," and "lies on" or some synonymous phrase. Likewise, the proof of the theorem is a series of assertions each containing these same terms and each justified directly or indirectly by one or more of our axioms. Suppose now that in each step of the proof we were to interchange the words "point" and "line." Each of these new statements would have as its final justification one or more assertions obtained from the axioms by similarly interchanging "point" and "line." But as we have just verified, the statements obtained from our axioms by interchanging "point" and "line" are in fact true in our system. Hence, merely by interchanging the words "point" and "line" in each step of our proof we have, in a purely mechanical fashion, constructed the proof of a new theorem, namely, the theorem obtained from the original one by interchanging the terms "point" and "line" in its statement.

The preceding observations make it clear that up to this point, at least, the principle of duality is valid in our system. In other words, for every theorem provable from Axioms 1 to 6 there is another which differs from the first only in the interchange of the dual terms "point" and "line" and whose proof differs from the proof of the first only in the consistent interchange of "point" and "line" in every step. The principle of duality in our present system thus depends upon a symmetry in the axioms themselves, and not, as in E_2^+ and Π_2, on the particular nature of the points, lines, and the incidence condition.

It is, of course, not automatically true that as we add additional axioms the principle of duality will remain valid in our system. In fact, if we wish to retain the principle of duality (as of course we do), then each time we adopt a new axiom we must not only show that the new system is still consistent, but we must also ensure that the principle of duality is preserved, either by adopting the dual of the new axiom also or else proving the dual as a theorem. The latter is actually what we shall do in each case.

Exercises

1. Prove Theorems 1 to 4.

2. Prove that the following theorem is true in any system satisfying Axioms 1 to 6: If there exists one line containing exactly n points, then:
(a) Every line contains exactly n points.
(b) Every point has exactly n lines passing through it.
(c) The system contains exactly $n^2 - n + 1$ points.
(d) The system contains exactly $n^2 - n + 1$ lines.

3. Show that if any line, together with the points which lie on it, is removed from a system satisfying Axioms 1 to 6, the resulting system still satisfies Axioms 1, 2, 5, and 6 but, instead of Axiom 3, now satisfies Axiom 3′: If l is a line and P is a point which is not on l, then there is one and only one line which contains P and has no point in common with l; and, instead of Axiom 4, now satisfies Axiom 4′: Every line contains at least two points. (A system satisfying Axioms 1, 2, 3′, 4′, 5, and 6 and containing only a finite number of points is sometimes called a **finite euclidean geometry**.)

4. Prove that the following theorem is true in any system satisfying the modified set of axioms described in Exercise 3: If there exists one line containing exactly n points, then

(a) Every line contains exactly n points.

(b) Every point has exactly $n + 1$ lines passing through it.

(c) The system contains exactly n^2 points.

(d) The system contains exactly $n(n + 1)$ lines.

5. Prove that the spherical model described in this section is isomorphic to E_2^+.

6. Can a system isomorphic to E_2 be obtained from the spherical model described in this section? How?

6.3 The Projectivity Axiom

Among the many important topics we discussed in our investigation of the geometry of E_2^+ and Π_2 were perspectivities and projectivities:

Definition 1

A one-to-one correspondence between the points of two ranges on distinct lines which has the property that the joins of corresponding points are concurrent is called a perspectivity.

Definition 2

A one-to-one correspondence between the points of two ranges which is the composition of a finite sequence of perspectivities is called a projectivity.

The fundamental property of projectivities was proved in Theorem 3, Sec. 4.6: a projectivity is uniquely determined by the assignment of three pairs of corresponding points. The proof we gave was naturally based upon the algebraic properties of E_2^+ and Π_2; and for all we know, the theorem is valid only in these two systems. In particular, this theorem is not a consequence of Axioms 1 to 6, because in Appendix 2 a finite geometry is described which satisfies Axioms 1 to 6 but in which different projectivities may have as many as six pairs of corresponding points in common! The property asserted by this theorem is so convenient to work with, however, and leads to so many important results, that since we cannot prove it, we shall incorporate it into our system as an axiom:

Axiom 7   The projectivity axiom

A projectivity between two ranges is uniquely determined by the assignment of three pairs of corresponding points.

The question of the consistency of the enlarged set of axioms, Axioms 1 to 7, is easily settled, for Π_2 provides us with a specific model in which we have already verified that Axioms 1 to 7 are valid. The other point which must be settled is whether or not the principle of duality continues to hold in the system after Axiom 7 has been added. If Axiom 7 were self-dual, this would be no problem. Since it is not, we must either assume its dual as a new axiom (subject, of course, to the usual requirement of consistency) or prove its dual as a theorem if the principle of duality is to be preserved. In this case, the dual of the axiom we have added is easy to prove:

Theorem 1

A projectivity between two pencils is uniquely determined by the assignment of three pairs of corresponding lines.

Proof Let a_1, b_1, c_1 be three lines of a pencil with vertex L_1, and let a_n, b_n, c_n be three lines of a pencil with vertex L_n. Now, contrary to the theorem, let us suppose that T and T' are two different projectivities each of which maps a_1 into a_n, b_1 into b_n, and c_1 into c_n. This means that there is at least one line, d_1, on L_1 whose images under T and T' are different, say d_n and d'_n. Thus we have the two chains of perspectivities

$$T: L_1(a_1,b_1,c_1,d_1) \overset{v_1}{\wedge} L_2 \overset{v_2}{\wedge} \cdots \overset{v_{n-2}}{\wedge} L_{n-1} \overset{v_{n-1}}{\wedge} L_n(a_n,b_n,c_n,d_n)$$

$$T': L_1(a_1,b_1,c_1,d_1) \overset{v_1'}{\wedge} L_2' \overset{v_2'}{\wedge} \cdots \overset{v_{m-2}'}{\wedge} L_{m-1}' \overset{v_{m-1}'}{\wedge} L_n(a_n,b_n,c_n,d_n')$$

Now let A_1, B_1, C_1, D_1 be the intersections of the lines a_1, b_1, c_1, d_1 with an arbitrary line x not passing through L_1, and let A_n, B_n, C_n, D_n, D'_n be the intersections of the lines a_n, b_n, c_n, d_n, d'_n with an arbitrary line y not passing through L_n. Then, under T,

$$x(A_1,B_1,C_1,D_1) \overset{L_1}{\wedge} v_1 \overset{L_2}{\wedge} \cdots \overset{L_{n-1}}{\wedge} v_{n-1} \overset{L_n}{\wedge} y(A_n,B_n,C_n,D_n)$$

while, under T',

$$x(A_1,B_1,C_1,D_1) \overset{L_1}{\wedge} v_1' \overset{L_2'}{\wedge} \cdots \overset{L_{m-1}'}{\wedge} v_{m-1}' \overset{L_n}{\wedge} y(A_n,B_n,C_n,D_n')$$

Thus from the assumption that there are two different projectivities between the points L_1 and L_n each of which maps a_1 into a_n, b_1 into b_n, and c_1 into c_n, we have reached the conclusion that there are two different projectivities between the lines x and y each of which maps A_1 into A_n, B_1 into B_n, and C_1 into C_n. However, this contradicts Axiom 7, and therefore the theorem is established.

Our next theorem tells us that another important result in E_2^+ and Π_2 is also true in any system which satisfies Axioms 1 to 7.

Theorem 2   The perspectivity theorem

Any projectivity between distinct lines in which the point of intersection of the lines is self-corresponding is a perspectivity.

Proof   Let P be the intersection of two lines, l_1 and l_2, let A_1 and B_1 be two points on l_1 each distinct from P, and let A_2 and B_2 be the images of A_1 and B_1, respectively, under a projectivity between l_1 and l_2 which maps P into itself. Then

$$l_1(P,A_1,B_1,\ldots) \sim l_2(P,A_2,B_2,\ldots)$$

Now if V is the intersection of A_1A_2 and B_1B_2, then

$$l_1(P,A_1,B_1,\ldots) \overset{V}{\wedge} l_2(P,A_2,B_2,\ldots)$$

However, by Axiom 7 there is a unique projectivity between l_1 and l_2 with the property that it maps A_1 into A_2, B_1 into B_2, and P into itself. Since the perspectivity from the center V is *a* projectivity which accomplishes this mapping, it is therefore the only one, and the theorem is established.

Corollary 1

If l_1, l_2, l_3 are three concurrent lines, and if $l_1 \overset{U}{\wedge} l_2 \overset{V}{\wedge} l_3$, then there exists a point W, collinear with U and V, such that $l_1 \overset{W}{\wedge} l_3$.

Proof   Let l_1, l_2, l_3 be three lines concurrent in a point P, and let U and V be centers of perspectivities such that $l_1 \overset{U}{\wedge} l_2 \overset{V}{\wedge} l_3$. Clearly, the point P which is common to l_1, l_2, and l_3 is self-corresponding under each of the perspectivities relating l_1 to l_3. Hence P is self-corresponding in the resultant projectivity between l_1 and l_3, and therefore, by Theorem 2, there is a point W such that

$$l_1 \overset{W}{\wedge} l_3$$

It remains now to prove that the center, W, of the last perspectivity is collinear with U and V. To do this, let us suppose first that the line UV does not pass through P. Then if Q_1, Q_2, Q_3 are, respectively, the intersections of UV and l_1, l_2, l_3 (Fig. 6.2), we have $l_1(Q_1,\ldots) \overset{U}{\wedge} l_2(Q_2,\ldots) \overset{V}{\wedge} l_3(Q_3,\ldots)$ and hence $l_1(Q_1,\ldots) \overset{W}{\wedge} l_3(Q_3,\ldots)$. Therefore, since Q_1 and Q_3 are corresponding points in this perspectivity, the center of the perspectivity, W, must lie on the line Q_1Q_3. Since this line is, by definition, the same as the line UV, we have thus shown that when UV does not pass through P, the centers of the three perspectivities are collinear.

If P lies on the line UV, then Q_1, Q_2, and Q_3 coincide at P, and our proof breaks down. In this case, we may consider, instead, the projectivity

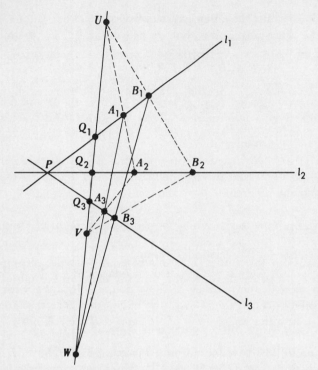

Fig. 6.2 The composition of perspectivities between concurrent lines.

defined by $l_1 \overset{W}{\wedge} l_3 \overset{V}{\wedge} l_2$, which, of course, is equivalent to the perspectivity $l_1 \overset{U}{\wedge} l_2$. The preceding argument can now be used to prove that U lies on the line VW, unless V and W are collinear with P. But if this happens, then U, V, and W are all collinear with P, and our proof is complete.

With Theorem 2 and its corollary, it is now easy to prove the theorem of Desargues:

Theorem 3   Desargues' theorem

If two triangles are centrally perspective, they are also axially perspective.

Proof   Let $\triangle A_1 A_2 A_3$ and $\triangle B_1 B_2 B_3$ be two triangles so situated that $l_1 : A_1 B_1$, $l_2 : A_2 B_2$, and $l_3 : A_3 B_3$ are concurrent in a point P, and let:

V_1 be the intersection of $A_2 A_3$ and $B_2 B_3$
V_2 be the intersection of $A_1 A_3$ and $B_1 B_3$
V_3 be the intersection of $A_1 A_2$ and $B_1 B_2$

Since V_1, V_2, V_3 are the intersections of corresponding sides of the given triangles, the theorem requires that we prove that V_1, V_2, V_3 are collinear. Now

$$l_1(P,A_1,B_1) \overset{V_3}{\wedge} l_2(P,A_2,B_2) \overset{V_1}{\wedge} l_3(P,A_3,B_3)$$

Hence, by Corollary 1, Theorem 2, it follows that there exists a point V'_2, collinear with V_1 and V_3, such that

$$l_1(P,A_1,B_1) \overset{V_2'}{\wedge} l_3(P,A_3,B_3)$$

Since (A_1,A_3) and (B_1,B_3) are corresponding points in this perspectivity, it follows that V'_2 is the intersection of A_1A_3 and B_1B_3. In other words, $V'_2 = V_2$, and the theorem is proved.

We are now in a position to prove the theorem of Pappus.

Theorem 4   The theorem of Pappus

Let A, B, C be three points on a line l and let A', B', C' be three points on a second line l', none of these points coinciding with the intersection of l and l'. Then the intersection of BC' and $B'C$, the intersection of CA' and $C'A$, and the intersection of AB' and $A'B$ are collinear.

Proof   Let l and l' be two lines intersecting in a point O, let A, B, C be three points on l distinct from O, let A', B', C' be three points on l' distinct from O, let:

A'' be the intersection of BC' and $B'C$
B'' be the intersection of CA' and $C'A$
C'' be the intersection of AB' and $A'B$;

let:

P be the intersection of BC' and AB'
Q be the intersection of $B'C$ and AC';

and let:

p be the line containing A, B', C'', P
q be the line containing A, B'', C', Q (Fig. 6.3).

Then

$$p(C'',B',P,A) \overset{B}{\wedge} l'(A',B',C',O) \overset{C}{\wedge} q(B'',Q,C',A)$$

and hence

$$p(C'',B',P,A) \sim q(B'',Q,C',A)$$

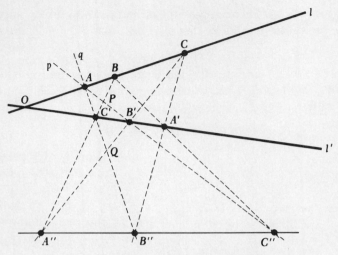

Fig. 6.3   Figure illustrating the theorem of Pappus.

Clearly, this is a projectivity between the distinct lines p and q in which the intersection of the lines, A, is self-corresponding. By Theorem 2 it is therefore a perspectivity, and the lines $C''B''$, $B'Q$, and PC' are necessarily concurrent. This implies that $B''C''$ must pass through the intersection of $B'Q$ and PC', namely, A''. Hence A'', B'', C'' are collinear, as asserted.

The line determined by the points A'', B'', C'' is commonly called the **pappus line** of the two triads, (A,B,C) and (A',B',C').

In Sec. 4.5 we showed that in both E_2^+ and Π_2, any projectivity between distinct lines could be expressed as the composition of at most two perspectivities. As a final result in this section we shall prove that this is valid not only in E_2^+ and Π_2 but also in any geometry satisfying Axioms 1 to 7. To do this it is convenient to begin by proving the following lemmas.

Lemma 1

If a, b, c are three nonconcurrent lines, and if

$$a(A_1,A_2,A_3,\ldots) \overset{V_1}{\wedge} b(B_1,B_2,B_3,\ldots) \overset{V_2}{\wedge} c(C_1,C_2,C_3,\ldots)$$

then for any line, b', concurrent with, but distinct from, a and b and not passing through V_2, there exists a point V_1', collinear with V_1 and V_2, such that

$$a(A_1,A_2,A_3,\ldots) \overset{V_1'}{\wedge} b'(B_1',B_2',B_3',\ldots) \overset{V_2}{\wedge} c(C_1,C_2,C_3,\ldots)$$

Proof Let a, b, c be three nonconcurrent lines, let V_1 and V_2 be the centers of perspectivities such that

$$a(A_1,A_2,A_3,...) \overset{V_1}{\wedge} b(B_1,B_2,B_3,...) \overset{V_2}{\wedge} c(C_1,C_2,C_3,...)$$

let b' be any line concurrent with, but distinct from, a and b and not passing through V_2, and let B_1', B_2', B_3', . . . be the points on b' which are perspective from V_2 with C_1, C_2, C_3, . . . (Fig. 6.4); i.e., let B_1', B_2', B_3', . . . be points on b' such that

(1) $$b'(B_1',B_2',B_3',...) \overset{V_2}{\wedge} c(C_1,C_2,C_3,...)$$

Then, clearly, $$b(B_1,B_2,B_3,...) \overset{V_2}{\wedge} b'(B_1',B_2',B_3',...)$$

Moreover, by hypothesis, $a(A_1,A_2,A_3,...) \overset{V_1}{\wedge} b(B_1,B_2,B_3,...)$. Hence, since a, b, and b' are distinct, concurrent lines, it follows from Corollary 1, Theorem 2, that there exists a point, V_1', collinear with V_1 and V_2, such that

$$a(A_1,A_2,A_3,...) \overset{V_1'}{\wedge} b'(B_1',B_2',B_3',...)$$

This, coupled with (1), proves the assertion of the lemma.

In exactly the same fashion, we can prove the companion result:

Lemma 2

If a, b, c are three nonconcurrent lines, and if

$$a(A_1,A_2,A_3,...) \overset{V_1}{\wedge} b(B_1,B_2,B_3,...) \overset{V_2}{\wedge} c(C_1,C_2,C_3,...)$$

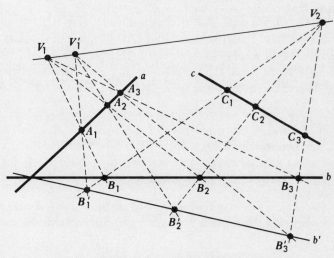

Fig. 6.4 Figure used in the proof of Lemma 1.

then for any line b', concurrent with, but distinct from, b and c and not passing through V_1, there exists a point V_2', collinear with V_1 and V_2, such that

$$a(A_1,A_2,A_3,...) \overset{V_1}{\wedge} b'(B_1',B_2',B_3',...) \overset{V_2'}{\wedge} c(C_1,C_2,C_3,...)$$

We can now turn our attention to the theorem itself.

Theorem 5

Any projectivity between distinct lines can be expressed as the composition of at most two perspectivities.

Proof At the outset, we observe that in order to prove the theorem it is sufficient to show that three successive perspectivities can always be reduced to two; for if this is proved, then repeated applications of this property will allow us to "telescope" any number of perspectivities until only two remain. Accordingly, let us suppose that we have a sequence of three perspectivities,

$$(2) \quad a(A_1,A_2,A_3,...) \overset{V_1}{\wedge} b(B_1,B_2,B_3,...) \overset{V_2}{\wedge} c(C_1,C_2,C_3,...) \overset{V_3}{\wedge} d(D_1,D_2,D_3,...)$$

There are five cases to consider:

1. a, b, c are concurrent.
2. b, c, d are concurrent.
3. No three of the lines a, b, c, d are concurrent.
4. a, b, d are concurrent.
5. a, c, d are concurrent.

Cases 1 and 2 can be disposed of immediately. In fact, in case 1, the two perspectivities relating a to b and b to c can, by Corollary 1, Theorem 2, be reduced to a single perspectivity; and in case 2, the two perspectivities relating b to c and c to d can similarly be reduced to a single perspectivity. Hence in each case the three given perspectivities can be reduced to two, as asserted.

In case 3, let P_1 be the intersection of a and b, and let P_2 be the intersection of c and d. Since no three of the lines a, b, c, d are concurrent, P_1 and P_2 are necessarily distinct points. Let x be the line P_1P_2, and let us suppose as a first subcase that x does not pass through V_2 (Fig. 6.5a). Then if X_1, X_2, X_3, ... are the points on x which are perspective with C_1, C_2, C_3, ... from V_2, we have

$$(3) \qquad a(A_1,A_2,A_3,...) \overset{V_1}{\wedge} b(B_1,B_2,B_3,...) \overset{V_2}{\wedge} x(X_1,X_2,X_3,...)$$

and

$$(4) \qquad x(X_1,X_2,X_3,...) \overset{V_2}{\wedge} c(C_1,C_2,C_3,...) \overset{V_3}{\wedge} d(D_1,D_2,D_3,...)$$

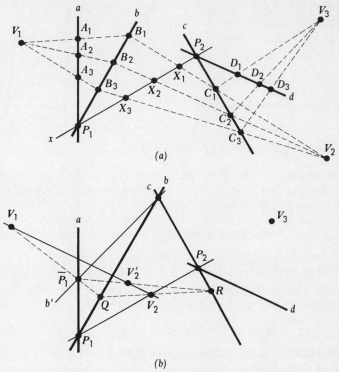

Fig. 6.5   Figures used in the proof of Theorem 5.

Since a, b, and x are concurrent, it follows from (3) and Corollary 1, Theorem 2, that there exists a point W_1 on V_1V_2 such that

$$(5) \qquad a(A_1,A_2,A_3,\ldots) \overset{W_1}{\overline{\wedge}} x(X_1,X_2,X_3,\ldots)$$

Similarly, since x, c, and d are concurrent, it follows from (4) that there exists a point W_2 on V_2V_3 such that

$$(6) \qquad x(X_1,X_2,X_3,\ldots) \overset{W_2}{\overline{\wedge}} d(D_1,D_2,D_3,\ldots)$$

Relations (5) and (6) together show that the three perspectivities (2) originally defining the mapping of a onto d can be reduced to two, as asserted.

If the line $x = P_1P_2$ passes through V_2, the points X_1, X_2, X_3, ... all coincide at V_2 and the preceding argument fails. However, if this happens, we can proceed as follows. Let b' be an arbitrary line on the point bc, distinct from b and c and not passing through either V_1 or ad. Let $\bar{P}_1$ be the

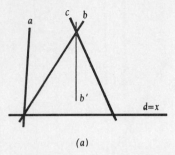 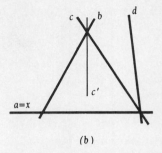

(a)   (b)

Fig. 6.6   Figures used in the discussion of cases 4 and 5, Theorem 5.

intersection of a and b', let Q be the intersection of $V_1\bar{P}_1$ and b, and let R be the intersection of V_2Q and c (Fig. 6.5b). Then under the first two of the given perspectivities, we have

$$a(P_1,\bar{P}_1,\ldots) \overset{V_1}{\wedge} b(P_1,Q,\ldots) \overset{V_2}{\wedge} c(P_2,R,\ldots)$$

Clearly, P_2 and R, being images of the distinct points P_1 and $\bar{P}_1$, are distinct. Now since b' is concurrent with b and c, it follows by Lemma 2 that there exists a point, V_2', on V_1V_2 but obviously not on b', such that the original chain of perspectivities can be replaced by the chain

(7) $$a(P_1,\bar{P}_1,\ldots) \overset{V_1}{\wedge} b' \overset{V_2'}{\wedge} c(P_2,R,\ldots) \overset{V_3}{\wedge} d$$

This implies that V_2' is the intersection of V_1V_2 and the line $\bar{P}_1R$; hence V_2' cannot lie on $\bar{P}_1P_2$. Therefore, since no three of the lines a, b', c, d are concurrent, and since V_2' does not lie on the line determined by the points $\bar{P}_1 = ab'$ and $P_2 = cd$, the three perspectivities in the equivalent chain (7) meet all the requirements of the first part of our proof in case 3 and can therefore be reduced to two by an application of our earlier argument.

In case 4, since x and d are the same line, the argument we used in case 3 breaks down. However, we can still reduce this case to case 3 as follows. Let b' be any line on the point bc which is distinct from b and c and does not pass through V_1 (Fig. 6.6a). Then by Lemma 2 there exists a point V_2' on V_1V_2 such that

$$a(A_1,A_2,A_3,\ldots) \overset{V_1}{\wedge} b'(B_1',B_2',B_3',\ldots) \overset{V_2'}{\wedge} c(C_1,C_2,C_3,\ldots) \overset{V_3}{\wedge} d(D_1,D_2,D_3,\ldots)$$

The lines a, b', c, and d are such that no three of them are concurrent. Hence the argument in case 3 can now be applied, and the reduction to two perspectivities can be accomplished.

In case 5, a and x are the same line, and again the general argument breaks down. However, this case can also be reduced to case 3. To do this, let c' be any line on the intersection of b and c which is distinct from b and c and does not pass through V_3 (Fig. 6.6b). Then by Lemma 1 there exists a point V_2' on V_2V_3 such that

$$a(A_1,A_2,A_3,\ldots) \overset{V_1}{\wedge} b(B_1,B_2,B_3,\ldots) \overset{V_2'}{\wedge} c'(C_1',C_2',C_3',\ldots) \overset{V_3}{\wedge} d(D_1,D_2,D_3,\ldots)$$

The lines a, b, c', and d are now such that no three of them are concurrent. Hence the argument in case 3 can be applied, and the proof of the theorem completed.

Corollary 1

Any projectivity can be reduced to at most three perspectivities.

Exercises

1. Prove Corollary 1, Theorem 5.

2. Illustrate by means of particular examples that Desargues' theorem is valid in the finite projective geometries described by Tables 6.2 and 6.3. Is this true in the system described by Table 6.1?

3. Illustrate by particular examples that the theorem of Pappus is valid in the finite projective geometries described by Tables 6.2 and 6.3. Is this true in the system described by Table 6.1?

4. In the finite projective geometry described by Table 6.2, express each of the following projectivities in terms of the minimum number of perspectivities:

(a) $l_3(P_2,P_9,P_4) \sim l_{11}(P_{13},P_6,P_4)$ (b) $l_1(P_1,P_{10},P_7) \sim l_{12}(P_7,P_9,P_8)$

(c) $l_2(P_8,P_3,P_5) \sim l_8(P_7,P_5,P_2)$ (d) $l_5(P_2,P_6,P_8) \sim l_5(P_6,P_2,P_{11})$

5. In the finite projective geometry described by Table 6.3, express each of the following projectivities as the composition of at most three perspectivities:

(a) $l_7(P_4,P_6,P_9) \sim l_7(P_4,P_6,P_{15})$ (b) $l_4(P_2,P_7,P_9) \sim l_4(P_7,P_2,P_{17})$

Is either of these projectivities an involution?

6. If $a(A_1,A_2,A_3,\ldots) \overset{U}{\wedge} b(B_1,B_2,B_3,\ldots) \overset{V}{\wedge} c(C_1,C_2,C_3,\ldots)$ and if a, b, and c are nonconcurrent, give the conditions, if any, under which there exists a point W such that

$$a(A_1,A_2,A_3,\ldots) \overset{W}{\wedge} c(C_1,C_2,C_3,\ldots)$$

7. Does Corollary 1, Theorem 2, establish a one-to-one correspondence between the lines of the pencil on P and the points of the range on UV if l_3, and therefore W, is considered variable? How can the line l_3 corresponding to a particular W on UV be determined?

8. If $\triangle A_1B_1C_1$ and $\triangle A_2B_2C_2$ are centrally perspective from the point O, and if P_1, P_2, P_3 are, respectively, the intersections of the sides (B_1C_1,B_2C_2),

(C_1A_1,C_2A_2), (A_1B_1,A_2B_2), show that the following pairs of triangles are also perspective and determine the center and axis of perspective in each case:

(a) $\triangle C_1C_2P_2$, $\triangle B_1B_2B_3$ (b) $\triangle A_1A_2P_3$, $\triangle C_1C_2P_1$ (c) $\triangle B_1B_2P_1$, $\triangle A_1A_2P_2$

(d) $\triangle OB_1C_1$, $\triangle A_2P_3P_2$ (e) $\triangle OC_1A_1$, $\triangle B_2P_1P_3$ (f) $\triangle OA_1B_1$, $\triangle C_2P_2P_1$

(g) $\triangle OB_2C_2$, $\triangle A_1P_3P_2$ (h) $\triangle OC_2A_2$, $\triangle B_1P_1P_3$ (i) $\triangle OA_2B_2$, $\triangle C_1P_2P_1$

9. If $\triangle ABC$ and $\triangle A'B'C'$ are perspective with center P and axis p, and if $\triangle BCD$ and $\triangle B'C'D'$ are perspective with center P and axis p, show that the triangles in each of the pairs $(\triangle ABD, \triangle A'B'D')$ and $(\triangle ACD, \triangle A'C'D')$ are also perspective with center P and axis p.

10. If a, b, c are three concurrent lines, and if U, V, W are three collinear points such that

$$a(A_1,A_2,A_3,\ldots) \overset{U}{\wedge} b(B_1,B_2,B_3,\ldots) \qquad b(B_1,B_2,B_3,\ldots) \overset{V}{\wedge} c(C_1,C_2,C_3,\ldots)$$

and $a(A_1,A_2,A_3,\ldots) \overset{W}{\wedge} b(B_1',B_2',B_3',\ldots)$, show that there is a point T, collinear with U, V, W, such that $b(B_1',B_2',B_3',\ldots) \overset{T}{\wedge} c(C_1,C_2,C_3,\ldots)$.

11. Show that if $a(A_1,A_2,A_3,\ldots) \overset{U}{\wedge} b(B_1,B_2,B_3,\ldots)$ and $b(B_1,B_2,B_3,\ldots) \overset{V}{\wedge} c(C_1,C_2,C_3,\ldots)$, and if b' is any line distinct from a, b, and UV, then there exist points U' and V' on UV and points B_1', B_2', B_3', $\ldots$ on b' such that $a(A_1,A_2,A_3,\ldots) \overset{U'}{\wedge} b'(B_1',B_2',B_3',\ldots)$ and $b'(B_1',B_2',B_3',\ldots) \overset{V'}{\wedge} c(C_1,C_2,C_3,\ldots)$.

12. If (A,B,C) and (A',B',C') are two sets of collinear points such that AA', BB', CC' are concurrent, prove that the pappus line of the two sets is concurrent with the lines which contain the two sets.

13. Prove the converse of the result of Exercise 12.

14. Given a projectivity, T, between cobasal ranges. Show that if a fixed point of T is known, then T can be expressed as the composition of two perspectivities.

15. Given a projectivity between cobasal ranges on a line l, with distinct fixed points F_1 and F_2, in which (A,A') and (B,B') are two pairs of mates. Show that there is also a projectivity between cobasal ranges on l, with fixed points F_1 and F_2, in which (A,B) and (A',B') are two pairs of mates. To what results in Π_2 does this correspond?

6.4 Further Relations among the Fundamental Theorems

In the last section we added the projectivity axiom to our system, and then from it we deduced the perspectivity theorem, Desargues' theorem, the theorem of Pappus, and the reduction theorem for chains of perspectivities.

With these results, we are now in a position to move on to new topics. However, "One man's theorem is another man's axiom," and it may help us to a better appreciation of the logical structure we are trying to build if we take time to reconsider the work of the last section in the light of this possibility. Specifically, in this section we shall show that the theorem (axiom) of Pappus is logically equivalent to the projectivity axiom (theorem). In other words, we shall show that either implies the other and all its consequences, and hence either could have been taken as an axiom. As a first step in the proof of this fact we shall show that the theorem of Pappus, without the projectivity axiom but with Axioms 1 to 6, of course, implies a special case of the perspectivity theorem.

Theorem 1

Independent of the projectivity axiom, the theorem of Pappus implies the following theorem: If l_1, l_2, l_3 are nonconcurrent lines, and if $l_1 \overset{V_1}{\wedge} l_2 \overset{V_2}{\wedge} l_3$ is a projectivity in which the intersection of l_1 and l_3 is self-corresponding, then this projectivity is a perspectivity.

Proof Let l_1, l_2, and l_3 be nonconcurrent lines, let:

P_1 be the intersection of l_2 and l_3
P_2 be the intersection of l_3 and l_1
P_3 be the intersection of l_1 and l_2

let $l_1 \overset{V_1}{\wedge} l_2 \overset{V_2}{\wedge} l_3$ be a projectivity between l_1 and l_3 in which the intersection, P_2, of l_1 and l_3 is self-corresponding, and let:

Q be the intersection of V_1P_2 and l_2
R be the intersection of V_1P_1 and V_2P_3

(Fig. 6.7). Now, by hypothesis, the given projectivity, $T: l_1 \overset{V_1}{\wedge} l_2 \overset{V_2}{\wedge} l_3$ leaves P_2 invariant, while the first of the two perspectivities in T, namely, $l_1 \overset{V_1}{\wedge} l_2$, sends P_2 into Q. Hence the second perspectivity, $l_2 \overset{V_2}{\wedge} l_3$, must send Q back into P_2, which is possible only if V_1 and V_2 are collinear with P_2 and Q. Suppose now that X_1 is an arbitrary point on l_1, and let X_2 and X_3 be its images on l_2 and l_3 under the perspectivities defining T. Then if we apply the theorem of Pappus, which we are assuming as part of our hypothesis, to the collinear triads

$$P_3 \quad X_2 \quad P_1$$
$$V_1 \quad P_2 \quad V_2$$

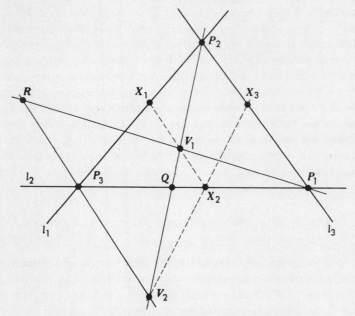

Fig. 6.7 Figure used in the proof of Theorem 1.

it follows that:

The intersection of P_3P_2 and X_2V_1, namely, X_1,
The intersection of P_3V_2 and P_1V_1, namely, R, and
The intersection of X_2V_2 and P_1P_2, namely, X_3,

are collinear. In other words, an arbitrary point X_1 on l_1 and its image X_3 under the given projectivity from l_1 to l_3 are always collinear with the point R. Since R is clearly a fixed point, independent of the position of X_1, it follows that the given projectivity is a perspectivity, as asserted.

We next prove that the theorem of Pappus implies Desargues' theorem.

Theorem 2

The theorem of Pappus, independent of the projectivity axiom, implies Desargues' theorem.

Proof Let l_1: A_1B_1, l_2: A_2B_2, l_3: A_3B_3 be three lines concurrent in a point O, so that $\triangle A_1A_2A_3$ and $\triangle B_1B_2B_3$ are perspective from O. Let:

P_1 be the intersection of A_2A_3 and B_2B_3
P_2 be the intersection of A_3A_1 and B_3B_1
P_3 be the intersection of A_1A_2 and B_1B_2

and let:

Q_1 be the intersection of P_1P_3 and l_1
Q_2 be the intersection of P_1P_3 and l_2
Q_3 be the intersection of P_1P_3 and l_3

(Fig. 6.8). To prove the theorem we must, of course, prove that P_1, P_2, P_3 are collinear. As a first case, let us suppose that P_1P_3 does not pass through O. At the outset, we note that if P_3 lies on l_1 or l_2, then either A_1 and B_1 or A_2 and B_2 coincide, and Desargues' theorem is trivially true. Similarly, if P_1 lies on l_2 or l_3, then either A_2 and B_2 or A_3 and B_3 coincide, and again Desargues' theorem is trivially true. Hence we may properly consider the perspectivities

$$l_1(A_1,B_1,Q_1,O,\ldots) \ \overset{P_3}{\wedge}\ l_2(A_2,B_2,Q_2,O,\ldots) \ \overset{P_1}{\wedge}\ l_3(A_3,B_3,Q_3,O,\ldots)$$

These perspectivities define a projectivity, T, between l_1 and l_3 in which O is a self-corresponding point. However, the lines l_1, l_2, l_3 are concurrent, and so Theorem 1 cannot be invoked to prove that T is a perspectivity, as we would like to show. Instead, we must proceed as follows. Let l' be any line through Q_1 distinct from l_1 and P_1P_3, and let A_2', B_2', and O' be the points

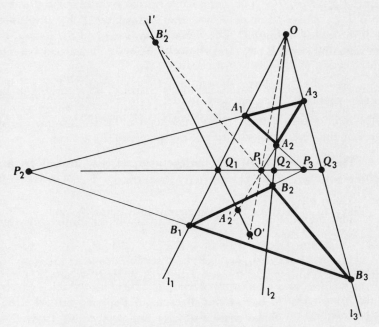

Fig. 6.8   Figure used in the proof of Theorem 2.

on l' which are perspective from P_1 with A_2, B_2, and O. Then T can be expressed in the form

$$(1) \quad l_1(A_1,B_1,Q_1,O,...) \overset{P_3}{\wedge} l_2(A_2,B_2,Q_2,O,...) \overset{P_1}{\wedge} l'(A_2',B_2',Q_1,O',...)$$
$$\overset{P_1}{\wedge} l_3(A_3,B_3,Q_3,O,...)$$

Now under the first two perspectivities in the chain (1), the point Q_1 is self-corresponding. Moreover, l_1, l_2, and l' are nonconcurrent. Hence by Theorem 1 these two perspectivities are equivalent to a single perspectivity, whose center, P', is of no concern to us. Thus T can be expressed in the form

$$l_1(A_1,B_1,Q_1,O,...) \overset{P'}{\wedge} l'(A_2',B_2',Q_1,O',...) \overset{P_1}{\wedge} l_3(A_3,B_3,Q_3,O,...).$$ In this form of T, the lines l_1, l', and l_3 are nonconcurrent and, of course, O is a self-corresponding point. Hence Theorem 1 can be invoked again, leading to the conclusion, which we were unable to reach earlier, that T is a perspectivity. Thus A_1A_3, B_1B_3, and Q_1Q_3 are concurrent. In other words, the intersection of A_1A_3 and B_1B_3, namely, P_2, lies on Q_1Q_3, that is, the line P_1P_3. Thus P_1, P_2, and P_3 are collinear, as required.

If P_1P_3 passes through O, then Q_1, Q_2, Q_3 all coincide at O, and the preceding argument fails. However, the three points P_1, P_2, P_3 and the three lines P_2P_3, P_3P_1, P_1P_2 play completely similar roles in our configuration. Hence if P_1P_3 passes through O, we may instead use P_1P_2, say, unless of course it too passes through O. But if this is the case, then P_1P_2O and P_1P_3O are the same line, and P_1, P_2, and P_3 are therefore collinear, as asserted.

Now that we have shown that the theorem of Pappus implies Desargues' theorem, it is an easy matter to prove that the theorem of Pappus implies the corollary of the perspectivity theorem:

Theorem 3

The theorem of Pappus, independent of the projectivity axiom, implies the corollary of the perspectivity theorem.

Proof Let l_1, l_2, l_3 be three lines concurrent in a point O, let

$$T: l_1(A_1,B_1,...) \overset{U}{\wedge} l_2(A_2,B_2,...) \overset{V}{\wedge} l_3(A_3,B_3,...)$$

be a projectivity between l_1 and l_3 in which the intersection, O, of l_1 and l_3 is self-corresponding, and let us apply Desargues' theorem (which we have just shown to be a consequence of the theorem of Pappus) to $\triangle A_1A_2A_3$ and $\triangle B_1B_2B_3$ (Fig. 6.9). Since these triangles are perspective from O, they must also be axially perspective. Moreover, since A_1A_2 and B_1B_2 intersect

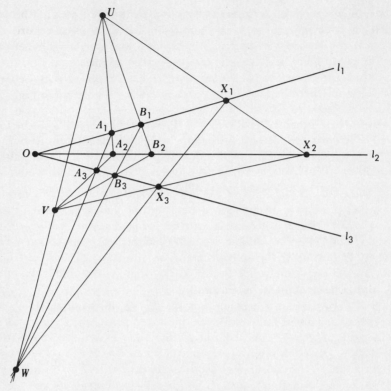

Fig. 6.9   Figure used in the proof of Theorem 3.

in U, and since A_2A_3 and B_2B_3 intersect in V, it follows that the intersection of the remaining sides, A_1A_3 and B_1B_3, say W, must be collinear with U and V. Now let X_1 be an arbitrary point on l_1, and let X_2 and X_3 be its images on l_2 and l_3 under the perspectivities defining T. Then $\triangle A_1A_2A_3$ and $\triangle X_1X_2X_3$ are also centrally perspective, and hence the intersections of their corresponding sides must be collinear. Now the intersection of A_1A_2 and X_1X_2 is U, and the intersection of A_2A_3 and X_2X_3 is V. Therefore the intersection of the remaining sides, A_1A_3 and X_1X_3, must lie on UV. However, from the first part of our proof, the intersection of A_1A_3 and UV is the point W. Hence X_1X_3 must also pass through W, which shows that a general point on l_1 and its image on l_3 under T are always collinear with the fixed point W. Thus T is a perspectivity, and our proof is complete.

If we now review the proof of the reduction theorem for chains of perspectivities, it is clear that we used only the corollary of the perspectivity theorem (which we have just shown to be a consequence of Desargues' theorem), independent of the projectivity axiom or the perspectivity theorem itself. Hence, since the theorem of Pappus implies Desargues' theorem, it follows that the theorem of Pappus implies that any projectivity between distinct lines can be reduced to at most two perspectivities.

Let us suppose, in particular, that we have a projectivity between distinct lines in which the intersection of the lines is self-corresponding. Like any other projectivity, it can of course be reduced to at most two perspectivities. Then either by Theorem 3 or Theorem 1, according as the lines involved are concurrent or nonconcurrent, it follows that the projectivity is in fact a perspectivity and hence is uniquely determined when two pairs of corresponding points, distinct from the intersection of the two lines, are given. These observations serve to establish the following theorem.

Theorem 4

The theorem of Pappus, independent of the projectivity axiom, implies the perspectivity theorem.

Our final result in this section is a rather surprising one, since it amounts to a special case of a theorem implying the theorem itself.

Theorem 5

The perspectivity theorem implies the projectivity axiom.

Proof Let A_1, B_1, C_1 be three points on a line l_1, let A_2, B_2, C_2 be three points on a line l_2, and let us suppose that there are two different projectivities, T and T', between l_1 and l_2 each of which maps A_1 into A_2, B_1 into B_2, and C_1 into C_2. This means that there is at least one point, X_1, on l_1 whose images under T and T', say X_2 and X_2', are different. Thus

$$T: l_1(A_1,B_1,C_1,X_1) \sim l_2(A_2,B_2,C_2,X_2)$$

$$T': l_1(A_1,B_1,C_1,X_1) \sim l_2(A_2,B_2,C_2,X_2')$$

Now let l and $\bar{l}$ be two lines intersecting in a point A, let B and C be two points on l distinct from A, and let $\bar{B}'$ and $\bar{C}'$ be two points on $\bar{l}$ distinct from A. From the construction described in Sec. 4.6 (Fig. 4.10) it is clear that there is at least one projectivity between l and l_1 of the form

$$T_1: l(A,B,C,\ldots) \sim l_1(A_1,B_1,C_1,\ldots)$$

Let X be the preimage of X_1 in this projectivity. Also there is at least one projectivity between l_2 and $\bar{l}$ of the form

$$T_2: l_2(A_2,B_2,C_2,\ldots) \sim \bar{l}(A,\bar{B},\bar{C},\ldots)$$

Let $\bar{X}$ and $\bar{X}'$ be the images of X_2 and X_2', respectively, under this projectivity. Obviously, since a projectivity is a one-to-one mapping, and since $\dot{X}_2$ and X_2' are distinct points, so too are $\bar{X}$ and $\bar{X}'$.

Now consider the sequence of projectivities

$$T_2 T T_1: l(A,B,C,X) \sim l_1(A_1,B_1,C_1,X_1) \sim l_2(A_2,B_2,C_2,X_2) \sim \bar{l}(A,\bar{B},\bar{C},\bar{X})$$

or $\qquad T_2 T T_1: l(A,B,C,X) \sim \bar{l}(A,\bar{B},\bar{C},\bar{X})$

and

$$T_2 T' T_1: l(A,B,C,X) \sim l_1(A_1,B_1,C_1,X_1) \sim l_2(A_2,B_2,C_2,X_2') \sim \bar{l}(A,\bar{B},\bar{C},\bar{X}')$$

or $\qquad T_2 T' T_1: l(A,B,C,X) \sim \bar{l}(A,\bar{B},\bar{C},\bar{X}')$

Since both $T_2 T T_1$ and $T_2 T' T_1$ have A as a self-corresponding point, it follows from the perspectivity theorem that both $T_2 T T_1$ and $T_2 T' T_1$ are perspectivities. Moreover, each maps B onto $\bar{B}$ and C onto $\bar{C}$; hence they must be the same perspectivity. This is impossible, however, since they assign distinct images, $\bar{X}$ and $\bar{X}'$, to the same point, X. Thus we have reached a contradiction which forces us to abandon the assumption that there can be two projectivities between l_1 and l_2 with three pairs of corresponding points in common, and the theorem is established.

Corollary 1

The theorem of Pappus implies the projectivity axiom.

Exercise

1. In the proof of Theorem 1 show that V_1 cannot lie on l_3 and V_2 cannot lie on l_1.

6.5 Applications of the Theorem of Pappus

According to Axiom 7, or equivalently as a consequence of the theorem of Pappus, a projectivity is uniquely determined when three pairs of corresponding points are given. Moreover, in Sec. 4.6 we described a construction by which the image of an arbitrary point under such a projectivity could be found. Although this construction arose during our discussion of E_2^+ and Π_2, it did not depend in any way on properties peculiar to these two systems and is, in fact, valid in any system satisfying Axioms 1 to 7. However, a simpler means of finding the image of an arbitrary point under a projectivity defined by three pairs of corresponding points is provided by the theorem of Pappus:

Theorem 1

Let A, B, C be three points on a line l, let A', B', C' be three points on a distinct line l', and let T be the projectivity which maps A into A', B into B', and C into C'. Then if X is an arbitrary point on l, and if X'' is the intersection of the line $A'X$ and the pappus line of the two triads (A,B,C) and (A',B',C'), the image of X under T is the point of intersection of AX'' and l'.

Proof Let A, B, C be three points on a line l, let A', B', C' be three points on a second line l', let:

A'' be the intersection of BC' and $B'C$
B'' be the intersection of CA' and $C'A$
C'' be the intersection of AB' and $A'B$

let l'' be the pappus line, $A''B''C''$, of the two triads (A,B,C) and (A',B',C'), let X be an arbitrary point on l, let X'' be the intersection of $A'X$ and l'', let $\bar{A}$ be the intersection of AA' and l'', and let X' be the intersection of AX'' and l' (Fig. 6.10a). Then

$$l(A,B,C,X) \overset{A'}{\wedge} l''(\bar{A},C'',B'',X'') \overset{A}{\wedge} l'(A',B',C',X')$$

Hence $l(A,B,C,X) \sim l'(A',B',C',X')$. In other words, X' is the image of X in the projectivity which maps A into A', B into B', and C into C', as asserted.

To find X' from X when A, B, C and A', B', C' are given, the figure required for the proof of the construction can be simplified to that shown in Fig. 6.10b. Obviously, B and B' or C and C' could be used just as well as A and A' in carrying out the construction.

The next two theorems, which are known as the **permutation theorems,** are reminiscent of certain of the results we discovered in the section on cross ratio, although here, since we have not yet introduced coordinates into our axiomatic structure, the concept of the cross ratio is still undefined.

Theorem 2

If $l(A,B,C,D) \sim l'(A',B',C',D')$, then also

$$l(A,B,C,D) \sim l'(B',A',D',C')$$

Proof Let T' be the projectivity $l(A,B,C,D) \sim l'(A',B',C',D')$, let U be any point which is not on l, let l'' be any line distinct from l which does not contain U, let A'', B'', C'', D'' be the points on l'' which are perspective from

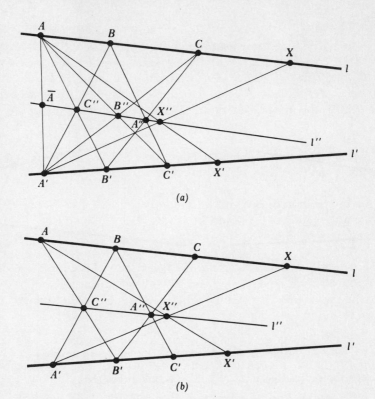

Fig. 6.10   The determination of the image of a point in a given projectivity by means of the theorem of Pappus.

U with A, B, C, D, and let V be the intersection of CA'' and BD'' (Fig. 6.11a). Then the pappus line of the two collinear triads

$$
\begin{array}{ccc}
A & B & C \\
B'' & A'' & D''
\end{array}
$$

is the line UV. Now in the projectivity, T'', between l and l'' in which $A \to B''$, $B \to A''$, $C \to D''$, the image of the point D can be found by the construction provided by Theorem 1: Join D to any one of the three given points, B'', A'', D'', on the image line, say D'' since the line DD'' is one about which we already have some information. Then join the intersection of this line and the pappus line, namely, U, to the preimage of D'' in T'', namely, C. This line, UC, then intersects l'' in the image of D under T'', namely, C''. Hence

(1) $l(A,B,C,D) \sim l''(B'',A'',D'',C'')$

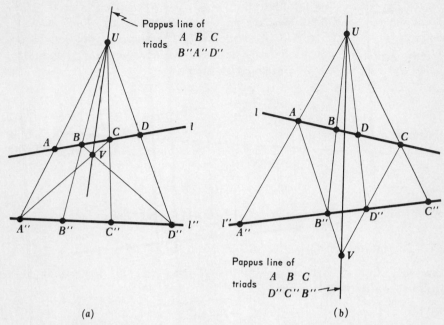

Fig. 6.11 Figures used in the proof of the permutation theorems.

Also, from the definition of A'', B'', C'', D'', it follows that

$$l(A,B,C,D) \overset{U}{\wedge} l''(A'',B'',C'',D'')$$

and therefore, simply by listing these same image pairs in a different order, we have

(2) $$l''(B'',A'',D'',C'') \overset{U}{\wedge} l(B,A,D,C)$$

Similarly, listing the various image pairs under the given projectivity, T', in a different order, it follows that

(3) $$l(B,A,D,C) \sim l'(B',A',D',C')$$

Hence, combining the projectivities (1), (2), and (3), we have

$$l(A,B,C,D) \sim l''(B'',A'',D'',C'') \overset{U}{\wedge} l(B,A,D,C) \sim l'(B',A',D',C')$$

or $l(A,B,C,D) \sim l'(B',A',D',C')$, as asserted.

Theorem 3

If $l(A,B,C,D) \sim l'(A',B',C',D')$, then also

$$l(A,B,C,D) \sim l'(D',C',B',A')$$

Proof   Let T' be the projectivity $l(A,B,C,D) \sim l'(A',B',C',D')$, let U be any point which is not on l, let l'' be any line distinct from l which does not contain U, let A'', B'', C'', D'' be the points on l'' which are perspective from U with A, B, C, D, and let V be the intersection of AB'' and CD'' (Fig. 6.11b). Then the pappus line of the two collinear triads

$$
\begin{array}{ccc}
A & B & C \\
D'' & C'' & B''
\end{array}
$$

is the line UV. Now in the projectivity, T'', between l and l'' in which $A \to D''$, $B \to C''$, $C \to B''$, the image of D is found, by the construction of Theorem 1, to be the point A''. Hence

(4) $$l(A,B,C,D) \sim l''(D'',C'',B'',A'')$$

Also, from the definition of A'', B'', C'', D'', it follows that

(5) $$l''(D'',C'',B'',A'') \overset{U}{\wedge} l(D,C,B,A)$$

Similarly, from the given projectivity, T', by listing the image-pairs in a different order, it follows that

(6) $$l(D,C,B,A) \sim l'(D',C',B',A')$$

Hence, combining (4), (5), and (6), we have

$$l(A,B,C,D) \sim l''(D'',C'',B'',A'') \overset{U}{\wedge} l(D,C,B,A) \sim l'(D',C',B',A')$$

or $l(A,B,C,D) \sim l'(D',C',B',A')$, as asserted.

Corollary 1
$$l(A,B,C,D) \sim l(B,A,D,C) \sim l(D,C,B,A) \sim l(C,D,A,B)$$

In Sec. 4.7 we showed that the cross ratio of four points (A,B,C,D) in the four orders (AB,CD), (BA,DC), (DC,BA), (CD,AB) were the same, and we also proved that there is a projectivity which sends four collinear points into four other collinear points if and only if the cross ratios of the two sets of points are equal. Thus the last corollary is a restatement of a result which we found to be true in E_2^+ and Π_2. Now, however, we have proved it to be true not only in E_2^+ and Π_2 but in any system satisfying Axioms 1 to 7.

Exercises

1. Given the projectivity $l(A,B,C,D,E,F) \sim l'(A',B',C',D',E',F')$. Show that the pappus line of the two sets (A,B,C), (A',B',C') is the same as the pappus line of the two sets (D,E,F), (D',E',F'). (Because of this property, the pappus

line of three points and their images under a given projectivity between distinct lines is often called the **axis** of the projectivity.)

2. If the axis of a projectivity (see Exercise 1) is given, how many pairs of corresponding points can be specified before the projectivity is completely determined?

3. Let T be a projectivity between distinct lines, l and l', intersecting in a point P. If P is considered first as a point of l and then as a point of l', what is its image and its preimage under T?

4. If A, B, C are three collinear points in E_2 and if P, Q, R are three other points such that $AQ \parallel BR$, $BP \parallel CQ$, and $CR \parallel AP$, prove that P, Q, and R are collinear.

5. Let $\triangle ABC$ be an arbitrary triangle in E_2, let P be an arbitrary point on BC, let l_1 and l_2 be lines on P such that $l_1 \parallel AB$ and $l_2 \parallel AC$, and let R and S be, respectively, the points of intersection of l_1 and l_2 with an arbitrary line on A. Show that $BS \parallel CR$.

6.6 Conclusion

In this chapter we began our axiomatic development of projective geometry by introducing the first seven of our axioms. Six of these were simple statements about the extent of our system and about the way in which points and lines are related. With the exception of Axiom 3, these six axioms were valid in euclidean as well as projective geometry. Axiom 3 denied the existence of parallel lines by asserting that two lines always intersected and thus ensured that we were no longer working in euclidean geometry. Axiom 7 dealt with projectivities between lines and was powerful enough to enable us to prove Desargues' theorem, the theorem of Pappus, and the reduction theorem for chains of perspectivities. We also discovered that if the theorem of Pappus was assumed, then the projectivity axiom could be proved as a theorem, which shows that the theorem of Pappus could have been chosen as our seventh axiom had we so desired. It is interesting that Desargues' theorem is not powerful enough to be used as an axiom in place of the projectivity axiom or the theorem of Pappus in the development we have in mind. A famous theorem of Wedderburn in the algebraic theory of finite fields shows that Desargues' theorem implies the theorem of Pappus and the projectivity axiom in finite projective geometries, but this is not true in projective geometries with infinitely many elements.

From a consideration of the structure of the first six axioms and their simple consequences, we inferred that every assertion they made about points and lines was equally true when the terms "point" and "line" were interchanged. This was not obvious after Axiom 7 was added to our system. However, we were able to prove its dual, and as a consequence it is clear that the principle of duality holds at least up to this point in our work. Whether or not this continues to be the case depends, of course, on whether or not we can consistently assume or prove the duals of such additional axioms as our development requires.

The chapter concluded with several important deductions from the theorem of Pappus. The first of these gave us an easy way of constructing the image of an arbitrary point under a projectivity defined by three pairs of corresponding points. The others identified for us certain permutations of four collinear points which were always projective with each other, even though in general there is no projectivity which maps four collinear points onto four assigned collinear image points.

seven

THE COMPLETE FOUR=POINT AND COMPLETE FOUR=LINE

7.1 Introduction

In this chapter we shall investigate the properties of two simple but exceedingly important configurations, the complete four-point and its dual, the complete four-line. The study of these figures will lead us to the notions of harmonic tetrads and involutions between cobasal ranges, which we encountered in Secs. 4.7 and 4.8. This time, however, our analysis will be based not upon the properties of a particular system (the algebraic representation) but rather upon the axioms we have thus far assumed, together with one new one, and our conclusions will therefore be valid in any system satisfying these axioms.

7.2 Definitions and Fundamental Properties

By a **simple four-point** we mean any set of four points no three of which are collinear. Since four points can be paired in six different ways, the points of a simple

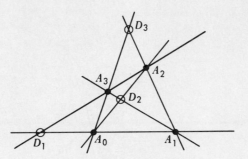

Fig. 7.1   A complete four-point.

four-point determine six lines, called the **sides** of the four-point. Two sides of a simple four-point whose intersection is not one of the four given points are said to be **opposite sides.** The intersections of the lines in the three pairs of opposite sides are called the **diagonal points** of the four-point.

The preceding observations lead us to the notion of a **complete four-point:**

Definition 1

The configuration consisting of the four points, six sides, and three diagonal points of a simple four-point is called a complete four-point.

Figure 7.1 shows a complete four-point, labeled in a way that we will often find convenient. The four points are identified by the subscripts 0, 1, 2, 3 attached to whatever carrier symbol we choose to use, say A. One and only one of the two lines in each pair of opposite sides passes through the point named by the subscript 0. The diagonal point on such a side is then identified by assigning to its carrier symbol, say D, the subscript of the other point of the four-point which lies on this side.

Dually, of course, beginning with the notion of a **simple four-line** as a set of four lines no three of which are concurrent, we are led to the concept of a **complete four-line:**

Definition 2

The configuration consisting of the four lines, six points, and three diagonal lines of a simple four-line is called a complete four-line.

Figure 7.2 shows a complete four-line, labeled in the same convenient way that we labeled the complete four-point in Fig. 7.1.

Figure 7.1 suggests a number of simple properties of a complete four-point, though of course it does not prove them. However, using the one

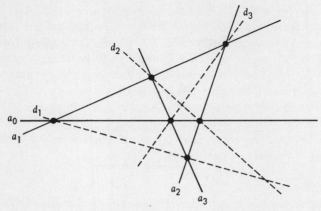

Fig. 7.2 A complete four-line.

fundamental fact that no three of the points of a simple four-point are collinear, it is easy to establish the following theorem:

Theorem 1

In any complete four-point:

1. No two of the sides can coincide.
2. No four of the sides can be concurrent.
3. No three of the sides can pass through the same diagonal point.
4. No two of the diagonal points can coincide.
5. No diagonal point can coincide with one of the four given points.
6. None of the four given points can be collinear with two diagonal points.

Dually, remembering that no three of the lines of a simple four-line are concurrent, it is easy to show that every complete four-line has the properties described in Theorem 2:

Theorem 2

In any complete four-line:

1. No two of the six points can be coincident.
2. No four of the six points can be collinear.
3. No three of the points can lie on the same diagonal line.
4. No two of the diagonal lines can coincide.
5. No diagonal line can coincide with one of the four given lines.
6. None of the four given lines can be concurrent with two diagonal lines.

Theorem 1 listed a number of properties of a complete four-point which were both obvious and true. No less obvious, but actually false, is another observation suggested by Fig. 7.1, namely, that the diagonal points of a complete four-point are not collinear. To show that this statement is false it is sufficient to exhibit one complete four-point whose diagonal points are collinear. No such example can be found in either E_2^+ or Π_2 because, as we shall soon see, the property in question is in fact true in these two systems. However, in the finite projective geometry defined by Table 6.3, *every* complete four-point has collinear diagonal points. For instance, the pairs of opposite sides of the four-point (P_1,P_2,P_5,P_6) are

$$P_1P_2 = l_{17} \quad \text{and} \quad P_5P_6 = l_{18}$$
$$P_1P_5 = l_{13} \quad \text{and} \quad P_2P_6 = l_{14}$$
$$P_1P_6 = l_{12} \quad \text{and} \quad P_2P_5 = l_{11}$$

and the intersections of the lines of these pairs, namely, the diagonal points P_{21}, P_{20}, P_{19}, all lie on l_{21}! This curious property can also be verified in the system defined by Table 6.1 and in infinitely many other finite projective geometries, as the following theorem, whose proof we must postpone to a later chapter, assures us.

Theorem 3

For every value of n of the form $2^k + 1$, $k = 1, 2, 3, \ldots$, there is a PG_n in which the diagonal points of every complete four-point are collinear.

Since we are primarily interested in establishing an axiomatic foundation for systems, in particular the algebraic representation, in which the diagonal points of every complete four-point are noncollinear, and since we cannot deduce this property from the axioms we have thus far adopted, we must postulate it. The simplest way to do this is to assume, as we shall, the following axiom.

Axiom 8

There is no complete four-point whose diagonal points are collinear.

Actually, this is a stronger assumption that we need, because, using only Axioms 1 to 7, it is possible to prove the following theorem.

Theorem 4

If the diagonal points of one complete four-point are collinear, then the diagonal points of every complete four-point are collinear.

Hence it is possible to ensure that there will be *no* complete four-point with collinear diagonal points by assuming simply that there is at least one complete four-point whose diagonal points are noncollinear. However, since we shall not digress to prove Theorem 4 (see Exercises 11 and 12), we shall adopt Axiom 8 even though it is unnecessarily strong.

We are now faced, as usual, with the twin obligations of showing that our latest axiom is consistent with those we have previously adopted and verifying that its dual is a valid assertion in our evolving system. The first of these requirements will be satisfied if we can exhibit a particular system in which Axioms 1 to 8 are all fulfilled. As before, we shall use the algebraic representation as our example; but since in our previous work, we did not discuss complete four-points, we cannot simply point to earlier results (as we did in establishing the consistency of Axiom 7) but must now verify in detail that Axiom 8 is true in Π_2.

Theorem 5

In the algebraic representation, there is no complete four-point whose diagonal points are collinear.

Proof Let $A_0:(x_0,y_0,z_0)$, $A_1:(x_1,y_1,z_1)$, $A_2:(x_2,y_2,z_2)$, $A_3:(x_3,y_3,z_3)$ be the vertices of an arbitrary complete four-point in Π_2. Then numbers $\lambda_0, \lambda_1, \lambda_2, \lambda_3$ exist such that

(1)
$$\lambda_0 x_0 + \lambda_1 x_1 + \lambda_2 x_2 + \lambda_3 x_3 = 0$$
$$\lambda_0 y_0 + \lambda_1 y_1 + \lambda_2 y_2 + \lambda_3 y_3 = 0$$
$$\lambda_0 z_0 + \lambda_1 z_1 + \lambda_2 z_2 + \lambda_3 z_3 = 0$$

In fact, by Corollary 1, Theorem 6, Sec. 3.7, $\lambda_{j-1} = (-1)^{j-1} c M_j$, where c is an arbitrary constant and M_j is the determinant of the (3,3) submatrix remaining when the jth column is deleted from the matrix

$$\begin{Vmatrix} x_0 & x_1 & x_2 & x_3 \\ y_0 & y_1 & y_2 & y_3 \\ z_0 & z_1 & z_2 & z_3 \end{Vmatrix}$$

Moreover, since no three of the points A_0, A_1, A_2, A_3 are collinear, none of the λ's is zero.

Now the point $(\lambda_0 x_0 + \lambda_1 x_1, \lambda_0 y_0 + \lambda_1 y_1, \lambda_0 z_0 + \lambda_1 z_1)$ is clearly collinear with A_0 and A_1. Likewise, the point

$$(\lambda_2 x_2 + \lambda_3 x_3, \lambda_2 y_2 + \lambda_3 y_3, \lambda_2 z_2 + \lambda_3 z_3)$$

is collinear with A_2 and A_3. Moreover, since Eqs. (1) can be written in the form

$$\lambda_0 x_0 + \lambda_1 x_1 = -(\lambda_2 x_2 + \lambda_3 x_3)$$
$$\lambda_0 y_0 + \lambda_1 y_1 = -(\lambda_2 y_2 + \lambda_3 y_3)$$
$$\lambda_0 z_0 + \lambda_1 z_1 = -(\lambda_2 z_2 + \lambda_3 z_3)$$

it follows that these two points are the same and must therefore be the point common to the lines A_0A_1 and A_2A_3, namely, the diagonal point D_1. Thus for D_1 we have either of the coordinate triples

$$D_1:(\lambda_0 x_0 + \lambda_1 x_1, \lambda_0 y_0 + \lambda_1 y_1, \lambda_0 z_0 + \lambda_1 z_1):$$
$$(\lambda_2 x_2 + \lambda_3 x_3, \lambda_2 y_2 + \lambda_3 y_3, \lambda_2 z_2 + \lambda_3 z_3)$$

and, similarly, for D_2 and D_3 we have

$$D_2:(\lambda_0 x_0 + \lambda_2 x_2, \lambda_0 y_0 + \lambda_2 y_2, \lambda_0 z_0 + \lambda_2 z_2):$$
$$(\lambda_1 x_1 + \lambda_3 x_3, \lambda_1 y_1 + \lambda_3 y_3, \lambda_1 z_1 + \lambda_3 z_3)$$

$$D_3:(\lambda_0 x_0 + \lambda_3 x_3, \lambda_0 y_0 + \lambda_3 y_3, \lambda_0 z_0 + \lambda_3 z_3):$$
$$(\lambda_1 x_1 + \lambda_2 x_2, \lambda_1 y_1 + \lambda_2 y_2, \lambda_1 z_1 + \lambda_2 z_2)$$

Let us now suppose, contrary to the theorem, that D_1, D_2, and D_3 are collinear on some line $[l,m,n]$. Then the coordinates of D_1, D_2, and D_3 must satisfy the incidence condition with the triple $[l,m,n]$. For D_1 we thus have the two equations

$$(\lambda_0 x_0 + \lambda_1 x_1)l + (\lambda_0 y_0 + \lambda_1 y_1)m + (\lambda_0 z_0 + \lambda_1 z_1)n = 0$$
$$(\lambda_2 x_2 + \lambda_3 x_3)l + (\lambda_2 y_2 + \lambda_3 y_3)m + (\lambda_2 z_2 + \lambda_3 z_3)n = 0$$

or, rearranging,

$$\lambda_0(lx_0 + my_0 + nz_0) + \lambda_1(lx_1 + my_1 + nz_1) = 0$$
$$\lambda_2(lx_2 + my_2 + nz_2) + \lambda_3(lx_3 + my_3 + nz_3) = 0$$

with similar relations for D_2 and D_3. For convenience let us put

$$s_i = lx_i + my_i + nz_i \qquad i = 0, 1, 2, 3$$

Then the incidence conditions that we have just obtained can be written in the form

$\lambda_0 s_0 + \lambda_1 s_1 = 0$	$\lambda_2 s_2 + \lambda_3 s_3 = 0$	for	D_1
$\lambda_0 s_0 + \lambda_2 s_2 = 0$	$\lambda_1 s_1 + \lambda_3 s_3 = 0$	for	D_2
$\lambda_0 s_0 + \lambda_3 s_3 = 0$	$\lambda_1 s_1 + \lambda_2 s_2 = 0$	for	D_3

From the first two equations in the first column and the last equation in the second column, we conclude at once that $\lambda_0 s_0 = \lambda_1 s_1 = \lambda_2 s_2 = 0$. Now we have already observed that none of the λ's is zero; hence these equations imply that $s_0 = s_1 = s_2 = 0$. However, it is impossible for any of the s's to be zero, because the vanishing of s_i implies that A_i lies on the line $[l,m,n]$ which contains the three diagonal points, whereas according to Theorem 1, no vertex of a complete four-point can be collinear with even two of the diagonal points. Hence we have reached a contradiction which forces us to abandon the supposition that the diagonal points are collinear. Thus the theorem is proved, and the consistency of Axiom 8 with Axioms 1 to 7 is established.

Corollary 1

In the extended euclidean plane there is no complete four-point whose diagonal points are collinear.

That the principle of duality remains valid after Axiom 8 is added to our system is guaranteed by the following theorem.

Theorem 6

There is no complete four-line whose diagonal lines are concurrent.

Proof Let a_0, a_1, a_2, a_3 be the four lines of an arbitrary four-line, and let d_1, d_2, d_3 be its diagonal lines. Consider now the four points

A_0: the intersection of a_1 and a_3
A_1: the intersection of a_1 and a_2
A_2: the intersection of a_0 and a_2
A_3: the intersection of a_0 and a_3

No three of these can be collinear, because if this were the case, at least two of the lines a_0, a_1, a_2, a_3 would have to be coincident, which is obviously impossible since they are the sides of a complete four-line. Hence A_0, A_1, A_2, A_3 can be considered the vertices of a complete four-point. The diagonal points of this four-point are then (Fig. 7.3)

D_1: the intersection of $A_0A_1 = a_1$ and $A_2A_3 = a_0$
D_2: the intersection of $A_0A_2 = d_2$ and $A_1A_3 = d_3$
D_3: the intersection of $A_0A_3 = a_3$ and $A_1A_2 = a_2$

Suppose now, contrary to the theorem, that the diagonal lines d_1, d_2, d_3 are concurrent. Since we have just observed that d_2 and d_3 intersect in the point D_2, this would imply that d_1 also passed through D_2. However, by definition, d_1 passes through the intersection of a_0 and a_1, that is, D_1, and the intersection of a_2 and a_3, that is, D_3. Hence if d_1 passed through D_2, it would follow that D_1, D_2, and D_3 were collinear, which is impossible by Axiom 8. Thus the assumption that d_1, d_2, and d_3 are concurrent must be abandoned, and the theorem is established.

Exercises

1. Prove Theorem 1.

2. Prove Theorem 2.

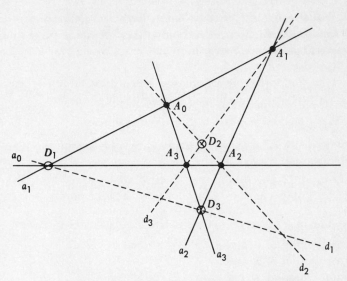

Fig. 7.3   Figure used in proving the dual of Axiom 8.

3. How many triangles are there in PG_n?

4. How many simple four-points are there in PG_n?

5. (a) Prove that in PG_3 the diagonal points of every complete four-point are collinear.
(b) Prove that every set of three collinear points in PG_3 is the set of diagonal points of a unique four-point.

6. Prove that in PG_4 there is no complete four-point whose diagonal points are collinear.

7. Show that every triangle in PG_4 is the diagonal triangle of a unique four-point.

8. If A_0, D_1, D_2, D_3 are four points in PG_4 no three of which are collinear, show that there is at most one complete four-point having A_0 as one of its vertices and $\triangle D_1 D_2 D_3$ as its diagonal triangle.

9. Show that in PG_5 two complete four-points cannot have the same diagonal triangle. Hence show that the diagonal points of every complete four-point in PG_5 are collinear.

10. Under the assumption that each triangle in PG_n is the diagonal triangle of the same number of complete four-points, show that there exist no PG_n's for odd values of n unless n is of the form $2^k + 1$.

11. Prove that if two complete four-points have two points and a diagonal point collinear with these points in common, and if one of these four-points has noncollinear diagonal points, then the other four-point also has non-collinear diagonal points.

12. Using the result of Exercise 11, prove that if there exists one complete four-point whose diagonal points are not collinear, then every complete four-point has diagonal points which are not collinear.

13. A four-point in E_2^+ consists of three finite points and one ideal point. What is its diagonal triangle?

14. A four-point in E_2^+ consists of two finite points and two ideal points. What is its diagonal triangle?

15. A four-line in E_2^+ consists of three finite lines and the ideal line. What is its diagonal triangle?

7.3 Harmonic Tetrads

In any complete four-point the line $D_i D_j$ meets two sides of the four-point at D_i and two other sides of the four-point at D_j. The intersections of $D_i D_j$ with the remaining two sides of the four-point cannot be vertices of the four-point since Theorem 1 assures us that no vertex can be collinear with two diagonal points. Likewise, neither of these intersections can be the third diagonal point, D_k, since Axiom 8 guarantees that the diagonal points of every four-point are noncollinear. Hence these intersections (Fig. 7.4) are additional points which we have not yet noted in our study of the complete four-point:

Definition 1

The intersection of any side of a complete four-point with the line determined by the two diagonal points which do not lie on that side is called a harmonic point of the four-point.

From Fig. 7.4 it is apparent that the configuration of a complete four-point with its six harmonic points is relatively complicated, and it is important that we have an efficient notation to identify the harmonic points as well as the diagonal points. To do this, we shall denote the two harmonic points on the line $D_i D_j$ by the symbols H_k and H_k', with the understanding that H_k refers to the harmonic point cut by the line $A_0 A_k$ and H_k' refers to the harmonic point cut by $A_i A_j$. An examination of Fig. 7.4 should make clear the pattern in this notation.

Dually, of course, we have the following definition.

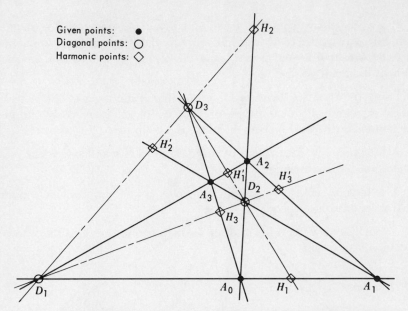

Given points: ●
Diagonal points: ○
Harmonic points: ◇

Fig. 7.4   A complete four-point with its diagonal points and harmonic points.

Definition 2

The join of any point of a complete four-line and the point of intersection of the two diagonal lines which do not pass through that point is called a harmonic line of the four-line.

Figure 7.5 shows the six harmonic lines of a typical complete four-line and the notation we shall use to describe them.

Our first theorem describes an interesting relation between complete four-lines and complete four-points.

Theorem 1

The six harmonic points of any complete four-point are the points of a complete four-line.

Proof Let A_0, A_1, A_2, A_3 be a complete four-point with diagonal points D_1, D_2, D_3 and harmonic points H_1, H_1', H_2, H_2', H_3, H_3'. By definition, the lines A_0D_3, A_1H_2', A_2H_1' all pass through A_3 (Fig. 7.4). Hence $\triangle A_0A_1A_2$ and $\triangle D_3H_2'H_1'$ are centrally perspective and therefore, by Desargues' theorem, they are also axially perspective. Now the intersection of the corresponding sides A_0A_1 and D_3H_2' is D_1, and the intersection of the corresponding sides A_0A_2 and D_3H_1' is D_2. Thus the intersection of the remaining sides, A_1A_2 and $H_2'H_1'$, must lie on D_1D_2. Since A_1A_2 intersects D_1D_2 in the point H_3', it

Given lines: ———
Diagonal lines: — · —
Harmonic lines: - - - - -

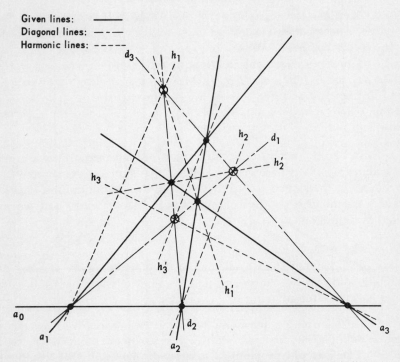

Fig. 7.5   A complete four-line with its diagonal lines and harmonic lines.

therefore follows that $H_2'H_1'$ also passes through H_3'; in other words, H_1', H_2', H_3' are collinear.

Similarly, A_1H_1, A_2H_2, A_3D_3 all pass through A_0, and therefore $\triangle A_1A_2A_3$ and $\triangle H_1H_2D_3$ are both centrally and axially perspective. Now the intersection of A_1A_3 and H_1D_3 is D_2, and the intersection of A_2A_3 and H_2D_3 is D_1. Thus the intersection of the remaining sides, A_1A_2 and H_1H_2, must be collinear with D_1 and D_2. Therefore, since A_1A_2 intersects D_1D_2 in the point H_3', it follows that H_1H_2 passes through H_3'; that is, H_1, H_2, H_3' are collinear.

Also, A_1H_1, A_2D_2, A_3H_3 all pass through A_0, and therefore $\triangle A_1A_2A_3$ and $\triangle H_1D_2H_3$ are both centrally and axially perspective. Moreover, the intersection of A_1A_2 and H_1D_2 is D_3, and the intersection of A_2A_3 and D_2H_3 is D_1. Thus the intersection of the sides A_1A_3 and H_1H_3 must be collinear with D_1 and D_3. Therefore, since A_1A_3 intersects D_1D_3 in the point H_2', it follows that H_1, H_2', H_3 are collinear.

Finally, A_1D_1, A_2H_2, A_3H_3 all pass through A_0, and hence $\triangle A_1A_2A_3$ and $\triangle D_1H_2H_3$ are both centrally and axially perspective. Furthermore, the intersection of A_1A_2 and D_1H_2 is D_3, and the intersection of A_1A_3 and

D_1H_3 is D_2. Thus the intersection of A_2A_3 and H_2H_3 must lie on D_2D_3. Therefore, since the intersection of A_2A_3 and D_2D_3 is H_1', it follows that H_1', H_2, H_3 are collinear.

Thus the six harmonic points lie by threes on four lines: a_0: $H_1'H_2'H_3'$, a_1: $H_1'H_2H_3$, a_2: $H_1H_2'H_3$, a_3: $H_1H_2H_3'$. Moreover, it is clear that these lines are all distinct and no three of them are concurrent. Hence a_0, a_1, a_2, a_3 is a four-line whose six points are H_1, H_1', H_2, H_2', H_3, H_3', and the theorem is established.

In Sec. 4.7 we introduced the notion of a *harmonic tetrad* as a set of four collinear points whose cross ratio was -1. Since we have, as yet, no coordinates in the system we are developing, this definition of a harmonic tetrad is meaningless in the context of our present work, and we must replace it with another:

Definition 3

Two pairs of points consisting of the diagonal points of a complete four-point and the two harmonic points collinear with them are said to form a harmonic tetrad.

In describing a harmonic tetrad, it is customary to list the two diagonal points as the first pair and the two harmonic points as the second pair. Since Definition 1 provides no basis for distinguishing between the two diagonal points or between the two harmonic points, it is clear that if (D_iD_j,H_kH_k') is a harmonic tetrad, so too is (D_jD_i,H_kH_k'), $(D_iD_j,H_k'H_k)$, and $(D_jD_i,H_k'H_k)$.

It is not obvious that the objects we called harmonic tetrads in Sec. 4.7 and the objects we are here calling harmonic tetrads are the same. Whenever two definitions are given for what is asserted to be the same concept, it is necessary to prove their equivalence, and after we have introduced coordinates in our axiomatic development we shall do this. Meanwhile, we should find it easy to believe that the two definitions do describe the same objects, as we observe how we can deduce from our second definition all the properties which we proved in Sec. 4.7.

Although there is no basis for distinguishing between the points within the respective pairs which form a harmonic tetrad, the pairs themselves are clearly distinguishable: one consists of diagonal points, the other of harmonic points. Hence, although (D_iD_j,H_kH_k') is a harmonic tetrad, one cannot conclude without proof that (H_kH_k',D_iD_j) is also a harmonic tetrad. This is the case, however, as our next theorem assures us.

Theorem 2

If (XY,LM) is a harmonic tetrad, so too is (LM,XY).

Proof The hypothesis that (XY,LM) is a harmonic tetrad means that there exists a complete four-point for which X and Y are two diagonal points, say D_1 and D_2, and L and M are the corresponding harmonic points, H_3 and H_3'. To prove the theorem we must show that there exists another four-point for which H_3 and H_3' are diagonal points and D_1 and D_2 are the corresponding harmonic points. To do this, let A_0, A_1, A_2, A_3 be the vertices of the four-point which contains the given harmonic tetrad, $(XY,LM) = (D_1D_2,H_3H_3')$, and consider the four points H_1, H_1', H_2, H_2' (Fig. 7.4). Clearly no three of these four points are collinear, for if this were the case, it would follow from the collinearity relations proved in Theorem 1 that all six of the harmonic points of the four-point $A_0A_1A_2A_3$ are collinear, which is impossible. Hence H_1, H_1', H_2, H_2' form a four-point, two of whose diagonal points are the intersection of H_1H_2' and H_2H_1', namely, H_3, and the intersection of H_1H_2 and $H_1'H_2'$, namely, H_3'. Moreover, the harmonic points which are collinear with the diagonal points H_3 and H_3' in the second four-point are the intersection of H_3H_3' and H_2H_2', namely, D_1, and the intersection of H_3H_3' and H_1H_1', namely, D_2. Thus (H_3H_3',D_1D_2) is a harmonic tetrad by definition, and the theorem is proved.

Corollary 1

If (XY,LM) is a harmonic tetrad, then (XY,ML), (YX,LM), (YX,ML), (LM,XY), (LM,YX), (ML,XY), and (ML,YX) are also harmonic tetrads.

Although our interest in harmonic tetrads suggests that there is something special about them, we have not yet proved this; and for all we know at present, every set of four collinear points is a harmonic tetrad. However, the next theorem makes it clear that this is not the case.

Theorem 3

If X, Y, and L are distinct collinear points, there is a unique point M collinear with them such that (XY,LM) is a harmonic tetrad.

Proof We shall prove this theorem by exhibiting a construction for the required point M. In other words, we shall show how to construct a complete four-point having X and Y as two diagonal points and L as a collinear harmonic point, and we shall prove that the second harmonic point thus determined is unique. To do this, let X, Y, and L be distinct collinear points, let A_0 be an arbitrary point which is not collinear with X, Y, and L, let A_1 and A_2 be points on A_0X and A_0Y, respectively, which are collinear with L, and let A_3 be the intersection of A_1Y and A_2X (Fig. 7.6). Then $A_0A_1A_2A_3$ is a four-point in which two of the diagonal points, D_1 and D_2, are, respectively,

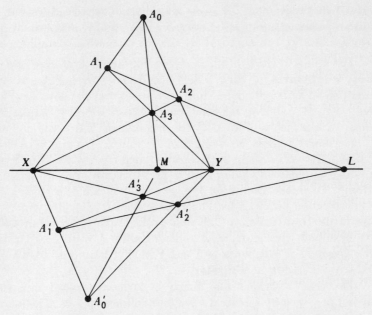

Fig. 7.6   Figure used in the proof of Theorem 3.

X and Y, and in which the harmonic point H_3' is L. The second harmonic point on XY, which is the required point M, is therefore the intersection of XY and the line A_0A_3.

To prove that the required fourth point $\dot{M}$ is unique, let A_0' be an arbitrary point distinct from A_0 and not collinear with X, Y, and L, and let A_1', A_2', and A_3' be defined in terms of A_0' just as A_1, A_2, A_3 were defined in terms of A_0. Then $A_0'A_1'A_2'A_3'$ is also a four-point in which $D_1' = X$, $D_2' = Y$, and $H_3' = L$, and in which H_3 is some point, M', which we hope to show is the same as M. To do this, we observe first that $\triangle A_0A_1A_2$ and $\triangle A_0'A_1'A_2'$ are axially perspective on the line XY. Therefore, by the dual of Desargues' theorem, they are centrally perspective; i.e., the lines A_0A_0', A_1A_1', and A_2A_2' are concurrent in some point, O. Likewise, $\triangle A_1A_2A_3$ and $\triangle A_1'A_2'A_3'$ are axially perspective on XY, and hence are also centrally perspective from some point, O'. Moreover, since O and O' are each on both A_1A_1' and A_2A_2', it follows that they are the same point. In other words, A_0A_0', A_1A_1', A_2A_2', and A_3A_3' *all* pass through the same point O $(= O')$. From this we conclude that $\triangle A_0A_1A_3$ and $\triangle A_0'A_1'A_3'$ are centrally perspective, and hence axially perspective. Now the intersection of A_0A_1 and $A_0'A_1'$ is X, and the intersection of A_1A_3 and $A_1'A_3'$ is Y. Hence the axis of the last perspectivity is XY, and so A_0A_3 and $A_0'A_3'$ must intersect XY in the same point. Therefore, since A_0A_3

intersects XY in M and $A_0'A_3'$ intersects XY in M', it follows that M and M' are the same point, which shows that M is unique and completes our proof.

The points of either of the pairs of a harmonic tetrad are said to be **harmonic conjugates** of each other with respect to the points of the remaining pair. The last theorem thus guarantees that *the harmonic conjugate of a given point with respect to two other points collinear with it is unique*, and it provides us with a means of actually constructing the harmonic conjugate.

Harmonic tetrads, as defined in Definition 3, perforce consist of distinct points. However, there are occasions when it is necessary to consider the harmonic conjugate of a point with respect to a pair of points to which it itself belongs. To extend Definition 3 to cover such a case, it is convenient to use the construction provided by Theorem 3 when L coincides with Y, say. The steps in the construction are all well defined, and it turns out that the required harmonic conjugate, M, coincides with L and Y; that is, *the harmonic conjugate of Y with respect to X and Y is Y*. A harmonic tetrad in this extended sense is said to be **singular.**

Theorem 3 showed that being a harmonic tetrad is a special property not possessed by a general set of four collinear points. The following theorems show that this property is preserved under an arbitrary projectivity; i.e., it is a projective property.

Theorem 4

If (XY,LM) is a nonsingular harmonic tetrad, then $(XY,LM) \sim (XY,ML) \sim (YX,LM) \sim (YX,ML)$.

Proof Let (XY,LM) be a harmonic tetrad, and let $A_0A_1A_2A_3$ be a four-point in which $D_1 = X$, $D_2 = Y$, $H_3 = L$, $H_3' = M$. Then (Fig. 7.4),

$$(D_1D_2,H_3H_3') \overset{H_1'}{\wedge} (D_1D_3,H_2H_2') \quad \text{and} \quad (D_1D_3,H_2H_2') \overset{H_1}{\wedge} (D_1D_2,H_3'H_3)$$

Hence $(D_1D_2,H_3H_3') \sim (D_1D_2,H_3'H_3)$; that is, $(XY,LM) \sim (XY,ML)$. The other assertions of the theorem follow from the permutation theorems (Theorems 2 and 3, Sec. 6.6).

Theorem 5

If $(XY,LM) \sim (XY,ML)$, then (XY,LM) is a harmonic tetrad.

Proof Let (XY,LM) be four collinear points such that $(XY,LM) \sim (XY,ML)$, and let us suppose, contrary to the theorem, that (XY,LM) is not a harmonic tetrad. Then by Theorem 3 there exists a unique point, Y', such that (XY',LM) *is* a harmonic tetrad, and by Theorem 4, $(XY',LM) \sim (XY',ML)$.

Now in the given projectivity, as well as in the last one, we have the mapping

$$X \to X \qquad L \to M \qquad M \to L$$

Hence, by Axiom 7, since these two projectivities have three pairs of corresponding points in common, they must be the same projectivity. Thus, listing in a single statement the images they assign, we have $(X,Y,Y',L,M) \sim (X,Y,Y',M,L)$. This projectivity has three self-corresponding points, X, Y, and Y', and therefore must be the identity, yet neither L nor M correspond to themselves. Hence we have a contradiction which forces us to abandon the supposition that (XY,LM) is not a harmonic tetrad, and our proof is complete.

Theorem 5 is essentially the converse of Theorem 4. Together the two theorems assert that (XY,LM) *is a harmonic tetrad if and only if* $(XY,LM) \sim (XY,ML)$.

Theorem 6

A tetrad projective with a harmonic tetrad is a harmonic tetrad.

Proof Let (XY,LM) be a harmonic tetrad, and let $(X'Y',L'M')$ be a tetrad such that $(X'Y',L'M') \sim (XY,LM)$. Equally well, of course, by listing the image pairs in a different order, this can be written in the form

$$(X'Y',M'L') \sim (XY,ML)$$

Now by Theorem 4 $(XY,LM) \sim (XY,ML)$. Hence, combining this with the two forms of the given projectivity, we have $(X'Y',L'M') \sim (X'Y',M'L')$. Therefore, by Theorem 5, $(X'Y',L'M')$ is a harmonic tetrad, as asserted.

Theorem 7

Any two harmonic tetrads are projective.

Proof Let (XY,LM) and $(X'Y',L'M')$ be two harmonic tetrads, and let us consider the projectivity in which $X \to X'$, $Y \to Y'$, $L \to L'$. Let M^* be the image of M under this projectivity, so that

$$(1) \qquad\qquad (XY,LM) \sim (X'Y',L'M^*)$$

Then, by the last theorem, $(X'Y',L'M^*)$ is a harmonic tetrad, since it is projective with the harmonic tetrad (XY,LM). However, we are given that $(X'Y',L'M')$ is a harmonic tetrad. Hence, by Theorem 3, $M^* = M$, and therefore, from (1), we have $(XY,LM) \sim (X'Y',L'M')$, as asserted.

Example 1

In the algebraic representation, a harmonic tetrad was *defined* to be a set of four collinear points, X, Y, L, M, such that $R(XY, LM) = -1$. Using the definitions and theorems of this section, *prove* this result.

In the algebraic representation, let

$$A_0 = \begin{Vmatrix} x_0 \\ y_0 \\ z_0 \end{Vmatrix} \quad A_1 = \begin{Vmatrix} x_1 \\ y_1 \\ z_1 \end{Vmatrix} \quad A_2 = \begin{Vmatrix} x_2 \\ y_2 \\ z_2 \end{Vmatrix} \quad A_3 = \begin{Vmatrix} x_3 \\ y_3 \\ z_3 \end{Vmatrix}$$

be the vertices of an arbitrary four-point. Then, as we observed in Sec. 7.2, the diagonal points D_2 and D_3 have the coordinate vectors

$$D_2 = \begin{Vmatrix} \lambda_0 x_0 + \lambda_2 x_2 \\ \lambda_0 y_0 + \lambda_2 y_2 \\ \lambda_0 z_0 + \lambda_2 z_2 \end{Vmatrix} \quad \text{and} \quad D_3 = \begin{Vmatrix} \lambda_0 x_0 + \lambda_3 x_3 \\ \lambda_0 y_0 + \lambda_3 y_3 \\ \lambda_0 z_0 + \lambda_3 z_3 \end{Vmatrix}$$

where λ_0, λ_1, λ_2, λ_3 are numbers, none of which is zero, such that

$$\begin{aligned} \lambda_0 x_0 + \lambda_1 x_1 + \lambda_2 x_2 + \lambda_3 x_3 &= 0 \\ \lambda_0 y_0 + \lambda_1 y_1 + \lambda_2 y_2 + \lambda_3 y_3 &= 0 \\ \lambda_0 z_0 + \lambda_1 z_1 + \lambda_2 z_2 + \lambda_3 z_3 &= 0 \end{aligned}$$

(2)

Adding these vectors, we have

$$D_2 + D_3 = \begin{Vmatrix} 2\lambda_0 x_0 + \lambda_2 x_2 + \lambda_3 x_3 \\ 2\lambda_0 y_0 + \lambda_2 y_2 + \lambda_3 y_3 \\ 2\lambda_0 z_0 + \lambda_2 z_2 + \lambda_3 z_3 \end{Vmatrix}$$

or, using Eqs. (2),

(3)
$$D_2 + D_3 = \begin{Vmatrix} \lambda_0 x_0 - \lambda_1 x_1 \\ \lambda_0 y_0 - \lambda_1 y_1 \\ \lambda_0 z_0 - \lambda_1 z_1 \end{Vmatrix} = \lambda_0 A_0 - \lambda_1 A_1$$

Obviously, the point $D_2 + D_3$ is a point on the line $D_2 D_3$. Moreover, from its representation as a linear combination of A_0 and A_1, it is clear that it is also a point on the line $A_0 A_1$. In other words, it is the intersection of $D_2 D_3$ and $A_0 A_1$, that is, the harmonic point H_1. Similarly, we have

(4)
$$D_2 - D_3 = \begin{Vmatrix} \lambda_2 x_2 - \lambda_3 x_3 \\ \lambda_2 y_2 - \lambda_3 y_3 \\ \lambda_2 z_2 - \lambda_3 z_3 \end{Vmatrix} = \lambda_2 A_2 - \lambda_3 A_3$$

which is the coordinate-vector of the intersection of $D_2 D_3$ and $A_2 A_3$, that is, the harmonic point H_1'.

From (3) and (4), it is clear that if the line $D_2 D_3$ is parametrized in terms of D_2 and D_3 as base points, then the parameters of H_1 are $(1,1)$ and the parameters of H_1' are $(1,-1)$. Hence, by Lemma 2, Sec. 4.7,

$\text{R}(D_2D_3,H_1H_1') = -1$. Finally, if (XY,LM) is an arbitrary harmonic tetrad, then by Theorem 6, $(XY,LM) \sim (D_2D_3,H_1H_1')$ and therefore, by Theorem 5, Sec. 4.7, $\text{R}(XY,LM) = \text{R}(D_2D_3,H_1H_1') = -1$, as asserted.

Exercises

1. In Π_2 find the coordinates of the diagonal points and the harmonic points of the four-point $A_0A_1A_2A_3$, given:

(a) $A_0:(1,1,1)$, $A_1:(1,0,0)$, $A_2:(0,1,0)$, $A_3:(0,0,1)$
(b) $A_0:(1,1,1)$, $A_1:(0,1,1)$, $A_2:(1,0,1)$, $A_3:(1,1,0)$
(c) $A_0:(1,1,1)$, $A_1:(-1,1,1)$, $A_2:(1,-1,1)$, $A_3:(1,1,-1)$

2. In the complete four-point $A_0A_1A_2A_3$, show that (A_0A_1,D_1H_1) and (A_2A_3,D_1H_1') are both harmonic tetrads. Using this result, show that there are nine harmonic tetrads associated with a given complete four-point.

3. Let x and y be two lines intersecting in a point O, let X_1, X_2, X_3 be three points on x each distinct from O, let Y_1, Y_2, Y_3 be three points on y each distinct from O, and let the triads (X_1,X_2,X_3) and (Y_1,Y_2,Y_3) be perspective from a point V. If Z_i is the harmonic conjugate of V with respect to X_i and Y_i, show that Z_1, Z_2, Z_3, and O are collinear. (The line containing Z_1, Z_2, Z_3, O is called the **polar line** of V with respect to the lines x and y.)

4. Given $\triangle ABC$ and a point V which is not on any side of $\triangle ABC$. Show that the triangle formed by the polar lines of V with respect to the pairs of sides of $\triangle ABC$ is perspective from V with $\triangle ABC$.

5. Show, via the construction of Theorem 3, that the harmonic conjugate of an ideal point L with respect to two finite points X and Y collinear with it is the midpoint of the segment $\overline{XY}$.

6. Show that in E_2 a point, L, and its harmonic conjugate, M, with respect to two points X and Y are always such that one of the pair (L,M) is a point of the segment $\overline{XY}$ and the other is not.

7. Let L_1, L_2, L_3, L_4 be four points on a line XY, and let M_i be the harmonic conjugate of L_i with respect to X and Y. Show that $(L_1,L_2,L_3,L_4) \sim (M_1,M_2,M_3,M_4)$ and that the fixed points of this projectivity are X and Y.

8. On the sides of $\triangle ABC$, let A' and A'' be a pair of harmonic conjugates with respect to B and C, let B' and B'' be a pair of harmonic conjugates with respect to C and A, and let C' and C'' be a pair of harmonic conjugates with respect to A and B. Show that corresponding sides of $\triangle ABC$, $\triangle A'B'C'$, and $\triangle A''B''C''$ are concurrent.

9. Let $A_0A_1A_2A_3$ be a complete four-point, and let $a_0a_1a_2a_3$ be the complete four-line determined by the six harmonic points of $A_0A_1A_2A_3$. Show that $A_0A_1A_2A_3$ and $a_0a_1a_2a_3$ have the same diagonal triangle.

10. Show that the diagonal triangle of any four-point is perspective with each triangle whose vertices are three of the four points of the given four-point.

11. Let O be any point which is not on a side of $\triangle ABC$, let A' be the intersection of OA and BC, let B' be the intersection of OB and CA, let C' be the intersection of OC and AB, let A'' be the harmonic conjugate of A' with respect to B and C, let B'' be the harmonic conjugate of B' with respect to C and A, and let C'' be the harmonic conjugate of C' with respect to A and B. Show that A'', B', C' are collinear and that there are two similar sets with this property. Show that AA'', BB', CC' are concurrent and that there are two similar sets with this property. Finally, show that A'', B'', C'' are collinear.

12. Show that (AC,BD) is a harmonic tetrad if and only if $(A,B,C,D) \sim (B,C,D,A)$.

13. Show that if two harmonic tetrads on distinct lines have a point in common, they are perspective from two different points.

14. Define a harmonic line-tetrad.

15. Given four points A_0, D_1, D_2, D_3 no three of which are collinear. Discuss the problem of constructing a four-point $A_0A_1A_2A_3$ which will have D_1, D_2, D_3 as its diagonal points.

7.4 Involutory Hexads

In this section we shall discuss still another class of point sets associated with a complete four-point, namely, involutory hexads, which will lead us to the study of involutions, a topic first introduced in Sec. 4.8.

Definition 1

Let l be an arbitrary line not passing through any vertex or any diagonal point of a complete four-point. Then any arrangement of the six points of intersection of l and the sides of the four-point into two triples of points such that corresponding points in the two triples lie on opposite sides of the four-point is called an involutory hexad.

If A_0, A_1, A_2, A_3 are the vertices of the four-point from which an involutory hexad is cut by a line l, it is convenient to denote by X_1, X_2, and X_3 the intersections of l and the sides A_0A_1, A_0A_2, A_0A_3 and to denote by X_1', X_2', X_3' the intersections of l and the sides opposite to A_0A_1, A_0A_2, A_0A_3.

With this notation, the involutory hexad is conveniently described by the symbol

$$(X_1 X_2 X_3, X_1' X_2' X_3')$$

Equally well, of course, it can be denoted by other symbols such as

$$(X_1 X_3 X_2, X_1' X_3' X_2'), \qquad (X_1 X_2 X_3', X_1' X_2' X_3), \qquad \cdots$$

as long as they satisfy the definitive requirement that corresponding points in the two triples lie on opposite sides of the four-point.

The study of involutory hexads raises many of the same questions we considered in our investigation of harmonic tetrads in the last section. Is there really anything special about involutory hexads, or can any six collinear points be arranged to form an involutory hexad? What are necessary and sufficient conditions that six collinear points form an involutory hexad? How many points can be given before an involutory hexad is uniquely determined? Is a hexad projective with an involutory hexad also an involutory hexad? Are any two involutory hexads projective with each other? The following theorems provide us with answers to these questions.

Theorem 1

If X_1, X_2, X_3, X_1', X_2' are five distinct collinear points, there is a unique point, X_3', such that $(X_1 X_2 X_3, X_1' X_2' X_3')$ is an involutory hexad.

Proof   Let X_1, X_2, X_3, X_1', X_2' be five points on a line x, let A_0 be an arbitrary point which is not on x, let A_1 be any point on $A_0 X_1$ distinct from A_0 and X_1, let A_3 be the intersection of $A_0 X_3$ and $A_1 X_2'$, and let A_2 be the intersection of $A_0 X_2$ and $A_3 X_1'$ (Fig. 7.7). Finally, let X_3' be the intersection of x and $A_1 A_2$. Then x intersects the sides of the four-point $A_0 A_1 A_2 A_3$ in the involutory hexad $(X_1 X_2 X_3, X_1' X_2' X_3')$, the first five points of which are the points given in the statement of the theorem. To complete our proof we must now show that the same sixth point, X_3', is obtained no matter how the construction is carried out.

To do this, let A_0' be a second arbitrary point which is not on x, let A_1', A_2', A_3' be determined from A_0' in the same way that A_1, A_2, A_3 were obtained from A_0, and let X_3^* be the intersection of x and the line $A_1' A_2'$. Then x intersects the four-point $A_0' A_1' A_2' A_3'$ in the involutory hexad

$$(X_1 X_2 X_3, X_1' X_2' X_3^*)$$

Now proceeding very much as in the proof of Theorem 3, Sec. 7.3, we note that $\triangle A_0 A_1 A_3$ and $\triangle A_0' A_1' A_3'$ are axially perspective on the line x and hence are also centrally perspective from some point, O. Likewise, $\triangle A_0 A_2 A_3$ and

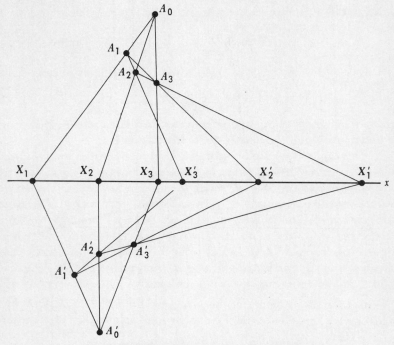

Fig. 7.7   Figure used in the proof of Theorem 1.

$\triangle A_0' A_2' A_3'$ are axially perspective on x and hence are also centrally perspective from some point O'. Now O is the intersection of the lines $A_0 A_0'$, $A_1 A_1'$, $A_3 A_3'$, and O' is the intersection of $A_0 A_0'$, $A_2 A_2'$, $A_3 A_3'$. Therefore O and O' are the same point, since each is the intersection of $A_0 A_0'$ and $A_3 A_3'$. Thus $A_0 A_0'$, $A_1 A_1'$, $A_2 A_2'$, $A_3 A_3'$ all pass through O $(= O')$, and therefore $\triangle A_0 A_1 A_2$ is perspective with $\triangle A_0' A_1' A_2'$ from O. It follows that these triangles are also axially perspective. Moreover, since the intersection of $A_0 A_1$ and $A_0' A_1'$ is X_1, and since the intersection of $A_0 A_2$ and $A_0' A_2'$ is X_2, the axis of perspectivity is the line x. Hence $A_1 A_2$ and $A_1' A_2'$ must intersect x in the same point, that is, $X_3' = X_3^*$, and the sixth point which forms an involutory hexad with the five given points in the indicated order is unique, as asserted.

Theorem 2

A necessary and sufficient condition that six collinear points X_1, X_2, X_3, X_1', X_2', X_3' should form an involutory hexad in which (X_1, X_1'), (X_2, X_2'), and (X_3, X_3') are pairs of corresponding points is that

$$(X_1, X_2, X_3, X_1') \sim (X_1', X_2', X_3', X_1)$$

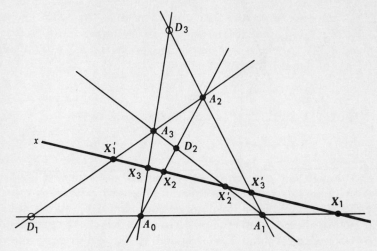

Fig. 7.8   Figure used in the proof of Theorem 2.

Proof   To prove the necessity of the given condition, let $(X_1X_2X_3,X_1'X_2'X_3')$ be an involutory hexad cut from a complete four-point $A_0A_1A_2A_3$ by a line x (Fig. 7.8). Then

$$(X_1,X_2,X_3,X_1') \overset{A_0}{\overline{\wedge}} (D_1,A_2,A_3,X_1') \overset{A_1}{\overline{\wedge}} (X_1,X_3',X_2',X_1')$$

and therefore $(X_1,X_2,X_3,X_1') \sim (X_1,X_3',X_2',X_1')$. Also, by the second permutation theorem (Theorem 3, Sec. 6.6), we have $(X_1,X_3',X_2',X_1') \sim (X_1',X_2',X_3',X_1)$. Therefore, combining the last two projectivities, we obtain $(X_1,X_2,X_3,X_1') \sim (X_1',X_2',X_3',X_1)$ as a necessary consequence of the assumption that

$$(X_1X_2X_3,X_1'X_2'X_3')$$

is an involutory hexad.

To prove the sufficiency of the given condition, let us suppose that X_1, X_2, X_3, X_1', X_2', X_3' are six collinear points such that $(X_1,X_2,X_3,X_1') \sim (X_1',X_2',X_3',X_1)$. Then, considering only the five points X_1, X_2, X_3, X_1', X_2', there is, by Theorem 1, a unique sixth point, X_3^*, such that

$$(X_1X_2X_3,X_1'X_2'X_3^*)$$

is an involutory hexad. Therefore, by the first part of the present theorem, we have $(X_1,X_2,X_3,X_1') \sim (X_1',X_2',X_3^*,X_1)$. Now this projectivity has three pairs of corresponding points in common with the projectivity we are given, namely, $(X_1,X_2,X_3,X_1') \sim (X_1',X_2',X_3',X_1)$. Hence the two projectivities are the same, and therefore each must assign the same image to X_3. Thus $X_3^* = X_3'$, which proves that $(X_1X_2X_3,X_1'X_2'X_3')$ is an involutory hexad, and our proof is complete.

Corollary 1

A necessary and sufficient condition that six collinear points, X_1, X_2, X_3, X_1', X_2', X_3', should form an involutory hexad in which (X_1, X_1'), (X_2, X_2'), and (X_3, X_3') are pairs of corresponding points is that $(X_1, X_2, X_3, X_2') \sim (X_1', X_2', X_3', X_2)$ or equally well $(X_1, X_2, X_3, X_3') \sim (X_1', X_2', X_3', X_3)$.

Corollary 2

If $(X_1 X_2 X_3, X_1' X_2' X_3')$ is an involutory hexad, then

$$(X_1 X_2 X_3, X_1' X_2' X_3') \sim (X_1' X_2' X_3', X_1 X_2 X_3)$$

Although any two harmonic tetrads are projective, it is not true that two involutory hexads are necessarily projective with each other. In fact, if $(X_1 X_2 X_3, X_1' X_2' X_3')$ is an involutory hexad, and if Y_1, Y_2, Y_3 are three collinear points, there is a unique projectivity in which Y_1 corresponds to X_1, Y_2 corresponds to X_2, and Y_3 corresponds to X_3, and in this projectivity the images Y_1', Y_2', Y_3' of X_1', X_2', X_3' are uniquely determined. On the other hand, Y_1' and Y_2' can be chosen arbitrarily, and a sixth point, Y_3', can always be found such that $(Y_1 Y_2 Y_3, Y_1' Y_2' Y_3')$ is an involutory hexad. Hence, in general two involutory hexads will not be projective with each other. We do, however, have the following theorem.

Theorem 3

If $(X_1 X_2 X_3, X_1' X_2' X_3')$ and $(Y_1 Y_2 Y_3, Y_1' Y_2' Y_3')$ are involutory hexads, and if $(X_1, X_2, X_3, X_1', X_2') \sim (Y_1, Y_2, Y_3, Y_1', Y_2')$, then

$$(X_1 X_2 X_3, X_1' X_2' X_3') \sim (Y_1 Y_2 Y_3, Y_1' Y_2' Y_3')$$

Proof Let $(X_1 X_2 X_3, X_1' X_2' X_3')$ and $(Y_1 Y_2 Y_3, Y_1' Y_2' Y_3')$ be two involutory hexads such that

(1) $$(X_1, X_2, X_3, X_1', X_2') \sim (Y_1, Y_2, Y_3, Y_1', Y_2')$$

Then by Theorem 2, $(X_1, X_2, X_3, X_1') \sim (X_1', X_2', X_3', X_1)$ and

$$(Y_1, Y_2, Y_3, Y_1') \sim (Y_1', Y_2', Y_3', Y_1)$$

Moreover, from (1) it follows that $(X_1, X_2, X_3, X_1') \sim (Y_1, Y_2, Y_3, Y_1')$. Hence, combining the last three projectivities, we have

(2) $$(X_1', X_2', X_3', X_1) \sim (Y_1', Y_2', Y_3', Y_1)$$

Now (1) and (2) are projectivities which have three pairs of corresponding points in common. Therefore they must be the same projectivity, and hence the mate of X_3' in (2), namely, Y_3', must also be the mate of X_3' in (1). Thus, adding this pair of corresponding points to (1), we have

$$(X_1X_2X_3, X_1'X_2'X_3') \sim (Y_1Y_2Y_3, Y_1'Y_2'Y_3')$$

as asserted.

Theorem 4

Any hexad projective with an involutory hexad is an involutory hexad.

Proof   Let $(X_1X_2X_3, X_1X_2'X_3')$ be an involutory hexad, and let

$$(Y_1, Y_2, Y_3, Y_1', Y_2', Y_3')$$

be a hexad such that

(3) $$(X_1, X_2, X_3, X_1', X_2', X_3') \sim (Y_1, Y_2, Y_3, Y_1', Y_2', Y_3')$$

Then from Theorem 2, $(X_1, X_2, X_3, X_1') \sim (X_1', X_2', X_3', X_1)$. Moreover, abstracting the appropriate pairs of points from the projectivity (3), we have $(X_1, X_2, X_3, X_1') \sim (Y_1, Y_2, Y_3, Y_1')$ and $(X_1', X_2', X_3', X_1) \sim (Y_1', Y_2', Y_3', Y_1)$. Hence, combining the last three projectivities, we obtain

$$(Y_1, Y_2, Y_3, Y_1') \sim (Y_1', Y_2', Y_3', Y_1)$$

which proves, according to Theorem 2, that $(Y_1Y_2Y_3, Y_1'Y_2'Y_3')$ is an involutory hexad, as asserted.

Definition 1 required that the line which cut an involutory hexad from a given four-point should not pass through any vertex or any diagonal point of the four-point. If the sectioning line does pass through a vertex or a diagonal point, we obtain what we shall call a **singular involutory hexad.** Specifically, if the sectioning line passes through D_1, the resulting singular hexad contains only five (distinct) points, since now $X_1 = X_1' = D_1$. Similarly, if the sectioning line passes through both D_1 and D_2, the resulting singular hexad contains only four (distinct) points, since now $X_1 = X_1' = D_1$ and $X_2 = X_2' = D_2$. In this case, the hexad reduces to a harmonic tetrad, which thus appears as a special case of an involutory hexad. If the sectioning line passes through one of the vertices, say A_1, again the hexad contains only four (distinct) points, for now $X_1 = X_2' = X_3' = A_1$. If the sectioning line passes through two vertices, say A_0 and A_1, the hexad is indeterminate, since now X_1 is any point of the line A_0A_1.

According to Axiom 7, a projectivity is uniquely determined when three pairs of corresponding points, say (X_1, X_1'), (X_2, X_2'), (X_3, X_3'), are given. When these six points are such that $(X_1X_2X_3, X_1'X_2'X_3')$ is an involutory hexad, the resulting projectivity is known as an **involution,** and has special properties which set it apart from other projectivities:

Definition 2

A projectivity between cobasal ranges which contains three pairs of corresponding points which form an involutory hexad is called an involution.

The following theorems describe some of the more interesting properties of involutions:

Theorem 5

The projectivity in which (X_1,X_1'), (X_2,X_2'), (X_3,X_3') are three pairs of corresponding points is an involution if and only if $(X_1,X_2,X_3,X_1') \sim (X_1',X_2',X_3',X_1)$.

Proof Let us suppose first that the given projectivity is an involution, so that it contains three pairs of corresponding points which form an involutory hexad, say $(Y_1Y_2Y_3,Y_1'Y_2'Y_3')$. Then by Corollary 2, Theorem 2, we have $(Y_1Y_2Y_3,Y_1'Y_2'Y_3') \sim (Y_1'Y_2'Y_3',Y_1Y_2Y_3)$ or, incorporating the information we have about the X's and their images under the same projectivity,

$$(Y_1,Y_2,Y_3,Y_1',Y_2',Y_3',X_1,X_2,X_3) \sim (Y_1',Y_2',Y_3',Y_1,Y_2,Y_3,X_1',X_2',X_3')$$

Now this projectivity and its inverse have three pairs of corresponding points in common, namely, (Y_1,Y_1'), (Y_2,Y_2'), (Y_3,Y_3'). Hence, by Axiom 7, this transformation and its inverse are identical. Therefore, since the inverse transformation maps X_1' into X_1, X_2' into X_2, and X_3' into X_3, the original transformation must effect this mapping also. Hence from the direct transformation we can abstract the relation $(X_1,X_2,X_3,X_1') \sim (X_1',X_2',X_3',X_1)$ which proves the "only if," or necessity, assertion of the theorem.

To prove the "if," or sufficiency, assertion of the theorem, we note that if $(X_1,X_2,X_3,X_1') \sim (X_1',X_2',X_3',X_1)$, then, by Theorem 2, $(X_1X_2X_3,X_1'X_2'X_3')$ is an involutory hexad, and the given projectivity is an involution by Definition 2.

Corollary 1

If a projectivity is an involution, then any three points and their images form an involutory hexad.

Corollary 2

If a projectivity between cobasal ranges contains one pair of distinct points which correspond reciprocally, i.e., if there is one pair of points, (X_1,X_1'), such that $(X_1,X_1',\ldots) \sim (X_1',X_1,\ldots)$, then the projectivity is an involution and every point and its image correspond reciprocally.

In our study of projectivities between cobasal ranges in the algebraic representation (Sec. 4.8), we found that every involution has two distinct self-corresponding points. This property cannot be derived as a consequence of the axioms we have thus far adopted, and in fact it is false in each of the finite geometries we have used as examples and also false in E_2^+. In the next chapter we shall introduce another axiom which will settle this question. Meanwhile, our theorems will be of a conditional nature, with the existence of one or more self-corresponding points a part of the hypothesis. Our first result shows that though there may be involutions without two self-corresponding points, there is a large class of involutions which do have two self-corresponding points.

Theorem 6

If F_1 and F_2 are any two distinct points, then the pairs of points which are harmonic conjugates with respect to F_1 and F_2 are mates in an involution having F_1 and F_2 as self-corresponding points.

Proof   Let F_1 and F_2 be distinct points, let X_1 and X_1' be any points which are harmonic conjugates with respect to F_1 and F_2, and let X_1'' be the mate of X_1' in the projectivity which maps F_1 into itself, F_2 into itself, and X_1 into X_1'. We then have the projectivity

$$(4) \qquad (F_1,F_2,X_1,X_1') \sim (F_1,F_2,X_1',X_1'')$$

From this it follows that $(F_1F_2,X_1'X_1'')$ is a harmonic tetrad since it is projective with a harmonic tetrad. Hence X_1' and X_1'' are harmonic conjugates with respect to F_1 and F_2, and therefore $X_1'' = X_1$, since the harmonic conjugate of a point with respect to two other points is unique. Thus X_1 and X_1' are reciprocally corresponding points in the projectivity (4) and, by Corollary 2, Theorem 5, this projectivity is an involution which, moreover, has F_1 and F_2 as fixed points.

If (X_2,X_2') is another pair of harmonic conjugates with respect to F_1 and F_2, it is clear from the preceding discussion that they too are mates in *an* involution having F_1 and F_2 as fixed points, but it is not obvious that this involution is the one in which X_1 and X_1' are mates. To show that X_2' is actually the mate of X_2 in the first involution, let us suppose that the first involution maps X_2 into X_2'' and therefore also maps X_2'' into X_2. Then we have $(F_1F_2,X_2X_2'') \sim (F_1F_2,X_2''X_2)$, which shows that X_2'' is the harmonic conjugate of X_2 with respect to F_1 and F_2. Finally, since the harmonic conjugate of a point with respect to two given points is unique, it follows that $X_2'' = X_2'$ as asserted.

Corollary 1

If an involution has a pair of distinct self-corresponding points, then every pair of mates in the involution is a pair of harmonic conjugates with respect to the fixed points of the involution.

Although we are not yet able to assert that every involution has exactly two distinct self-corresponding points, we can prove that no involution has just one self-corresponding point:

Theorem 7

If an involution has one self-corresponding point, then it has a second self-corresponding point distinct from the first.

Proof Let X_1 and X_1' be an arbitrary pair of mates in an involution having a fixed point, F_1, let F_2 be the harmonic conjugate of F_1 with respect to X_1 and X_1', and let F_2' be the mate of F_2 in the given involution. Since neither X_1 nor X_1' coincides with F_1, it follows that F_2 is distinct from F_1. In the given involution we thus have $(F_1,F_2,X_1,X_1') \sim (F_1,F_2',X_1',X_1)$. Moreover. since (F_1F_2,X_1X_1') is a harmonic tetrad, so is $(F_1F_2',X_1'X_1)$, since the two tetrads are projective. Therefore, both (F_1,F_2) and (F_1,F_2') are harmonic conjugates with respect to X_1 and X_1', which implies that $F_2' = F_2$. Thus F_2 is its own image in the given involution, which thus has two distinct self-corresponding points, as asserted.

If we know that a projectivity is an involution, then if we are given one pair of distinct corresponding points, say (X_1,X_1'), we are actually given a second pair because we also know that X_1 corresponds to X_1'. Hence by Axiom 7, the involution will be completely determined by the specification of one further point and its image, whether that point is self-corresponding or not. If the points in both of the given pairs are self-corresponding, then we have only two pairs of corresponding points given, and Axiom 7 cannot be invoked. In this case, however, we are given the two fixed points of the involution, and Theorem 6 and its corollary guarantee that the involution is uniquely determined by the two points. Thus we have established the following theorem.

Theorem 8

An involution is uniquely determined when two pairs of corresponding points are given.

We have usually thought of projectivities in what might be called an *active* or *dynamic* sense; i.e., we have regarded them as mappings which transformed, or carried, certain points into certain other points. Equally well, however, we can think of projectivities in a *passive* or *static* sense; i.e., we can consider them to be simply a set of ordered pairs of points, the first point in each pair being a point of the object range and the second point being its correspondent in the image range. If we have two projectivities on the same line, and if we take the second, or passive, point of view, i.e., if we regard each of them simply as a set of ordered pairs of points, it is natural to ask whether these two sets have anything in common, and if so, what? When the projectivities are both involutions, the question has a neat answer which is given in the following theorem.

Theorem 9

If every involution has two distinct self-corresponding points, then two involutions on the same line always have one and only one pair of mates in common.

Proof   Let I_1 and I_2 be two involutions on the same line, let F_1 and F_1' be the fixed points of I_1, and let F_2 and F_2' be the fixed points of I_2. Now consider the involution, I_3, in which (F_1, F_1') and (F_2, F_2') are two pairs of corresponding points. By hypothesis, every involution has two distinct self-corresponding points; hence I_3 has two distinct fixed points, say X and X'. Then according to Corollary 1, Theorem 6, both $(F_1 F_1', XX')$ and $(F_2 F_2', XX')$ are harmonic tetrads. Hence X and X' are harmonic conjugates with respect to both (F_1, F_1') and (F_2, F_2'), and therefore, by Theorem 6 itself, X and X' are mates in both I_1 and I_2. Finally, it is clear from Theorem 8 that I_1 and I_2 can have no more than one pair of mates in common. Hence the theorem is established.

Exercises

1. Prove Corollary 1, Theorem 6.

2. Let (X, X') and (Y, Y') be two pairs of mates in an involution whose fixed points are F_1 and F_2. Show that (X, Y'), (X', Y), and (F_1, F_2) are three pairs of mates in another involution.

3. Given $x(X_1, X_2, X_3, \dots) \overset{U}{\wedge} y(Y_1, Y_2, Y_3, \dots) \overset{V}{\wedge} x(X_1', X_2', X_3', \dots)$. Find a necessary and sufficient condition that the projectivity $x(X_1, X_2, X_3, \dots) \sim x(X_1', X_2', X_3', \dots)$ should be an involution.

4. If $(A, B, C, D, F_1, F_2) \sim (B, C, D, A, F_1, F_2)$, show that (A, C) and (B, D) are two pairs of mates in the involution whose self-corresponding points are F_1 and F_2.

5. Given $x(X_1,X_2,X_3,...) \overset{U}{\sim} y(Y_1,Y_2,Y_3,...) \wedge x(X_1',X_2',X_3',...)$. Show that the projectivity $x(X_1,X_2,X_3,...) \sim x(X_1',X_2',X_3',...)$ is an involution if and only if U is on the pappus line of the projectivity

$$x(X_1,X_2,X_3,...) \sim y(Y_1,Y_2,Y_3,...)$$

6. Define an involutory line hexad.

7. In Π_2, A_0, A_1, A_2, A_3 and P are, respectively, $(1,1,1)$, $(1,0,0)$, $(0,1,0)$, $(0,0,1)$, and (p_1,p_2,p_3). Show that there is a unique line on P which cuts the four-point $A_0A_1A_2A_3$ in an involutory hexad belonging to an involution for which P is a fixed point, and find the coordinates of the second fixed point, P', in the involution on this line.

8. In Π_2 let (X_1,X_1'), (X_2,X_2'), (X_3,X_3') be three pairs of points on a line x whose parameters are given, respectively, by the roots of the quadratic equations $Q_1: a_1\lambda^2 + 2b_1\lambda\mu + c_1\mu^2 = 0$, $Q_2: a_2\lambda^2 + 2b_2\lambda\mu + c_2\mu^2 = 0$, Q_3: $a_3\lambda^2 + 2b_3\lambda\mu + c_3\mu^2 = 0$. Show that $(X_1X_2X_3,X_1'X_2'X_3')$ is an involutory hexad if and only if

$$\begin{vmatrix} a_1 & b_1 & c_1 \\ a_2 & b_2 & c_2 \\ a_3 & b_3 & c_3 \end{vmatrix} = 0$$

9. Using the results of Exercise 8, show that the involution determined by the points defined by the equations $Q_1 = 0$ and $Q_2 = 0$ consists of precisely those pairs of points defined by the family of equations $k_1Q_1 + k_2Q_2 = 0$.

10. In Π_2 what is the equation of the involution whose fixed points are the points defined by the equation $A\lambda^2 + 2B\lambda\mu + C\mu^2 = 0$?

7.5 Further Results

There are a number of famous theorems in elementary geometry which assert the concurrence of three particular lines, such as the medians, altitudes, or angle bisectors, passing through the respective vertices of a general triangle. Such results obviously involve the notion of measurement and hence are not theorems of projective geometry. However, there are projective theorems, dealing with the concurrence of lines and the collinearity of points associated with a general triangle, from which these familiar euclidean theorems can be obtained as metrical specializations. In particular, the following and its dual lead to many important euclidean results.

Theorem 1

Let A_1, A_2, A_3 be the vertices of an arbitrary triangle, let a_1, a_2, a_3 be lines passing, respectively, through A_1, A_2, A_3, let x be any line distinct from

a_1, a_2, a_3 and the sides of $\triangle A_1A_2A_3$, let X_1, X_2, X_3 be, respectively, the intersections of x and the lines a_1, a_2, a_3, and let X_1', X_2', X_3' be, respectively, the intersections of x and the lines A_2A_3, A_3A_1, A_1A_2. Then a_1, a_2, a_3 are concurrent if and only if $(X_1X_2X_3,X_1'X_2'X_3')$ is an involutory hexad.

Proof   With the notation defined in the statement of the theorem, let us suppose first that the lines a_1, a_2, a_3 on the respective vertices A_1, A_2, A_3 are concurrent in a point A_0 (Fig. 7.9). Then it follows immediately that $(X_1X_2X_3,X_1'X_2'X_3')$ is an involutory hexad, since it is an appropriately ordered set of six points cut from the sides of the four-point $A_0A_1A_2A_3$ by the line x.

   Next let us suppose that $(X_1X_2X_3,X_1'X_2'X_3')$ is an involutory hexad. Let A_0 be the intersection of a_1 and a_2, let a_3^* be the line A_0A_3, and let X_3^* ($= a_3^*$) be the intersection of A_0A_3 ($= a_3^*$) and x. Then by the first part of our proof, $(X_1X_2X_3^*,X_1'X_2'X_3')$ is an involutory hexad. However, we are given that $(X_1X_2X_3,X_1'X_2'X_3')$ is also an involutory hexad. Hence, by Theorem 1, Sec. 7.4, it follows that $X_3^* = X_3$. Thus a_3^* and a_3 are the same line, which proves that a_1, a_2, a_3 are concurrent, as asserted.

Theorem 2

   Let a_1, a_2, a_3 be the sides of an arbitrary triangle, let A_1, A_2, A_3 be points lying, respectively, on a_1, a_2, a_3, let X be any point distinct from A_1,

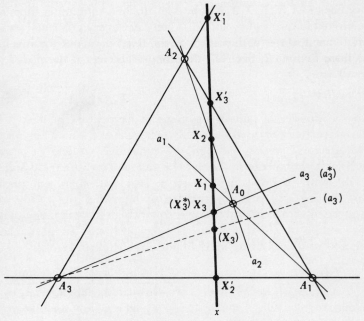

Fig. 7.9   Figure used in the proof of Theorem 1.

A_2, A_3 and the vertices of $\triangle a_1a_2a_3$, let x_1, x_2, x_3 be, respectively, the joins of X and the points A_1, A_2, A_3, and let x_1', x_2', x_3' be, respectively, the joins of X and the points a_2a_3, a_3a_1, a_1a_2. Then A_1, A_2, A_3 are collinear if and only if $(x_1x_2x_3, x_1'x_2'x_3')$ is an involutory (line) hexad.

In order to obtain euclidean specializations of Theorem 1 and its dual, Theorem 2, we shall need the following two lemmas.

Lemma 1

In the algebraic representation, $(X_1X_2X_3, X_1'X_2'X_3')$ is an involutory hexad if and only if

$$R(X_2'X_3', X_1X_1')R(X_3'X_1', X_2X_2')R(X_1'X_2', X_3X_3') = -1$$

or, equally well,

$$R(X_2X_3, X_1X_1')R(X_3X_1, X_2X_2')R(X_1X_2, X_3X_3') = -1$$

Proof Let X_1, X_2, X_3, X_1', X_2', X_3' be six collinear points in Π_2, and let us suppose first that $(X_1X_2X_3, X_1'X_2'X_3')$ is an involutory hexad. Then (X_1, X_1'), (X_2, X_2'), (X_3, X_3') are pairs of mates in an involution. Moreover, although we do not yet have an axiomatic basis for asserting this in general, we have shown that in Π_2 every involution has two distinct self-corresponding points. Let F_1 and F_2 be the fixed points of the involution determined by the involutory hexad $(X_1X_2X_3, X_1'X_2'X_3')$, and let the line containing these points be parametrized in terms of F_1 and F_2. Then since every point and its image are harmonic conjugates with respect to the fixed points of the involution, it follows from Lemma 2, Sec. 4.7, that the parameters of the points of the given hexad are

$$X_1:(\lambda_1, \mu_1) \qquad X_2:(\lambda_2, \mu_2) \qquad X_3:(\lambda_3, \mu_3)$$
$$X_1':(\lambda_1, -\mu_1) \qquad X_2':(\lambda_2, -\mu_2) \qquad X_3':(\lambda_3, -\mu_3)$$

Hence

$$R(X_2'X_3', X_1X_1')R(X_3'X_1', X_2X_2')R(X_1'X_2', X_3X_3')$$

$$= \frac{\begin{vmatrix} \lambda_2 & \lambda_1 \\ -\mu_2 & \mu_1 \end{vmatrix} \cdot \begin{vmatrix} \lambda_3 & \lambda_1 \\ -\mu_3 & -\mu_1 \end{vmatrix}}{\begin{vmatrix} \lambda_2 & \lambda_1 \\ -\mu_2 & -\mu_1 \end{vmatrix} \cdot \begin{vmatrix} \lambda_3 & \lambda_1 \\ -\mu_3 & \mu_1 \end{vmatrix}} \frac{\begin{vmatrix} \lambda_3 & \lambda_2 \\ -\mu_3 & \mu_2 \end{vmatrix} \cdot \begin{vmatrix} \lambda_1 & \lambda_2 \\ -\mu_1 & -\mu_2 \end{vmatrix}}{\begin{vmatrix} \lambda_3 & \lambda_2 \\ -\mu_3 & -\mu_2 \end{vmatrix} \cdot \begin{vmatrix} \lambda_1 & \lambda_2 \\ -\mu_1 & \mu_2 \end{vmatrix}}$$

$$\times \frac{\begin{vmatrix} \lambda_1 & \lambda_3 \\ -\mu_1 & \mu_3 \end{vmatrix} \cdot \begin{vmatrix} \lambda_2 & \lambda_3 \\ -\mu_2 & -\mu_3 \end{vmatrix}}{\begin{vmatrix} \lambda_1 & \lambda_3 \\ -\mu_1 & -\mu_3 \end{vmatrix} \cdot \begin{vmatrix} \lambda_2 & \lambda_3 \\ -\mu_2 & \mu_3 \end{vmatrix}} = -1$$

as asserted.

Now let us suppose that X_1, X_2, X_3, X_1', X_2', X_3' are six collinear points such that $R(X_2'X_3',X_1X_1')R(X_3'X_1',X_2X_2')R(X_1'X_2',X_3X_3') = -1$. By Theorem 8, Sec. 7.4, there is a unique involution in which (X_1,X_1') and (X_2,X_2') are pairs of mates. Let F_1 and F_2 be the fixed points of this involution, and let the line containing the six given points be parametrized in terms of F_1 and F_2. Then X_1, X_2, X_3, X_1', X_2' have the same parameters they had in the first part of our proof. However, since we do not yet know that X_3' is the mate of X_3 in the involution we have just defined, we cannot assert that its parameters are $(\lambda_3, -\mu_3)$. Instead, we must at present regard its parameters as arbitrary, say (λ_3', μ_3'). Now, evaluating the cross-ratio expression which is our hypothesis in this portion of the proof, using the parameters we have just identified, we obtain, after straightforward simplifications,

$$\frac{(\lambda_3'\mu_2 - \lambda_2\mu_3')(\lambda_1\mu_3 + \lambda_3\mu_1)}{(\lambda_1\mu_3' - \lambda_3'\mu_1)(\lambda_2\mu_3 + \lambda_3\mu_2)} = -1$$

or, cross multiplying, collecting terms, and factoring,

$$(\lambda_3'\mu_3 + \lambda_3\mu_3')(\lambda_1\mu_2 - \lambda_2\mu_1) = 0$$

Since X_1 and X_2 are distinct points, it follows that $\lambda_1\mu_2 - \lambda_2\mu_1 \neq 0$. Hence $\lambda_3'\mu_3 + \lambda_3\mu_3' = 0$, which implies that $\lambda_3' = k\lambda_3$ and $\mu_3' = -k\mu_3$. Therefore, by Lemma 2, Sec. 4.7, X_3' is the harmonic conjugate of X_3 with respect to the fixed points F_1 and F_2, and thus $(X_1X_2X_3,X_1'X_2'X_3')$ is an involutory hexad, as asserted. The necessity and sufficiency of the second assertion of the theorem follow in exactly the same fashion.

Lemma 2

If P_1, P_2, Q_1, Q_2, Q_3 are five collinear points in Π_2, then

$$R(P_1P_2,Q_1Q_2) = -R(P_1P_2,Q_1Q_3)$$

if and only if Q_2 and Q_3 are harmonic conjugates with respect to P_1 and P_2.

Proof   Let P_1, P_2, Q_1, Q_2, Q_3 be five collinear points. Then from the multiplication theorem for cross ratios (Exercise 13, Sec. 4.7) we have

$$R(P_1P_2,Q_1Q_2)R(P_1P_2,Q_2Q_3) = R(P_1P_2,Q_1Q_3)$$

From this it follows immediately that $R(P_1P_2,Q_1Q_2) = -R(P_1P_2,Q_1Q_3)$ if and only if $R(P_1P_2,Q_2Q_3) = -1$, which is true if and only if Q_2 and Q_3 are harmonic conjugates with respect to P_1 and P_2, as asserted.

Using Lemma 1, we can now obtain, in the algebraic representation, the following modification of Theorem 1.

Theorem 3

Let A_1, A_2, A_3 be the vertices of an arbitrary triangle, let x be any line which does not pass through a vertex of $\triangle A_1A_2A_3$, let P_1', P_2', P_3' be,

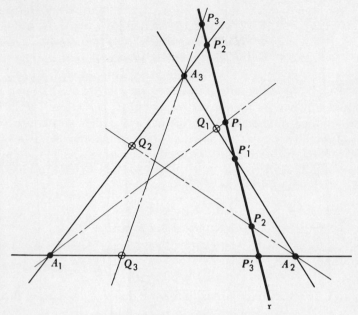

Fig. 7.10   Figure used in the proof of Theorem 3.

respectively, the points in which x intersects A_2A_3, A_3A_1, A_1A_2, and let Q_1, Q_2, Q_3 be three points located, respectively, on A_2A_3, A_3A_1, A_1A_2. Then the lines joining Q_1, Q_2, Q_3 to the corresponding vertices of $\triangle A_1A_2A_3$ are concurrent if and only if

$$\mathrm{R}\,(A_3A_2,Q_1P_1')\mathrm{R}\,(A_1A_3,Q_2P_2')\mathrm{R}\,(A_2A_1,Q_3P_3') = -1$$

Proof   Let $\triangle A_1A_2A_3$ be an arbitrary triangle, let Q_1, Q_2, Q_3 be three points located, respectively, on A_2A_3, A_3A_1, A_1A_2, let x be an arbitrary line not passing through any vertex of $\triangle A_1A_2A_3$, let P_1', P_2', P_3' be the points in which x intersects A_2A_3, A_3A_1, A_1A_2, and let P_1, P_2, P_3 be the points in which x intersects A_1Q_1, A_2Q_2, A_3Q_3 (Fig. 7.10). Then, by Theorem 1 and Lemma 1, the lines A_1Q_1, A_2Q_2, A_3Q_3 are concurrent if and only if

$$\mathrm{R}\,(P_2'P_3',P_1P_1')\mathrm{R}\,(P_3'P_1',P_2P_2')\mathrm{R}\,(P_1'P_2',P_3P_3') = -1$$

However,

$$(P_2',P_3',P_1,P_1') \overset{A_1}{\wedge} (A_3,A_2,Q_1,P_1')$$

$$(P_3',P_1',P_2,P_2') \overset{A_2}{\wedge} (A_1,A_3,Q_2,P_2')$$

$$(P_1',P_2',P_3,P_3') \overset{A_3}{\wedge} (A_2,A_1,Q_3,P_3')$$

Hence

$$\text{R}(P_2'P_3',P_1P_1') = \text{R}(A_3A_2,Q_1P_1')$$
$$\text{R}(P_3'P_1',P_2P_2') = \text{R}(A_1A_3,Q_2P_2')$$
$$\text{R}(P_1'P_2',P_3P_3') = \text{R}(A_2A_1,Q_3P_3')$$

and therefore the necessary and sufficient condition for the concurrence of A_1Q_1, A_2Q_2, A_3Q_3 becomes

$$\text{R}(A_3A_2,Q_1P_1')\text{R}(A_1A_3,Q_2P_2')\text{R}(A_2A_1,Q_3P_3') = -1$$

as asserted.

If we now identify x as the ideal line in a euclidean specialization of Π_2, we obtain the following theorem.

Theorem 4

If $\triangle A_1A_2A_3$ is an arbitrary triangle, and if Q_1, Q_2, Q_3 are points on A_2A_3, A_3A_1, and A_1A_2, respectively, then A_1Q_1, A_2Q_2, and A_3Q_3 are concurrent if and only if

$$\frac{(A_1Q_3)}{(Q_3A_2)}\frac{(A_2Q_1)}{(Q_1A_3)}\frac{(A_3Q_2)}{(Q_2A_1)} = 1$$

Proof   From Exercise 11, Sec. 4.7, we know that if W is the ideal point on the line containing the finite points X, Y, Z, then

$$\text{R}(XY,ZW) = -\frac{(XZ)}{(ZY)}$$

where, as usual, (UV) denotes the length of the segment $\overline{UV}$. Hence, taking x to be the ideal line, so that P_1', P_2', P_3' are, respectively, the ideal points on the lines A_2A_3, A_3A_1, A_1A_2, and applying this evaluation to the three cross ratios in the formula of Theorem 3, we have

$$\left[-\frac{(A_3Q_1)}{(Q_1A_2)}\right]\left[-\frac{(A_1Q_2)}{(Q_2A_3)}\right]\left[-\frac{(A_2Q_3)}{(Q_3A_1)}\right] = -1$$

or, rearranging slightly,

$$\frac{(A_1Q_3)}{(Q_3A_2)}\frac{(A_2Q_1)}{(Q_1A_3)}\frac{(A_3Q_2)}{(Q_2A_1)} = 1$$

as asserted.

Theorem 4 is known as the **theorem of Ceva,** after its discoverer, the Italian mathematician Giovanni Ceva (c. 1647–1736). It is of great importance in advanced euclidean geometry.

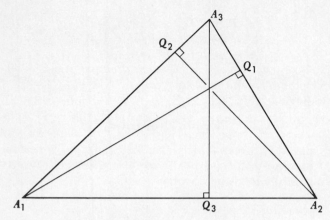

Fig. 7.11   Figure used in Example 1.

Example 1

Using Ceva's theorem, prove that the altitudes of a triangle are concurrent.

Let Q_1, Q_2, Q_3 be the feet of the perpendiculars from the vertices A_1, A_2, A_3 of an arbitrary triangle to the opposite sides (Fig. 7.11). Then

$$\triangle A_1 A_2 Q_2 \approx \triangle A_1 A_3 Q_3 \qquad \text{and therefore} \qquad \frac{(A_1 Q_3)}{(A_1 Q_2)} = \frac{(A_1 A_3)}{(A_1 A_2)}$$

$$\triangle A_2 A_3 Q_3 \approx \triangle A_2 A_1 Q_1 \qquad \text{and therefore} \qquad \frac{(A_2 Q_1)}{(A_2 Q_3)} = \frac{(A_2 A_1)}{(A_2 A_3)}$$

$$\triangle A_3 A_1 Q_1 \approx \triangle A_3 A_2 Q_2 \qquad \text{and therefore} \qquad \frac{(A_3 Q_2)}{(A_3 Q_1)} = \frac{(A_3 A_2)}{(A_3 A_1)}$$

Now, multiplying the last three equations, we have

$$\frac{(A_1 Q_3)}{(Q_3 A_2)} \frac{(A_2 Q_1)}{(Q_1 A_3)} \frac{(A_3 Q_2)}{(Q_2 A_1)} = 1$$

Hence, by Ceva's theorem, $A_1 Q_1$, $A_2 Q_2$, and $A_3 Q_3$ are concurrent, as asserted.

Using Lemmas 1 and 2, we can now obtain the following modification of Theorem 2.

Theorem 5

Let A_1, A_2, A_3 be the vertices of an arbitrary triangle, let x be any line which does not pass through a vertex of $\triangle A_1 A_2 A_3$, let P_1, P_2, P_3 be the intersections of x and the lines $A_2 A_3$, $A_3 A_1$, $A_1 A_2$, and let Q_1, Q_2, Q_3 be arbitrary points on $A_2 A_3$, $A_3 A_1$, $A_1 A_2$, respectively. Then Q_1, Q_2, Q_3 are collinear if and only if

$$\mathrm{R}\,(A_2A_3,Q_1P_1)\mathrm{R}\,(A_3A_1,Q_2P_2)\mathrm{R}\,(A_1A_2,Q_3P_3) = 1$$

Proof   Let A_1, A_2, A_3 be the vertices of an arbitrary triangle, let P_1, P_2, P_3 be collinear points on the respective sides of $\triangle A_1A_2A_3$, let Q_1, Q_2, Q_3 be arbitrary points on A_2A_3, A_3A_1, A_1A_2, let X be the intersection of Q_2A_2 and Q_3A_3, let $\bar{Q}_1$ be the intersection of A_1X and A_2A_3, let x_1, x_2, x_3 be the lines joining X to P_1, P_2, P_3, respectively, and let x_1', x_2', x_3' be the lines joining X to A_1, A_2, A_3 (Fig. 7.12). Let us suppose first that Q_1, Q_2, Q_3 are collinear. Then, since P_1, P_2, P_3 are collinear, it follows from Theorem 2 and Lemma 1, restated for line hexads, that

$$\mathrm{R}\,(x_2'x_3',x_1x_1')\mathrm{R}\,(x_3'x_1',x_2x_2')\mathrm{R}\,(x_1'x_2',x_3x_3') = -1$$

However, from Theorem 4, Sec. 4.7,

$$\mathrm{R}\,(x_2'x_3',x_1x_1') = \mathrm{R}\,(A_2A_3,P_1\bar{Q}_1)$$
$$\mathrm{R}\,(x_3'x_1',x_2x_2') = \mathrm{R}\,(A_3A_1,P_2Q_2)$$
$$\mathrm{R}\,(x_1'x_2',x_3x_3') = \mathrm{R}\,(A_1A_2,P_3Q_3)$$

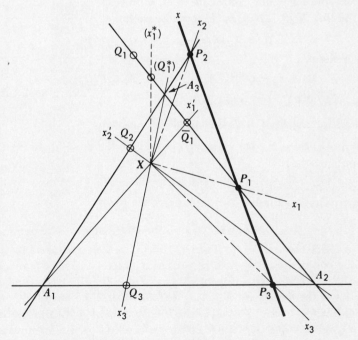

Fig. 7.12   Figure used in the proof of Theorem 5.

and therefore

$$R(A_2A_3,P_1\bar{Q}_1)R(A_3A_1,P_2Q_2)R(A_1A_2,P_3Q_3) = -1$$

Now from the properties of the four-point $XA_1A_2A_3$, whose diagonal points are $\bar{Q}_1$, Q_2, Q_3, it follows that $\bar{Q}_1$ is the harmonic conjugate of Q_1 with respect to A_2 and A_3. Hence, by Lemma 2, we have

$$R(A_2A_3,P_1\bar{Q}_1) = -R(A_2A_3,P_1Q_1)$$

and therefore

$$R(A_2A_3,P_1Q_1)R(A_3A_1,P_2Q_2)R(A_1A_2,P_3Q_3) = 1$$

Finally, since $R(RS,TU) = 1/R(RS,UT)$, the last equation implies that

$$R(A_2A_3,Q_1P_1)R(A_3A_1,Q_2P_2)R(A_1A_2,Q_3P_3) = 1$$

as asserted.

Now suppose that

$$R(A_2A_3,Q_1P_1)R(A_3A_1,Q_2P_2)R(A_1A_2,Q_3P_3) = 1$$

or, equally well,

$$R(A_2A_3,P_1Q_1)R(A_3A_1,P_2Q_2)R(A_1A_2,P_3Q_3) = 1$$

Let Q_1^* be the harmonic conjugate of Q_1 with respect to A_2 and A_3, and let x_1^* be the line XQ_1^*. Then, by Lemma 2, we have

$$R(A_2A_3,P_1Q_1) = -R(A_2A_3,P_1Q_1^*)$$

and therefore

$$R(A_2A_3,P_1Q_1^*)R(A_3A_1,P_2Q_2)R(A_1A_2,P_3Q_3) = -1$$

By Theorem 4, Sec. 4.7, this implies that

$$R(x_2'x_3',x_1'x_1^*)R(x_3'x_1',x_2'x_2')R(x_1'x_2',x_3'x_3') = -1$$

Furthermore, since P_1, P_2, P_3 are collinear, we have, as in the first part of our proof,

$$R(x_2'x_3',x_1'x_1')R(x_3'x_1',x_2'x_2')R(x_1'x_2',x_3'x_3') = -1$$

Hence, comparing the last two equations, we have

$$R(x_2'x_3',x_1'x_1^*) = R(x_2'x_3',x_1'x_1')$$

which implies that $x_1^* = x_1'$. Thus $Q_1^* = \bar{Q}_1$, which means that Q_1 is the harmonic conjugate of $\bar{Q}_1$ with respect to A_2 and A_3, since Q_1^* and Q_1 were harmonic conjugates with respect to A_2 and A_3. Finally, from the harmonic properties of the four-point $XA_1A_2A_3$, it follows that the harmonic conjugate of $\bar{Q}_1$ with respect to A_2 and A_3 is the intersection of A_2A_3 and the diagonal line Q_2Q_3. Hence Q_1, Q_2, Q_3 are collinear, the "if," or sufficiency, assertion of the theorem is established, and our proof is complete.

If we now identify x as the ideal line in a euclidean specialization of Π_2, we obtain the following theorem.

Theorem 6

If $\triangle A_1 A_2 A_3$ is an arbitrary triangle, and if Q_1, Q_2, Q_3 are points on $A_2 A_3$, $A_3 A_1$, $A_1 A_2$, respectively, then Q_1, Q_2, Q_3 are collinear if and only if

$$\frac{(A_1 Q_3)}{(Q_3 A_2)} \frac{(A_2 Q_1)}{(Q_1 A_3)} \frac{(A_3 Q_2)}{(Q_2 A_1)} = -1$$

Proof   From Exercise 11, Sec. 4.7, we know that if W is the ideal point on the line containing the finite points X, Y, Z, then

$$\mathrm{R}(XY, ZW) = -\frac{(XZ)}{(ZY)}$$

Hence, taking the collinear points P_1, P_2, P_3 to be the ideal points on the lines $A_2 A_3$, $A_3 A_1$, $A_1 A_2$, and using this formula to evaluate the cross ratios in the formula of Theorem 5, we obtain

$$\left[-\frac{(A_2 Q_1)}{(Q_1 A_3)} \right]\left[-\frac{(A_3 Q_2)}{(Q_2 A_1)} \right]\left[-\frac{(A_1 Q_3)}{(Q_3 A_2)} \right] = 1$$

or
$$\frac{(A_1 Q_3)}{(Q_3 A_2)} \frac{(A_2 Q_1)}{(Q_1 A_3)} \frac{(A_3 Q_2)}{(Q_2 A_1)} = -1$$

as asserted.

Theorem 6 is known as the **theorem of Menelaus,** after the Greek mathematician, Menelaus of Alexandria (A.D. c. 100), who discovered it. Like Ceva's theorem, it is of fundamental importance in advanced euclidean geometry.

Example 2

Let a_0, a_1, a_2, a_3 be the sides of an arbitrary complete four-line, let P_1, P_2, P_3, P_1', P_2', P_3' be, respectively, the vertices $a_0 a_1$, $a_0 a_2$, $a_0 a_3$, $a_2 a_3$, $a_1 a_3$, $a_1 a_2$, and let M_1, M_2, M_3 be, respectively, the midpoints of the segments determined by the pairs of opposite vertices (P_1, P_1'), (P_2, P_2'), (P_3, P_3'). Using the theorem of Menelaus, prove that in the euclidean plane M_1, M_2, M_3 are collinear.

In addition to the points identified in the statement of the problem, let Q_1, Q_2, Q_3 be, respectively, the midpoints of the segments $\overline{P_2' P_3}$, $\overline{P_3 P_1}$, $\overline{P_1 P_2}$ (Fig. 7.13). Since M_1, Q_2, Q_3 are the midpoints of the segments $\overline{P_1 P_1'}$, $\overline{P_1 P_3}$, $\overline{P_1 P_2'}$, it follows from elementary geometry that M_1, Q_2, Q_3 are collinear

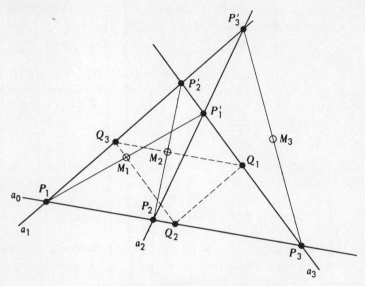

Fig. 7.13 Figure used in Example 2.

on a line parallel to a_3. Hence

$$\frac{(Q_3 M_1)}{(M_1 Q_2)} = \frac{(P_2' P_1')}{(P_1' P_3)}$$

Similarly, M_2, Q_1, Q_3 are collinear on a line parallel to a_0, and M_3, Q_1, Q_2 are collinear on a line parallel to a_1, and therefore

$$\frac{(Q_1 M_2)}{(M_2 Q_3)} = \frac{(P_3 P_2)}{(P_2 P_1)} \qquad \text{and} \qquad \frac{(Q_2 M_3)}{(M_3 Q_1)} = \frac{(P_1 P_3')}{(P_3' P_2')}$$

Combining these equalities, we have

$$\frac{(Q_2 M_3)}{(M_3 Q_1)} \frac{(Q_3 M_1)}{(M_1 Q_2)} \frac{(Q_1 M_2)}{(M_2 Q_3)} = \frac{(P_1 P_3')}{(P_3' P_2')} \frac{(P_2' P_1')}{(P_1' P_3)} \frac{(P_3 P_2)}{(P_2 P_1)}$$

Now P_1', P_2, P_3' are collinear points on the respective sides of $\triangle P_1 P_2' P_3$. Hence, by the theorem of Menelaus, the right-hand side of the last equation is equal to 1. Thus the left-hand side is also equal to -1, and therefore, by the theorem of Menelaus again, the points M_1, M_2, M_3 on the respective sides of $\triangle Q_1 Q_2 Q_3$ are collinear, as asserted.

Exercises

1. Using Fig. 7.14a, give a strictly euclidean proof of the theorem of Ceva.

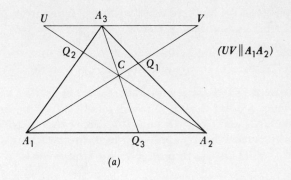

$(UV \| A_1A_2)$

(a)

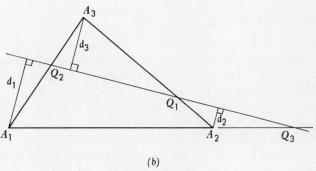

(b)

Fig. 7.14

2. Using Fig. 7.14*b*, give a strictly euclidean proof of the theorem of Menelaus.

3. Prove the following generalization of the theorem of Menelaus. If Q_1, Q_2, Q_3, Q_4 are, respectively, the points in which a transversal, t, intersects the sides A_1A_2, A_2A_3, A_3A_4, A_4A_1, of a quadrilateral $A_1A_2A_3A_4$, then

$$\frac{(A_1Q_1)}{(Q_1A_2)} \frac{(A_2Q_2)}{(Q_2A_3)} \frac{(A_3Q_3)}{(Q_3A_4)} \frac{(A_4Q_4)}{(Q_4A_1)} = 1$$

4. Using Ceva's theorem, prove that: (*a*) The medians of a triangle are concurrent.

(*b*) The bisectors of the internal angles of a triangle are concurrent.

5. Using the theorem of Menelaus, prove that the bisectors of the external angles of a triangle intersect the opposite sides in three collinear points.

6. Prove that the lines joining the vertices of a triangle to the points of contact of the opposite sides with the inscribed circle are concurrent. (The

point of concurrence in this case is known as the **Gergonne point,** after the French mathematician J. D. Gergonne.)

7. If Γ is the circumscribed circle of $\triangle ABC$, show that the tangents to Γ at A, B, and C intersect the opposite sides of $\triangle ABC$ in three collinear points.

8. (A euclidean proof of Desargues' theorem.) Let $\triangle ABC$ and $\triangle A'B'C'$ be perspective from a point O and, as usual, let A'', B'', C'' be the intersections of BC and $B'C'$, CA and $C'A'$, AB and $A'B'$. Show that $\triangle ABC$ and $\triangle A'B'C'$ are also axially perspective by applying the theorem of Menelaus to $\triangle BCO$, $\triangle CAO$, and $\triangle ABO$ using, respectively, the transversals $B'C'A''$, $C'A'B''$, and $A'B'C''$.

7.6 Conclusion

This chapter has been devoted primarily to a discussion of the properties of complete four-points, sometimes called **complete quadrangles,** and complete four-lines, sometimes called **complete quadrilaterals.** In the course of this investigation, we introduced another new axiom, namely, the assertion that there is no complete four-point with collinear diagonal points. Subsequently, we encountered in a general setting the important notions of harmonic tetrads, involutory hexads, and involutions on a line which we first met in our study of E_2^+ and Π_2. Finally, we developed projective criteria for the concurrence of lines and the collinearity of points located, respectively, on the vertices and on the sides of a general triangle, and by suitably specializing these results we obtained the important theorems of Ceva and Menelaus, which play a fundamental role in advanced euclidean geometry.

eight ‖ *CONICS*

8.1 Introduction

Both from our work in analytic geometry and our earlier study of E_2^+ and Π_2 in this book, we already have considerable familiarity with conics and their fundamental properties. However, this past experience has been acquired exclusively in coordinatized systems, where the curves we studied could be described by equations and analyzed via algebraic techniques. Now we propose to discuss conics in a more general setting. Naturally, this will require new definitions, eventually to be reconciled with the old, and axiomatic rather than algebraic arguments. It is interesting, however, as we shall see in this chapter, that all the nonmetrical properties of conics can be derived as consequences of our axioms, without recourse to any equations.

8.2 Definitions and Fundamental Properties

In euclidean geometry, where the principle

of duality is not valid, we almost invariably regard a conic as a set of points satisfying certain definitive conditions, and lines related to the curve—such as tangents and polars—have only a secondary status. In projective geometry, where duality ensures that points are no more fundamental than lines, it is presumably (and actually) possible to study conics either as sets of points having interesting related lines or as sets of lines having interesting related points, i.e., either as **loci** or as **envelopes**:

Definition 1

A locus is a set of points to which an arbitrary point does or does not belong, according as it does or does not satisfy a specified condition.

Definition 2

An envelope is a set of lines to which an arbitrary line does or does not belong, according as it does or does not satisfy a specified condition.

Definition 3

The locus of the points of intersection of corresponding lines of two projective pencils is called a point-conic.

The vertices of the two pencils which generate a point-conic are called the **generating bases** of the conic (Fig. 8.1*a*). If the generating bases of a point-conic are distinct, and if the line joining these bases viewed as a line of one pencil does not correspond to itself viewed as a line of the other pencil, the point-conic is said to be **nonsingular.** All other point-conics are said to be **singular.** Singular conics, as here defined, are also singular in the usual sense; i.e., they consist of the points of one or more straight lines. In fact if U and V are distinct generating bases of a point-conic, and if the line UV as a line of the pencil on U corresponds to itself as a line of the pencil on V, then from Definition 3 it follows that every point of UV is a point of the conic. Moreover, in this case, since the projectivity between the pencil on U and the pencil on V has a self-corresponding line, namely, UV, the projectivity must be a perspectivity. Since corresponding lines in a perspectivity intersect on the axis of the perspectivity, the points of this axis are also points of the conic, which therefore consists of the points of two distinct lines (Fig. 8.2*a*). On the other hand, if the vertices of the two pencils coincide, then the single generating base is clearly a point of the conic. Furthermore, if the projectivity between the given cobasal pencils has one or more

self-corresponding lines,[1] then the points of every such line are also points of
the conic, and again the conic is singular in the ordinary sense (Fig. 8.2*b*).

Definition 4

The envelope of the lines joining corresponding points of two
projective ranges is called a line-conic.

The lines bearing the two ranges which generate a line-conic are
called the **generating bases** of the conic (Fig. 8.1*b*). If the generating bases
of a line-conic are distinct, and if their point of intersection is not self-
corresponding in the projectivity between the two ranges, the line-conic is

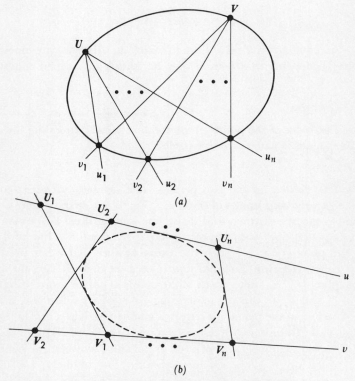

Fig. 8.1   The generation of nonsingular point-conics and line-conics.

[1] In the algebraic representation, we found that every projectivity between cobasal pencils,
or cobasal ranges, had two self-corresponding elements, either distinct or coincident. However, as we observed in our discussion of the fixed points of involutions in Sec. 7.4, we cannot
deduce this fact from the axioms we have thus far adopted. In the next section we shall add
another axiom which will guarantee the existence of two self-corresponding elements, either
distinct or coincident, in any projectivity between cobasal pencils or cobasal ranges.

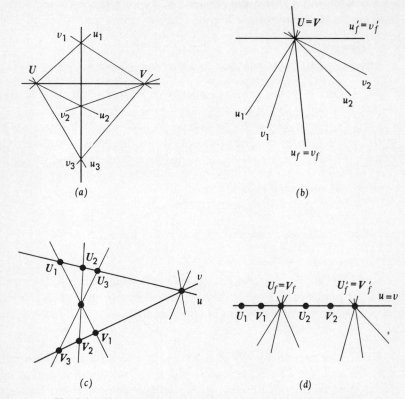

Fig. 8.2   The generation of singular point-conics and line-conics.

said to be **nonsingular.** All other line-conics are said to be **singular.** If u and v are distinct generating bases of a line-conic, and if the intersection of u and v is self-corresponding, then from Definition 4 it is clear that any line on the point uv is a line of the line-conic. Moreover, in this case the projectivity is a perspectivity, and every line on the center of perspectivity is also a line of the line-conic. Thus the conic consists of the lines of two distinct pencils (Fig. 8.2c). On the other hand, if the lines carrying the two ranges coincide, then clearly the single generating base, $u = v$, is a line of the conic. Furthermore, if the projectivity between the given cobasal ranges has one or more self-corresponding points, then all the lines on each of these points satisfy the condition of Definition 4 and are also lines of the line-conic (Fig. 8.2d).

It is clear (with the qualification noted in the last footnote) that every singular point-conic consists of the points of one or more ranges and every singular line-conic consists of the lines of one or more pencils. However, it is not obvious that nonsingular conics may not also consist of entire ranges or entire pencils. That this is not the case is an immediate consequence of our first few simple theorems.

Theorem 1

Every point-conic contains its generating bases.

Proof Let U and V be the generating bases of a point-conic, and consider the line UV as a line of the pencil on U. Obviously, it has for its image in the given projectivity *some* line of the pencil on V, and clearly it intersects this line in the point V, which is therefore a point of the conic. Similarly, considering UV as a line of the pencil on V, it has a corresponding line in the pencil on U which it intersects in the point U. Hence U is also a point of the point-conic, and our proof is complete.

Theorem 2

Every line-conic contains its generating bases.

Theorem 3

If U, V, A, B, C are five points no four of which are collinear, and if no two of the points A, B, C are collinear with either U or V, then there is a unique point-conic which contains A, B, C and has U and V as generating bases.

Proof Under the hypotheses of the theorem, it is clear that of the six lines UA, UB, UC and VA, VB, VC no two in either set can coincide and at most one line in the first set can coincide with a line in the second set. Hence the projectivity

$$U(A,B,C,\ldots) \sim V(A,B,C,\ldots)$$

is uniquely specified, and moreover, it defines a conic which contains A, B, C and has U and V as generating bases, as required.

Theorem 4

If u, v, a, b, c are five lines no four of which are concurrent, and if no two of the lines a, b, c are concurrent with either u or v, then there is a unique line-conic which contains a, b, c and has u and v as generating bases.

Theorem 5

Two distinct points of a nonsingular point-conic cannot be collinear with a generating base of the conic.

Proof Let U and V be the generating bases of a nonsingular point-conic, let A and B be any other points on the conic, and let us suppose first, contrary to the assertion of the theorem, that A is collinear with U and V. Then since A must be the intersection of a pair of corresponding lines in the projectivity between the pencil on U and the pencil on V, it follows that UA and UV must be corresponding lines in this projectivity. However, the existence of a self-corresponding line in the projectivity which generates a point-conic means that the conic is singular. Since our hypothesis is that the given conic is nonsingular, we thus have reached a contradiction, which proves that A cannot be collinear with the two generating bases, U and V.

Now suppose that A and B are collinear with one of the generating bases, say U. Then since A and B are both on the conic, the line l, which contains A, B, and U, must correspond both to the line VA and to the line VB, which is impossible since a projectivity is a one-to-one correspondence in which no element has two distinct images. Hence A and B cannot be collinear with a generating base, and our proof is complete.

Theorem 6

Two distinct lines of a nonsingular line-conic cannot be concurrent with a generating base of the conic.

Theorem 5 makes it clear that nonsingular point-conics do not consist of one or more ranges, and Theorem 6 makes it clear that nonsingular line-conics do not consist of one or more pencils.

From the definition of a point-conic, it appears that the generating bases, which by Theorem 1 are points of the conic, have special properties not possessed by other points of the conic. This is not the case, however, as the next theorem assures us.

Theorem 7

Any two points of a nonsingular point-conic can serve as generating bases of the conic.

Proof   Let U and V be the generating bases of a nonsingular point-conic, let A, B, C, D be any other points on the conic, let x be the line CD, and let:

A' be the intersection of UA and CD
B' be the intersection of UB and CD
A'' be the intersection of VA and CD
B'' be the intersection of VB and CD

(Fig. 8.3). Then since two distinct points on a nonsingular point-conic cannot be collinear with a generating base, it follows that neither U nor V

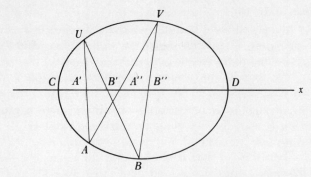

Fig. 8.3   Figure used in the proof of Theorem 7.

can lie on the line $x = CD$ and that of the six points A', B', A'', B'', C, D, the only ones which can possibly coincide are (A',B'') and (A'',B').

Now from the definition of a point-conic,

$$U(A,B,C,D) \sim V(A,B,C,D)$$

Also

$$U(A,B,C,D) \; \overline{\wedge} \; x(A',B',C,D) \qquad \text{and} \qquad V(A,B,C,D) \; \overline{\wedge} \; x(A'',B'',C,D)$$

Hence, $x(A',B',C,D) \sim x(A'',B'',C,D)$, and further, by the first permutation theorem, $x(A',B',C,D) \sim x(B'',A'',D,C)$. Now this projectivity contains a pair of reciprocally corresponding points, namely, (C,D), and therefore it is an involution in which, necessarily, every point and its image are also reciprocally corresponding. Thus, since A'' corresponds to B', it follows that B' corresponds to A'', and we can write $x(A',A'',C,D) \sim x(B'',B',D,C)$. From this projectivity, again using the first permutation theorem, we have

$$x(A',A'',C,D) \sim x(B',B'',C,D)$$

Now $x(A',A'',C,D) \; \overline{\wedge} \; A(U,V,C,D)$, and $x(B',B'',C,D) \; \overline{\wedge} \; B(U,V,C,D)$. Hence, combining these two perspectivities with the last projectivity, we obtain

$$A(U,V,C,D) \sim B(U,V,C,D)$$

Similarly, of course, if E is any other point on the conic, we can show that

$$A(U,V,C,E) \sim B(U,V,C,E)$$

Moreover, since the last two projectivities have three pairs of corresponding lines in common, they must be the same projectivity. Hence we can write

$$A(U,V,C,D,E,...) \sim B(U,V,C,D,E,...)$$

which proves that A and B, which were any two points on the given conic, can be used as generating bases, as asserted.

Corollary 1

Any two points on a singular point-conic consisting of two ranges can be used as generating bases, provided the points belong to the same range and neither one is the point common to the two ranges.

Corollary 2

A singular point-conic consisting of a single range cannot have two generating bases, but any one of its points can serve as the single generating base of the conic.

Corollary 3

There is a unique point-conic containing five given points no four of which are collinear.

Corollary 4

No three points of a nonsingular point-conic can be collinear.

Theorem 8

Any two lines of a nonsingular line-conic can serve as generating bases of the conic.

According to Corollary 3 of Theorem 7, there is, except in very special cases, a unique conic containing five given points. Hence, if six points are given, there is in general no conic which contains them all. There may be, however, and the next theorem gives us a useful test for determining when this will be the case.

Theorem 9

Six points, A, B, C, D, E, F, no three of which are collinear, will all belong to the same nonsingular point-conic if and only if

$$A(C,D,E,F) \sim B(C,D,E,F)$$

Proof Let A, B, C, D, E, F be six points no three of which are collinear, and let us suppose first that there is a point-conic which contains them all. Then, since any two points of a nonsingular point-conic can be used as generating bases, it follows that $A(C,D,E,F) \sim B(C,D,E,F)$, which proves the "only if" assertion of the theorem.

Next, let us suppose that the six points are such that $A(C,D,E,F) \sim B(C,D,E,F)$, but, contrary to the theorem, let us suppose that there is no conic which contains them all. In any event, however, there is a unique conic, Γ, which contains the five points A, B, C, D, E. Moreover, the line AF

has *some* line of the pencil on B as its image in the projectivity which defines Γ, and these lines intersect in a point $F' \neq F$ belonging to Γ (Fig. 8.4). Thus in the projectivity defining Γ, we have $A(C,D,E,F') \sim B(C,D,E,F')$. This, however, has three pairs of corresponding lines in common with the projectivity we are given, and hence these two projectivities are the same. Therefore, since AF' and AF are the same line, and since, by hypothesis, BF is the line which corresponds to AF, it follows that BF' and BF are the same line. Thus we have the lines $AF (= AF')$ and $BF (= BF')$ intersecting in the points F and F', one of which is on Γ, the other of which is not. This contradiction forces us to abandon the supposition that the sixth point, F, does not belong to Γ. In other words, there is a conic, Γ, containing each of the six given points, and the "if" assertion of the theorem is also established.

Corollary 1

If A, B, C, D are four points in E_2^+ or in Π_2 no three of which are collinear, the locus of a point P with the property that the cross ratio of the lines joining it to A, B, C, D is a constant is a nonsingular point-conic containing A, B, C, D.

Theorem 10

Six lines, a, b, c, d, e, f, no three of which are concurrent, will all belong to the same nonsingular line-conic if and only if $a(c,d,e,f) \sim b(c,d,e,f)$.

Example 1

$A:(1,0)$, $B:(-1,0)$, $C:(0,1)$, $D:(0,-1)$ are four points in the euclidean plane. Find the locus of a point P if the cross ratio of the four lines PA, PB, PC, PD is a constant, r.

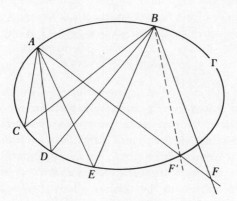

Fig. 8.4 Figure used in the proof of Theorem 9.

Let the coordinates of a general point P be (x_0, y_0). Then $x - x_0 = 0$ and $y - y_0 = 0$ are two distinct lines on P, and any line of the pencil on P has an equation of the form $\lambda(x - x_0) + \mu(y - y_0) = 0$. In particular,

The equation of PA is $\quad y_0(x - x_0) + (1 - x_0)(y - y_0) = 0$.
The equation of PB is $-y_0(x - x_0) + (1 + x_0)(y - y_0) = 0$.
The equation of PC is $\quad (1 - y_0)(x - x_0) + x_0(y - y_0) = 0$.
The equation of PD is $\quad (1 + y_0)(x - x_0) - x_0(y - y_0) = 0$.

Computing the cross ratio of the four lines PA, PB, PC, PD from their parameters, as read from the last four equations, we have

$$r = \frac{\begin{vmatrix} y_0 & 1 - y_0 \\ 1 - x_0 & x_0 \end{vmatrix} \cdot \begin{vmatrix} -y_0 & 1 + y_0 \\ 1 + x_0 & -x_0 \end{vmatrix}}{\begin{vmatrix} y_0 & 1 + y_0 \\ 1 - x_0 & -x_0 \end{vmatrix} \cdot \begin{vmatrix} -y_0 & 1 - y_0 \\ 1 + x_0 & x_0 \end{vmatrix}} = \frac{1 - (x_0 + y_0)^2}{1 - (x_0 - y_0)^2}$$

Hence, simplifying and then dropping the subscripts, the locus of P is the curve whose equation is

(1) $$x^2 + 2\,\frac{1 + r}{1 - r}\,xy + y^2 = 1$$

This is a hyperbola if $r > 0$, a parabola if $r = 0$, and an ellipse if $r < 0$. It is a circle if and only if $r = -1$, that is, if and only if the lines PA, PB, PC, PD form a harmonic tetrad, in that order.

If $r = 0$, the equation of the conic factors into

$$(x + y - 1)(x + y + 1) = 0$$

and the conic thus consists of the lines AC and BD. If $r = 1$, the equation reduces to $xy = 0$, and the conic consists of the lines AB and CD. If r is infinite, the equation of the conic factors into $(x - y - 1)(x - y + 1) = 0$, and the conic consists of the lines AD and BC. Thus the values of r corresponding to singular tetrads (Sec. 4.7) also correspond to the singular conics in the family (1). For every value of r, however, the conic (1) contains the four given points A, B, C, D.

If we are given three pairs of corresponding lines in a projectivity between two pencils, it is possible, as we know (see Exercise 6, Sec. 4.6) to construct the line of the second pencil which corresponds to any given line of the first pencil. Using this construction to implement Definition 3, it is clearly possible to construct as many points as desired on the unique non-singular conic determined by five points no three of which are collinear. Other ways for the point-by-point construction of the conic on five given points are also available, and we shall investigate two of them in subsequent

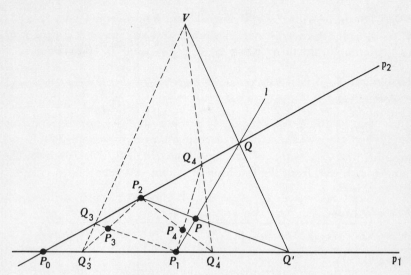

Fig. 8.5 Maclaurin's conic construction.

sections of this chapter. The most convenient of all these constructions, however, is probably the one due to Colin Maclaurin (1698–1746), which we shall now describe.

Let P_0, P_1, P_2, P_3, P_4 be five points no three of which are collinear, let any three of these points, say P_0, P_1, P_2, be selected, and let p_1 and p_2 be the lines P_0P_1 and P_0P_2, respectively. Let Q_3 and Q_3' be the points in which P_1P_3 and P_2P_3 intersect p_2 and p_1, respectively (Fig. 8.5). Similarly, let Q_4 and Q_4' be the points in which P_1P_4 and P_2P_4 intersect p_2 and p_1, respectively. Finally, let V be the intersection of Q_3Q_3' and Q_4Q_4', let Q be the intersection of p_2 and an arbitrary line, l, on P_1, and let Q' be the intersection of VQ and p_1. Then the intersection, P, of P_2Q' and $l = P_1Q$ is the second intersection of l and the conic determined by the five given points. To justify this construction, we note that

$$P_1(P_0,Q_3,Q_4,Q) \overset{p_2}{\wedge} V(P_0,Q_3',Q_4',Q') \overset{p_1}{\wedge} P_2(P_0,Q_3',Q_4',Q')$$

that is, $P_1(P_0,Q_3,Q_4,Q,\ldots) \sim P_2(P_0,Q_3',Q_4',Q',\ldots)$. By Definition 3, the locus of the intersections of corresponding lines in this projectivity is a conic on the generating bases, P_1 and P_2. Moreover, the intersection of P_1P_0 and P_2P_0 is P_0, the intersection of P_1Q_3 and P_2Q_3' is P_3, and the intersection of P_1Q_4 and P_2Q_4' is P_4. Hence this conic also contains P_0, P_3, P_4. Therefore the point of intersection of P_1Q and P_2Q', namely, P, is also on the conic determined by the five given points, as required. Additional properties of this construction will be found in the exercises.

Exercises

1. Show that if every point-conic which contains four given points is singular, then at least three of the points are collinear.

2. How many singular point-conics are there which contain the vertices of a given four-point?

3. Show that in PG_n there are exactly n points on every nonsingular point-conic. How many points are there on a singular point-conic in PG_n?

4. Prove Corollary 1, Theorem 7.

5. Prove Corollary 2, Theorem 7.

6. Prove Corollary 3, Theorem 7.

7. Prove Corollary 4, Theorem 7.

8. Prove Corollary 1, Theorem 9.

9. Given the points $A:(1,2)$, $B:(1,1)$, $C:(2,0)$, $U:(0,1)$, and $V:(0,0)$ in E_2. Letting the pencil on U be parametrized in terms of the lines $x = 0$ and $y - 1 = 0$ and the pencil on V be parametrized in terms of the lines $x = 0$ and $y = 0$, find the equation of the projectivity in which $UA \sim VA$, $UB \sim VB$, $UC \sim VC$. Using this, find the equation of the conic which is determined by the points A, B, C, U, V.

10. Work Exercise 9 if A, B, C are, respectively, $(2,1)$, $(1,0)$, $(-1,3)$.

11. Work Example 1 if A, B, C, D are, respectively, the points $(0,0)$, $(1,1)$, $(3,0)$, $(0,-1)$.

12. Show that in E_2 every conic of the family

$$\lambda(x^2 + 3y^2 - x + 3y - 6) + \mu(x^2 + 2xy - 3x) = 0$$

contains the points $A:(0,1)$, $B:(1,1)$, $C:(3,0)$, $D:(0,-2)$. If P is an arbitrary point on the conic corresponding to the values (λ,μ), and if p_a, p_b, p_c, p_d are, respectively, the lines PA, PB, PC, PD, verify that $R(p_a p_b, p_c p_d)$ is independent of P and find its value in terms of λ and μ.

13. In Maclaurin's construction in E_2, let P_0, P_1, and P_2 be the points $(0,0)$, $(a,0)$, and (b,c), and let the point V determined from the five points P_0, P_1, P_2, P_3, P_4 have coordinates (u,v). Show that the equation of the conic determined by P_0, P_1, P_2, P_3, P_4 is

$$c^2vx^2 + (ac^2 - 2bcv)xy + (b^2v + acu - abv - abc)y^2$$
$$- ac^2vx + (2abcv - ac^2u)y = 0$$

14. Using the results of Exercise 13, show that the conic Γ determined by five points, P_0, P_1, P_2, P_3, P_4, is an ellipse, a parabola, or a hyperbola according as the point V lies inside, on, or outside the hyperbola whose equation is $4bv^2 - 4cuv + ac^2 = 0$. Determine the position of V when Γ is the circle on P_0, P_1, P_2, and devise a construction for V in this case.

15. Show that the locus of the points V corresponding to the conics on four points, P_0, P_1, P_2, P_3, is a straight line. Show, further, that if this line intersects the hyperbola defined in Exercise 14 in two points, there are two different parabolas on P_0, P_1, P_2, P_3 and devise a construction for the direction of the axes of these parabolas. What can be said about the conics on P_0, P_1, P_2, P_3 if the locus of V is tangent to the hyperbola of Exercise 14? If the locus of V does not intersect the hyperbola?

8.3 The Intersections of a Line and a Conic

If P is a point on a nonsingular conic in the euclidean plane, it is a familiar fact that there is one and only one line through P which has no other point in common with the conic, i.e., is tangent to the conic. That this is also a property of conics in any system satisfying Axioms 1 to 8 is guaranteed by the next theorem.

Theorem 1

Through every point of a nonsingular point-conic there is one and only one line which meets the conic in no other point.

Proof Let P and P' be any two points on a nonsingular conic Γ. Then, since any two points on a nonsingular point-conic can be generating bases, there is a projectivity between the pencil on P and the pencil on P' such that corresponding lines in the projectivity intersect in points of Γ. Now consider the line $p' = P'P$ as a line of the pencil on P'. To it, of course, there corresponds some line of the pencil on P, say p, and let us suppose that p intersects Γ in a second point, Q, distinct from P (Fig. 8.6). Then by definition, the line $P'Q$ must be the line of the pencil on P' which corresponds to the line p of the pencil on P. But by hypothesis the line on P' which corresponds to p is $P'P$. Moreover, $P'P$ and $P'Q$ are distinct lines, since P and Q are distinct points and $P'P$ cannot have a third intersection with Γ. Thus there are two lines corresponding to p in the projectivity which generates Γ, which is impossible. Hence we must abandon the supposition that the line p, corresponding to $P'P$, meets Γ in a second point. Since P was an arbitrary point on Γ, this proves that at any point of a nonsingular point-conic there is at least one line which is tangent to the conic.

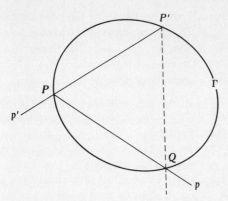

Fig. 8.6 Figure used in the proof of
Theorem 1.

To prove that at each point of a nonsingular conic there is at most
one tangent to the conic, let us assume that at some point P of a nonsingular
conic Γ there are two such lines, say p_1 and p_2, and let us take P and any
other point, P', as generating bases of Γ. Then, either from the preceding
argument or from the definition of a conic, it follows that both p_1 and p_2
must correspond to the line $P'P$ in the projectivity which generates Γ. Since
this is impossible, we must abandon the supposition that there are two tangents to Γ at P, and our proof is complete.

Corollary 1

If A, B, C, D, and P are five points on a nonsingular point-conic Γ,
and if A', B', C', D' are, respectively, points on the tangents to Γ at A, B, C, D
which are distinct from A, B, C, D, then

$$P(A,B,C,D) \sim A(A',B,C,D) \sim B(A,B',C,D) \sim C(A,B,C',D) \sim D(A,B,C,D')$$

Corollary 2

If A, B, C, D are four points no three of which are collinear, and if
t is a line on A which does not pass through B, C, or D, there is a unique conic
which passes through B, C, and D and is tangent to t at A.

Corollary 3

If A, B, C are three noncollinear points, if t_a is a line on A which
does not pass through B or C, and if t_b is a line on B which does not pass
through A or C, there is a unique conic which passes through C, is tangent
to t_a at A, and is tangent to t_b at B.

Theorem 2

On every line of a nonsingular line-conic there is one and only one point which is on no other line of the conic.

Just as a line which passes through one and only one point of a nonsingular point-conic is called a tangent line of the point-conic, so a point which is on one and only one line of a nonsingular line-conic is called a **tangent point** of the line-conic.

We have already seen (Corollary 4, Theorem 8, Sec. 8.2) that no line can intersect a nonsingular point-conic in more than two points. Moreover, from the definition of a conic it is obvious that there are lines which meet a nonsingular conic in exactly two points, and from Theorem 1 it follows that there exist lines which have exactly one point in common with a nonsingular point-conic. This raises the question of whether there are lines which have no point in common with a given conic, as in E_2 and E_2^+, or whether, as in Π_2, every line meets every nonsingular point-conic in at least one point. Since both E_2^+ and Π_2 satisfy Axioms 1 to 8, the answer is immediate. *There are some systems satisfying Axioms 1 to 8 in which every line meets every nonsingular point-conic in at least one point, and there are other systems satisfying Axioms 1 to 8 for which this is not the case.* Thus if we are interested, as we are, in developing a system in which every line will intersect every conic in at least one point, it is necessary to ensure it by adopting an appropriate new axiom. The one we add to our system for this purpose makes no reference to conics, but, as we shall soon see, it suffices to guarantee the intersection property we want.

Axiom 9

Every projectivity between cobasal ranges has at least one self-corresponding point.

The consistency of this axiom with the others we have adopted is clear from a consideration of Π_2. In fact, in Π_2 the general projectivity between cobasal ranges is defined by an equation of the form

$$a\lambda\lambda' + b\lambda\mu' + c\mu\lambda' + d\mu\mu' = 0$$

and the self-corresponding points of this projectivity are determined by the condition $(\lambda',\mu') = (k\lambda,k\mu)$, $k \neq 0$. This leads at once to the quadratic equation

$$a\lambda^2 + (b + c)\lambda\mu + d\mu^2 = 0$$

which always has at least one solution in Π_2, where coordinates may take on complex values.

To make sure that the principle of duality is still valid after the addition of Axiom 9, we must now prove the dual of our new axiom.

Theorem 3

Every projectivity between cobasal pencils has at least one self-corresponding line.

Proof Let $P(a_1,a_2,a_3,...)$ and $P(b_1,b_2,b_3,...)$ be two cobasal pencils which are projective with each other, and let p be any line which is not on P. Let A_1, A_2, A_3, ... be the points in which p intersects a_1, a_2, a_3, ... , and let B_1, B_2, B_3, ... be the points in which p intersects b_1, b_2, b_3, ... Then

$$p(A_1,A_2,A_3,...) \barwedge P(a_1,a_2,a_3,...) \sim P(b_1,b_2,b_3,...) \barwedge p(B_1,B_2,B_3,...)$$

Hence $p(A_1,A_2,A_3,...) \sim p(B_1,B_2,B_3,...)$. By Axiom 9, this projectivity between cobasal ranges on p has at least one self-corresponding point, F, and clearly the line PF is self-corresponding in the projectivity between the two pencils on P. Hence the given projectivity has at least one self-corresponding line, as asserted.

With Axiom 9 available, we can now prove that every line meets every nonsingular conic in at least one point:

Theorem 4

Every line has at least one point in common with every non-singular point-conic.

Proof Let Γ be an arbitrary nonsingular point-conic, let x be an arbitrary line, let U and U' be two points of Γ which are not points of x, and let U and U' be taken as generating bases of Γ. Then there is a projectivity $U(u_1,u_2,u_3,...) \sim U'(u_1',u_2',u_3',...)$ such that the intersections of corresponding lines (u_1,u_1'), (u_2,u_2'), (u_3,u_3'), ... are points of Γ. Let X_1, X_2, X_3, ... be the points in which x intersects the lines u_1, u_2, u_3, ... , and let X_1', X_2', X_3', ... be the points in which x intersects the lines u_1', u_2', u_3', ... (Fig. 8.7). Then

$$x(X_1,X_2,X_3,...) \barwedge U(u_1,u_2,u_3,...) \sim U'(u_1',u_2',u_3',...) \barwedge x(X_1',X_2',X_3',...)$$

and hence $x(X_1,X_2,X_3,...) \sim x(X_1',X_2',X_3',...)$. By Axiom 9, this projectivity between cobasal ranges has at least one self-corresponding point. Moreover, if a point X_i, say, is self-corresponding, it must be the intersection of u_i and u_i'; that is, it must be a point of Γ. Hence x, which was an arbitrary line, must have at least one point in common with Γ, as asserted.

Corollary 1

Every line has at least one point in common with every point-conic, singular or nonsingular.

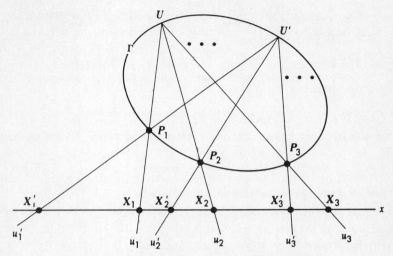

Fig. 8.7 Figure used in the proof of Theorem 4.

Theorem 5

Every point has at least one line in common with every nonsingular line-conic.

It is interesting to note that with the introduction of Axiom 9, all finite projective geometries have been eliminated from the collection of systems we are studying. To see this, we recall from Exercise 3, Sec. 8.2, that in any PG_n satisfying Axioms 1 to 8 every nonsingular point-conic contains exactly n points. Hence, pairing these points in all possible ways, there are exactly $n(n-1)/2$ lines which meet a given conic in two distinct points. In addition to these lines, there are, by Theorem 1, n other lines which meet the conic in a single point. Therefore the number of lines in PG_n which meet a given nonsingular point-conic is $n(n-1)/2 + n = n(n+1)/2$. However, according to Exercise 2, Sec. 6.2, the total number of lines in PG_n is $n^2 - n + 1$. Hence, subtracting, we find that there are

$$n^2 - n + 1 - n(n+1)/2 = [(n-1)(n-2)]/2$$

lines of PG_n which have nothing in common with a given nonsingular point-conic. Since this number is positive if $n \geq 3$, we thus have a contradiction of Theorem 4 in every PG_n satisfying Axioms 1 to 8, which shows that in none of them is Axiom 9 satisfied.

Exercises

1. Prove Corollary 1, Theorem 1.

2. Prove Corollary 2, Theorem 1.

3. Prove Corollary 3, Theorem 1.

4. Prove Theorem 2 without using the principle of duality.

5. State the duals of Corollaries 1, 2, and 3 of Theorem 1.

6. Prove Theorem 5 without using the principle of duality.

7. Prove that every involution has exactly two distinct self-corresponding points.

8. Prove that two involutions on the same line or in the same pencil always have one and only one pair of mates in common.

9. Given $A:(0,1)$, $B:(0,0)$, $C:(1,1)$, $D:(2,0)$, and $t_a: x - y + 1 = 0$ in E_2. Letting the pencil on A be parametrized in terms of the lines $x = 0$ and $y - 1 = 0$ and the pencil on B be parametrized in terms of the lines $x = 0$ and $y = 0$, find the equation of the projectivity in which $AC \sim BC$, $AD \sim BD$, $t_a \sim BA$. From this, find the equation of the conic which passes through B, C, D and is tangent to t_a at A.

10. Given $A:(1,2)$, $B:(3,1)$, $C:(0,0)$, $t_a: x - y + 1 = 0$, and $t_b: x - 3 = 0$ in E_2. Letting the pencil on A be parametrized in terms of the lines $x - 1 = 0$ and $y - 2 = 0$ and the pencil on B be parametrized in terms of the lines $x - 3 = 0$ and $y - 1 = 0$, find the equation of the projectivity in which $AC \sim BC$, $t_a \sim BA$, $t_b \sim AB$. From this, find the equation of the conic which passes through C, is tangent to t_a at A, and is tangent to t_b at B.

8.4 Desargues' Conic Theorem

We have already noted the fundamental importance in projective geometry of Desargues' theorem on perspective triangles. Another result due to Desargues, known as **Desargues' conic theorem,** is also of great utility:

Theorem 1

If A_0, A_1, A_2, A_3 are the vertices of a complete four-point, and if x is any line which does not contain a vertex of this four-point, then every point-conic which passes through A_0, A_1, A_2, A_3 intersects x in points which are mates in the same involution on x.

Proof Let A_0, A_1, A_2, A_3 be the vertices of a complete four-point, and let x be any line which does not pass through A_0, A_1, A_2, or A_3. With our usual notation for involutory hexads, let X_1, X_2, X_3 be, respectively, the intersections of x and A_0A_1, A_0A_2, A_0A_3, and let X_1', X_2', X_3' be, respectively, the

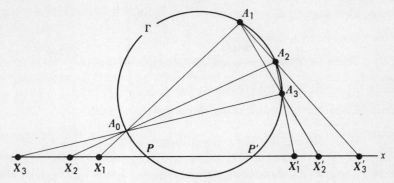

Fig. 8.8   Figure used in the proof of Desargues' conic theorem.

intersections of x and A_2A_3, A_1A_3, A_1A_2. Finally, let P and P' be the inter-sections of x and any point-conic, Γ, on A_0, A_1, A_2, A_3 (Fig. 8.8). Then, taking A_0 and A_2 as generating bases of Γ, we have

$$A_0(A_1,A_3,P,P') \sim A_2(A_1,A_3,P,P')$$

Also, $\qquad\qquad A_0(A_1,A_3,P,P') \overline{\wedge} x(X_1,X_3,P,P')$

and $A_2(A_1,A_3,P,P') \overline{\wedge} x(X'_3,X'_1,P,P') \sim x(X'_1,X'_3,P',P)$, the last step following by the first permutation theorem. Hence, combining these projectivities, we obtain $x(X_1,X_3,P,P') \sim x(X'_1,X'_3,P',P)$. Since the last projectivity has a pair of reciprocally corresponding points, it follows that it is an involution. Therefore P and P' are mates in the unique involution, T, determined by the two pairs of corresponding points (X_1,X'_1) and (X_3,X'_3). Since X_1, X'_1, X_3, X'_3 are independent of the particular conic on A_0, A_1, A_2, A_3, it follows simi-larly that the intersections of x and any other nonsingular conic on these four points are also mates in T. Furthermore, the three singular conics on A_0, A_1, A_2, A_3, namely, the conics consisting of the pairs of lines (A_0A_1,A_2A_3), (A_0A_2,A_1A_3), (A_0A_3,A_1A_2), intersect x in the pairs of points (X_1,X'_1), (X_2,X'_2), (X_3,X'_3), which are obviously pairs of mates in T. This completes our proof.

Corollary 1

If A_0, A_1, A_2, A_3 are the vertices of a complete four-point, and if x is any line which does not pass through any of these points, then any point and its mate in the involution determined on x by the involutory hexad in which x intersects the given four-point lie on a point-conic which contains A_0, A_1, A_2, A_3.

Corollary 2

If A_0, A_1, A_2, A_3 are four points no three of which are collinear, and if x is any line which does not pass through any of these points, there are two and only two nonsingular point-conics on A_0, A_1, A_2, A_3 which are tangent to x.

If two of the four points involved in Desargues' conic theorem, say A_0 and A_1, coincide while at the same time the line A_0A_1 remains defined, we have an interesting special case of the theorem:

Theorem 2

If A_1, A_2, A_3 are three noncollinear points, if t is any line on A_1 which does not pass through A_2 or A_3, and if x is any line which does not contain A_1, A_2, or A_3, then every point-conic on A_2 and A_3 which is tangent to t at A_1 intersects x in points which are pairs of mates in the same involution on x.

Proof Let A_1, A_2, A_3 be three noncollinear points, let t be a line on A_1 but not on A_2 or A_3, let A_1' be a point on t distinct from A_1, and let x be a line which does not pass through A_1, A_2, or A_3. Also let X_1 be the intersection of x and t, let X_1', X_2', X_3' be, respectively, the intersections of x and the lines A_2A_3, A_1A_3, A_1A_2, and let P and P' be the intersections of x and any conic passing through A_2 and A_3 and tangent to t at A_1 (Fig. 8.9). Then from Corollary 1, Theorem 1, Sec. 8.3, we have

$$A_1(A_1',A_3,P,P') \sim A_2(A_1,A_3,P,P')$$

Also $\qquad A_1(A_1',A_3,P,P') \mathbin{\overline{\wedge}} x(X_1,X_2',P,P')$

and $\qquad A_2(A_1,A_3,P,P') \mathbin{\overline{\wedge}} x(X_3',X_1',P,P') \sim x(X_1',X_3',P',P)$

the last step following from the first permutation theorem. Hence, combining

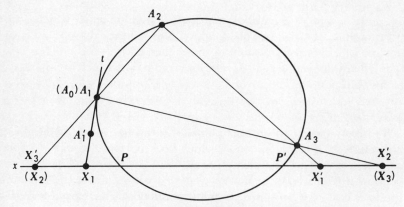

Fig. 8.9   Figure used in the proof of Theorem 2.

these projectivities, we obtain

$$x(X_1,X_2',P,P') \sim x(X_1',X_3',P',P)$$

which shows that P and P' are mates in the unique involution, T, in which (X_1,X_1') and (X_2',X_3') are pairs of corresponding points. This time there are only two singular conics in the family of conics defined by the conditions of the theorem, namely, the conics, consisting of the points on the pairs of lines (t,A_2A_3) and (A_1A_2,A_1A_3). Since the intersections of these conics and the line x, namely, the pairs of points (X_1,X_1') and (X_3',X_2') are obviously pairs of mates in T, the theorem is established.

Corollary 1

If A_1, A_2, A_3 are three noncollinear points, if t is any line on A_1 which does not pass through A_2 or A_3, and if x is any line which does not pass through A_1, A_2, or A_3, then any point and its mate in the involution on x in which the intersections of x with t and A_2A_3 and the intersections of x with A_1A_2 and A_1A_3 are pairs of mates lie on a point-conic which passes through A_2 and A_3 and is tangent to t at A_1.

Corollary 2

If A_1, A_2, A_3 are three noncollinear points, if t is any line on A_1 which does not pass through A_2 or A_3, and if x is any line which is not on A_1, A_2, or A_3, then there are two and only two nonsingular point-conics which contain A_2 and A_3, are tangent to t at A_1, and are tangent to x.

By an argument very much like the one we used in the proof of the last theorem, the following theorem can be established.

Theorem 3

If A_1 and A_2 are distinct points, if t_1 is a line on A_1 which does not pass through A_2, if t_2 is a line on A_2 which does not pass through A_1, and if x is any line which does not contain A_1 or A_2, then every conic which is tangent to t_1 at A_1 and tangent to t_2 at A_2 intersects x in points which are mates in the same involution on x.

Corollary 1

If A_1 and A_2 are distinct points, if t_1 is a line on A_1 which does not pass through A_2, if t_2 is a line on A_2 which does not pass through A_1, and if x is any line which is not on either A_1 or A_2, then any point and its mate in the involution on x in which the intersections of x with t_1 and t_2 correspond and the intersection of x and A_1A_2 is self-corresponding lie on a conic which touches t_1 at A_1 and touches t_2 at A_2.

Corollary 2

If A_1 and A_2 are distinct points, if t_1 is a line on A_1 which is not on A_2, if t_2 is a line on A_2 which is not on A_1, and if x is any line which is on neither A_1 nor A_2, then there is one and only one nonsingular conic which touches t_1 at A_1, touches t_2 at A_2, and is tangent to x.

Example 1

A, B, C, I, J are five points no three of which are collinear. A', B', C' are, respectively, the points in which BC, CA, AB intersect the line IJ, and A'', B'', C'' are the harmonic conjugates of A', B', C' with respect to I and J. If P is an arbitrary point, and if A_1, B_1, C_1 are, respectively, the points of intersection of the pairs of lines (PA'',BC), (PB'',AC), (PC'',AB), prove that A_1, B_1, C_1 are collinear if and only if P is a point of the conic determined by the points A, B, C, I, J.

Let Γ be the unique nonsingular point-conic which contains the five given points A, B, C, I, J, and let us suppose first that P is a point of Γ. Then if A_2, B_2, C_2 are, respectively, the points in which PA, PB, PC intersect IJ, it follows, by applying Desargues' conic theorem to the four-point $PABC$ and the line IJ, that (A',A_2), (B',B_2), (C',C_2) are pairs of mates in an involution on IJ. That is,

$$(A',B',C',I,J) \sim (A_2,B_2,C_2,J,I)$$

Also, it follows from the definition of A'', B'', C'' that (A',A''), (B',B''), (C',C'') are pairs of mates in an involution on IJ whose fixed points are I and J. Thus

$$(A',B',C',I,J) \sim (A'',B'',C'',I,J)$$

Hence, combining these two projectivities, we have

$$(A'',B'',C'',I,J) \sim (A_2,B_2,C_2,J,I)$$

Moreover, this projectivity has (I,J) as a pair of reciprocally corresponding points and is therefore an involution. Hence $(A''B''C'',A_2B_2C_2)$ is an involutory hexad, and therefore, by Theorem 2, Sec. 7.5, A_1, B_1, C_1 are collinear, as asserted.

On the other hand, if P is not a point of Γ, then I and J are not reciprocally corresponding points in the involution which is cut on IJ by the conics on A, B, C, P. In particular, let K $(\neq J)$ be the second intersection of IJ and the conic on A, B, C, P, I, let L $(\neq I)$ be the second intersection of IJ and the conic on A, B, C, P, J, and let M be the harmonic conjugate of K with respect to I and J. Then (A',A_2), (B',B_2), (C',C_2), (I,K), (J,L) are pairs of mates in an involution on IJ, and therefore

$$(A',B',C',I,J,K) \sim (A_2,B_2,C_2,K,L,I)$$

Also, we still have the involution

$$(A',B',C',I,J,K) \sim (A'',B'',C'',I,J,M)$$

Hence, combining these two projectivities, we have

$$(A'',B'',C'',I,J,M) \sim (A_2,B_2,C_2,K,L,I)$$

In this projectivity, K corresponds to I, but I does not correspond to K. Hence this projectivity is not an involution, $(A''B''C'',A_2B_2C_2)$ is not an involutory hexad, and therefore A_1, B_1, C_1 are not collinear.

This result has an interesting euclidean specialization if we take I and J to be the circular points at infinity. In this case the conic on A, B, C, I, J becomes the circle on A, B, C, the lines PA'', PB'', PC'' become the perpendiculars from P to the sides BC, CA, AB of $\triangle ABC$ (see Exercise 26, Sec. 4.7), and we have **Simpson's theorem:** *The feet of the perpendiculars from a point P to the sides of an arbitrary triangle are collinear if and only if P is on the circle determined by the vertices of the triangle.*

Among other things, Desargues' conic theorem provides us with another method for the point-by-point construction of a nonsingular conic when five points on the conic are given. Although not as simple as Maclaurin's construction, it is nonetheless worth noting:

Let A_0, A_1, A_2, A_3 and P be five points no three of which are collinear, let x be an arbitrary line on P, and let it be required to find the second intersection of x and the nonsingular conic, Γ, determined by A_0, A_1, A_2, A_3, and P. If X_1 and X_1' are, respectively, the intersections of x with A_0A_1 and A_2A_3, and if X_2 and X_2' are, respectively, the intersections of x with A_0A_2 and A_1A_3, then (X_1,X_1'), (X_2,X_2'), together with P and the required second intersection of x and Γ, say P', are pairs of mates in an involution on x. Thus $(X_1X_2P,X_1'X_2'P')$ is an involutory hexad, and since the five points X_1, X_2, P, X_1', X_2' are known, the sixth point, P', can be found by Theorem 1, Sec. 7.4.

Exercises

1. Prove Corollary 1, Theorem 1.

2. Prove Corollary 2, Theorem 1.

3. Prove Corollary 1, Theorem 2.

4. Prove Corollary 2, Theorem 2.

5. Prove Theorem 3.

6. Prove Corollary 1, Theorem 3.

7. Prove Corollary 2, Theorem 3.

8. Let T be an involution on a line x, and let A_0, A_1, A_2 be three non-collinear points none of which is on x. Show that all point-conics on A_0, A_1, A_2 and a pair of mates in T are also on a fourth point, A_3.

9. Show that there is no set of four points in E_2 through which a unique parabola can be passed. Is it possible to have four points in E_2 through which no parabola can be passed?

10. If A_0, A_1, A_2, A_3 are four points of E_2 no three of which are collinear, and if l is a line which does not pass through any of these points, is there always a conic containing A_0, A_1, A_2, A_3 which is tangent to l?

11. Let $A_0A_1A_2A_3$ and $A_0'A_1'A_2'A_3'$ be two simple four-points, and let x be a line which contains none of these points. Show that there is one and only one pair of points, (P,Q), on x such that simultaneously A_0, A_1, A_2, A_3, P, Q lie on a point-conic and A_0', A_1', A_2', A_3', P, Q also lie on a point-conic.

12. Let D_1 and D_2 be two diagonal points of a complete four-point $A_0A_1A_2A_3$. Show that any point-conic on A_0, A_1, A_2, A_3 intersects D_1D_2 in a pair of points which are harmonic conjugates with respect to D_1 and D_2.

13. Let A, B, C, D be four points no three of which are collinear, let a be a line on A which is not on B, C, or D, and let d be a line which is on D but not on A, B, or C. Devise a construction for the second intersection of d and the conic which contains B, C, D and is tangent to a at A.

14. Let A, B, C be three noncollinear points, let a be a line on A but not on B or C, let b be a line on B but not on A or C, let c be a line on C but not on A or B, and let Γ be the unique conic which passes through C, is tangent to a at A, and is tangent to b at B. Devise a construction for the second intersection of c and Γ.

15. Show that in any complete four-point, A_0, A_1, H_2, H_3, H_2', H_3' all lie on a nonsingular point-conic. Show also that A_2, A_3, H_2, H_3, H_2', H_3' all lie on a nonsingular point-conic.

16. Give an algebraic proof of Desargues' conic theorem in Π_2.

17. Given a complete four-point $A_0A_1A_2A_3$ and an arbitrary point P. Is there a line on P which cuts this four-point in an involutory hexad belonging to an involution for which P is a fixed point?

8.5 Pascal's Theorem

The theorem of Pascal is one of the most famous theorems in mathematics, not only because of its elegance and importance but because of the circumstances surrounding its discovery. When Blaise Pascal (1623–1662) was a boy of twelve, he asked his father what mathematics was, and his father, apparently identifying geometry with all of mathematics, told him that it was "the method of making figures with exactness and of finding out what proportions they relatively had to one another." With no other clue than this, and with no books or teachers to guide him, Pascal formulated the necessary definitions and axioms and proved the first 25 or 30 theorems in Euclid's elements! When his father discovered what his son was doing, he gave him a copy of Euclid, which the boy mastered easily. With this background, young Pascal by the time he was sixteen had written a treatise on conics which culminated in the theorem which still bears his name.[1]

Theorem 1   Pascal's theorem

If A, B, C, A', B', C' are six points on a nonsingular point-conic, then the intersection of BC' and $B'C$, the intersection of CA' and $C'A$, and the intersection of AB' and $A'B$ are collinear.

Proof   Let A, B, C, A', B', C' be six points on a nonsingular conic, Γ, let p be the line AB', let q be the line AC', let B_1 be the intersection of BC' and p, and let C_1 be the intersection of CB' and q (Fig. 8.10). Further, let A'' be the intersection of BC' and $B'C$, let B'' be the intersection of CA' and $C'A$, and let C'' be the intersection of AB' and $A'B$. Then $B(A',B',C',A) \overline{\wedge} p(C'',B',B_1,A)$, and $C(A',B',C',A) \overline{\wedge} q(B'',C_1,C',A)$. However, from the fact that any two points on a nonsingular conic can be generating bases it follows that $B(A',B',C',A) \sim C(A',B',C',A)$. Hence

$$p(C'',B',B_1,A) \sim q(B'',C_1,C',A)$$

This projectivity between ranges on distinct lines has a self-corresponding point, A, and therefore it must be a perspectivity. Thus $B''C'''$ must pass through the point of intersection of $B'C_1$ and B_1C', namely, A'', and the theorem is proved.

Pascal's theorem is often stated in the following more descriptive but less precise form: *the intersections of the opposite sides of a hexagon inscribed in a conic are collinear.* The line containing the intersections of the opposite sides of the inscribed hexagon is usually called the **Pascal line** of the hexagon.

---

[1] At the age of nineteen, Pascal invented the first successful calculating machine. Later, he made important contributions to the theory of probability and to the calculus. By his middle twenties, however, persistent ill health and a naturally mystical temperament led him to abandon most of his scientific studies for the pursuit of theology and philosophy, and today he is known to nonmathematicians primarily for his religious essays.

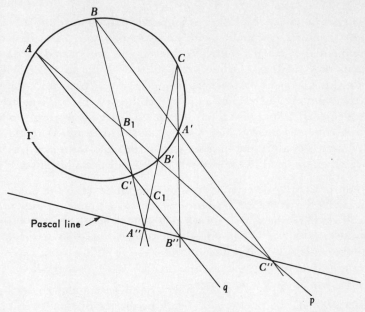

Fig. 8.10   Pascal's theorem.

Most of the applications of Pascal's theorem actually involve its converse:

Theorem 2

If A, B, C, A', B', C' are six points no three of which are collinear, and if the intersection of BC' and $B'C$, the intersection of CA' and $C'A$, and the intersection of AB' and $A'B$ are collinear, then A, B, C, A', B', C' all lie on a nonsingular point-conic.

Proof   Let A, B, C, A',B',C' be six points no three of which are collinear, let p be the line AB', let q be the line AC', let B_1 be the intersection of BC' and p, and let C_1 be the intersection of CB' and q (Fig. 8.10). Further, let A'' be the intersection of BC' and $B'C$, let B'' be the intersection of CA' and $C'A$, and let C'' be the intersection of AB' and $A'B$. Then, since A'', B'', C'' are collinear, by hypothesis, we have $p(C'',B',B_1,A) \overset{A''}{\barwedge} q(B'',C_1,C',A)$. Moreover

$$p(C'',B',B_1,A) \;\overline{\wedge}\; B(A',B',C',A)$$

and $q(B'',C_1,C',A) \;\overline{\wedge}\; C(A',B',C',A)$. Therefore $B(A',B',C',A) \sim C(A',B',C',A)$, and hence, by Theorem 9, Sec. 8.2, A, B, C, A', B', C' all lie on a non-singular point-conic, as asserted.

There are a number of interesting singular cases of Pascal's theorem which arise when various ones of the six given points coincide and the lines

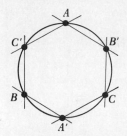

Fig. 8.11   The standard labeling
of the points in Pascal's theorem.

they would ordinarily determine are given as tangents to the conic. If we continue with the standard reference order we adopted in the proof of Pascal's theorem, namely, that shown in Fig. 8.11 (even though the points may not appear to our eye to follow this sequence around the conic), we have the six cases described in the following theorems. The first two of these theorems are trivially true; the last four require proof, though we shall leave all but one of the proofs as exercises.

Theorem 3

Pascal's theorem is true if two points which are not consecutive in the standard ordering are coincident.

Figure 8.12*a* illustrates Theorem 3 when $A = C$.

Theorem 4

Pascal's theorem is true if three points which are consecutive in the standard ordering are coincident, provided that the lines normally determined by the two pairs of consecutive points in such a triple are each given as the tangent to the conic at the point of coincidence.

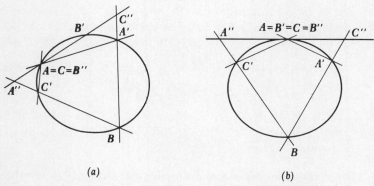

(a)   (b)

Fig. 8.12   Two trivial special cases of Pascal's theorem.

Figure 8.12*b* illustrates Theorem 4 when $A = B' = C$.

Theorem 5

Pascal's theorem is true if any side of the inscribed hexagon is replaced by a tangent to the given conic.

Proof   Let $A(= B')$, C, A', B, C' be five points on a nonsingular conic, Γ, let AB' denote the tangent to Γ at the point $A = B'$, let p be the line $A'C$, let q be the line $B'C$, let A_1 be the intersection of p and AB', and let B_1 be the intersection of q and $A'B$ (Fig. 8.13). Also, let A'' be the intersection of BC' and $B'C$, let B'' be the intersection of CA' and $C'A$, and let C'' be the intersection of AB' and $A'B$. Then $A(A',B',C',C) \, \overline{\wedge} \, p(A',A_1,B'',C)$, and

$$B(A',B',C',C) \, \overline{\wedge} \, q(B_1,B',A'',C)$$

Moreover, from the projective definition of Γ, using A and B as generating bases, we have $A(A',B',C',C) \sim B(A',B',C',C)$, and therefore, combining this with the preceding two projectivities, we obtain

$$p(A',A_1,B'',C) \sim q(B_1,B',A'',C)$$

This projectivity between distinct lines has a self-corresponding point, namely, C, and therefore it must be a perspectivity. Hence $A''B''$ must pass through the intersection of $A'B_1$ and A_1B', namely, C''. Thus A'', B'', C'' are collinear, and the theorem is established.

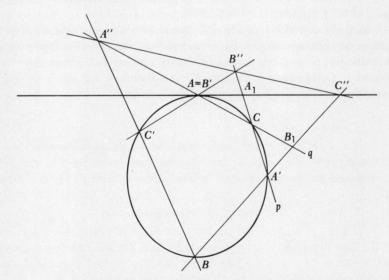

Fig. 8.13   Figure used in the proof of Theorem 5.

Theorem 6

Pascal's theorem is true if two sides of the inscribed hexagon which are neither opposite nor adjacent in the standard ordering are replaced by tangents to the given conic.

Theorem 7

Pascal's theorem is true if two sides of the inscribed hexagon which are opposite in the standard ordering are replaced by tangents to the given conic.

Theorem 8

Pascal's theorem is true if three sides of the inscribed hexagon no two of which are adjacent in the standard ordering are replaced by tangents to the given conic.

Corollary 1

A triangle circumscribed about a nonsingular point-conic is perspective with the triangle whose vertices are the points in which the sides of the first triangle touch the given conic.

In most applications, it is the converses of these theorems which are used, but we shall leave their statements and proofs as exercises.

The dual of Pascal's theorem is known as **Brianchon's theorem,** in honor of C. J. Brianchon (1785–1864), who discovered it while he was a student at the École Polytechnique in Paris. It is interesting that although Brianchon himself appears to have had serious doubts about the validity of the principle of duality, originally discovered by J. V. Poncelet (1788–1867) and J. D. Gergonne (1771–1859), his theorem did much to impress mathematicians with the importance of the principle:

Theorem 9

If a, b', c, a', b, c' are six lines of a nonsingular line-conic, then the line determined by the points bc' and $b'c$, the line determined by the points ca' and $c'a$, and the line determined by the points ab' and $a'b$ are concurrent.

Theorem 10   The converse of Brianchon's theorem

If a, b', c, a', b, c' are six lines no three of which are concurrent, and if the line determined by the points bc' and $b'c$, the line determined by the points ca' and $c'a$, and the line determined by the points ab' and $a'b$ are concurrent, then a, b', c, a', b, c' all belong to a nonsingular line-conic.

Using Pascal's theorem, or rather its converse, it is possible to carry out a great variety of interesting constructions, not only in the projective plane but also in the euclidean plane:

Example 1

Given five points no three of which are collinear and an arbitrary line passing through one of the points, find the second intersection of this line and the conic determined by the given points.

We have already noted how Maclaurin's construction and Desargues' conic theorem enable us to carry out the construction required in this example. Using Pascal's theorem, we proceed as follows. Let A, B', C, A', and B be five points no three of which are collinear, and let x be an arbitrary line on the point B. Since our aim is to determine the second intersection, say C', of x and the conic on the five given points, the line x is actually the line BC'. Hence A'' can be found at once as the intersection of $x = BC'$ and $B'C$. Similarly, C'' can be found as the intersection of AB' and $A'B$, since each of these lines is available to us (Fig. 8.14a). Thus the pascal line of the configuration is determined. Next we can locate B'' as the intersection of $A'C$ and the pascal line. Finally, C' can be located as the intersection of $BC' = x$ and AB'', and by Theorem 2 as the required intersection.

Example 2

Let P, Q, R be three noncollinear points, let p be a line on P but not on Q or R, let q be a line on Q but not on P or R, and let r be a line on R but not on P or Q. Find the second intersection of r and the conic which contains P, Q, R and is tangent to p and q at P and Q, respectively.

To systematize our procedure, let us introduce our standard notation by writing $P = A = B'$, $Q = C = A'$, and $R = B$. The line AB' is then the given tangent p, the line $A'C$ is the given tangent q, and the line BC' is the given line r on which we are required to locate the second intersection, C' (Fig. 8.14b). Since $AB' = p$ and $A'B$ are both available to us, we can locate C'' at once. Similarly, with BC' and $B'C$ known, we can find A''. This determines the pascal line of the configuration and permits us to find B'' as the intersection of $A'C = q$ and the pascal line. Finally, C' can be located as the intersection of $BC' = r$ and AB''. By the converse of Theorem 6, the point C' which we have thus constructed is on the conic defined by the given data.

Various euclidean constructions can be carried out by means of Pascal's theorem if we keep in mind the relation between the ideal line in E_2^+ and the various types of conics in E_2. Specifically:

A parabola is a conic which is tangent to the ideal line, and the direction of its axis identifies the ideal point at which the ideal line touches the parabola.

A hyperbola is a conic which intersects the ideal line in distinct real points. The directions of the asymptotes identify these ideal points, and the asymptotes themselves are the tangents to the hyperbola at these ideal points.

Example 3

Given four finite points on a hyperbola and the direction of one asymptote, show how to construct the asymptote itself.

Being given the direction of an asymptote means that we are given an ideal point on the hyperbola. Hence we are in fact given five points on a conic and are asked to construct the tangent to the conic at a specified one of these points. To systematize our work, let A, B', C, and A' be the four finite points we are given, and let B be the ideal point we are given, via the

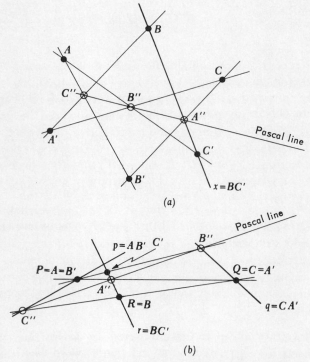

Fig. 8.14 Two constructions based on Pascal's theorem.

direction which defines it. Since we want to construct the tangent to the hyperbola at B, it follows that B is also to be considered the point C'. Now from the definition of ideal points, it follows that $A'B$ is simply the line through A' which is parallel to the line which defines the direction of the asymptote, and therefore the ideal point B. The point C'' can then be found at once as the intersection of this line through A' and the line AB'. Similarly, by finding the intersection of CA' and the line through A which is parallel to the line defining $B = C'$, we can find B''. Thus the pascal line is determined, and A'' can be found as the intersection of $B'C$ and the pascal line. With A'' located, the required asymptote, namely, the line BC', can be constructed at once as the line through A'' in the direction of B.

Exercises

1. Prove Theorem 6.

2. Prove Theorem 7.

3. Prove Theorem 8.

4. Prove Corollary 1, Theorem 8.

5. Let P, Q, R be three collinear points, and let U and V be two points neither of which is on the line containing P, Q, R. Show that the other six intersections of the lines UP, UQ, UR, VP, VQ, VR all lie on a nonsingular point-conic.

6. Let Γ be a conic which contains the vertices of a four-point $A_0A_1A_2A_3$, let p and q be the tangents to Γ at any two of the points A_0, A_1, A_2, A_3, and let U be the intersection of p and q. Show that U is collinear with two diagonal points of the given four-point.

7. Let Γ be a point-conic containing the vertices of a four-point $A_0A_1A_2A_3$, and let a_0, a_1, a_2, a_3 be, respectively, the tangents to Γ at A_0, A_1, A_2, A_3. Show that the six points A_0, A_1, A_2, A_3, a_0a_1, a_2a_3 all lie on a nonsingular point-conic. Using this fact, show that a_0a_1 and a_2a_3 are harmonic conjugates with respect to D_2 and D_3.

8. Using Pascal's theorem, show that three tangents to a nonsingular point-conic cannot be concurrent.

9. Let Γ be the conic determined by the five points P_1, P_2, P_5, P_6, P_{17} in the finite projective geometry defined by Table 6.3. Show that all the tangents to Γ are concurrent. How can this result be reconciled with the result of Exercise 8?

10. Let X_1, X_2, X_3, X_1', X_2', X_3' be an involutory hexad associated with a given four-point, and let Y_1, Y_2, Y_3, Y_1', Y_2', Y_3' be, respectively, the harmonic conjugates of X_1, X_2, X_3, X_1', X_2', X_3' with respect to the two points of the

four-point with which each is collinear. Show that Y_1, Y_2, Y_3, Y_1', Y_2', Y_3' are all on a point-conic. Under what conditions will this point-conic be singular?

11. Let $\triangle ABC$ and $\triangle A'B'C'$ be centrally perspective. Show that the intersections of the sides of one triangle with the noncorresponding sides of the other all lie on a nonsingular point-conic. What is the dual of this result?

12. Show that six points on a nonsingular conic determine 60 pascal lines when all possible orders of the points are considered. Show, further, that the 15 lines determined by the 6 given points intersect in 45 additional points through each of which pass 4 of these pascal lines.

13. Devise a point-by-point construction for the conic determined by each of the following sets of data:
(*a*) Four points no three of which are collinear and the tangent to the conic at one of these points
(*b*) Three points on a hyperbola and the direction of the asymptotes
(*c*) One point on a hyperbola and the two asymptotes
(*d*) Three points on a parabola and the direction of the axis
(*e*) Two points on a parabola, the tangent to the parabola at one of these points, and the direction of the axis

14. Devise a line-by-line construction for the conic determined by each of the following sets of data:
(*a*) Five lines no three of which are concurrent
(*b*) Four lines no three of which are concurrent and the tangent point of one of the lines
(*c*) The asymptotes of a hyperbola and one line which is tangent to the hyperbola

15. Given three points on a parabola and the direction of the axis, devise a construction for the axis, the vertex, the focus, and the directrix.

16. Given two points on a parabola and the tangents at these points, devise a construction for the direction of the axis.

17. Given one point and the two asymptotes of a hyperbola, devise a construction for the tangent to the hyperbola at the given point. Thence show that the segment intercepted by the asymptotes on any tangent to a hyperbola is bisected by the point of contact of the tangent.

18. Prove that the tangents to a nonsingular point-conic in any finite projective geometry satisfying Axioms 1 to 7 are all concurrent if and only if the diagonal points of every four-point are collinear. *Hint:* Use Pascal's theorem.

8.6 Polar Theory

In Sec. 4.9 when we first introduced the concept of the polar line of a point P with respect to a conic Γ, we had, a priori, a line singled out for special consideration, namely, the line defined by the equation $P^T A X = 0$, where $X^T A X = 0$ was the equation of Γ. In our axiomatic development, however, there is no line suggested for investigation until our first theorem identifies it.

Theorem 1

The locus of the harmonic conjugates of a point with respect to the intersections of a nonsingular point-conic and the lines of the pencil on that point is a range of points.

Proof Let Γ be an arbitrary nonsingular point-conic, let P be an arbitrary point, and let us suppose first that P is not a point of Γ. Let (A,A') be a pair of points on Γ which are collinear with P, and let A_1 be the harmonic conjugate of P with respect to A and A'. Similarly, let (B,B') be a second pair of points on Γ which are collinear with P, and let B_1 be the harmonic conjugate of P with respect to B and B'. Obviously P is one of the diagonal points of the four-point $AA'BB'$; let Q and R be the other two. Finally, let T be the intersection of the tangents to Γ at A and A' (Fig. 8.15). Now consider the special case of Pascal's theorem defined by the correspondence

$$
\begin{array}{cccccc}
A & B' & C & A' & B & C' \\
A & A & B & A' & A' & B'
\end{array}
$$

where, of course, the lines AB' $(= AA)$ and $A'B$ $(= A'A')$ are, respectively, the tangents to Γ at A and A'. From this it follows that Q, R, and T are collinear. Furthermore, from the properties of the harmonic points of a complete four-point, we know that Q, R, A_1, B_1 are collinear. Hence Q, R, T, A_1, B_1 are all collinear.

Now if the pair of points (B,B') is replaced by any other pair of points on Γ which are collinear with P, say (C,C'), if C_1 is the harmonic conjugate of P with respect to C and C', and if Q' and R' are the other diagonal points of the four-point $AA'CC'$, it follows similarly that Q', R', T, A_1, C_1 are all collinear. Moreover, since T and A_1 are not changed when the pair (B,B') is replaced by the pair (C,C'), we have, in fact, the extended collinearity relation $QQ'RR'TA_1B_1C_1$, which proves that the harmonic conjugate of P with respect to *any* pair of points on Γ which are collinear with P lies on the line TA_1.

The line TA_1 clearly cannot pass through P, for if this were the case, then the three diagonal points of each of the four-points $AA'BB'$, $AA'CC'$, ... would be collinear, contrary to Axiom 8. Furthermore, TA_1 cannot be

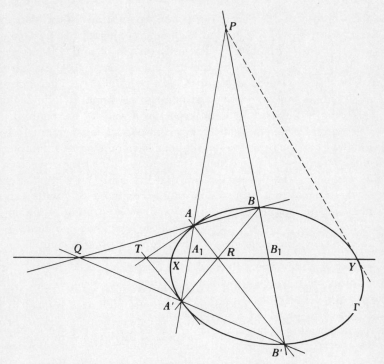

Fig. 8.15 Figure used in the proof of Theorem 1.

tangent to Γ. In fact, since A_1 is distinct from A and A' on PA_1, it follows that TA, TA', and TA_1 are distinct lines. Hence if TA_1 were tangent to Γ, there would be three tangents from T to Γ, which is impossible, by Exercise 8, Sec. 8.5. Thus TA_1, since it is not tangent to Γ, has two distinct points, say X and Y, in common with Γ. Now consider the line joining P to either of the intersections of TA_1 and Γ, say Y. If this line is not tangent to Γ at Y, it must have a second point, Y', in common with Γ. In this case the harmonic conjugate of P with respect to Y and Y' is a point distinct from Y, and therefore the line PY has both Y and this harmonic conjugate in common with TA_1, which is impossible. Thus the lines joining P to each of the intersections of TA_1 and Γ are tangent to Γ at these intersections, and by Exercise 8, Sec. 8.5, these are the only tangents to Γ which pass through P. From the properties of singular harmonic tetrads, it follows that on each of these tangents the harmonic conjugate of P with respect to the coincident intersections of Γ and the tangent is the point of tangency. Since we have just seen that each of these points of tangency lies on TA_1, we have thus shown that on *every* line through P, the harmonic conjugate of P with respect to the intersections of that line and Γ lies on TA_1.

To complete this part of our proof, we must now show, conversely, that every point, K_1, of TA_1 is the harmonic conjugate of P with respect to the intersections of PK_1 and Γ. To verify this, let K and K' be the points common to Γ and PK_1. Then by the first part of our proof, the harmonic conjugate of P with respect to K and K' is a point of TA_1. Therefore, since PK_1 and TA_1 have one and only one point in common, it follows that K_1 is the harmonic conjugate of P with respect to K and K', as required.

Finally, let us suppose that the given point P is a point of Γ. In this case it follows from the properties of singular harmonic tetrads that on every line through P, except the one which is tangent to Γ, the harmonic conjugate of P with respect to the intersections of that line and Γ is P itself. Furthermore, on the tangent to Γ at P, the harmonic conjugate of P with respect to the coincident intersections of Γ and the tangent is *any* point of the tangent. Hence, when P is a point of Γ, the locus of the harmonic conjugates of P is still a line, namely, the tangent to Γ at P, and our proof is complete.

Definition 1

If P is any point, and if Γ is any nonsingular point-conic, the line, p, which is the locus of the harmonic conjugates of P with respect to pairs of points of Γ which are collinear with P, is called the polar line or polar of P with respect to Γ, and P is called the pole of the line p with respect to Γ.

It should be noted that the terms *pole* and *polar* which appear in Definition 1 are *not* duals of each other, since each refers to a point-conic.

Certain observations we made in the proof of Theorem 1 are important enough to warrant restatement as corollaries of the theorem.

Corollary 1

A point lies on its polar with respect to a given nonsingular point-conic if and only if the point lies on the conic.

Corollary 2

Through any point which is not on a nonsingular point-conic there pass exactly two lines which are tangent to the conic.

Corollary 3

The polar of a point with respect to a nonsingular point-conic which does not contain the point is the line determined by the points of contact of the tangents to the conic from the given point.

Corollary 4

The polar of a point with respect to a nonsingular point-conic which contains the point is the tangent to the conic at the given point.

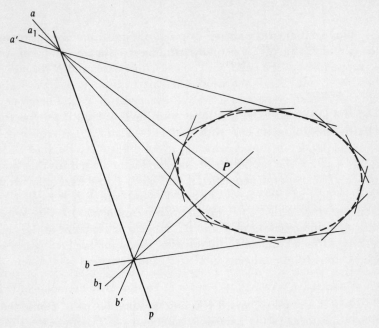

Fig. 8.16   The pole of a line with respect to a nonsingular line-conic.

Corollary 5

If P and P' are distinct points, their polars with respect to a given nonsingular point-conic are distinct.

Theorem 2

The envelope of the harmonic conjugates of a line with respect to the lines of a nonsingular line-conic which pass through the respective points of that line is a point.

The meaning of Theorem 2 is illustrated in Fig. 8.16.

Definition 2

If p is any line, and if Φ is any nonsingular line-conic, the point P which is the envelope of the harmonic conjugates of p with respect to the pairs of lines of Φ which are concurrent with p is called the pole of p with respect to Φ, and p is called the polar of P with respect to Φ.

Again it should be noted that the terms *pole* and *polar* appearing in Definition 2 are not duals of each other. However, the phrase "pole and polar with respect to a nonsingular point-conic" *is* the dual of the phrase "polar and pole with respect to a nonsingular line-conic."

Our next theorem shows that a property of polars which we first proved only in E_2^+ and Π_2 is actually valid in any system satisfying Axioms 1 to 9.

Theorem 3

If a point Q lies on the polar of a point P with respect to a nonsingular point-conic Γ, then P lies on the polar of Q with respect to Γ.

Proof Let p be the polar of a point P with respect to a nonsingular point-conic Γ, let Q be a point on p, and let us suppose first that neither P nor Q lies on Γ. Then if R and S are the (distinct) points in which the line PQ intersects Γ, it follows that Q is the harmonic conjugate of P with respect to R and S. But this implies that P is also the harmonic conjugate of Q with respect to R and S, which, in turn, means that P is on the polar of Q with respect to Γ.

If P is not a point of Γ and Q is, then Q must be the point of contact of one of the tangents to Γ from P. In this case, the polar of Q is the tangent to Γ at Q, and this obviously passes through P. If P is a point of Γ but Q is not, then Q must lie on the tangent to Γ at P. In this case, the polar of Q is the line determined by P and the point of contact of the second tangent from Q to Γ, and this line obviously passes through P. Under the hypotheses of the theorem, P and Q cannot both be points of Γ unless they are the same point, in which case the theorem is trivially true. Hence our proof is complete.

Corollary 1

If a line q passes through the pole of a line p with respect to a nonsingular point-conic Γ, then p passes through the pole of q with respect to Γ.

Various interesting results now follow easily from Theorem 3:

Theorem 4

If $P, Q, R, S, \ldots$ are collinear points, their polars with respect to a nonsingular point-conic are concurrent.

Proof Let $P, Q, R, S, \ldots$ be points on a line x, let $p, q, r, s, \ldots$ be the polars of $P, Q, R, S, \ldots$ with respect to a nonsingular point-conic Γ, and let X be the pole of x with respect to Γ. Then since the points $P, Q, R, S, \ldots$ lie on the polar of X, it follows from Theorem 3 that their polars, $p, q, r, s, \ldots$, all pass through X. In other words, $p, q, r, s, \ldots$ are concurrent, as asserted.

Corollary 1

If $p, q, r, s, \ldots$ are concurrent lines, their poles with respect to a nonsingular point-conic are collinear.

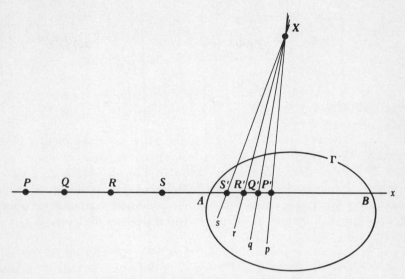

Fig. 8.17 Figure used in the proof of Theorem 5.

Theorem 5

If P, Q, R, S are four points on a line x, and if p, q, r, s are the polars of these points with respect to a nonsingular point-conic Γ, then $x(P,Q,R,S) \sim X(p,q,r,s)$, where X is the pole of x with respect to Γ.

Proof Let Γ be a nonsingular point-conic, let P, Q, R, S be four points on a line x, let p, q, r, s be the polars of these points with respect to Γ, and let X be the pole of the line x. Then, by the preceding theorem, p, q, r, s all pass through X. We now have two possibilities to consider according as x is or is not tangent to Γ. If x is not tangent to Γ, let A and B be the points in which x intersects Γ, and let P', Q', R', S' be the harmonic conjugates of P, Q, R, S with respect to A and B. Then P', Q', R', S' are on p, q, r, s, respectively, and therefore (Fig. 8.17) $x(P',Q',R',S') \overline{\wedge} X(p,q,r,s)$. Furthermore, (P,P'), (Q,Q'), (R,R'), (S,S') are pairs of mates in the involution on x whose fixed points are A and B. Hence $x(P,Q,R,S) \sim x(P',Q',R',S')$. Thus, combining these two projectivities, we have $x(P,Q,R,S) \sim X(p,q,r,s)$, as asserted.

If x is tangent to Γ, the preceding argument breaks down because A, B, P', Q', R', S', and X all coincide at the point of tangency. In this case we first project P, Q, R, S from x onto a line, x_1, which is not tangent to Γ. Let V be the center of this perspectivity, let P_1, Q_1, R_1, S_1 be the images of P, Q, R, S, on x_1, let X_1 be the pole of x_1 with respect to Γ, and let p_1, q_1, r_1, s_1, and v be, respectively, the polars of P_1, Q_1, R_1, S_1, and V. Then, by the first part of the theorem, $x_1(P_1,Q_1,R_1,S_1) \sim X_1(p_1,q_1,r_1,s_1)$. Moreover, from the definition of

P_1, Q_1, R_1, S_1, we have $x(P,Q,R,S) \overset{V}{\wedge} x_1(P_1,Q_1,R_1,S_1)$. Also, since the points (P,P_1), (Q,Q_1), (R,R_1), (S,S_1) are collinear with V, the polar lines (p,p_1), (q,q_1), (r,r_1), (s,s_1) must be concurrent with v. Hence

$$X(p,q,r,s) \overset{v}{\wedge} X_1(p_1,q_1,r_1,s_1)$$

Finally, combining these three projectivities, we have $x(P,Q,R,S) \sim X(p,q,r,s)$, and our proof is complete.

Corollary 1

If $x(X_1,X_2,X_3,...) \sim y(Y_1,Y_2,Y_3,...)$, then

$$X(x_1,x_2,x_3,...) \sim Y(y_1,y_2,y_3,...)$$

where X and Y are the poles of x and y with respect to any nonsingular point-conic, Γ, and x_1, x_2, x_3, $...$, y_1, y_2, y_3, $...$ are the polars of X_1, X_2, X_3, $...$, Y_1, Y_2, Y_3, $...$ with respect to Γ.

Corollary 2

If XR, XS, XT, XU are four lines on a point X, and if r, s, t, u are the polars of R, S, T, U with respect to a nonsingular point-conic, Γ, then $X(R,S,T,U) \sim x(r,s,t,u)$, where x is the polar of X with respect to Γ.

Theorem 6

If P, Q, R, S, T, U are six points on a nonsingular point-conic, their polars with respect to any nonsingular point-conic are six lines of a nonsingular line-conic.

Proof   Let P, Q, R, S, T, U be six points on a nonsingular point-conic, and let p, q, r, s, t, u be the polars of these points with respect to a nonsingular point-conic, Γ. Then from the definition of a point-conic, we have $P(R,S,T,U) \sim Q(R,S,T,U)$. Also, from Corollary 2, Theorem 5, we have $P(R,S,T,U) \sim p(r,s,t,u)$ and $Q(R,S,T,U) \sim q(r,s,t,u)$. Hence $p(r,s,t,u) \sim q(r,s,t,u)$, which, by Theorem 10, Sec. 8.2, proves that p, q, r, s, t, u are all lines of a line-conic. Moreover, since the point-conic containing P, Q, R, S, T, U is nonsingular, the projectivity $P(r,s,t,u) \sim Q(R,S,T,U)$ is not a perspectivity; i.e., it can have no self-corresponding line. Hence the projectivity $p(r,s,t,u) \sim q(r,s,t,u)$ can have no self-corresponding point, i.e., it is not a perspectivity, and therefore the line-conic which it defines is nonsingular. This completes our proof.

Corollary 1

The polars of the points of a given nonsingular point-conic with respect to any nonsingular point-conic are all lines of the same nonsingular line-conic, and conversely.

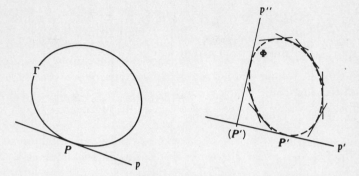

Fig. 8.18   Figure used in the proof of Corollary 2, Theorem 6.

Corollary 2

If p is tangent to a nonsingular point-conic Γ at a point P, and if Φ is the line-conic which contains the polars of the points of Γ with respect to a given nonsingular point-conic, then the pole of p is the tangent-point of the polar of P considered as a line of Φ.

Proof   Let p be the tangent to a nonsingular point-conic Γ at a point P, let P' and p' be, respectively, the pole of p and the polar of P with respect to a given nonsingular point-conic, and let Φ be the line-conic to which the polars of the points of Γ belong. Then by Theorem 3, P' lies on the line p'. If P' is not the tangent-point of p' as a line of Φ, then a second line, p'', of Φ must pass through P' (Fig. 8.18). Moreover, by Corollary 1, Theorem 6, the pole of p'' must be a point, P'', of Γ which is also on p. But this is impossible, since p, being a tangent to Γ, cannot contain two distinct points of Γ. Thus the assertion of the corollary is established.

Corollary 3

If p is the tangent to a nonsingular point-conic Γ at a point P, if Φ is the line-conic which contains the polars of the points of Γ with respect to a given nonsingular point-conic, and if p is the polar of P in this polarity, then P is the tangent point of p as a line of Φ.

Our last three theorems bring point-conics and line-conics together into a single self-dual configuration:

Theorem 7

The set of all tangents to a nonsingular point-conic is a nonsingular line-conic.

Proof Let Γ be a nonsingular point-conic. By Corollary 1, Theorem 6, the polars of the points of Γ with respect to *any* nonsingular point-conic are lines of a nonsingular line-conic. If, in particular, we consider the polars of the points of Γ with respect to Γ itself, they become simply the tangents to Γ, which thus all belong to a nonsingular line-conic. Moreover, there can be no line of this line-conic which is not a tangent to Γ. In fact, if there were such a line, p, it would have to meet Γ in two distinct points, say P_1 and P_2. The tangents to Γ at these points, say p_1 and p_2, are, of course, lines of the line-conic, and by Corollary 3, Theorem 6, P_1 and P_2 are the tangent-points of these lines as lines of the line-conic. But this is impossible, since a tangent-point is by definition a point through which passes only one line of a line-conic, whereas p and p_1 both pass through P_1, and p and p_2 both pass through P_2. Thus not only is every tangent to Γ a line of the line-conic, but, conversely, every line of the line-conic is a tangent to Γ, and the theorem is established.

Theorem 8

The set of all tangent-points of a nonsingular line-conic is a nonsingular point-conic.

In view of Theorems 7 and 8, we can now drop the adjectives *point* and *line*, and simply speak of conics, leaving it to the context to determine whether we are primarily interested in the points or in the lines of what is now a self-dual configuration.

Theorem 9

If P is any point, and if p is its polar with respect to the points of a nonsingular conic, then p is also the polar of P with respect to the lines of the conic.

Proof The theorem is obviously true if P is a point of the conic, because then, by either definition, the polar, p, is the tangent to the conic at P. Let us therefore suppose that P is not a point of the conic. Let p_1 and p_2 be the tangents to the conic which pass through P, and let P_1 and P_2 be their points of contact. Then the polar of P relative to the points of the conic is the line P_1P_2. On the other hand, p_1 and p_2 are two lines of the conic, and P_1 and P_2 are their tangent points. The pole of the line P_1P_2 is therefore the point p_1p_2, that is, P, which is the same thing as saying that P_1P_2 is the polar of P relative to the lines of the conic. Thus $p = P_1P_2$ is the polar of P by either definition, as asserted.

Theorem 10

If p is any line, and if P is its pole with respect to the lines of a nonsingular conic, then P is also the pole of p with respect to the points of the conic.

Definition 3

Two points each of which lies on the polar of the other with respect to a nonsingular conic, Γ, are said to be conjugate points with respect to Γ.

Definition 4

Two lines each of which passes through the pole of the other with respect to a nonsingular conic, Γ, are said to be conjugate lines with respect to Γ.

Exercises

1. Prove Corollary 5, Theorem 1.

2. Prove Corollary 3, Theorem 6.

3. What meaning, if any, can be given to the notion of the polar of a point with respect to a singular conic?

4. Let P_1, P_2, and Q be three points on a nonsingular conic, Γ, let P be the pole of P_1P_2, let t be any line on P, let R_1 be the intersection of t and P_1Q, and let R_2 be the intersection of t and P_2Q. Show that R_1 and R_2 are conjugate points with respect to Γ.

5. A, B, C, D, and P are five points on a nonsingular conic. Show that the necessary and sufficient condition that AB and CD be conjugate lines with respect to the conic is that $P(AB,CD)$ be a harmonic line tetrad.

6. Show that if a complete four-point is inscribed in a nonsingular conic, the polar of each diagonal point with respect to the conic is the line determined by the other two diagonal points.

7. Let $ABCD$ be a four-point inscribed in a nonsingular conic, Γ, and let a, b, c, d be, respectively, the tangents to Γ at A, B, C, D. Show that the four-point $ABCD$ and the four-line $abcd$ have the same diagonal triangle.

8. Given five points, A, B, C, D, E, no three of which are collinear, and an arbitrary point, P. Devise a construction for the polar of P with respect to the conic determined by A, B, C, D, E.

9. Given five points, A, B, C, D, E, no three of which are collinear, and an arbitrary line p. Devise a construction for the pole of p with respect to the conic determined by A, B, C, D, E.

10. Given a nonsingular conic Γ and a point P which is not on Γ. Show that there are infinitely many triangles which have P as one vertex and are self-polar with respect to Γ, that is, have the property that the polar of each vertex with respect to Γ is the opposite side. Is it possible to find a triangle which is self-polar with respect to Γ if two of its vertices are assigned arbitrarily?

11. If a point varies along a line, show that its polars with respect to two given conics intersect in points which all lie on a conic.

12. In E_2^+ show that the pole of the ideal line with respect to a central conic, Γ, is the center of Γ.

13. In E_2^+ show that the diagonals of a parallelogram either inscribed in, or circumscribed to, a central conic are diameters of the conic.

14. In E_2^+ show that if a parallelogram is inscribed in a central conic, the tangents to the conic at the vertices of the parallelogram form a second parallelogram, and conversely.

15. Devise a construction for the center of a conic when five of its points are given. Is it possible to construct the center of a conic when five of its lines are given?

8.7 Projectivities on a Conic

In this section we shall extend the concepts of perspectivity and projectivity to include correspondences involving the points and lines of nonsingular conics. In doing this, we shall find it convenient to broaden the meaning of the terms *range* and *pencil*:

> *Definition 1*
> The set of points belonging to a nonsingular conic is called a *range on the conic*.

> *Definition 2*
> The set of lines belonging to a nonsingular conic is called a pencil on the conic.

By analogy with our earlier notation for ranges on lines and pencils on points, we shall denote a range on a conic Γ by the symbol $\Gamma(A,B,C,...)$, where A, B, C, ... are points of Γ, and a pencil on Γ by the symbol

$$\Gamma(a,b,c,...)$$

where $a, b, c, \ldots$ are lines of Γ. We shall also extend the use of the symbols $\overline{\wedge}$ and $\sim$ to perspectivities and projectivities involving ranges and pencils on conics, as described in the following definitions.

Definition 3

A range $\Gamma(A,B,C,\ldots)$ on a nonsingular conic Γ is said to be perspective with a range $x(A',B',C',\ldots)$ on a line x if and only if there exists a point V on Γ but not on x such that the lines AA', BB', CC', $\ldots$ all pass through V.

Definition 4

A range $\Gamma(A,B,C,\ldots)$ on a nonsingular conic Γ is said to be projective with a range $x(A',B',C',\ldots)$ on a line x if and only if there exists a range $y(A'',B'',C'',\ldots)$ such that $\Gamma(A,B,C,\ldots) \overline{\wedge} y(A'',B'',C'',\ldots)$ and

$$y(A'',B'',C'',\ldots) \sim x(A',B',C',\ldots)$$

Definition 5

Two ranges on the same or different nonsingular conics are said to be projective if and only if each is projective with the same range on some line.

Definition 6

A pencil $\Gamma(a,b,c,\ldots)$ on a nonsingular conic Γ is said to be perspective with a pencil $X(a',b',c',\ldots)$ on a point X if and only if there exists a line v on Γ but not on X such that the points aa', bb', cc', $\ldots$ all lie on v.

Definition 7

A pencil $\Gamma(a,b,c,\ldots)$ on a nonsingular conic Γ is said to be projective with a pencil $X(a',b',c',\ldots)$ on a point X if and only if there exists a pencil $Y(a'',b'',c'',\ldots)$ such that $\Gamma(a,b,c,\ldots) \overline{\wedge} Y(a'',b'',c'',\ldots)$ and

$$Y(a'',b'',c'',\ldots) \sim X(a',b',c',\ldots)$$

Definition 8

Two pencils on the same or different nonsingular conics are said to be projective if and only if each is projective with the same pencil on some point.

Definitions 3 and 6 are illustrated in Fig. 8.19.

The data required to determine a projectivity between a range on a conic and a range on a line or a projectivity between ranges on two conics are described in our first two theorems, which follow immediately from Axiom 7 and Theorem 9, Sec. 8.2.

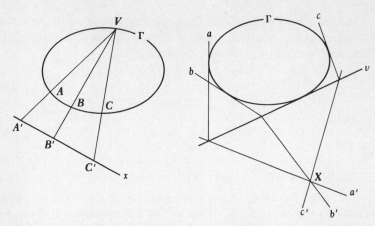

Fig. 8.19   A perspectivity between a range on a conic and a range on a line, and a perspectivity between a pencil on a conic and a pencil on a point.

Theorem 1

A projectivity between a range on a conic and a range on a line is uniquely determined when three pairs of corresponding points are given.

Corollary 1

A perspectivity between a range on a conic and a range on a line is uniquely determined when one pair of corresponding points is given.

Theorem 2

A projectivity between two ranges on the same or different conics is uniquely determined when three pairs of corresponding points are given.

Corollary 1

If a projectivity between ranges on the same conic has more than two self-corresponding points, it is the identity.

Our next theorem provides us with an elegant way of determining the fixed points of a projectivity between ranges on the same conic.

Theorem 3

The fixed points of the projectivity $\Gamma(A,B,C,...) \sim \Gamma(A',B',C',...)$ on a conic Γ are the points in which the pascal line of the set A, B', C, A', B, C' intersects Γ.

Proof   Let (A,A'), (B,B'), (C,C') be three pairs of corresponding points in a projectivity on a conic Γ, let A'' be the intersection of BC' and $B'C$, let B'' be

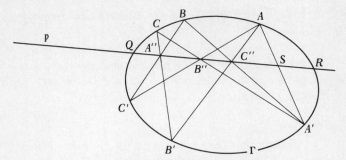

Fig. 8.20   Figure used in the proof of Theorem 3.

the intersection of CA' and $C'A$, let C'' be the intersection of AB' and $A'B$, let p be the pascal line of the points A, B', C, A', B, C', let Q and R be the intersections of p and Γ, and let S be the intersection of p and the line AA' (Fig. 8.20). Then $\Gamma(Q,A,B,C,R) \stackrel{A'}{\overline{\wedge}} p(Q,S,C'',B'',R)$, and

$$\Gamma(Q,A',B',C',R) \stackrel{A}{\overline{\wedge}} p(Q,S,C'',B'',R)$$

Hence

$$\Gamma(Q,A,B,C,R) \sim \Gamma(Q,A',B',C',R)$$

Therefore Q and R are fixed points in the projectivity determined on Γ by the three pairs of corresponding points (A,A'), (B,B'), (C,C'). Moreover, by Corollary 1, Theorem 2, the projectivity can have no other fixed points. Hence the theorem is proved.

Corollary 1

In a projectivity $\Gamma(A,B,C,...) \sim \Gamma(A',B',C',...)$ on a nonsingular conic Γ, the pascal line of the two triples (A,B,C) and (A',B',C') is the same as the pascal line of any other three points and their images under the given projectivity.

By means of Theorem 3, we can now construct the fixed points of a projectivity between ranges on the same line, provided we are given a single conic:

Theorem 4

If $x(A,B,C,...) \sim x(A',B',C',...)$ is a projectivity on a line x, if V, A_1, B_1, C_1, A_1', B_1', C_1' are points on a nonsingular conic, Γ, such that $\Gamma(A_1,B_1,C_1) \stackrel{V}{\overline{\wedge}} x(A,B,C)$ and $\Gamma(A_1',B_1',C_1') \stackrel{V}{\overline{\wedge}} x(A',B',C')$, and if Q_1 and R_1 are the fixed points of the projectivity $\Gamma(A_1,B_1,C_1,...) \sim \Gamma(A_1',B_1',C_1',...)$, then the fixed points of the given projectivity on x are the projections of Q_1 and R_1 from V onto x.

Not only can the notion of a projectivity between cobasal ranges be extended to include projectivities between ranges on the same conic, but so can the notion of an involution:

Definition 9

A projectivity between two ranges on the same nonsingular conic in which there is a pair of reciprocally corresponding points is said to be an involution on the conic.

Theorem 5

Every pair of corresponding points in an involution on a nonsingular conic is a pair of reciprocally corresponding points.

Proof Let (A,B) be a pair of reciprocally corresponding points in an involution on a conic Γ, let (C,D) be any other pair of corresponding points, and let X be the image of D, so that (D,X) is also a pair of corresponding points. Then $\Gamma(A,B,C,D,...) \sim \Gamma(B,A,D,X,...)$. Moreover, by the first permutation theorem, we have $\Gamma(B,A,D,C,...) \sim \Gamma(A,B,C,D,...)$. Hence, combining these two projectivities, we obtain

$$\Gamma(B,A,D,C,...) \sim \Gamma(B,A,D,X,...)$$

Since this projectivity has three pairs of self-corresponding points, it is the identity, and therefore $X = C$. Thus not only does D correspond to C, but C also corresponds to D. This completes our proof, since (C,D) was any pair of corresponding points.

Theorem 6

The lines joining corresponding points in an involution on a nonsingular conic all pass through a common point.

Proof Let (A,A') be a pair of mates in an involution on a conic, Γ, let p be the line determined by the fixed points, Q and R, of the involution, let S be the intersection of p and the line AA', let P be the harmonic conjugate of S with respect to A and A', and let (B,B') be a second pair of mates in the given involution (Fig. 8.21). Then $\Gamma(A,B,A',B',...) \sim \Gamma(A',B',A,B,...)$, and since, by Corollary 1, Theorem 3, p is the pascal line of any two triples of corresponding points, it follows that the intersection of AB' and $A'B$, say U, and the intersection of AB and $A'B'$, say V, are both on p. Thus U and V are two of the three diagonal points of the four-point $AA'BB'$. From the

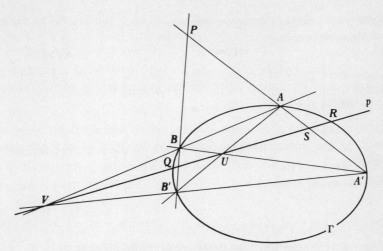

Fig. 8.21 Figure used in the proof of Theorem 6.

harmonic properties of a complete four-point, the third diagonal point must be P. Hence BB' must also pass through P. Similarly, any other pair of corresponding points, (C,C'), must be collinear with P, since P depends only on the pair (A,A') and the pascal line p.

Theorem 7

If $\Gamma(A,B,C,...) \sim \Gamma(A',B',C',...)$ is a projectivity on a nonsingular conic, and if AA', BB', and CC' are concurrent, the projectivity is an involution.

Proof Let $\Gamma(A,B,C,...) \sim \Gamma(A',B',C',...)$ be a projectivity on a nonsingular conic, Γ, and let AA', BB', CC' be concurrent in a point V. Now consider the unique involution on Γ in which (A,A') and (B,B') are pairs of reciprocally corresponding points, and let C'' be the mate of C in this involution. Then by the last theorem, C and C'' are collinear with V. Hence $C' = C''$, and therefore the given projectivity is an involution since it has three pairs of mates in common with the involution $\Gamma(A,B,C,...) \sim \Gamma(A',B',C',...)$. This completes our proof.

Exercises

1. Prove Corollary 1, Theorem 1.

2. Prove Theorem 1.

3. Prove Theorem 2.

4. Prove Theorem 4.

5. Given three pairs of corresponding points in a projectivity between ranges on the same conic. Devise a construction for the image of an arbitrary point on the conic.

6. Given three pairs of corresponding points in a projectivity between a conic and a line. Devise a construction for finding the image of an arbitrary point on either the conic or the line.

7. Given three pairs of corresponding points in a projectivity between two conics. Devise a construction for finding the image of an arbitrary point on either conic.

8. Can the fixed points of a projectivity between ranges on the same conic ever be coincident?

9. Show that the fixed points of an involution on a conic are always distinct.

10. Show that the permutation theorems are valid for projectivities involving ranges on conics.

11. Show that any projectivity on a conic is the composition of two involutions.

12. Given five points no three of which are collinear. Devise a construction for the intersection of an arbitrary line and the conic determined by the five points.

13. Given five finite points which are known to lie on a hyperbola. Devise a construction for the asymptotes of the hyperbola.

14. Given three nonconcurrent lines and three noncollinear points none of which lies on any of the given lines. Does there exist a triangle whose sides pass through the given points and whose vertices lie on the given lines? If there is such a triangle, devise a construction for locating its vertices.

15. Given four lines, a_0, a_1, a_2, a_3, and a triangle none of whose vertices lies on any of the given lines. Does there exist a four-point $A_0 A_1 A_2 A_3$ whose vertices lie, respectively, on a_0, a_1, a_2, a_3 which has the given triangle as diagonal triangle? If such a four-point exists, devise a construction for locating its vertices.

8.8 Further Results

The properties of conics have been investigated in almost unbelievable detail, and entire books have been written about them.[1] We shall conclude this chapter by presenting a few typical examples.

[1] See, for instance, George Salmon, "A Treatise on Conic Sections," Chelsea Publishing Company, New York, 1954.

Theorem 1

If two four-points have the same diagonal points, then there is a conic which contains the vertices of both four-points.

Proof Let $A_0 A_1 A_2 A_3$ and $B_0 B_1 B_2 B_3$ be two four-points each of which has the diagonal points D_1, D_2, D_3, and let H_1, H_1', . . . , H_3' be the harmonic points of $B_0 B_1 B_2 B_3$. We must now consider two cases, according as no vertex of the second four-point lies on a side of the first or at least one vertex of the second four-point lies on a side of the first. In the second case, let us suppose for definiteness that B_0 lies on the line $A_0 A_1$. Then from the properties of the four-point $B_0 B_1 B_2 B_3$ it follows that $B_2 B_3$ is the harmonic conjugate of $B_0 B_1$ with respect to $D_1 D_2$ and $D_1 D_3$. Moreover, $B_0 B_1$ is the line $A_0 A_1$, since B_1 is collinear with B_0 and D_1 both of which are on $A_0 A_1$. Therefore, since $A_2 A_3$ is the harmonic conjugate of $A_0 A_1$ with respect to $D_1 D_2$ and $D_1 D_3$, it follows that $B_2 B_3$ and $A_2 A_3$ are the same line. In other words, A_0, A_1, A_2, A_3, B_0, B_1, B_2, B_3 all lie on the singular conic consisting of the lines $A_0 A_1$ and $A_2 A_3$, and the theorem is trivially true.

If no vertex of the second four-point is on a side of the first, we can consider the nonsingular conic, Γ, determined by A_0, A_1, A_2, A_3, B_0. By the result of Exercise 6, Sec. 8.6, $D_2 D_3$ is the polar of D_1 with respect to Γ. Moreover, in the four-point $B_0 B_1 B_2 B_3$ the line $B_0 D_1$ intersects $D_2 D_3$ in the harmonic point H_1. Therefore, if X is the second intersection of $B_0 D_1$ and Γ, it follows that $(D_1 H_1, B_0 X)$ is a harmonic tetrad. However, from the harmonic properties of the four-point $B_0 B_1 B_2 B_3$, we know that $(D_1 H_1, B_0 B_1)$ is also a harmonic tetrad. Hence $X = B_1$, which proves that B_1 is a point of Γ. Similarly, B_2 and B_3 are also points of Γ, and the theorem is proved.

Theorem 2

If P is a point which does not lie on any side of a four-point $A_0 A_1 A_2 A_3$, then the polars of P with respect to the conics which contain A_0, A_1, A_2, A_3 are all concurrent.

Proof Let P be a point which does not lie on any side of a four-point $A_0 A_1 A_2 A_3$, let Γ be the unique nonsingular conic determined by A_0, A_1, A_2, A_3, P, and let t be the tangent to Γ at P. In other words, t is the polar of P with respect to Γ. Now by Desargues' conic theorem, the conics of the pencil on A_0, A_1, A_2, A_3 intersect t in pairs of points which are mates in an involution on t, one of whose fixed points is P. If Q is the other fixed point of this involution, the conic Φ determined by A_0, A_1, A_2, A_3, Q is tangent to t at Q, and therefore the polar of P with respect to Φ passes through Q. Now let Ψ be any conic on A_0, A_1, A_2, A_3 other than Γ and Φ, and let R and S be the points in which t intersects Ψ. Then (PQ, RS) is a harmonic tetrad, and therefore the polar of P with respect to Ψ' also passes through Q. Since Ψ' was any conic on A_0, A_1, A_2, A_3 other than Γ and Φ, and since we have already verified that the polars of P with respect to Γ and Φ pass through Q, our proof is complete.

Theorem 3

If $\triangle ABC$ and $\triangle A'B'C'$ are two triangles so related that the sides $B'C'$, $C'A'$, and $A'B'$ of $\triangle A'B'C'$ are, respectively, the polars of the vertices A, B, and C of $\triangle ABC$ with respect to a nonsingular conic, Γ, then $\triangle ABC$ and $\triangle A'B'C'$ are perspective.

Proof Let Γ be a nonsingular conic, and let $\triangle ABC$ and $\triangle A'B'C'$ be so related that:

$B'C' = a'$ is the polar of A with respect to Γ.
$C'A' = b'$ is the polar of B with respect to Γ.
$A'B' = c'$ is the polar of C with respect to Γ.

And, perforce,

$BC = a$ is the polar of A' with respect to Γ.
$CA = b$ is the polar of B' with respect to Γ.
$AB = c$ is the polar of C' with respect to Γ.

Further, let B'' be the point bb', let B_1 be the point ab', let C_1 be the point ac', and let c_1 be the line $A'C$. Since the polar of A' is $BC = a$ and the polar of C is $A'B' = c'$, it follows that the pole of c_1 is the intersection of a and c', that is, the point C_1. Now consider the concurrent lines BA', BB', BA, and BC_1. The poles of these lines are the points in which the polar of B, namely, b', is met by the polars of A', B', A, and C_1, namely, a, b, a', and c_1. Moreover, from Theorem 5, Sec. 8.6, these lines are projective with their poles, that is, $B(A',B',A,C_1) \sim b'(a,b,a',c_1)$. Now $b'a = B_1$, $b'b = B''$, $b'a' = C'$, and $b'c_1 = A'$. Hence the last projectivity can be written $B(A',B',A,C_1) \sim b'(B_1,B'',C',A')$. Also, $b'(B_1,B'',C',A') \ \overline{\wedge}\ C(B_1,B'',C',A')$, and, by the second permutation theorem, $C(B_1,B'',C',A') \sim C(A',C',B'',B_1)$. Furthermore, the line CB'' is the same as the line CA. Hence from the last two projectivities we have $b'(B_1,B'',C',A') \sim C(A',C',A,B_1)$ and, finally,

$$B(A',B',A,C_1) \sim C(A',C',A,B_1)$$

or, since $BC_1 = BC$ and $CB_1 = CB$,

$$B(A',B',A,C) \sim C(A',C',A,B)$$

These two projective pencils have a self-corresponding line, namely, the line BC; hence they must be perspective. Clearly, the axis of this perspectivity is the line AA', and therefore the corresponding lines BB' and CC'

must also intersect on AA'. Thus AA', BB', CC' are concurrent, which shows that $\triangle ABC$ and $\triangle A'B'C'$ are perspective and completes our proof.

Corollary 1   Hesse's theorem

If two pairs of opposite sides of a four-point are conjugate with respect to a nonsingular conic, the remaining two sides are also conjugate with respect to the conic.

Proof Let $ABCD$ be a four-point such that (AB,CD) and (AC,BD) are pairs of conjugate lines with respect to a nonsingular conic Γ. Let $B'C'$, $C'A'$, and $A'B'$ be, respectively, the polars of A, B, and C, so that, perforce, BC, CA, and AB are also the polars of A', B', and C'. Then since CD and AB are conjugate lines, and since C' is the pole of AB, it follows that CD must pass through C'. Likewise, since BD and AC are conjugate lines, and since B' is the pole of AC, it follows that BD must pass through B'. Thus D must be the intersection of BB' and CC'. However, by Theorem 3, we know that AA' is concurrent with BB' and CC', that is, D also lies on AA'. Finally, since A' is the pole of BC, and since AD passes through A', it follows that AD and BC are also conjugate lines with respect to Γ, as asserted.

If two triangles, such as those described in the preceding theorem, have the property that each side of either triangle is the polar of the corresponding vertex of the other triangle with respect to a nonsingular conic Γ, each triangle is said to be the **polar** of the other with respect to Γ. It may happen, of course, that a triangle is its own polar with respect to a conic Γ, and when this is the case, the triangle is said to be **self-polar** with respect to Γ. Our final theorem gives us an interesting property of two triangles each of which is self-polar with respect to the same conic.

Theorem 4

If two triangles are each self-polar with respect to a nonsingular conic, then their six vertices all lie on another conic.

Proof Let Γ be a nonsingular conic, and let $\triangle ABC$ and $\triangle A'B'C'$ be two triangles each of which is self-polar with respect to Γ. Then:

$BC = a$ is the polar of A with respect to Γ.

$CA = b$ is the polar of B with respect to Γ.

$AB = c$ is the polar of C with respect to Γ.

$B'C' = a'$ is the polar of A' with respect to Γ.

$C'A' = b'$ is the polar of B' with respect to Γ.

$A'B' = c'$ is the polar of C' with respect to Γ.

Further, let B_1, C_1, B'', and C'' be, respectively, the points of intersection of $BC = a$ and the lines $A'C' = b'$, $A'B' = c'$, AB', and AC'. Clearly, since B'' lies on the line AB', its polar is concurrent with the polars of A and B'; that is, it passes through the intersection of BC and $A'C'$, namely, B_1. Similarly, the polar of C'' passes through C_1. Now from the definition of B'' and C'', it follows that $A(B,C,B',C') \overline{\wedge} a(B,C,B'',C'')$. Also, from Theorem 5, Sec. 8.6, the points B, C, B'', C'', which are collinear on $BC = a$, are projective with their polars, which are concurrent in the point A, the polar of BC. Thus $a(B,C,B'',C'') \sim A(C,B,B_1,C_1)$. Hence $A(B,C,B',C') \sim A(C,B,B_1,C_1)$. However, since B, C, B_1, C_1 are collinear on a, it is clear that

$$A(C,B,B_1,C_1) \overset{a}{\overline{\wedge}} A'(C,B,B_1,C_1) \sim A'(B,C,C_1,B_1)$$

the last step following from the first permutation theorem. Therefore, $A(B,C,B',C') \sim A'(B,C,C_1,B_1)$. Finally, since $A'C_1 = A'B'$ and $A'B_1 = A'C'$, the last projectivity can be written

$$A(B,C,B',C') \sim A'(B,C,B',C')$$

which, by Theorem 9, Sec. 8.2, proves that A, B, C and A', B', C' are all on a conic, as asserted.

Exercises

1. Given a nonsingular conic Γ and a triangle none of whose vertices is on Γ. Show that there are two four-points whose vertices are on Γ and which have the given triangle as diagonal triangle, and devise a construction for locating the vertices of these two four-points.

2. Given a nonsingular conic Γ and a triangle none of whose vertices is on Γ. Determine whether or not there exists a triangle with vertices on Γ each of whose sides passes through a vertex of the given triangle. If such a triangle exists, devise a construction for locating its vertices.

3. Given seven points A, B, P, Q, R, U, V such that no three of the points A, B, P, Q, R are collinear and no three of the points P, Q, R, U, V are collinear. Devise a construction for the fourth point common to the conic on A, B, P, Q, R and the conic on P, Q, R, U, V.

4. Show that the center and axis of the perspectivity established in Theorem 3 are pole and polar with respect to Γ.

5. Given three noncollinear points and two lines neither of which contains any of the given points. Show that there are four and only four conics which pass through the three given points and are tangent to the two given lines.

6. Given four points no three of which are collinear and a line which does not pass through any of these points. Show that there are two and only two conics which pass through the four given points and are tangent to the given line.

7. If A, B, C, A', B', C' are six points on a nonsingular conic, show that there is one and only one conic with respect to which $\triangle ABC$ and $\triangle A'B'C'$ are simultaneously self-polar.

8. If A, B, C, A', B', C' are six points on a nonsingular conic, show that the six sides of $\triangle ABC$ and $\triangle A'B'C'$ are all lines of a second conic.

9. Let Γ be a nonsingular conic, let A, B, P be three noncollinear points none of which is on Γ, and let Q and R be two points on Γ which are collinear with P. Show that there is a second pair of points (Q',R') on Γ such that Q' and R' are collinear with P and such that A, B, Q, Q', R, R' are all on a conic.

10. Let $(X_1X_2X_3,X_1'X_2'X_3')$ be the involutory hexad cut from a four-point $A_0A_1A_2A_3$ by a line x which does not pass through any vertex of the four-point, and let Y_1, Y_2, Y_3, Y_1', Y_2', Y_3' be, respectively, the harmonic conjugates of the points of this hexad with respect to the two vertices of the four-point with which each is collinear. Show that the locus of the poles of x with respect to the pencil of conics on A_0, A_1, A_2, A_3 is a conic, and show, further, that this conic contains the diagonal points of the four-point, the fixed points of the involution determined on x by the given involutory hexad, and the six points Y_1, Y_2, $\ldots$, Y_3'. (For obvious reasons, this conic is called the **eleven-point conic** of the given configuration.)

8.9 Conclusion

In this chapter we have developed the fundamental properties of conics, starting with the definition of a point-conic as the locus of intersections of corresponding lines of two projective pencils and the definition of a line-conic as the envelope of the joins of corresponding points of two projective ranges. In doing this we found it necessary to adopt as another axiom the assertion that every projectivity between cobasal ranges has at least one fixed point. With this, together with our earlier axioms, we were able to prove such familiar properties as the existence of a unique nonsingular conic containing five given points no three of which are collinear, the existence of a unique tangent at each point of a nonsingular conic, the existence of either one or two intersections of a line and a conic, and various of the polar properties we first encountered in Sec. 4.9.

As essentially new results, we developed the concept of a conic as a self-dual configuration consisting simultaneously of a locus and an envelope, we established Desargues' conic theorem and the theorem of Pascal, and we devised a method of constructing the fixed points of any projectivity between cobasal ranges. Using these theorems, we found that we were able to carry out a variety of interesting constructions involving conics both in projective planes satisfying our axioms and, with suitable specializations, in the euclidean plane as well.

nine

THE INTRODUCTION OF COORDINATES

9.1 Introduction

Our objective in this chapter is the introduction of coordinates, both nonhomogeneous and homogeneous, in the presently uncoordinatized structure we have evolved from Axioms 1 to 9. As a first step, we shall investigate briefly the properties of generalized number systems, or *fields*, which are abstract mathematical systems with two binary operations analogous to the processes of addition and multiplication in ordinary arithmetic. Then, after we have defined the sum and product of two points, we shall develop an arithmetic of collinear points satisfying all the field axioms. When this has been accomplished, we shall be able to introduce our final axiom, which will postulate that the set of points on an arbitrary line, with a single point excluded, is isomorphic to the field of complex numbers. Having thus established the possibility of a nonhomogeneous coordinate system on any line, we shall identify two lines as axes and extend the coordinatization into the entire projective plane, with the exception of the points of a single line analogous to the ideal line in E_2^+. Finally, we shall introduce homogeneous coordinates for every point in the projective plane, without exception.

9.2 Fields

A **field** is an abstract mathematical system whose properties, as set forth in the following definition, are analogous to, and in fact were suggested by, the properties of the real-number system.

Definition 1

A field, $\mathscr{F}$, is a set of elements $\{a,b,c,...\}$ with two binary operations, referred to as "addition" and "multiplication" and denoted by the symbols $+$ and $\cdot$, which has the following properties:

1. $\mathscr{F}$ is closed under addition; i.e., for any elements, a and b in $\mathscr{F}$, there exists an element c in $\mathscr{F}$ such that $a + b = c$.

2. The operation of addition is commutative; i.e., if a and b are any elements of $\mathscr{F}$, then $a + b = b + a$.

3. The operation of addition is associative; i.e., if a, b, and c are any elements of $\mathscr{F}$, then $(a + b) + c = a + (b + c)$.

4. $\mathscr{F}$ contains an additive identity; i.e., there exists at least one element, z, in $\mathscr{F}$ with the property that $x + z = x$ for every element x in $\mathscr{F}$.

5. Every element in $\mathscr{F}$ has an additive inverse; i.e., for each element x and each additive inverse z in $\mathscr{F}$, there exists at least one element, $-x$, such that $x + (-x) = z$.

6. $\mathscr{F}$ is closed under multiplication; i.e., for any elements p and q in $\mathscr{F}$, there exists an element r of $\mathscr{F}$ such that $p \cdot q = r$.

7. The operation of multiplication is commutative; i.e., if p and q are any elements of $\mathscr{F}$, then $p \cdot q = q \cdot p$.

8. The operation of multiplication is associative; i.e., if p, q, and r are any elements of $\mathscr{F}$, then $(p \cdot q) \cdot r = p \cdot (q \cdot r)$.

9. $\mathscr{F}$ contains a multiplicative identity; i.e., there exists at least one element, u, in $\mathscr{F}$ with the property that $y \cdot u = y$ for every element y in $\mathscr{F}$.

10. Every element of $\mathscr{F}$ which is not an additive identity has a multiplicative inverse; i.e., for each element y in $\mathscr{F}$ which is not an additive identity, z, and for each multiplicative identity, u, there exists at least one element y^{-1} such that $y \cdot y^{-1} = u$.

11. The operation of multiplication is distributive over addition; i.e., if a, b, and p are elements of $\mathscr{F}$, then $p \cdot (a + b) = p \cdot a + p \cdot b$.

It is interesting to note that corresponding statements in the set 1 to 5 and the set 6 to 10 in Definition 1 make comparable assertions about the operation of addition and the operation of multiplication, while statement 11 serves to relate the two operations.

The real-number system, with addition and multiplication having their usual meanings, is undoubtedly the most familiar example of a field. Other well-known examples are the set of all rational numbers and the set of all complex numbers. It is also easy to verify that with addition and multiplication defined in the usual way, the set of all real numbers of the form $a + b\sqrt{2}$, where a and b are rational, or, more generally, the set of all real numbers of the form $a + b\sqrt{n}$, where a and b are rational and n is a given integer, is a field.

In each of the preceding examples there was only one additive identity, namely the number 0, and only one multiplicative identity, namely the number 1. This suggests that an arbitrary field has a unique additive identity and a unique multiplicative identity, and although this is not asserted in the definition of a field, it can be proved to be the case:

Theorem 1
A field has a unique additive identity.

Proof Let $\mathscr{F}$ be an arbitrary field, and let us suppose that both z and z' are additive identities in $\mathscr{F}$. Then, by property 4, $x + z = x$ for each element x in $\mathscr{F}$. In particular, if we take x to be the second additive identity, z', we have $z' + z = z'$. From this, using the commutative property of addition, it follows that $z + z' = z'$. Now, by hypothesis, z' is an additive identity; hence by property 4 we have $z + z' = z$. Therefore $z = z'$, since each is the element $z + z'$; and $\mathscr{F}$ has only one additive identity, as asserted.

In essentially the same way, we can establish the companion theorem for multiplication:

Theorem 2
A field has only one multiplicative identity.

With Theorems 1 and 2 available, it is easy to prove the uniqueness of the additive and multiplicative inverses described in properties 5 and 10:

Theorem 3
An arbitrary element of a field has a unique additive inverse.

Proof   Let a be an arbitrary element of a field $\mathscr{F}$, and let us suppose that both $-a$ and $-a'$ are additive inverses of a. Then by property 5, $a + (-a) = z$, and also $a + (-a') = z$. Hence $a + (-a) = a + (-a')$, since each is the (unique) element z. Therefore, $(-a') + [a + (-a)] = (-a') + [a + (-a')]$. Now using properties 2, 3, and 4, we have for the respective members of the last equality,

$$(-a') + [a + (-a)] = [(-a') + a] + (-a) = z + (-a) = -a$$

and $(-a') + [a + (-a')] = [(-a') + a] + (-a') = z + (-a') = -a'$

Hence $-a = -a'$; and therefore a has only one additive inverse, as asserted.

Theorem 4

Each element of a field except the additive identity has a unique multiplicative inverse.

By analogy with elementary arithmetic, the additive identity of a field is usually referred to as the **zero element** of the field, and the multiplicative identity is usually referred to as the **unit element** of the field. Also, the additive inverse of an element is often called the **negative** of that element, and a sum of the form $a + (-b)$ is often written simply as $a - b$. Similarly, the multiplicative inverse of an element is often called the **reciprocal** of that element, and the product $p \cdot (q^{-1})$ is often written simply as p/q. Although the elements of a field need not be numbers, it is customary to use numerical coefficients to abbreviate the sums which arise when an element of a field is added to itself a number of times. Thus we write $a + a = 2a$, $a + a + a = 3a, \ldots$. Similarly, we use the exponential notation to denote the products which arise when an element of a field is multiplied by itself a number of times. Thus we write $a \cdot a = a^2$, $a \cdot a \cdot a = a^3, \ldots$.

If a is an element of a field, the smallest value of n for which $na = z$ is called the **order** of a. If there is no integer n for which $na = z$, the element a is said to be of **infinite order.**

The following theorems are now easy to establish, though we shall leave their proofs as exercises.

Theorem 5   *The cancellation law for addition*

If a, b, and c are elements of a field, and if $a + b = a + c$, then $b = c$.

Theorem 6   *The cancellation law for multiplication*

If p, q, and r are elements of a field, and if p is not the additive identity, then $p \cdot q = p \cdot r$ implies $q = r$.

Corollary 1

If z is the additive identity of a field, then the relation $p \cdot q = z$ implies that either p or q is the element z.

Theorem 7

In any field, each of the equations $a + x = b$ and $x + a = b$ has the unique solution $x = b + (-a) = b - a$.

Theorem 8

In any field, each of the equations $px = q$ and $xp = q$ has the unique solution $x = qp^{-1} = q/p$, provided p is not the additive identity.

The examples of fields which we gave above were all infinite fields; i.e., each contained an infinite number of elements. There are also fields with only a finite number of elements, and since these play a fundamental role in the construction of finite geometries such as we discussed in Sec. 6.2, we shall illustrate here how such fields may be obtained. In brief, the process is the following: First we choose any prime number p. Then we take the integers $0, 1, 2, \ldots, p - 1$ as the elements of our field, $\mathscr{F}$, and we define the operations of addition and multiplication in the usual way, except that at the end of every calculation we reduce each result modulo p; that is, we retain only the remainder when the actual result is divided by p. With these definitions it is possible to verify that each of the field properties holds in the system $\mathscr{F}$. For instance, if $p = 3$, so that the elements of our field, $\mathscr{F}$, are the three numbers 0, 1, 2, we have the following as the tables which define the operations of addition and multiplication in $\mathscr{F}$:

+	0	1	2
0	0	1	2
1	1	2	0
2	2	0	1

·	0	1	2
0	0	0	0
1	0	1	2
2	0	2	1

On the other hand, if $p = 5$, the elements in $\mathscr{F}$ are the numbers 0, 1, 2, 3, 4, and the tables defining addition and multiplication are

+	0	1	2	3	4
0	0	1	2	3	4
1	1	2	3	4	0
2	2	3	4	0	1
3	3	4	0	1	2
4	4	0	1	2	3

·	0	1	2	3	4
0	0	0	0	0	0
1	0	1	2	3	4
2	0	2	4	1	3
3	0	3	1	4	2
4	0	4	3	2	1

From the finite field arising from a given prime, p, it is possible to construct a finite projective geometry with exactly $p + 1$ points on each line in the following way. The points of the geometry are defined to be ordered triples (x_1, x_2, x_3) whose entries are elements of the field, i.e., integers between 0 and $p - 1$, inclusive. As usual, the triple $(0,0,0)$ is excluded, and it is understood that proportional triples identify the same point. Since there are p possibilities for each of the three entries, the number of distinct triples which can be formed with the numbers 0, 1, 2, ..., $p - 1$ is p^3. However, since the triple $(0,0,0)$ is inadmissible, there are actually only $p^3 - 1$ different triples to consider. Moreover, since each of the $p - 1$ triples obtained from any particular triple by multiplying it by the numbers 1, 2, ..., $p - 1$ corresponds to the same point, it follows that collectively the $p^3 - 1$ triples identify only

$$\frac{p^3 - 1}{p - 1} = p^2 + p + 1 = (p + 1)^2 - (p + 1) + 1$$

points. This, of course, checks the results of Exercise 2, Sec. 6.2, for the total number of points in a finite projective geometry having $n + 1$ points on every line.

The lines of our finite geometry can now be defined as the $p^2 + p + 1$ ordered triples $[l_1, l_2, l_3]$ constructed in precisely the same way that we formed the triples (x_1, x_2, x_3), with the incidence condition $l_1 x_1 + l_2 x_2 + l_3 x_3 = 0$, modulo p, as the condition which tells us whether a given point does or does not lie on a given line. For every value of p it can be verified that the system constructed in this manner satisfies Axioms 1 to 6, Sec. 6.2.

To illustrate the process we have just described, let us take $p = 3$ and construct the corresponding finite geometry. Clearly there are $3^3 - 1 = 26$ different ordered triples which we can form from the numbers 0, 1, 2, excluding the triple $(0,0,0)$. However, since each triple when multiplied by 2 yields a different triple corresponding to the same point or line, as the case may be, we actually will have only $26/2 = 13$ points and 13 lines in our system. It is now easy to list these explicitly:

$$
\begin{array}{ll}
P_1 : (1,0,0) : (2,0,0) & l_1 : [1,0,0] \\
P_2 : (0,1,0) : (0,2,0) & l_2 : [0,1,0] \\
P_3 : (0,0,1) : (0,0,2) & l_3 : [0,0,1] \\
P_4 : (0,1,1) : (0,2,2) & l_4 : [0,1,1] \\
P_5 : (1,0,1) : (2,0,2) & l_5 : [1,0,1] \\
P_6 : (1,1,0) : (2,2,0) & l_6 : [1,1,0] \\
P_7 : (0,1,2) : (0,2,1) & l_7 : [0,1,2] \\
P_8 : (1,0,2) : (2,0,1) & l_8 : [1,0,2] \\
P_9 : (1,2,0) : (2,1,0) & l_9 : [1,2,0]
\end{array}
$$

$$P_{10}:(2,1,1):(1,2,2) \qquad l_{10}:[2,1,1]$$
$$P_{11}:(1,2,1):(2,1,2) \qquad l_{11}:[1,2,1]$$
$$P_{12}:(1,1,2):(2,1,1) \qquad l_{12}:[1,1,2]$$
$$P_{13}:(1,1,1):(2,2,2) \qquad l_{13}:[1,1,1]$$

The calculations required to identify the points which lie on each line are now perfectly straightforward, and the incidence relations displayed in the following table are easily checked. For instance, P_{12} lies on l_{12} because

l_1	l_2	l_3	l_4	l_5	l_6	l_7	l_8	l_9	l_{10}	l_{11}	l_{12}	l_{13}
P_2	P_1	P_1	P_1	P_2	P_3	P_1	P_2	P_3	P_5	P_4	P_4	P_7
P_3	P_3	P_2	P_7	P_8	P_9	P_4	P_5	P_6	P_6	P_6	P_5	P_8
P_4	P_5	P_6	P_{11}	P_{10}	P_{10}	P_{10}	P_{11}	P_{12}	P_7	P_8	P_9	P_9
P_7	P_8	P_9	P_{12}	P_{12}	P_{11}	P_{13}	P_{13}	P_{13}	P_{10}	P_{11}	P_{12}	P_{13}

$1 \cdot 1 + 1 \cdot 1 + 2 \cdot 2 = 6 = 0$, mod 3, and P_{11} does not lie on l_{12} because $1 \cdot 1 + 2 \cdot 1 + 1 \cdot 2 = 5 = 2 \neq 0$, mod 3. It is easy to show that the system we have just constructed and the system described by Table 6.2 are isomorphic.

The preceding process for constructing a finite field with n elements breaks down if n is not a prime, p. For instance, the numbers 0, 1, 2, 3 do not form a field under the operations of addition and multiplication modulo 4 because, in particular, the element 2 has no multiplicative inverse. In fact, since $2 \cdot 0 = 0$, $2 \cdot 1 = 2$, $2 \cdot 2 = 4 = 0$ (mod 4) and $2 \cdot 3 = 6 = 2$ (mod 4), it is clear that there is no element, x, such that $2 \cdot x = 1$, as required by property 10 of Definition 1.

Actually, finite fields exist when and only when the number of elements, n, is a power of a prime, say $n = p^k$, $k = 1, 2, 3, \ldots$. When $k = 1$, they can be constructed by the process we described above. When $k > 1$, they can be constructed in the following way, as extensions of the field arising from p itself. Let $\mathscr{F}$ be the field whose elements are the numbers $0, 1, 2, \ldots, p - 1$ and in which addition and multiplication are understood to be carried out modulo p, where p is an arbitrary prime. Furthermore, let

$$(1) \qquad x^k + a_1 x^{k-1} + a_2 x^{k-2} + \cdots + a_{k-1} x + a_k = 0$$

be an equation whose coefficients are elements of $\mathscr{F}$ but whose left member cannot be resolved into factors whose coefficients are elements of $\mathscr{F}$. Such an equation is said to be **irreducible** over $\mathscr{F}$. Now consider the set, $\mathscr{F}^*$, consisting of all expressions of the form

$$b_0 + b_1 x + b_2 x^2 + \cdots + b_{k-2} x^{k-2} + b_{k-1} x^{k-1}$$

which can be constructed from the symbol x, using the elements of $\mathscr{F}$ as coefficients. Clearly, since there are k b's and p possible values for each b, the set $\mathscr{F}^*$ contains p^k elements. Finally, in $\mathscr{F}^*$ let addition and multiplication be defined in the "natural" way, using the addition and multiplication tables of the b's and relation (1) to reduce all calculations. With these definitions, it can be shown that $\mathscr{F}^*$ is a field.[1]

To illustrate this extension process, let us attempt to construct a finite field with four elements. Since $4 = 2^2$ and 2 is a prime, it should be possible to do this, beginning with the field $\mathscr{F}$ whose elements are the numbers 0 and 1. Our first step is to select a second-degree equation whose coefficients are elements of $\mathscr{F}$, that is, either 0 or 1, and which is irreducible over $\mathscr{F}$. The only possibilities are

$$x^2 = 0 \qquad x^2 + 1 = 0 \qquad x^2 + x = 0 \qquad x^2 + x + 1 = 0$$

Now in $\mathscr{F}$ the additive inverse of 1, namely, -1, is 1 itself. That is, in $\mathscr{F}$, $1 = -1$, and therefore the preceding equations can be rewritten in the form

$$x^2 = x \cdot x = 0 \qquad x^2 - 1 = (x + 1)(x - 1) = 0$$

$$x(x + 1) = 0 \qquad x^2 + x + 1 = 0$$

This shows that each of the first three is reducible over $\mathscr{F}$, and hence we have no choice but to continue with the equation

$$x^2 + x + 1 = x^2 - x - 1 = 0 \qquad \text{or} \qquad x^2 = x + 1$$

which is, in fact, irreducible (see Exercise 15). We now consider the set $\mathscr{F}^*$ of expressions of the form $b_0 + b_1 x$, where each of the b's is either 0 or 1. There are four such elements, namely, 0, 1, x, $1 + x$. Finally, using the relation $x^2 = x + 1$, we construct the addition and multiplication tables in $\mathscr{F}^*$:

$+$	0	1	x	$1 + x$
0	0	1	x	$1 + x$
1	1	0	$1 + x$	x
x	x	$1 + x$	0	1
$1 + x$	$1 + x$	x	1	0

[1] It is instructive to compare this procedure with the process by which the complex-number field is generated from the real field by adjoining to the latter an arbitrary symbol, i, satisfying the relation $i^2 + 1 = 0$, and then considering the set of all elements of the form $a + bi$, where a and b are elements of the real field, and all products are simplified by using the relation $i^2 + 1 = 0$, that is, $i^2 = -1$. The two procedures are, of course, the same.

·	0	1	x	$1 + x$
0	0	0	0	0
1	0	1	x	$1 + x$
x	0	x	$1 + x$	1
$1 + x$	0	$1 + x$	1	x

For instance,

$$x + (1 + x) = 1 + (x + x) = 1 + (1 + 1)x = 1 + 0 \cdot x = 1$$

and

$$(1 + x)(1 + x) = 1 + (x + x) + x^2 = 1 + 0 + x^2 = 1 + (x + 1) = x$$

Using $\mathscr{F}^*$, it is now possible to construct a finite projective geometry having five points on every line, although we shall leave this as an exercise.

As we mentioned above, there exists a finite field with n elements if and only if n is a power of a prime. Moreover, it can be shown[1] that for each such value of n there is essentially only one field, namely, the one whose construction we have just described. In other words, *any two finite fields with the same number of elements are isomorphic.* For any field with p^k elements, the prime p has an important significance which is made clear in the next two theorems.

Theorem 9

All nonzero elements of a field have the same order.

Proof Let a and b be any nonzero elements of a field, $\mathscr{F}$, and let a be of order n. By definition, this means that n is the smallest integer such that $na = z$, where z is the additive identity or zero element of $\mathscr{F}$. Now if $na = z$, then $(na)b = zb = z$. Also

$$(na)b = (a + a + \cdots + a)b = ab + ab + \cdots + ab$$
$$= ba + ba + \cdots + ba = (b + b + \cdots + b)a = (nb)a$$

Hence $(na)b = z$ implies $(nb)a = z$, and since $a \neq z$, this implies, by Corollary 1, Theorem 6, that $nb = z$. Moreover, n must be the smallest integer for which $nb = z$, for otherwise, by repeating the argument with the roles of a and b interchanged, we would reach a contradiction of the fact that the order of a is n. Thus we have shown that if one element of $\mathscr{F}$ is of order n, then every element of $\mathscr{F}$ is of order n, as asserted.

[1] See, for instance, G. Birkhoff and S. MacLane, "A Survey of Modern Algebra," pp. 445–447, The Macmillan Company, New York, 1953.

Corollary 1

If one element of a field is of infinite order, then every element of the field is of infinite order.

Theorem 10

The common value of the orders of the nonzero elements of a field with p^k elements is the prime p.

Proof   Since every field with p^k elements contains as a subfield the field $\mathscr{F}$ consisting of the p elements $0, 1, 2, \ldots, p - 1$, it is sufficient to consider this subfield and prove that it contains at least one element whose order is p. Now from the properties of addition modulo p, it follows that every element, x, in $\mathscr{F}$ has the property that $px = z$. If p is not the order of x, let us suppose that $p_1 < p$ is the smallest positive integer such that $p_1 x = z$. By the division algorithm, p can be expressed in the form $p = p_1 q + r$, where the remainder, r, is less than p_1. Hence the relation $px = z$ implies that $(p_1 q + r)x = z$ or $q(p_1 x) + rx = z$, or simply $rx = z$. However, since $r < p_1$, this contradicts the assumption that p_1 was the smallest integer such that $p_1 x = z$. Hence we must abandon the supposition that p is not the order of x, and the theorem is established.

Definition 2

The common value of the orders of the nonzero elements of a field $\mathscr{F}$ is called the characteristic of $\mathscr{F}$.

With Theorems 9 and 10 available, we can now give a proof of an assertion we made without proof in Sec. 7.2.

Theorem 11

If PG_n is a finite projective geometry constructed from a field of characteristic p, then the diagonal points of a four-point in PG_n are collinear if and only if $p = 2$, that is, if and only if the number of points on each line of PG_n is $2^k + 1$.

Proof   Let $A_0 : (x_0, y_0, z_0)$, $A_1 : (x_1, y_1, z_1)$, $A_2 : (x_2, y_2, z_2)$, $A_3 : (x_3, y_3, z_3)$ be the vertices of an arbitrary four-point in a PG_n constructed from a finite field, $\mathscr{F}$. Then as in Sec. 7.2, numbers $\lambda_0, \lambda_1, \lambda_2, \lambda_3$, none of which is zero, exist such that the diagonal points of the four-point are

$$D_1: \begin{matrix} (\lambda_0 x_0 + \lambda_1 x_1, \ \lambda_0 y_0 + \lambda_1 y_1, \ \lambda_0 z_0 + \lambda_1 z_1) \\ (\lambda_2 x_2 + \lambda_3 x_3, \ \lambda_2 y_2 + \lambda_3 y_3, \ \lambda_2 z_2 + \lambda_3 z_3) \end{matrix}$$

$$D_2: \begin{matrix} (\lambda_0 x_0 + \lambda_2 x_2, \ \lambda_0 y_0 + \lambda_2 y_2, \ \lambda_0 z_0 + \lambda_2 z_2) \\ (\lambda_1 x_1 + \lambda_3 x_3, \ \lambda_1 y_1 + \lambda_3 y_3, \ \lambda_1 z_1 + \lambda_3 z_3) \end{matrix}$$

$$D_3: \begin{matrix} (\lambda_0 x_0 + \lambda_3 x_3, \ \lambda_0 y_0 + \lambda_3 y_3, \ \lambda_0 z_0 + \lambda_3 z_3) \\ (\lambda_1 x_1 + \lambda_2 x_2, \ \lambda_1 y_1 + \lambda_2 y_2, \ \lambda_1 z_1 + \lambda_2 z_2) \end{matrix}$$

Now suppose that the field, $\mathscr{F}$, from which the coordinates are taken is of characteristic 2, and consider the determinant of the coordinates of D_1, D_2, D_3:

$$\begin{vmatrix} \lambda_2 x_2 + \lambda_3 x_3 & \lambda_2 y_2 + \lambda_3 y_3 & \lambda_2 z_2 + \lambda_3 z_3 \\ \lambda_1 x_1 + \lambda_3 x_3 & \lambda_1 y_1 + \lambda_3 y_3 & \lambda_1 z_1 + \lambda_3 z_3 \\ \lambda_1 x_1 + \lambda_2 x_2 & \lambda_1 y_1 + \lambda_2 y_2 & \lambda_1 z_1 + \lambda_2 z_2 \end{vmatrix}$$

If each of the first two rows is added to the third row, and if the result is simplified by using the fact that $x + x = 0$ for every element in a field of characteristic 2, each element in the new third row is zero. Hence the determinant is zero, and the diagonal points D_1, D_2, D_3 are collinear, as asserted.

Conversely, let us suppose that the diagonal points D_1, D_2, D_3 are collinear on a line $[l_1, l_2, l_3]$. Then if we let $s_i = l_1 x_i + l_2 y_i + l_3 z_i$, $i = 0, 1, 2, 3$, it follows, as in Sec. 7.2, that

$$\begin{aligned} \lambda_0 s_0 + \lambda_1 s_1 = 0 && \lambda_2 s_2 + \lambda_3 s_3 = 0 \\ \lambda_0 s_0 + \lambda_2 s_2 = 0 && \lambda_1 s_1 + \lambda_3 s_3 = 0 \\ \lambda_0 s_0 + \lambda_3 s_3 = 0 && \lambda_1 s_1 + \lambda_2 s_2 = 0 \end{aligned}$$

From the three equations in the second column, it follows at once that $-\lambda_1 s_1 - \lambda_1 s_1 = 0$. Hence in the field $\mathscr{F}$ from which the coordinates are taken, there is at least one element, $x = -\lambda_1 s_1$, such that $x + x = 2x = 0$. Therefore, by Theorem 9, the characteristic of $\mathscr{F}$ is 2, and our proof is complete.

Exercises

1. Prove Theorem 2.

2. Prove Theorem 4.

3. Prove Theorem 5.

4. Prove Theorem 6.

5. Prove Corollary 1, Theorem 6.

6. Prove Theorem 7.

7. Prove Theorem 8.

8. Show that the PG_4 obtained in this section is isomorphic to the system defined by Table 6.2.

9. Using the appropriate addition and multiplication tables developed in this section, construct a PG_5 and show that it is isomorphic to the system defined by Table 6.3.

10. Construct a PG_6.

11. (*a*) Construct a finite field containing eight elements by adding to the field with two elements an element x satisfying the equation $x^3 = x + 1$.

(*b*) Work part (*a*) using an element x satisfying the equation $x^3 = x^2 + 1$.
(*c*) Show that the fields constructed in parts (*a*) and (*b*) are isomorphic to
each other.
(*d*) Can part (*a*) be worked using an element x satisfying the equation
$x^3 = x^2 + x + 1$? Why?

12. Show that the set of numbers of the form $a + b\sqrt{2} + c\sqrt{3} + d\sqrt{6}$,
where a, b, c, d are rational, is a field under the operations of ordinary
addition and multiplication.

13. Show that if the coefficients in a quadratic equation are elements of a
field $\mathscr{F}$, the equation may be solved by the usual quadratic formula if and
only if the characteristic of $\mathscr{F}$ is different from 2.

14. Show that the characteristic of every finite field must be a prime.

15. Assuming the factor theorem of elementary algebra, show that a
polynomial is irreducible over a field $\mathscr{F}$ if and only if for each element in $\mathscr{F}$
the value of the polynomial is different from 0. Using this fact, show that
$x^2 + x + 1$ is indeed irreducible over the field whose elements are $(0,1)$.

9.3 The Arithmetic of Collinear Points

Each of the fields which we discussed in the last section was either completely,
or at least essentially, numerical in character. In this section, we propose to
define the sum and the product of two points in such a way that a field can
be constructed from the points of an arbitrary line *prior* to any coordinatization of the line. After this has been accomplished, coordinates can then be
introduced by postulating an isomorphism between the field of points on the
line and a suitable field of numbers.

To define the sum and the product of points on an arbitrary line, l,
we must first identify three points on l to be used in formulating our definitions. These points, which are arbitrary except for the requirement that
they be distinct, we shall label with the symbols L_0, L_1, and L_ω, suggesting,
respectively, the zero point, the unit point, and the ideal point on an
ordinary euclidean line. The points L_0, L_1, L_ω are, together, called the
gauge points on l. The set of points on l with the exception of L_ω is called
the **open set** on l.

Definition 1

If L_0, L_1, and L_ω are the gauge points on a line l, the sum of any
points L_x and L_y of the open set on l is the point, $L_x + L_y$, which is the mate
of L_0 in the involution on l in which L_x and L_y are a pair of mates and L_ω
is self-corresponding.

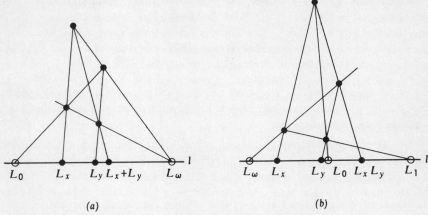

Fig. 9.1 The sum and the product of two points.

The familiar construction, based on Desargues' conic theorem, which yields the sum of two points relative to a given set of gauge points, is illustrated in Fig. 9.1a. It should be noted that the sum of two points is independent of L_1 and depends only on the points themselves and the gauge points L_0 and L_ω.

Definition 2

If L_0, L_1, and L_ω are the gauge points on a line l, and if L_x and L_y are any points of the open set on l distinct from L_0, then the product of L_x and L_y is the point, L_xL_y, which is the mate of L_1 in the involution on l in which L_x and L_y are a pair of mates and L_0 and L_ω are a pair of mates. If either L_x or L_y is the gauge point L_0, then the product of L_x and L_y is also the point L_0.

Figure 9.1b illustrates the construction which yields the product of two points in terms of a given set of gauge points.

We must now verify that the addition and multiplication of points, as we have defined these operations, satisfy the 11 requirements of Definition 1, Sec. 9.2.

At the outset, it is obvious that property 1 holds, because we know that an involution is completely determined when one self-corresponding point, L_ω, and one pair of reciprocally corresponding points, (L_x, L_y), are given, and in such an involution the point L_0 always has a unique image. Likewise, property 6 holds, because if either L_x or L_y is the point L_0, then by definition their product is also L_0; and if neither L_x nor L_y is the point L_0, then their product is the unique image of L_1 in the involution which is completely determined by the two pairs of corresponding points (L_x, L_y) and (L_0, L_ω).

Property 2 holds because in any involution every pair of mates is a pair of reciprocally corresponding points, and hence the roles of L_x and L_y can be interchanged without changing the image of L_0 in the involution which defines their sum. Similarly, property 7 holds because if either L_x or L_y is L_0, then, by definition, both $L_x L_y$ and $L_y L_x$ equal L_0; and if neither L_x nor L_y is L_0, then their roles can be interchanged without changing the image of L_1 in the involution which defines their product.

Property 4 holds because, in fact, L_0 is the unique additive identity. This follows trivially since $L_0 + L_x$ is the image of L_0 in an involution in which, by definition, the image of L_0 is L_x. Likewise, property 9 holds because, in fact, L_1 is the unique multiplicative identity. This also is obvious because if L_x is the point L_0, then $L_1 L_x = L_1 L_0 = L_0$; and if L_x is not L_0, then $L_1 L_x$ is the image of L_1 in an involution in which, by definition, L_x is the image of L_1.

To verify property 5, that is, to show that for each point L_x of the open set on l there is a point, $-L_x$, such that $L_x + (-L_x) = L_0$, we observe that the required point is, in fact, the image of L_x in the unique involution in which L_0 and L_ω are self-corresponding points. To verify property 10, that is, to show that for each point L_x of the open set which is distinct from L_0 there is a point L_x^{-1} such that $L_x L_x^{-1} = L_1$, we observe that the required point is, in fact, the image of L_x in the unique involution in which L_0 and L_ω are reciprocally corresponding points and L_1 is self-corresponding.

Preparatory to verifying the remaining requirements of Definition 1, Sec. 9.2, namely, properties 3, 8, and 11, it will be convenient to prove the following lemmas.

Lemma 1

If $L_x + L_a = L_p$, $L_x + L_b = L_q$, $L_x + L_c = L_r, \ldots$, or if $L_a + L_x = L_p$, $L_b + L_x = L_q$, $L_c + L_x = L_r, \ldots$ on a line l with gauge points L_0, L_1, L_ω, then

$$l(L_\omega, L_0, L_a, L_b, L_c, \ldots) \sim l(L_\omega, L_x, L_p, L_q, L_r, \ldots)$$

Proof   On a line l with gauge points L_0, L_1, L_ω, let $L_x + L_a = L_p$, $L_x + L_b = L_q$, $L_x + L_c = L_r, \ldots$, and let the construction which yields these sums be carried out using the same two lines, u and v, on L_ω, the same line, h, on L_0, and the same line, k, on L_x. Further, let O' be the intersection of h and u, and let O be the intersection of k and u. Then (Fig. 9.2),

$$l(L_\omega, L_0, L_a, L_b, L_c, \ldots) \overset{O'}{\barwedge} v(L_\omega, X, P, Q, R, \ldots) \overset{O}{\barwedge} l(L_\omega, L_x, L_p, L_q, L_r, \ldots)$$

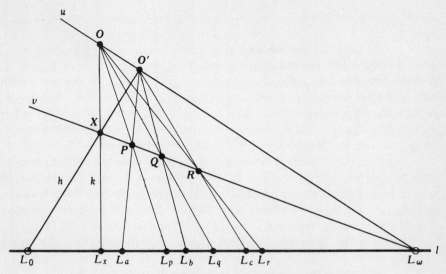

Fig. 9.2   Figure used in the proof of Lemma 1.

and therefore

$$l(L_\omega, L_0, L_a, L_b, L_c, \ldots) \sim l(L_\omega, L_x, L_p, L_q, L_r, \ldots)$$

The second assertion of the lemma follows at once since the construction shown in Fig. 9.2 defines equally well the sums $L_a + L_x, L_b + L_x, L_c + L_x, \ldots$.

Lemma 2

If $L_x L_a = L_p$, $L_x L_b = L_q$, $L_x L_c = L_r$, $\ldots$, or if $L_a L_x = L_p$, $L_b L_x = L_q$, $L_c L_x = L_r$, $\ldots$ on a line l with gauge points L_0, L_1, L_ω, then

$$l(L_\omega, L_0, L_1, L_a, L_b, L_c, \ldots) \sim l(L_\omega, L_0, L_x, L_p, L_q, L_r, \ldots)$$

Proof   On a line l with gauge points L_0, L_1, L_ω, let $L_x L_a = L_p$, $L_x L_b = L_q$, $L_x L_c = L_r$, $\ldots$, and let the constructions which yield these products be carried out using the same line u on L_x, the same line v on L_ω, the same line h on L_0, and the same line k on L_1, as shown in Fig. 9.3. Further, let O and O' be, respectively, the intersections of v with the lines u and k. Then

$$l(L_\omega, L_0, L_1, L_a, L_b, L_c, \ldots) \overset{O'}{\wedge} h(W, L_0, X, P, Q, R, \ldots) \overset{O}{\wedge} l(L_\omega, L_0, L_x, L_p, L_q, L_r, \ldots)$$

and therefore

$$l(L_\omega, L_0, L_1, L_a, L_b, L_c, \ldots) \sim l(L_\omega, L_0, L_x, L_p, L_q, L_r, \ldots)$$

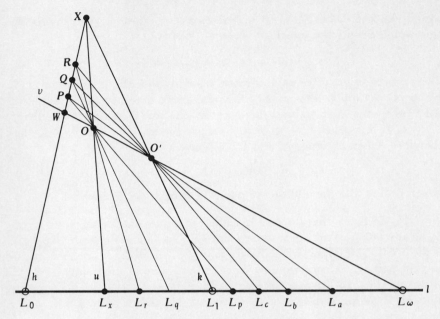

Fig. 9.3   Figure used in the proof of Lemma 2.

The second assertion of the lemma follows at once since the construction shown in Fig. 9.3 defines equally well the products L_aL_x, L_bL_x, L_cL_x, ...

To prove that the addition of points is associative, let L_x, L_y, L_z be any points of the open set on a line l, and let

$$L_x + L_y = L_u \qquad L_y + L_z = L_v \qquad L_u + L_z = L_s$$

Applying Lemma 1 to the last two of these relations, with $x = z$, $a = y$, $b = u$, $p = v$, and $q = s$, we have

(1) $$l(L_\omega, L_0, L_y, L_u) \sim l(L_\omega, L_z, L_v, L_s)$$

Also, from the involution which defines the sum $L_x + L_y = L_u$, we have $l(L_\omega, L_0, L_u, L_x, L_y) \sim l(L_\omega, L_u, L_0, L_yL_x)$, or, abstracting the four points which appear in the left side of the projectivity (1),

(2) $$l(L_\omega, L_0, L_y, L_u) \sim l(L_\omega, L_u, L_x, L_0)$$

Hence, combining (1) and (2), we have

(3) $$l(L_\omega, L_u, L_x, L_0) \sim l(L_\omega, L_z, L_v, L_s)$$

This is a projectivity in which L_ω is self-corresponding, L_v corresponds to L_x, and L_s is the mate of L_0. However, we do not yet know that this projectivity is an involution, and hence we cannot conclude at this stage (as we would like to do) that $L_x + L_v = L_s$. Instead, we continue with the involution which defines the sum $L_u + L_z = L_s$, namely,

$$l(L_\omega, L_0, L_s, L_u, L_z) \sim l(L_\omega, L_s, L_0, L_z, L_u)$$

This involution has three pairs of mates in common with the projectivity (3). Hence, by Axiom 7, they must be the same projectivity. In other words, (3) is an involution after all, and therefore we can now conclude from (3) that $L_x + L_v = L_s$. Thus $L_x + L_v = L_u + L_z$, since each is equal to L_s, and therefore, substituting for L_v and L_u, we have

$$L_x + (L_y + L_z) = (L_x + L_y) + L_z$$

which proves that the addition of points is associative, as asserted.

To prove that the multiplication of points is associative, let L_x, L_y, L_z be any points of the open set on a line l. If any one of the points L_x, L_y, L_z is the point L_0, it follows immediately that $(L_x L_y) L_z = L_x (L_y L_z)$, since each is just the point L_0. Hence we may suppose that none of the points L_x, L_y, L_z is the point L_0. In this case let $L_x L_y = L_u$, $L_y L_z = L_v$, $L_u L_z = L_s$. Applying Lemma 2 to the last two of these equations, with $x = z$, $a = u$, $b = y$, $p = s$, $q = v$, we have

(4) $$l(L_\omega, L_0, L_1, L_u, L_y) \sim l(L_\omega, L_0, L_z, L_s, L_v)$$

Also, since $L_x L_y = L_u$, we have from the definition of the product of two points,

(5) $$l(L_\omega, L_0, L_1, L_u, L_y) \sim l(L_0, L_\omega, L_u, L_1, L_x)$$

Hence, from (4) and (5) we have

$$l(L_0, L_\omega, L_u, L_1, L_x) \sim l(L_\omega, L_0, L_z, L_s, L_v)$$

Since this projectivity contains a pair of reciprocally corresponding points, (L_0, L_ω), it is an involution. Hence, since L_s is the mate of L_1 in this involution, it follows that $L_x L_v = L_s$. Thus $L_u L_z = L_x L_v$, since each is the point L_s. Therefore, substituting for L_u and L_v, we obtain

$$(L_x L_y) L_z = L_x (L_y L_z)$$

which proves that the multiplication of points is associative, as asserted.

Finally, to verify property 11 of Definition 1, Sec. 9.2, that is, to verify that the multiplication of points is distributive over addition, we note first that if L_x is the point L_0, then the assertion $L_x(L_y + L_z) = L_x L_y + L_x L_z$ is trivially true. Hence we may suppose that $L_x \neq L_0$. In this case, let

$$L_y + L_z = L_r \qquad L_x L_y = L_s \qquad L_x L_z = L_t \qquad L_x L_r = L_u$$

Then by Lemma 2 we have

(6) $$l(L_\omega,L_0,L_1,L_y,L_z,L_r) \sim l(L_\omega,L_0,L_x,L_s,L_t,L_u)$$

Also, from the fact that $L_y + L_z = L_r$, we have

(7) $$l(L_\omega,L_0,L_r,L_y,L_z) \sim l(L_\omega,L_r,L_0,L_z,L_y)$$

Now by rearranging the listing of images in (6), we obtain

(6a) $$l(L_\omega,L_0,L_r,L_y,L_z) \sim l(L_\omega,L_0,L_u,L_s,L_t)$$

and

(6b) $$l(L_\omega,L_r,L_0,L_z,L_y) \sim l(L_\omega,L_u,L_0,L_t,L_s)$$

Hence, combining the projectivities (6a) and (6b) with the projectivity (7), we obtain $l(L_\omega,L_0,L_u,L_s,L_t) \sim l(L_\omega,L_u,L_0,L_t,L_s)$. Since this projectivity contains a pair of reciprocally corresponding points, (L_s,L_t), it is an involution. Moreover, since L_ω is self-corresponding, and since L_u is the mate of L_0, this involution implies that $L_s + L_t = L_u$. Thus $L_x L_r = L_s + L_t$, since each is the point L_u. Therefore, substituting for L_r, L_s, and L_t, we have

$$L_x(L_y + L_z) = L_x L_y + L_x L_z$$

which proves that the multiplication of points is distributive over addition, as asserted.

We have now verified that each of the 11 requirements for a field holds in the open set of points on an arbitrary line when sums and products of points are defined as in Definitions 1 and 2. Thus the open set of points on a line is an example of a field. With this knowledge, we can now introduce a coordinate system on an arbitrary line, l, by postulating an isomorphism between the field of points in the open set on l and some appropriate number field. However, since our ultimate objective is the coordinatization of the entire projective plane, with the possible exception of the points of one special line, and since this will require the simultaneous coordinatization of several lines to serve as axes, it is important that we determine whether each line must be assumed isomorphic to the same number field or whether different lines can, at the same time, be respectively isomorphic to essentially different number fields. The next theorem answers this question by showing that the open sets on any two lines are isomorphic and hence cannot be isomorphic to number fields which are not themselves isomorphic.

Theorem 1

The open sets on any two lines of the projective plane are isomorphic.

Proof Let l and m be two lines of the projective plane, let (L_0, L_1, L_ω) and (M_0, M_1, M_ω) be the gauge points on l and m, and consider the projectivity

$$(8) \qquad l(L_0, L_1, L_\omega, \ldots) \sim m(M_0, M_1, M_\omega, \ldots)$$

By means of this projectivity, each point on l corresponds to a unique point on m, and vice versa. Hence the first requirement for an isomorphism between l and m is satisfied.

Now let L_x and L_y be any points of the open set on l, let $L_x + L_y = L_p$, and let $L_x L_y = L_q$. Furthermore, let M_u, M_v, M_r, M_s be, respectively, the images of L_x, L_y, L_p, L_q under the projectivity (8), so that

$$(8a) \qquad l(L_0, L_1, L_\omega, L_x, L_y, L_p, L_q) \sim m(M_0, M_1, M_\omega, M_u, M_v, M_r, M_s)$$

Now since $L_x + L_y = L_p$, it follows, from the definition of the sum of two points, that

$$(9) \qquad l(L_\omega, L_0, L_p, L_x, L_y) \sim l(L_\omega, L_p, L_0, L_y, L_x)$$

Hence, combining this with (8a), we have $m(M_\omega, M_0, M_r, M_u, M_v) \sim m(M_\omega, M_r, M_0, M_v, M_u)$. This is an involution on the line m in which M_ω is self-corresponding, (M_u, M_v) is a pair of reciprocally corresponding points, and M_r is the mate of M_0. Therefore $M_u + M_v = M_r$. In other words, the image of the sum of any two points on l is the sum of the images of these points; i.e., the one-to-one correspondence between l and m defined by (8) is sum-preserving.

Also, since $L_x L_y = L_q$, it follows from the definition of the product of two points that

$$l(L_\omega, L_0, L_1, L_q, L_x, L_y) \sim l(L_0, L_\omega, L_q, L_1, L_y, L_x)$$

Hence, combining this with (8a), we have

$$m(M_\omega, M_0, M_1, M_s, M_u, M_v) \sim m(M_0, M_\omega, M_s, M_1, M_v, M_u)$$

This is an involution on m in which (M_0, M_ω) and (M_u, M_v) are pairs of reciprocally corresponding points and M_s is the mate of M_1. Therefore $M_u M_v = M_s$. In other words, the image of the product of any two points of the open set on l is the product of the images of these points; i.e., the one-to-one correspondence between l and m defined by (8) is also product-preserving. Thus, (8) establishes an isomorphism between l and m, since it is a one-to-one correspondence preserving both sums and products.

We are now in a position to adopt our final axiom:

Axiom 10

The open set of points on any line in the projective plane is isomorphic with the field of complex numbers.

As usual, our first obligation, now that we have adopted a new axiom, is to show that it is consistent with the rest of our axioms. To do this, we shall turn again to the algebraic representation, where we have previously verified that Axioms 1 to 9 are valid, and show that Axiom 10 holds also. Let $L_0:(x_0,y_0,z_0)$ and $L_\omega:(x_\omega,y_\omega,z_\omega)$ be two points on an arbitrary line, l, in Π_2, let l be parametrized in terms of L_0 and L_ω, so that $L_0:(1,0)$ and $L_\omega:(0,1)$, and let L_1 be the point whose parameters are $(1,1)$. Then the coordinates of any point of the open set on l are of the form $(1,k)$, and we can set up the necessary one-to-one correspondence between the open set on l and the field of complex numbers by associating the general point $(1,k)$ with the complex number k, and conversely.

Now consider the involution on l in which L_ω is self-corresponding and (L_x,L_y) is a pair of mates. To find the equation of this involution, we must impose the conditions $(0,1) \leftrightarrow (0,1)$ and $(1,x) \leftrightarrow (1,y)$ on the general equation

$$(10) \qquad\qquad a\lambda\lambda' + b(\lambda\mu' + \lambda'\mu) + d\mu\mu' = 0$$

The first condition requires that $d = 0$. The second requires that

$$a + b(x + y) = 0$$

that is, $a = x + y$ and $b = -1$. Hence the equation of the involution is

$$(x + y)\lambda\lambda' - (\lambda\mu' + \lambda'\mu) = 0$$

Now in this involution, the mate of L_0, that is, the point whose parameters correspond to the pair $(1,0)$, is the point whose parameters are $(1, x + y)$. Thus $L_x + L_y = L_{x+y}$, which shows that sums are preserved in the correspondence we have set up between the open set on l and the field of complex numbers.

Next, consider the involution on l in which (L_0,L_ω) and (L_x,L_y) are two pairs of self-corresponding points. Imposing the first of these conditions on the equation of the general projectivity (10), we find $b = 0$, and imposing the second, we find $a = xy$ and $d = -1$. The equation of the required involution is therefore $xy\lambda\lambda' - \mu\mu' = 0$. In this involution, the mate of L_1, that is, the point whose parameters correspond to the pair $(1,1)$, is the point $(1,xy)$. Hence $L_xL_y = L_{xy}$, which shows that products are also preserved in the correspondence we have set up. The correspondence is therefore an isomorphism, and our verification of the consistency of Axiom 10 is complete.

Our final obligation is to show that the principle of duality remains valid after the adoption of Axiom 10, by proving the dual of Axiom 10. To do this, we must first define the gauge lines on a point and then the sum and product of two lines by dualizing Definitions 1 and 2. The duals of the arguments we presented earlier in this section would then show that the open set of lines on a point is a field and that the open sets of lines on any two points are isomorphic. Finally, we would show that the open set of lines on a point is isomorphic to the field of complex numbers by proving that the open set of lines on a point is isomorphic to the open set of points on a line, which in turn is isomorphic with the complex-number field. This last we would do as follows. Let l be a line with gauge points L_0, L_1, L_ω, let L be an arbitrary point, and let the gauge lines l_0, l_1, l_ω be the lines LL_0, LL_1, LL_ω. Then a one-to-one correspondence between the open set of lines on L and the open set of points on l is set up by the perspectivity

$$(11) \qquad L(l_0,l_1,l_\omega,l_x,l_y,\ldots) \;\overline{\wedge}\; l(L_0,L_1,L_\omega,L_x,L_y,\ldots)$$

Now consider the involution on l which defines the relation $L_x + L_y = L_u$: $l(L_\omega,L_0,L_x,L_y) \sim l(L_\omega,L_u,L_y,L_x)$. Combining this with the projectivity (11), we have $L(l_\omega,l_0,l_x,l_y) \sim L(l_\omega,l_u,l_y,l_x)$ which implies that $l_x + l_y = l_u$. In precisely the same way, by considering the involution which defines the relation $L_x L_y = L_v$ we can show that $l_x l_y = l_v$, which completes the verification that the correspondence set up between the open set of lines on L and the open set of points on l by the perspectivity (11) is an isomorphism. Since the open set of points on l is isomorphic with the field of complex numbers, by Axiom 10, and since the relation of being isomorphic is transitive, it follows that the open set of lines on an arbitrary point, L, is also isomorphic with the complex field; and the dual of Axiom 10 is verified.

An immediate consequence of the isomorphism between the open set of points on an arbitrary line and the field of complex numbers is the following simple, but useful, theorem.

Theorem 2

If l and m are lines with gauge points (L_0,L_1,L_ω) and (M_0,M_1,M_ω), respectively, then

$$l(L_0,L_1,L_\omega,L_x,L_y,\ldots) \sim m(M_0,M_1,M_\omega,M_x,M_y,\ldots)$$

Proof   Let l be a line with gauge points (L_0,L_1,L_ω), and let m be a line with gauge points (M_0,M_1,M_ω). From Axiom 10 we know that the open sets on l and m are both isomorphic to the field of complex numbers. Moreover, from the proof of Theorem 1, we know that the projectivity

$$(12) \qquad l(L_0,L_1,L_\omega,\ldots) \sim m(M_0,M_1,M_\omega,\ldots)$$

establishes an isomorphism between the open sets on l and m. Thus we have a chain of isomorphisms whose net result is to establish an isomorphism between the field of complex numbers and itself. Now suppose that the mate of L_x in the projectivity (12) is M_z. Then, beginning with the sum of the numbers 1 and x, we have the mappings indicated in the following scheme:

$$1 + x \;\rightarrow\; L_1 + L_x \rightarrow M_1 + M_z \;\rightarrow\; 1 + z$$

$$\text{complex field} \quad \text{open set on } l \quad \text{open set on } m \quad \text{complex field}$$

Therefore, since these successive isomorphisms are necessarily sum-preserving, it follows that $1 + x = 1 + z$ or $z = x$. Thus the theorem is established.

Corollary 1

The complex number corresponding to each point of the open set on an arbitrary line, l, is uniquely determined by the gauge points on l.

With the adoption of Axiom 10, it is no longer necessary to use labels such as L_0, L_1, L_x, L_y for the points of the open set on an arbitrary line. The only significant feature in each of these names is the subscript, that is, the complex number to which the point corresponds, and henceforth we shall usually identify points by the corresponding complex numbers. The point L_ω is left unlabeled by this process, since the subscript in the name L_ω is not a number. This, of course, does not imply that there is anything exceptional about the point L_ω, which in fact can be any point on the line in question, but merely evidences that L_ω has been selected to play an exceptional role in the coordinatization of the line.

Exercises

1. On a euclidean line, l, locate L_2, L_3, ..., given that L_ω is the ideal point on l. Are the points L_0, L_1, L_2, L_3, ... equally spaced on l?

2. In the arithmetic of points on a line l, what meaning, if any, can be assigned to the symbol $-L_x$? $1/L_x$? $2L_x$? $3L_x$? $\frac{1}{2}L_x$? $\frac{1}{3}L_x$? $\frac{3}{2}L_x$? $\frac{2}{3}L_x$? $\sqrt{L_x}$? L_x^2?

3. Devise a construction for the point denoted by each of the symbols in Exercise 2.

4. Given points A, B, C on a line l with gauge points L_0, L_1, L_ω. Is it possible to construct a point X such that:
(a) $X + A = L_0$ (b) $X + A = L_1$ (c) $X + A = L_\omega$ (d) $X + A = B$
(e) $AX = L_0$ (f) $AX = B$ (g) $AX + B = L_0$ (h) $AX + B = C$

5. Given points A, B, C on a line l with gauge points L_0, L_1, L_ω. Is it possible to construct a point X such that $AX^2 + BX + C = L_0$?

6. Discuss the possibility of establishing an arithmetic of points on a non-singular conic.

9.4 Projectivities in Nonhomogeneous Coordinates

To obtain the condition which characterizes pairs of mates in a projectivity, it is convenient to prove three lemmas first.

Lemma 1

If l and m are any coordinatized lines of the projective plane, there exists a projectivity such that

$$(L_\omega, L_0, L_x, L_y, L_z, \ldots) \sim m(M_\omega, M_t, M_{x+t}, M_{y+t}, M_{z+t}, \ldots)$$

Proof   If l and m are lines with gauge points (L_0, L_1, L_ω) and (M_0, M_1, M_ω), respectively, there are two cases to consider:

1. l and m are the same line with the same gauge points.
2. l and m are lines with different gauge points.

In case 1 we have $M_r = L_r$ for all values of r. Also

$$L_x + L_t = L_{x+t} \qquad L_y + L_t = L_{y+t} \qquad L_z + L_t = L_{z+t} \qquad \cdots$$

Hence, by Lemma 1, Sec. 9.3,

$$l(L_\omega, L_0, L_x, L_y, L_z, \ldots) \sim l(L_\omega, L_t, L_{x+t}, L_{y+t}, L_{z+t}, \ldots)$$

as asserted. In case 2, we first use Theorem 2, Sec. 9.3, to set up the projectivity $l(L_\omega, L_0, L_x, L_y, L_z, \ldots) \sim m(M_\omega, M_0, M_x, M_y, M_z, \ldots)$. Then, using case 1 of the present theorem, we have further

$$m(M_\omega, M_0, M_x, M_y, M_z, \ldots) \sim m(M_\omega, M_t, M_{x+t}, M_{y+t}, M_{z+t}, \ldots)$$

Finally, combining the last two projectivities, we obtain the assertion of the theorem in this case also.

Lemma 2

If l and m are any coordinatized lines of the projective plane, there exists a projectivity such that

$$l(L_\omega, L_0, L_1, L_x, L_y, L_z, \ldots) \sim m(M_\omega, M_0, M_t, M_{xt}, M_{yt}, M_{zt}, \ldots)$$

Proof   If l and m are lines with gauge points (L_0, L_1, L_ω) and (M_0, M_1, M_ω), respectively, there are again two cases to consider:

1. l and m are the same line with the same gauge points.
2. l and m are lines with different gauge points.

In case 1 we have $M_r = L_r$ for all values of r. Also $L_x L_t = L_{xt}$, $L_y L_t = L_{yt}$, $L_z L_t = L_{zt}, \ldots$. Hence, by Lemma 2, Sec. 9.3,

$$l(L_\omega, L_0, L_1, L_x, L_y, L_z, \ldots) \sim l(L_\omega, L_0, L_t, L_{xt}, L_{yi}, L_{tz}, \ldots)$$

as asserted. Case 2 we handle by first setting up the appropriate projectivity between l and m and then using case 1, just as we did in the proof of Lemma 1.

Lemma 3

If l and m are any coordinatized lines of the projective plane, there exists a projectivity such that

$$l(L_\omega, L_0, L_1, L_t, L_x, L_y, L_z, \ldots) \sim m(M_0, M_\omega, M_t, M_1, M_{t/x}, M_{t/y}, M_{t/z}, \ldots)$$

provided $t \neq 0$.

Proof Again, as in Lemmas 1 and 2, there are two cases to consider:

1. l and m are the same line with the same gauge points.
2. l and m are lines with different gauge points.

In case 1, $M_r = L_r$ for all values of r, and the assertion of the theorem is precisely the involution that defines the products

$$L_x L_{t/x} = L_t \qquad L_y L_{t/y} = L_t \qquad L_z L_{t/z} = L_t \qquad \ldots$$

Case 2 we handle, as in the proofs of Lemmas 1 and 2, by first setting up the appropriate projectivity between l and m and then using the result of case 1.

Using the preceding lemmas, we can now prove our main theorem on projectivities:

Theorem 1

If L_x, L_y, L_z, L_w are four points on a line l, each distinct from L_ω, and if $M_{x'}$, $M_{y'}$, $M_{z'}$, $M_{w'}$ are four points on a line m, each distinct from M_ω, then the necessary and sufficient condition that there should exist a projectivity such that $l(L_x, L_y, L_z, L_w) \sim m(M_{x'}, M_{y'}, M_{z'}, M_{w'})$ is that there should exist four numbers, a, b, c, d, such that $ad - bc \neq 0$ and such that the pairs of numbers (x,x'), (y,y'), (z,z'), (w,w') each satisfies the relation $u' = (au + b)/(cu + d)$.

Proof To prove the sufficiency of the condition of the theorem, let us suppose first that $c \neq 0$. Then by Lemma 1 there is a projectivity in which

$$(1) \qquad l(L_x, L_y, L_z, L_w) \sim m(M_{x_1}, M_{y_1}, M_{z_1}, M_{w_1})$$

where $\quad x_1 = x + \dfrac{d}{c} \quad y_1 = y + \dfrac{d}{c} \quad z_1 = z + \dfrac{d}{c} \quad w_1 = w + \dfrac{d}{c}$

Also, by Lemma 2 there exists a projectivity in which

$$(2) \qquad m(M_{x_1}, M_{y_1}, M_{z_1}, M_{w_1}) \sim m(M_{x_2}, M_{y_2}, M_{z_2}, M_{w_2})$$

where

$$x_2 = \frac{c^2}{bc - ad} x_1 \qquad y_2 = \frac{c^2}{bc - ad} y_1 \qquad z_2 = \frac{c^2}{bc - ad} z_1 \qquad w_2 = \frac{c^2}{bc - ad} w_1$$

Furthermore, using Lemma 3 with $t = 1$, we obtain the projectivity

$$(3) \qquad m(M_{x_2}, M_{y_2}, M_{z_2}, M_{w_2}) \sim m(M_{x_3}, M_{y_3}, M_{z_3}, M_{w_3})$$

where $\qquad x_3 = \dfrac{1}{x_2} \qquad y_3 = \dfrac{1}{y_2} \qquad z_3 = \dfrac{1}{z_2} \qquad w_3 = \dfrac{1}{w_2}$

Finally, using Lemma 1 again, we obtain the projectivity

$$(4) \qquad m(M_{x_3}, M_{y_3}, M_{z_3}, M_{w_3}) \sim m(M_{x_4}, M_{y_4}, M_{z_4}, M_{w_4})$$

where $\quad x_4 = x_3 + \dfrac{a}{c} \quad y_4 = y_3 + \dfrac{a}{c} \quad z_4 = z_3 + \dfrac{a}{c} \quad w_4 = w_3 + \dfrac{a}{c}$

Hence, combining projectivities (1) to (4), we therefore have

$$l(L_x, L_y, L_z, L_w) \sim m(M_{x_4}, M_{y_4}, M_{z_4}, M_{w_4})$$

Now from the relations connecting the successive x's, we have

$$x_4 = x_3 + \frac{a}{c} = \frac{1}{x_2} + \frac{a}{c} = \frac{bc - ad}{c^2 x_1} + \frac{a}{c} = \frac{bc - ad}{c(cx + d)} + \frac{a}{c} = \frac{ax + b}{cx + d}$$

Thus, *if* the numbers (x, x') satisfy the relation

$$(5) \qquad u' = \frac{au + b}{cu + d}$$

then x_4 is in fact x'. Similarly, if (y, y'), (z, z'), and (w, w') satisfy Eq. (5), then $y_4 = y'$, $z_4 = z'$, and $w_4 = w'$. Hence if (5) is satisfied by the appropriate pairs of coordinates, there is a projectivity in which $M_{x'}$, $M_{y'}$, $M_{z'}$, $M_{w'}$ are, respectively, the images of L_x, L_y, L_z, L_w, provided $c \neq 0$.

If $c = 0$, the condition $ad - bc \neq 0$ guarantees that neither a nor d is zero. In this case, the sufficiency of the given condition can be established by considering the following pair of projectivities:

(6) $$l(L_x, L_y, L_z, L_w) \sim m(M_{x_1}, M_{y_1}, M_{z_1}, M_{w_1})$$

where
$$x_1 = x + \frac{b}{a} \qquad y_1 = y + \frac{b}{a}$$

$$z_1 = z + \frac{b}{a} \qquad w_1 = w + \frac{b}{a}$$

(by Lemma 1)

(7) $$m(M_{x_1}, M_{y_1}, M_{z_1}, M_{w_1}) \sim m(M_{x_2}, M_{y_2}, M_{z_2}, M_{w_2})$$

where
$$x_2 = \frac{a}{d} x_1 \qquad y_2 = \frac{a}{d} y_1 \qquad z_2 = \frac{a}{d} z_1 \qquad w_2 = \frac{a}{d} w_1 \qquad \text{(by Lemma 2)}$$

Now from the relations connecting the successive x's, we have

$$x_2 = \frac{a}{d} x_1 = \frac{a}{d}\left(x + \frac{b}{a}\right) = \frac{ax + b}{d}$$

Thus if Eq. (5) holds with $c = 0$, then $x_2 = x'$. Similarly, if (y, y'), (z, z'), (w, w') satisfy (5), with $c = 0$, then $y_2 = y'$, $z_2 = z'$, $w_2 = w'$, and the proof of the sufficiency of the condition of the theorem is complete when $c = 0$ as well as when $c \neq 0$.

To prove the necessity of the condition of the theorem, we begin with the assumption that

(8) $$l(L_x, L_y, L_z, L_w) \sim m(M_{x'}, M_{y'}, M_{z'}, M_{w'})$$

Then numbers a, b, c, d can surely be found which will satisfy the three conditions

$$x' = \frac{ax + b}{cx + d} \qquad y' = \frac{ay + b}{cy + d} \qquad z' = \frac{az + b}{cz + d}$$

Moreover, these numbers must be such that $ad - bc \neq 0$, for otherwise we would have $a = kc$, $b = kd$ and the expression $u' = (au + b)/(cu + d)$ would reduce to the constant k. This, in turn, would imply that $x' = y' = z'$, which is impossible since $M_{x'}$, $M_{y'}$, $M_{z'}$ are known to be distinct points. Now suppose that

$$\frac{aw + b}{cw + d} = w''$$

Then by the first part of the theorem,

$$l(L_x, L_y, L_z, L_w) \sim m(M_{x'}, M_{y'}, M_{z'}, M_{w''})$$

Furthermore, this projectivity has three pairs of mates in common with the projectivity (8). Hence the two projectivities are the same, which means that $w'' = w'$. Therefore (w, w'), as well as the pairs (x, x'), (y, y'), (z, z') satisfy the condition of the theorem, and our proof of the necessity of the condition is complete.

Although the projectivity $l(L_x, L_y, L_z, L_w) \sim m(M_{x'}, M_{y'}, M_{z'}, M_{w'})$ perforce assigns a unique image to L_ω, this cannot be determined from (5) since it is meaningless to substitute the nonnumerical symbol ω into this formula in an attempt to identify the mate of L_ω. Likewise, (5) assigns no image to the point $L_{-d/c}$, although, of course, this point has an image in the projectivity. The images of L_ω and $L_{-d/c}$ can easily be found, however, by tracing these points through the component projectivities which we used in the proof of the sufficiency assertion of Theorem 1. The results are given in the following corollary.

Corollary 1

In the projectivity defined by the formula $u' = (au + b)/(cu + d)$, the images of L_ω and $L_{-d/c}$ are $M_{a/c}$ and M_ω, respectively, provided $c \neq 0$. If $c = 0$, the image of L_ω is M_ω; and there is no point, and hence no image, corresponding to the label $L_{-d/c}$.

Proof Let the formula $u' = (au + b)/(cu + d)$ define a projectivity built up from successive projectivities as in the proof of Theorem 1, and let us suppose first that $c \neq 0$. Then for L_ω we have the following sequence of images under the respective component projectivities:

$$L_\omega \to M_\omega \to M_\omega \to M_0 \to M_{a/c}$$

Similarly, for $L_{-d/c}$ we have the sequence of images

$$L_{-d/c} \to M_0 \to M_0 \to M_\omega \to M_\omega$$

On the other hand, if $c = 0$, then we have from the two component projectivities required to "synthesize" the given projectivity in this case

$$L_\omega \to M_\omega \to M_\omega$$

Obviously, when $c = 0$, the formula $u' = (au + b)/d$ assigns a well-defined image to every labeled point, without exception. Therefore since we have just determined the image of the only unlabeled point, our proof is complete.

Corollary 2

A projectivity is described by an equation of the form $u' = au + b$ if and only if the unlabeled points of the two ranges are mates in the projectivity.

Exercises

1. Prove Corollary 2, Theorem 1.

2. Find the equation of the projectivity in which:
(a) $L_2 \rightarrow M_1$, $L_{-1} \rightarrow M_2$, $L_0 \rightarrow M_3$
(b) $L_1 \rightarrow M_{-1}$, $L_2 \rightarrow M_2$, $L_3 \rightarrow M_1$
(c) $L_1 \rightarrow M_2$, $L_2 \rightarrow M_{-1}$, $L_\omega \rightarrow M_\omega$
(d) $L_1 \rightarrow M_\omega$, $L_\omega \rightarrow M_0$, $L_0 \rightarrow M_1$

3. Find the six projectivities which map L_0, L_1, L_ω into the six possible permutations of these points. Do these projectivities form a group?

4. Prove Theorem 1 using a different sequence of projectivities.

5. What is the necessary and sufficient condition that a projectivity between cobasal ranges should have a single self-corresponding point?

9.5 Homogeneous Coordinates on a Line

If x is the nonhomogeneous coordinate of a labeled point, L_x, on an arbitrary line, l, with gauge points L_0, L_1, L_ω, and if x_1 and x_2 are numbers such that $x_1/x_2 = x$, then (x_1,x_2) is said to be a pair of **homogeneous coordinates** for the point L_x relative to L_0, L_1, L_ω. Obviously, the second homogeneous coordinate of any labeled point must be different from zero. Moreover, if (x_1,x_2) is a pair of homogeneous coordinates for a labeled point, L_x, then so is (kx_1,kx_2), provided $k \neq 0$. The assignment of homogeneous coordinates to the points of l is completed by defining the homogeneous coordinates of the unlabeled point, L_ω, to be any ordered pair of numbers of the form $(k,0)$, $k \neq 0$.

Once homogeneous coordinates have been introduced on the lines of the projective plane, it becomes possible to restate the results of Theorem 1, Sec. 9.4, and its corollary as a single, completely symmetric relation:

Theorem 1

The necessary and sufficient condition that there should exist a projectivity between a line l and a line l' in which the image of an arbitrary point (x_1,x_2) on l is the point (x_1',x_2') on l' is that (x_1,x_2) and (x_1',x_2') should satisfy an equation of the form

$$\alpha x_1 x_1' + \beta x_1 x_2' + \gamma x_2 x_1' + \delta x_2 x_2' = 0 \qquad \alpha\delta - \beta\gamma \neq 0$$

Proof   If x_2, $x_2' \neq 0$, the condition of the theorem can be written in the form

$$\alpha \frac{x_1}{x_2} \frac{x_1'}{x_2'} + \beta \frac{x_1}{x_2} + \gamma \frac{x_1'}{x_2'} + \delta = 0$$

which is completely equivalent to the condition of Theorem 1, Sec. 9.4, if we make the identifications

$$\frac{x_1}{x_2} = u \qquad \frac{x_1'}{x_2'} = u' \qquad \alpha = c \qquad \beta = -a \qquad \gamma = d \qquad \delta = -b$$

In other words, for pairs of corresponding points each of which is a labeled point in the nonhomogeneous coordinate systems on l and l', the condition of the present theorem and the condition of Theorem 1, Sec. 9.4, are identical. Moreover, we know from Corollary 1, Theorem 1, Sec. 9.4, that if $\alpha \neq 0$, that is, $c \neq 0$, then the image of the (unlabeled) point $(1,0)$ on l is the point $(\beta, -\alpha)$, that is, $L_{a/c}'$, on l'; and it is easy to verify that the relation given in the theorem is satisfied by the pairs $(x_1, x_2) = (1,0)$ and $(x_1', x_2') = (\beta, -\alpha)$. Furthermore, if $\alpha \neq 0$, that is, $c \neq 0$, the preimage of the (unlabeled) point $(1,0)$ on l' is the point $(\gamma, -\alpha)$, that is, $L_{-d/c}$, on l; and again it is easy to check that the pairs $(x_1, x_2) = (\gamma, -\alpha)$ and $(x_1', x_2') = (1,0)$ satisfy the relation given in the theorem. Finally, it is clear that the (unlabeled) point $(1,0)$ on l' corresponds to the (unlabeled) point $(1,0)$ on l if and only if $\alpha = 0$, as required by Corollary 1, Theorem 1, Sec. 9.4.

Although the relation

(1)
$$\alpha x_1 x_1' + \beta x_1 x_2' + \gamma x_2 x_1' + \delta x_2 x_2' = 0$$

is very much like the relation

(2)
$$\alpha \lambda_1 \lambda_n + \beta \lambda_1 \mu_n + \gamma \mu_1 \lambda_n + \delta \mu_1 \mu_n = 0$$

of Theorem 2, Sec. 4.5, the significance of the variables in the two equations is quite different. In (1) the variables (x_1, x_2) and (x_1', x_2') are pairs of *homogeneous coordinates* with respect to the three gauge points on the respective lines. In (2) the variables (λ_1, μ_1) and (λ_n, μ_n) are pairs of *parameters* with respect to the two base points on the respective lines. In (1), the coordinates (x_1, x_2) and (x_1', x_2') have meaning apart from any coordinatization of the projective plane as a whole, and indeed we have not yet undertaken such a coordinatization. In (2), the parameters (λ_1, μ_1) and (λ_n, μ_n) are scalar multipliers of the coordinate vectors of the appropriate base points and have meaning only in relation to a coordinatization of the entire plane.

However, because (1) and (2) are algebraically identical, there are certain properties of Eq. (1) which can be inferred, without further proof, from the corresponding properties of Eq. (2). In particular, we have the following results, which were originally proved in Sec. 4.8.

Theorem 2

A projectivity $\alpha x_1 x_1' + \beta x_1 x_2' + \gamma x_2 x_1' + \delta x_2 x_2' = 0$ between cobasal ranges has either one or two self-corresponding points according as

$$(\beta + \gamma)^2 - 4\alpha\delta$$

is equal to or different from zero.

Theorem 3

A projectivity $\alpha x_1 x_1' + \beta x_1 x_2' + \gamma x_2 x_1' + \delta x_2 x_2' = 0$ between cobasal ranges is an involution if and only if $\beta = \gamma$.

We know from Axiom 7 that a projectivity is uniquely determined when three pairs of corresponding points are given. Hence if P_1, P_2, P_3, P_4 and P_1', P_2', P_3', P_4' are two sets of four collinear points, there is, in general, no projectivity in which $(P_1, P_2, P_3, P_4) \sim (P_1', P_2', P_3', P_4')$. To determine conditions under which there will exist a projectivity in which the image of $P_i:(x_i, y_i)$ is $P_i':(x_i', y_i')$, $i = 1, 2, 3, 4$, let $L_0:(0,1)$, $L_1:(1,1)$, $L_\omega:(1,0)$ be the gauge points on any convenient line, l, and let $Q:(q_1, q_2)$ be the image of P_4 in the unique projectivity in which L_0, L_ω, L_1 are, respectively, the images of P_1, P_2, P_3. Then

$$(3) \qquad (P_1, P_2, P_3, P_4) \sim (L_0, L_\omega, L_1, Q)$$

Therefore, since our object is to ensure the projectivity

$$(P_1, P_2, P_3, P_4) \sim (P_1', P_2', P_3', P_4')$$

we must also have

$$(4) \qquad (P_1', P_2', P_3', P_4') \sim (L_0, L_\omega, L_1, Q)$$

Now if the equation of the projectivity (3) is

$$\alpha u u' + \beta u v' + \gamma v u' + \delta v v' = 0$$

it follows, by substituting the coordinates of the pairs of corresponding points (P_1, L_0), (P_2, L_ω), (P_3, L_1), (P_4, Q), that

$$
\begin{aligned}
\beta x_1 && + \delta y_1 &= 0 \\
\alpha x_2 && + \gamma y_2 &= 0 \\
\alpha x_3 + \beta x_3 + \gamma y_3 + \delta y_3 &= 0 \\
\alpha x_4 q_1 + \beta x_4 q_2 + \gamma y_4 q_1 + \delta y_4 q_2 &= 0
\end{aligned}
$$

Hence, by Theorem 5, Sec. 3.6, it follows that

$$
\begin{vmatrix}
0 & x_1 & 0 & y_1 \\
x_2 & 0 & y_2 & 0 \\
x_3 & x_3 & y_3 & y_3 \\
x_4 q_1 & x_4 q_2 & y_4 q_1 & y_4 q_2
\end{vmatrix} = 0
$$

If this is expanded in terms of the elements of the last row and the resulting equation solved for q_2/q_1, we find without difficulty that

$$(5) \qquad \frac{q_2}{q_1} = \frac{(x_1 y_3 - x_3 y_1)(x_2 y_4 - x_4 y_2)}{(x_1 y_4 - x_4 y_1)(x_2 y_3 - x_3 y_2)}$$

Similarly, from the projectivity (4), we find that

$$(6) \qquad \frac{q_2}{q_1} = \frac{(x_1' y_3' - x_3' y_1')(x_2' y_4' - x_4' y_2')}{(x_1' y_4' - x_4' y_1')(x_2' y_3' - x_3' y_2')}$$

Hence, equating these two expressions for q_2/q_1, we obtain

$$(7) \qquad \frac{(x_1 y_3 - x_3 y_1)(x_2 y_4 - x_4 y_2)}{(x_1 y_4 - x_4 y_1)(x_2 y_3 - x_3 y_2)} = \frac{(x_1' y_3' - x_3' y_1')(x_2' y_4' - x_4' y_2')}{(x_1' y_4' - x_4' y_1')(x_2' y_3' - x_3' y_2')}$$

as a necessary condition that there should be a projectivity in which

$$(P_1, P_2, P_3, P_4) \sim (P_1', P_2', P_3', P_4')$$

Conversely, let us assume that condition (7) holds, and let us consider the two projectivities $(P_1, P_2, P_3, P_4) \sim (L_0, L_\omega, L_1, Q)$ and

$$(P_1', P_2', P_3', P_4') \sim (L_0, L_\omega, L_1, R)$$

where $Q:(q_1, q_2)$ and $R:(r_1, r_2)$ are simply the images of P_4 and P_4' in the projectivities determined by the first three points and their images in the respective cases. Then, exactly as before, the ratio of the coordinates of Q is given by (5), and the ratio of the coordinates of R is given by the fraction on the right of Eq. (6). Moreover, since (7) is assumed to hold, it follows that $q_2/q_1 = r_2/r_1$. Hence Q and R are the same point, and therefore

$$(P_1, P_2, P_3, P_4) \sim (P_1', P_2', P_3', P_4')$$

since each is projective with the tetrad (L_0, L_ω, L_1, Q). Thus, recognizing that the two fractions in Eq. (7) are just the cross ratios of the points P_1, P_2, P_3, P_4 and the points P_1', P_2', P_3', P_4', calculated from the coordinates of the points instead of the parameters of the points, as in Sec. 4.6, we have the counterpart of Theorem 5, Sec. 4.7:

Theorem 4

The cross ratio of four collinear points and the cross ratio of their images under any projectivity are equal.

Corollary 1

Four collinear points, P_1, P_2, P_3, P_4, are projective with four other collinear points, P_1', P_2', P_3', P_4', if and only if $R(P_1 P_2, P_3 P_4) = R(P_1' P_2', P_3' P_4')$.

Because the algebraic structure of the cross ratio of four collinear points is the same whether the cross ratio is calculated from the parameters or the coordinates of the points, it follows that the formal properties of the cross ratio are the same under either interpretation. In particular, the permutation properties given by Theorems 1 to 3, Sec. 4.7, are valid not only in E_2^+ and Π_2 but also in any projective plane satisfying the axioms we have adopted.

Also, using Theorem 4 and the permutation properties of the cross ratio, the following result, which was a definition in E_2^+ and Π_2, can be proved as a theorem in the projective plane.

Theorem 5

In the projective plane, four collinear points, P_1, P_2, P_3, P_4, form a harmonic tetrad if and only if $R(P_1P_2,P_3P_4) = -1$.

From Theorem 5, we obtain immediately the following equivalent of Lemma 2, Sec. 4.7.

Theorem 6

The harmonic conjugate of a general point $P:(x_1,x_2)$ on a line l with respect to the gauge points L_0 and L_ω on l is the point P' whose coordinates are $(x_1,-x_2)$.

Finally, for use when we attempt to introduce homogeneous coordinates into the projective plane as a whole, we note the following equivalent of the dual of Theorem 5, Sec. 7.5.

Theorem 7

If A, B, C, I are four points in the projective plane, no three of which are collinear, if:

A_1 is the intersection of AI and BC,
B_1 is the intersection of BI and CA,
C_1 is the intersection of CI and AB,

if:

The gauge points, A_0, A_1, A_ω on BC are, respectively, C, A_1, B,
The gauge points, B_0, B_1, B_ω on CA are, respectively, A, B_1, C,
The gauge points, C_0, C_1, C_ω on AB are, respectively, B, C_1, A,

and if A', B', C' are, respectively, arbitrary points on the lines BC, CA, AB, then AA', BB', CC' are concurrent if and only if

$$R(A_0A_\omega,A'A_1)R(B_0B_\omega,B'B_1)R(C_0C_\omega,C'C_1) = 1$$

Exercises

1. Prove Theorem 5.

2. Prove Theorem 6.

3. Verify that Theorem 7 is the dual of Theorem 5, Sec. 7.5.

4. Prove that the fixed points of every involution are distinct.

5. Prove algebraically that every projectivity between cobasal ranges can be expressed as the composition of two involutions.

6. Under what conditions, if any, is the composition of two involutions also an involution?

7. Prove algebraically that two involutions on the same line have one and only one pair of mates in common.

8. Let P and P' be the points determined by the roots of the equation $ax_1{}^2 + bx_1x_2 + cx_2{}^2 = 0$, and let Q and Q' be the points determined by the roots of the equation $dx_1{}^2 + ex_1x_2 + fx_2{}^2 = 0$. What is the equation of the involution determined by the pairs of mates (P,P') and (Q,Q')?

9.6 Nonhomogeneous Coordinates in the Projective Plane

Now that we know how to assign coordinates to the points of the open set on an arbitrary line, we introduce nonhomogeneous coordinates in the projective plane itself in the following fashion.

Let l and m be distinct lines of the projective plane, and let O be their intersection. On l choose gauge points X_0, X_1, X_ω in such a way that X_0 is the point O; and on m choose gauge points Y_0, Y_1, Y_ω in such a way that Y_0 is also the point O. The open sets on l and m we suppose to be identified with the field of complex numbers in the way described in Sec. 9.3.

Now let P be any point which is not on the line $X_\omega Y_\omega$. If P is a point of the line l, it coincides with some point X_x on l, and we assign to P the pair of coordinates $(x,0)$. If P is a point of the line m, it coincides with some point Y_y on m, and we assign to P the pair of coordinates $(0,y)$. If P does not lie on either l or m, let X_x be the intersection of l and the line PY_ω, and let Y_y be the intersection of m and the line PX_ω. In this case, we assign to P the pair of coordinates (x,y). Obviously, two points which are distinct cannot have the same pair of coordinates, and, conversely, points which do not have the same pair of coordinates are distinct.

By analogy with ordinary analytic geometry, the first entry in the ordered pair of coordinates of an arbitrary point is called the x **coordinate,** or **abscissa,** and the second entry is called the y **coordinate,** or **ordinate.**

The line l, from whose points the x coordinate of a general point is obtained, is called the x **axis.** The line m, from whose points the y coordinate of a general point is obtained, is called the y **axis.** The point $O = X_0 = Y_0$, which is the intersection of l and m, is called the **origin** of the coordinate system.

The foregoing procedure does not and cannot assign a pair of coordinates to a point which is on the line $X_\omega Y_\omega$. In fact, the most this process can do is to assign the *same* label, (ω,ω), to *each* point of $X_\omega Y_\omega$. Moreover, since ω is not a number, (ω,ω) cannot be considered a pair of coordinates at all. The "unlabeled line," $X_\omega Y_\omega$, is, of course, no different from any other line in the plane, and by a suitable choice of l, m, and the gauge points X_ω and Y_ω, any line can be made to play the role of the unlabeled line. The fact that the points on one line are left without coordinates by our present method of coordinatization is the result of a flaw in the method, which the introduction of homogeneous coordinates, in the next section, will eliminate.

It should come as no surprise to us that in the projective plane lines are described by linear equations, and conversely. However, our intuition and past experience notwithstanding, it is necessary that we prove this "obvious" fact in the context of our axiomatic development. The next two theorems confirm this natural expectation.

Theorem 1

Every line in the projective plane except the unlabeled line has an equation of the form $l_1 x + l_2 y + l_3 = 0$, where l_1 and l_2 are not both zero.

Proof   Consider first any line except the unlabeled line which passes through the unlabeled point Y_ω, on the y axis. From the definition of coordinates, it follows that every labeled point on such a line has an x coordinate which is a constant, say a. Furthermore, it is clear that any point whose x coordinate is a lies on this line. Hence $x = a$ is an equation of this line, and moreover this equation is of the asserted form. Similarly, of course, any line on X_ω, except the unlabeled line, has an equation $y = b$, which is of the asserted form.

Now let us consider an arbitrary line, p, which does not pass through either X_ω or Y_ω, and let $P_i:(x_i,y_i)$ be an arbitrary labeled point on p. Then, by definition, the intersection of the x axis and the line $P_i Y_\omega$ is the point X_{x_i}, and the intersection of the y axis and the line $P_i X_\omega$ is the point Y_{y_i} (Fig. 9.4). Thus

$$(1) \qquad p(P_1,P_2,P_3,\ldots,P_\omega) \overset{Y_w}{\wedge} l(X_{x_1},X_{x_2},X_{x_3},\ldots,X_\omega)$$

$$(2) \qquad p(P_1,P_2,P_3,\ldots,P_\omega) \overset{X_w}{\wedge} m(Y_{y_1},Y_{y_2},Y_{y_3},\ldots,Y_\omega)$$

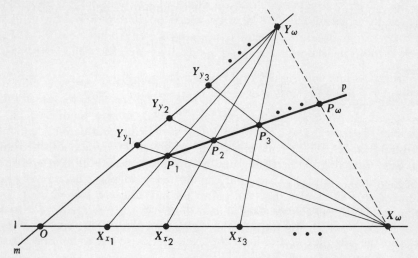

Fig. 9.4 Figure used in the proof of Theorem 1.

Hence, combining these projectivities, we have

(3) $$l(X_{x_1}, X_{x_2}, X_{x_3}, \ldots, X_\omega) \sim m(Y_{y_1}, Y_{y_2}, Y_{y_3}, \ldots, Y_\omega)$$

Moreover, it is obvious that X_ω and Y_ω are mates in this projectivity. Therefore, by Corollary 2, Theorem 1, Sec. 9.4, numbers a and b must exist such that $y_i = ax_i + b$, for all values of i. In other words, if (x, y) are the coordinates of any labeled point on the line p, then $y = ax + b$. Furthermore, from Theorem 1, Sec. 9.4, if $P_i : (x_i, y_i)$ is any point whose coordinates satisfy the equation $y = ax + b$, then X_{x_i} and Y_{y_i} are a pair of mates in the projectivity (3), and hence, from the perspectivities (1) and (2), it follows that P_i is on the line p whose equation is $y = ax + b$. Since this equation is of the form described in the theorem, our proof is complete.

Corollary 1

With the exception of the unlabeled line, every line in the projective plane can be represented parametrically by a pair of equations of the form

$$x = at + b \qquad y = ct + d \qquad a, c \text{ not both zero}$$

Theorem 2

For all values of l_1, l_2, l_3 such that l_1 and l_2 are not both zero, the locus of the equation $l_1 x + l_2 y + l_3 = 0$ is a straight line.

Proof Consider the locus defined by the equation $l_1 x + l_2 y + l_3 = 0$. If l_1 and l_2 are not both zero, it is obvious that there are infinitely many pairs of values of x and y which satisfy the equation. Let (x_1, y_1) and (x_2, y_2) be two such pairs, so that

$$(4) \qquad \begin{aligned} l_1 x_1 + l_2 y_1 + l_3 &= 0 \\ l_1 x_2 + l_2 y_2 + l_3 &= 0 \end{aligned}$$

Now there is a line determined by the points $P_1 : (x_1, y_1)$ and $P_2 : (x_2, y_2)$, and by Theorem 1 it has an equation of the form $m_1 x + m_2 y + m_3 = 0$, so that

$$(5) \qquad \begin{aligned} m_1 x_1 + m_2 y_1 + m_3 &= 0 \\ m_1 x_2 + m_2 y_2 + m_3 &= 0 \end{aligned}$$

From the pair of equations (4) and the pair of equations (5), it follows that the numbers in each of the triples $[l_1, l_2, l_3]$ and $[m_1, m_2, m_3]$ are proportional to the signed determinants from the matrix $\left\| \begin{matrix} x_1 & y_1 & 1 \\ x_2 & y_2 & 1 \end{matrix} \right\|$. Hence $[l_1, l_2, l_3]$ and $[m_1, m_2, m_3]$ are proportional triples, and therefore $l_1 x + l_2 y + l_3 = 0$ and $m_1 x + m_2 y + m_3 = 0$ are equivalent equations. In other words, they represent the same locus, and from the second of these equations that locus must be a line, as asserted.

Corollary 1

If a and c are not both zero, the equations $x = at + b$ and $y = ct + d$ are the parametric equations of a line.

Various familiar theorems on lines are now easy to establish. In particular, we have the following useful results.

Theorem 3

The equation of the line determined by the distinct points $P_1 : (x_1, y_1)$ and $P_2 : (x_2, y_2)$ can be written in the form $\begin{vmatrix} x & y & 1 \\ x_1 & y_1 & 1 \\ x_2 & y_2 & 1 \end{vmatrix} = 0.$

Theorem 4

A point is on the line determined by the distinct points $P_1 : (x_1, y_1)$ and $P_2 : (x_2, y_2)$ if and only if its coordinates can be expressed in the form

$$\left(\frac{\lambda x_1 + \mu x_2}{\lambda + \mu}, \frac{\lambda y_1 + \mu y_2}{\lambda + \mu} \right)$$

Just as we expect, conics in the projective plane are described by second-degree equations, and conversely:

Theorem 5

Every conic in the projective plane except the singular conic consisting of two ranges on the unlabeled line has an equation of the form $ax^2 + 2hxy + by^2 + 2gx + 2fy + c = 0.$

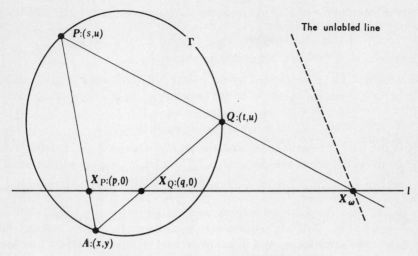

Fig. 9.5   Figure used in the proof of Theorem 5.

Proof   Let us suppose first that the conic, Γ, is singular. If it consists of the ranges on the lines $l_1x + l_2y + l_3 = 0$ and $m_1x + m_2y + m_3 = 0$, then, clearly, the coordinates of each of its labeled points satisfies the equation $(l_1x + l_2y + l_3)(m_1x + m_2y + m_3) = 0$, and whether the lines are distinct or not, this equation is of the asserted form. Moreover, every point whose coordinates satisfy this equation is on one or the other of the lines and hence is a point of the conic. Similarly, if Γ consists of the range on the line $l_1x + l_2y + l_3 = 0$ and the range on the unlabeled line, the coordinates of each of its labeled points, and no others, satisfy the equation $l_1x + l_2y + l_3 = 0$, which, again, is of the asserted form.

   If the conic Γ is nonsingular, let $P:(s,u)$ and $Q:(t,u)$ be two labeled points on Γ which, for convenience, we assume to be collinear with X_ω on a line distinct from the x axis. Furthermore, let $A:(x,y)$ be an arbitrary labeled point on Γ, and let $X_P:(p,0)$ and $X_Q:(q,0)$ be, respectively, the intersections of PA and QA with the line, l, which is the x axis (Fig. 9.5). Then if P and Q are used as generating bases for Γ, we have

$$P(A,\dots) \sim Q(A,\dots)$$

Also, $P(A,\dots) \ \overline{\wedge}\ l(X_P,\dots)$ and $Q(A,\dots) \ \overline{\wedge}\ l(X_Q,\dots)$

Hence, combining these projectivities, we have

(6) $$l(X_P,\dots) \sim l(X_Q,\dots)$$

Therefore, from the work of Sec. 9.4, we know that there exist numbers a, b, c, d such that the coordinates of corresponding points in the projectivity

(6) satisfy the relation

$$(7) \qquad apq + bp + cq + d = 0$$

Moreover, since P, A, X_P are collinear, we have

$$\begin{vmatrix} x & y & 1 \\ s & u & 1 \\ p & 0 & 1 \end{vmatrix} = 0 \qquad \text{or} \qquad p = \frac{ux - sy}{u - y}$$

and since Q, A, X_Q are collinear, we have

$$\begin{vmatrix} x & y & 1 \\ t & u & 1 \\ q & 0 & 1 \end{vmatrix} = 0 \qquad \text{or} \qquad q = \frac{ux - ty}{u - y}$$

Hence, substituting into Eq. (7) and clearing of fractions, we have

$$(8) \qquad a(ux - sy)(ux - ty) + b(ux - sy)(u - y) \\ + c(ux - ty)(u - y) + d(u - y)^2 = 0$$

which is obviously an equation of the asserted form. This derivation breaks down if $y = u$. However, in this case A is either P or Q, and by direct substitution it is easy to verify that the coordinates of $P:(s,u)$ and $Q:(t,u)$ satisfy Eq. (8).

Conversely, if x and y are any two numbers satisfying Eq. (8), then by reversing our steps it follows that X_P and X_Q are corresponding points in the projectivity (7) and therefore the point $A:(x,y)$ is a point on the conic. This completes our proof.

The general equation

$$(9) \qquad ax^2 + 2hxy + by^2 + 2gx + 2fy + c = 0$$

does not have the structure of Eq. (8), nor is it apparent how it can be put in the form (8); therefore we cannot work back from Eq. (9) to the projectivity (7) and thereby prove that the locus of Eq. (9) is a conic. However, by an argument very much like the one we used to prove Theorem 2, it can be shown that any equation of the form (9) is equivalent to an equation which is the equation of a conic. Thus, although we shall leave the details of the proof to the exercises, we do have the following theorem.

Theorem 6

The locus of any equation of the form

$$ax^2 + 2hxy + by^2 + 2gx + 2fy + c = 0$$

is a conic.

Exercises

1. Prove Corollary 1, Theorem 1.

2. Prove Corollary 1, Theorem 2.

3. Prove Theorem 3.

4. Prove Theorem 4.

5. Given $l: l_1x + l_2y + l_3 = 0$ and $m: m_1x + m_2y + m_3 = 0$. Show that any line on the intersection of l and m has an equation of the form

$$\lambda(l_1x + l_2y + l_3) + \mu(m_1x + m_2y + m_3) = 0$$

6. Show that if $l_1m_2 - l_2m_1 = 0$, the equations $l_1x + l_2y + l_3 = 0$ and $m_1x + m_2y + m_3 = 0$ have no solution. Does this contradict Axiom 3?

7. Prove Theorem 6.

8. Discuss the introduction of nonhomogeneous line coordinates into the projective plane.

9. State and prove the dual of Theorem 1.

9.7 Homogeneous Coordinates in the Projective Plane

Once nonhomogeneous coordinates have been introduced into the projective plane, it is natural to ask whether they can be used to define homogeneous coordinates· as we did in the euclidean plane. The answer is Yes, and the procedure is essentially the same.

Let (x',y') be the nonhomogeneous coordinates of an arbitrary labeled point, P, in the projective plane. Then any three numbers (x_1,x_2,x_3) such that $x_1/x_3 = x'$ and $x_2/x_3 = y'$ can be considered homogeneous coordinates of P. If O is the origin of the given nonhomogeneous system, the assignment of homogeneous coordinates is completed by giving to the unlabeled point on the line OP the coordinate-triple $(x_1,x_2,0)$. Since the equation of the line OP is $y'x - x'y = 0$, the nonhomogeneous coordinates of any point on OP are of the form kx', ky'. Hence the same set of homogeneous coordinates is assigned to the unlabeled point on OP no matter what point on the line OP is used in the process.

This method of assigning homogeneous coordinates to the points of the projective plane, though simple, is marred by the fact that it introduces the dissymmetry of an initially unlabeled line into what should be a completely symmetric process. Hence it seems desirable that we choose another, more symmetric procedure as our "official" way of establishing homogeneous coordinates in the projective plane.

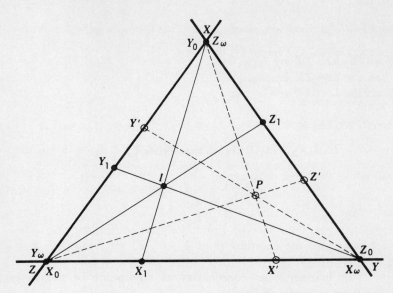

Fig. 9.6   The triangle of reference for a system of homogeneous coordinates in the projective plane.

To do this, we begin with an arbitrary four-point $XYZI$ whose diagonal points are:

The intersection, X_1, of XI and YZ,
The intersection, Y_1, of YI and ZX,
The intersection, Z_1, of ZI and XY.

On the sides of $\triangle XYZ$ we choose gauge points as follows:

On YZ let $X_0 = Z$, $X_1 = X_1$, $X_\omega = Y$,
On ZX let $Y_0 = X$, $Y_1 = Y_1$, $Y_\omega = Z$,
On XY let $Z_0 = Y$, $Z_1 = Z_1$, $Z_\omega = X$.

Triangle XYZ we shall call the **triangle of reference,** and the point I we shall call the **unit point** (Fig. 9.6).

Now let P be any point which is not on any side of the triangle of reference and let:

X' be the intersection of XP and YZ,
Y' be the intersection of YP and ZX,
Z' be the intersection of ZP and XY.

Without loss of generality we can suppose that the homogeneous coordinates:

Of X' on the line YZ are (x_2,x_3),
Of Y' on the line ZX are (x_3,x_1),
Of Z' on the line XY are (x_1,t).

Moreover, by Theorem 7, Sec. 9.5, we have

$$\mathrm{R}(X_0X_\omega,X'X_1)\mathrm{R}(Y_0Y_\omega,Y'Y_1)\mathrm{R}(Z_0Z_\omega,Z'Z_1) = 1$$

or, evaluating the three cross ratios,

$$\frac{x_2}{x_3}\frac{x_3}{x_1}\frac{x_1}{t} = 1$$

Hence $t = x_2$, and the coordinates of X', Y', Z' are, respectively, (x_2,x_3), (x_3,x_1), (x_1,x_2).

The homogeneous coordinates of any point in the projective plane are now defined as follows:

Definition 1

1. If P is any point on YZ, and if its coordinates relative to X_0, X_1, X_ω are (x_2,x_3), then its homogeneous coordinates in the plane are $(0,x_2,x_3)$.

2. If P is any point on ZX, and if its coordinates relative to Y_0, Y_1, Y_ω are (x_3,x_1), then its homogeneous coordinates in the plane are $(x_1,0,x_3)$.

3. If P is any point on XY, and if its coordinates relative to Z_0, Z_1, Z_ω are (x_1,x_2), then its homogeneous coordinates in the plane are $(x_1,x_2,0)$.

4. If P is any point which is not on a side of the triangle of reference, and if (x_2,x_3), (x_3,x_1), (x_1,x_2) are the homogeneous coordinates of its projections from X, Y, Z on YZ, ZX, XY, respectively, then its homogeneous coordinates in the plane are (x_1,x_2,x_3).

With the exception of the vertices of the triangle of reference, Definition 1 unambiguously assigns a triple of coordinates (x_1,x_2,x_3) to every point in the projective plane. It is not immediately clear, however, that the coordinates of the vertices X, Y, Z are uniquely determined, since each of these points lies on two sides of the triangle of reference and hence has coordinates assigned to it according to two different rules. To check this question, let us consider the point X, for example. As a point on XY, it plays the role of Z_ω. Hence its homogeneous coordinates on XY are $(1,0)$ and therefore, according to Definition 1, its homogeneous coordinates in the plane are $(1,0,0)$. On the other hand, as a point of ZX, the point X plays the role of Y_0. Its homogeneous coordinates on ZX are therefore $(0,1)$, and hence its homogeneous coordinates in the plane are again $(1,0,0)$. Thus both as a point of XY and as a point of ZX, the homogeneous coordinates of X are $(1,0,0)$. Similarly, it is easy to check that by either of the two relevant parts of Definition 1, the homogeneous coordinates of Y are $(0,1,0)$ and the homogeneous coordinates of Z are $(0,0,1)$.

   It is interesting now to verify that the two methods we have suggested for defining homogeneous coordinates give us the same coordinatization of the projective plane when the frames of reference of the non-homogeneous and homogeneous systems are suitably related. To see this, consider an arbitrary nonhomogeneous coordinate system with gauge points L_0, L_1, L_ω on the x axis and gauge points M_0, M_1, M_ω on the y axis, L_0 and M_0, of course, coinciding at the origin, O. Furthermore, let U be the intersection of the lines $L_\omega M_1$ and $M_\omega L_1$, and let N_1 be the intersection of the unlabeled line and the line OU. Finally, let $P:(x,y)$ be an arbitrary point which is on neither the x axis, the y axis, nor the unlabeled line, let L_x be the projection of P from M_ω onto the x axis, let M_y be the projection of P from L_ω onto the y axis, and let N be the projection of P from O onto the unlabeled line. Then the homogeneous coordinates of P are (x_1,x_2,x_3) (according to our first scheme of definition) if and only if $x = x_1/x_3$ and $y = x_2/x_3$, that is, if and only if the homogeneous coordinates of L_x with respect to L_0, L_1, L_ω are (x_1,x_3) and the homogeneous coordinates of M_y with respect to M_0, M_1, M_ω are (x_2,x_3).

   Now consider the homogeneous coordinate system for which

$X\ (= Y_0 = Z_\omega)$ is the point L_ω,
$Y\ (= Z_0 = X_\omega)$ is the point M_ω,
$Z\ (= X_0 = Y_\omega)$ is the point O,
I is the point U,
X_1 is the point M_1,
Y_1 is the point L_1,
Z_1 is the point N_1

(Fig. 9.7). Since $Y_0 = L_\omega$, $Y_1 = L_1$, $Y_\omega = L_0$, it follows from the permutation properties of the cross ratio (Theorem 2, Sec. 4.7) and the fact that

$$\mathrm{R}\,(L_0 L_\omega, L_x L_1) = x = x_1/x_3$$

that the homogeneous coordinates of L_x with respect to Y_0, Y_1, Y_ω, on the line ZX, are (x_3,x_1). On the other hand, since $X_0 = M_0$, $X_1 = M_1$, $X_\omega = M_\omega$, the homogeneous coordinates of M_y with respect to X_0, X_1, X_ω, on the line YZ, are still (x_2,x_3). Finally, by Theorem 7, Sec. 9.5, the homogeneous

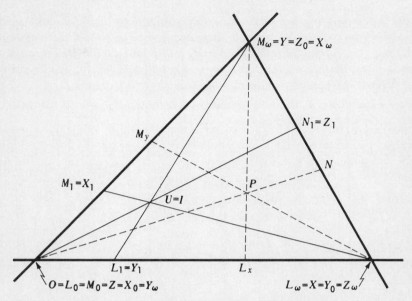

Fig. 9.7 A nonhomogeneous coordinate system and a homogeneous coordinate system superposed to yield identical homogeneous coordinates.

coordinates of N with respect to Z_0, Z_1, Z_ω, on the line XY, are (x_1,x_2). Hence, according to Definition 1, the homogeneous coordinates of P are (x_1,x_2,x_3), just as they were when they were derived from the nonhomogeneous coordinates of P. That the two methods of assigning homogeneous coordinates give the same coordinates to points on the lines YZ, ZX, and XY is almost obvious, and easily checked.

The equation of a line in homogeneous coordinates can be derived by an argument very much like the one we used in proving Theorem 1, Sec. 9.6. However, it is easier to obtain it by simply transforming the equation of the line from nonhomogeneous to homogeneous coordinates. Let us suppose, then, that l is an arbitrary line and that we desire its equation in a given homogeneous coordinate system. If we choose a nonhomogeneous coordinate system related to the given homogeneous system as shown in Fig. 9.7, we know that the two sets of coordinates are related by the equations $x = x_1/x_3$, $y = x_2/x_3$, and we also know that in the nonhomogeneous system the line has an equation of the form $l_1 x + l_2 y + l_3 = 0$. Hence, transforming the equation into homogeneous coordinates, we obtain

(1) $$l_1 x_1 + l_2 x_2 + l_3 x_3 = 0$$

as an equation satisfied by the homogeneous coordinates of every point on the given line, and no others, except possibly for points that were unlabeled in the nonhomogeneous coordinate system, i.e., points on the line XY in the homogeneous system. Now the homogeneous coordinates of the unlabeled point on l, say N, can be determined from the equation of the line through O, that is Z, which has no labeled intersection with l. Clearly, this line is the line defined (in nonhomogeneous coordinates) by the equation

(2) $$l_1 x + l_2 y = 0$$

Hence, from the first of our two equivalent methods of defining homogeneous coordinates, the homogeneous coordinates of N are $(l_2, -l_1, 0)$, since the point with nonhomogeneous coordinates $(l_2, -l_1)$ lies on the line defined by Eq. (2). Obviously the coordinates $(l_2, -l_1, 0)$ of N satisfy the homogeneous form of the equation of l. Conversely, the homogeneous coordinates of no other point on XY can satisfy the equation of l, since a triple of coordinates of the form $(a, b, 0)$ will satisfy Eq. (1) if and only if $l_1 a + l_2 b = 0$, that is, if and only if $(a, b, 0)$ is the point N. Finally, we observe that the line XY, which has no equation in nonhomogeneous coordinates, has the equation $x_3 = 0$ in homogeneous coordinates, and this, too, is of the asserted form. Thus we have established the following theorem.

Theorem 1

Every line in the projective plane has an equation of the form $l_1 x_1 + l_2 x_2 + l_3 x_3 = 0$.

The following theorems, which in most cases are just restatements of results that we have previously established, are now easy to prove.

Theorem 2

The locus of every equation of the form $l_1 x_1 + l_2 x_2 + l_3 x_3 = 0$ is a line.

Theorem 3

The equation of the line on the points (a_1, a_2, a_3) and (b_1, b_2, b_3) can be written in the form

$$\begin{vmatrix} x_1 & x_2 & x_3 \\ a_1 & a_2 & a_3 \\ b_1 & b_2 & b_3 \end{vmatrix} = 0$$

Theorem 4

If $A:(a_1, a_2, a_3)$, $B:(b_1, b_2, b_3)$, $C:(c_1, c_2, c_3)$, $D:(d_1, d_2, d_3)$ are four collinear points, then

$$\begin{aligned} R\,(AB, CD) &= R\,[(a_2, a_3)(b_2, b_3),\ (c_2, c_3)(d_2, d_3)] \\ &= R\,[(a_3, a_1)(b_3, b_1),\ (c_3, c_1)(d_3, d_1)] \\ &= R\,[(a_1, a_2)(b_1, b_2),\ (c_1, c_2)(d_1, d_2)] \end{aligned}$$

whenever these cross ratios are meaningful, and there is always at least one of these cross ratios which is meaningful.

Theorem 5

If $A:(a_1,a_2,a_3)$ and $B:(b_1,b_2,b_3)$ are distinct points, and if $P_i:\lambda_i A + \mu_i B$, $i = 1, 2, 3, 4$, are four points collinear with A and B, then

$$\mathrm{R}(P_1P_2,P_3P_4) = \mathrm{R}[(\lambda_1,\mu_1)(\lambda_2,\mu_2), (\lambda_3,\mu_3)(\lambda_4,\mu_4)]$$

In Sec. 9.5 we noted that the cross ratio of four collinear points arose in two contexts that at first seemed quite different, one involving the *coordinates* of the points, the other involving the *parameters* of the points. The last two theorems tell us that this distinction is of no significance, since the cross ratio is the same from either point of view.

The equation of a conic in homogeneous coordinates can be derived by an argument similar to that used to prove Theorem 5, Sec. 9.6. However, as in the preceding discussion of the equation of a line, it is now easier to obtain it by simply transforming the equation of the conic from nonhomogeneous to homogeneous coordinates. Let us suppose, then, that Γ is an arbitrary conic and that we desire its equation in a given homogeneous coordinate system. If we choose a nonhomogeneous system related to the given homogeneous system as shown in Fig. 9.7, we know that the two sets of coordinates are related by the equations $x = x_1/x_3$ and $y = x_2/x_3$, and we also know that in nonhomogeneous coordinates the conic has an equation of the form $ax^2 + 2hxy + by^2 + 2gx + 2fy + c = 0$, arising, in fact, from Eq. (8), Sec. 9.6, namely,

$$a(ux - sy)(ux - ty) + b(ux - sy)(u - y) \\ + c(ux - ty)(u - y) + d(u - y)^2 = 0$$

or, in homogeneous coordinates,

$$(3) \qquad a(ux_1 - sx_2)(ux_1 - tx_2) + b(ux_1 - sx_2)(ux_3 - x_2) \\ + c(ux_1 - tx_2)(ux_3 - x_2) \\ + d(ux_3 - x_2)^2 = 0$$

The last equation is thus a relation satisfied by the homogeneous coordinates of every point of the given conic, except possibly those points which were unlabeled in the nonhomogeneous system. To complete our discussion we must therefore show that:

1. The homogeneous coordinates of every point common to Γ and XY satisfy Eq. (3).

2. Every point on XY whose coordinates satisfy Eq. (3) is a point of Γ.

To verify requirement 1, we return to the proof of Theorem 5, Sec. 9.6, and observe that an unlabeled point on Γ, that is, a point common to Γ and the line XY, arises when and only when the lines PX_P and QX_Q intersect in an unlabeled point, i.e., a point whose third homogeneous coordinate is zero. Now the equations of PX_P and QX_Q are, respectively,

$$ux_1 + (p - s)x_2 - pux_3 = 0 \qquad \text{and} \qquad ux_1 + (q - t)x_2 - qux_3 = 0$$

and we find without difficulty that their intersection is the point whose homogeneous coordinates are

$$x_1 = pt - qs \qquad x_2 = u(p - q) \qquad x_3 = (p - q) - (s - t)$$

Hence we must verify that these coordinates satisfy Eq. (3) when $x_3 = (p - q) - (s - t) = 0$, subject, of course, to the condition $apq + bp + cq + d = 0$. The details of this verification are perfectly straightforward, and we leave them as an exercise.

To verify requirement 2, let $(x_1', x_2', 0)$ be any point on XY whose coordinates satisfy the equation of Γ. Equation (3) then gives us the condition

$$a(ux_1' - sx_2')(ux_1' - tx_2') + b(ux_1' - sx_2')(-x_2')$$
$$+ c(ux_1' - tx_2')(-x_2') + d(-x_2')^2 = 0$$

which implies that $p = (ux_1' - sx_2')/(-x_2')$ and $q = (ux_1' - tx_2')/(-x_2')$ are the nonhomogeneous coordinates of corresponding points, X_P and X_Q, in the projectivity on the x axis which arose in the proof of Theorem 5, Sec. 9.6. Hence, the intersection of PX_P and QX_Q is a point of Γ, and it is easy to check that it is, in fact, the point $(x_1', x_2', 0)$. In other words, any point of XY whose coordinates satisfy the equation of Γ is a point of Γ. If $x_2' = 0$, the preceding argument breaks down; but for this to happen, the coefficient a must be zero, which implies that X_P and X_Q coincide at the (unlabeled) point $(1,0,0)$ on the x axis. In this case, being the intersection of PX_P and QX_Q, this point is clearly a point of Γ. Finally, if Γ is a singular conic, it is clear from our discussion of the equation of a line that the equation of Γ is of the asserted form. Thus, summarizing, we have the following theorem.

Theorem 6

Every conic in the projective plane has an equation of the form

$$ax_1{}^2 + 2hx_1x_2 + bx_2{}^2 + 2gx_1x_3 + 2fx_2x_3 + cx_3{}^2 = 0$$

It is now a straightforward matter to prove the following theorems, most of which are essentially restatements of results we have already established.

Theorem 7

The locus of any equation of the form

$$ax_1{}^2 + 2hx_1x_2 + bx_2{}^2 + 2gx_1x_3 + 2fx_2x_3 + cx_3{}^2 = 0$$

is a conic.

Theorem 8

The equation

$$ax_1{}^2 + 2hx_1x_2 + bx_2{}^2 + 2gx_1x_3 + 2fx_2x_3 + cx_3{}^2 = 0$$

represents a singular conic if and only if

$$\begin{vmatrix} a & h & g \\ h & b & f \\ g & f & c \end{vmatrix} = 0$$

Theorem 9

If

$$ax_1{}^2 + 2hx_1x_2 + bx_2{}^2 + 2gx_1x_3 + 2fx_2x_3 + cx_3{}^2 = 0$$

is the equation of a nonsingular conic, Γ, then:

1. The equation of the polar of a point (y_1, y_2, y_3) with respect to Γ is

$$(ay_1 + hy_2 + gy_3)x_1 + (hy_1 + by_2 + fy_3)x_2 + (gy_1 + fy_2 + cy_3)x_3 = 0$$

If $Y:(y_1, y_2, y_3)$ is a point of Γ, this is the equation of the tangent to Γ at Y.

2. The equation of the singular conic consisting of the tangents to Γ which pass through a point (y_1, y_2, y_3) which is not on Γ is

$$(ax_1{}^2 + 2hx_1x_2 + bx_2{}^2 + 2gx_1x_3 + 2fx_2x_3 + cx_3{}^2)$$
$$\cdot (ay_1{}^2 + 2hy_1y_2 + by_2{}^2 + 2gy_1y_3 + 2fy_2y_3 + cy_3{}^2)$$
$$-[(ay_1 + hy_2 + gy_3)x_1 + (hy_1 + by_2 + fy_3)x_2 + (gy_1 + fy_2 + cy_3)x_3]^2 = 0$$

Exercises

1. Is every triple, (x_1, x_2, x_3), with at least one nonzero entry a set of homogeneous coordinates for some point of the projective plane?

2. Can a point have two nonproportional triples of homogeneous coordinates with respect to a given coordinate system in the projective plane?

3. Prove Theorem 4.

4. Prove Theorem 5.

5. Complete the proof of Theorem 6 by verifying that $x_1 = pt - qs$, $x_2 = u(p - q)$, $x_3 = (p - q) - (s - t)$ satisfy Eq. (3) when $apq + bp + cq + d = 0$ and $x_3 = 0$.

6. Verify Theorem 4, given $P_1:(1,2,1)$, $P_2:(1,1,-1)$, $P_3:(0,1,2)$, $P_4:(-1,1,5)$.

7. Verify Theorem 5 by computing the cross ratio of the points in Exercise 6 using their parameters in terms of:
(a) P_1 and P_2 (b) P_1 and P_3

8. Prove Theorem 9.

9. Given $\Gamma: x_1{}^2 - 2x_2x_3 = 0$ and the points $P:(2,2,1)$ and $Q:(-2,1,2)$ on Γ. If P and Q are used as generating bases of Γ, find the equation of the projectivity induced on the line $x_1 = 0$ by the projectivity which generates Γ.

10. Work Exercise 9, given $\Gamma: x_1{}^2 + x_2{}^2 - 2x_1x_3 = 0$, $P:(1,1,1)$, and $Q:(1,-1,1)$.

11. To what form does the equation of a general conic reduce if:
(a) The triangle of reference is self-polar with respect to the conic?
(b) The vertices of the triangle of reference are on the conic?
(c) The triangle of reference consists of two tangents to the conic and the line determined by the points of contact of these tangents?

12. Show that if two triangles are self-polar with respect to a given conic, their six vertices lie on a second conic. Is the converse true?

13. What is the locus of the poles of a given line with respect to the conics on four given points no three of which are collinear?

14. The sides, BC, CA, AB, of $\triangle ABC$ are tangent to a conic at the respective points A', B', C'. Prove algebraically that AA', BB', CC' are concurrent.

15. Show that if two nonsingular conics have four distinct points in common, there is one and only one triangle which is self-polar with respect to both conics.

9.8 Line-Coordinates in the Projective Plane

The process of introducing coordinates for the lines of the projective plane is just the dual of the process we have employed in defining point-coordinates.

First of all, we identify gauge lines p_0, p_1, p_ω in an arbitrary pencil of lines. Then, using the dual of Axiom 10, we introduce first nonhomogeneous and then homogeneous coordinates into the pencil. Then, by-passing the introduction of a nonhomogeneous coordinate system for the

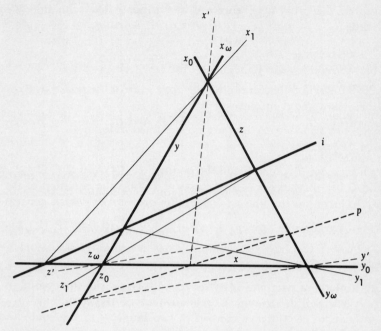

Fig. 9.8 The triangle of reference for a system of homogeneous line-coordinates in the projective plane.

lines of the projective plane, we assign homogeneous coordinates to the lines of the plane by the dual of Definition 1, Sec. 9.7.

First we choose an arbitrary four-line x, y, z, i whose diagonal lines are:

The join, x_1, of xi and yz,
The join, y_1, of yi and zx,
The join, z_1, of zi and xy.

On the vertices of $\triangle xyz$ we choose gauge lines as follows (Fig. 9.8):

On yz let $x_0 = z$, $x_1 = x_1$, $x_\omega = y$,
On zx let $y_0 = x$, $y_1 = y_1$, $y_\omega = z$,
On xy let $z_0 = y$, $z_1 = z_1$, $z_\omega = x$.

Now let p be any line which is not on any vertex of $\triangle xyz$, and let:

x' be the join of xp and yz,
y' be the join of yp and zx,
z' be the join of zp and xy.

Then without loss of generality we can suppose that the homogeneous coordinates

Of x' on the point yz are $[l_2, l_3]$,
Of y' on the point zx are $[l_3, l_1]$,
Of z' on the point xy are $[l_1, t]$.

and by Theorem 5, Sec. 7.5, we can show that $t = l_2$. The homogeneous coordinates of an arbitrary line are now defined as follows.

Definition 1

1. If p is any line on yz, and if its homogeneous coordinates relative to x_0, x_1, x_ω are $[l_2, l_3]$, then its homogeneous coordinates in the plane are $[0, l_2, l_3]$.

2. If p is any line on zx, and if its homogeneous coordinates relative to y_0, y_1, y_ω are $[l_3, l_1]$, then its homogeneous coordinates in the plane are $[l_1, 0, l_3]$.

3. If p is any line on xy, and if its homogeneous coordinates relative to z_0, z_1, z_ω are $[l_1, l_2]$, then its homogeneous coordinates in the plane are $[l_1, l_2, 0]$.

4. If p is any line which is not on a vertex of $\triangle xyz$, and if $[l_2, l_3]$, $[l_3, l_1]$, $[l_1, l_2]$ are the homogeneous coordinates of the lines which join its intersections with x, y, and z to the opposite vertices, yz, zx, and xy, then its homogeneous coordinates in the plane are $[l_1, l_2, l_3]$.

Under Definition 1, x, y, and z are assigned homogeneous coordinates according to two different schemes, since each of these lines is on two different vertices of the triangle of reference, $\triangle xyz$. However, it is easy to check that under either definition the homogeneous coordinates of x, y, and z are, respectively, $[1,0,0]$, $[0,1,0]$, $[0,0,1]$.

It is now easy to establish the following theorems which are, in fact, just the duals of theorems we have already obtained.

Theorem 1

Every point has an equation of the form

$$a_1 l_1 + a_2 l_2 + a_3 l_3 = 0$$

Theorem 2

The envelope of every equation of the form

$$a_1 l_1 + a_2 l_2 + a_3 l_3 = 0$$

is a point.

Theorem 3

The equation of the point of intersection of the line $[p_1,p_2,p_3]$ and the line $[q_1,q_2,q_3]$ can be written in the form $\begin{vmatrix} l_1 & l_2 & l_3 \\ p_1 & p_2 & p_3 \\ q_1 & q_2 & q_3 \end{vmatrix} = 0.$

Theorem 4

If $p:[p_1,p_2,p_3]$, $q:[q_1,q_2,q_3]$, $r:[r_1,r_2,r_3]$, $s:[s_1,s_2,s_3]$ are four concurrent lines, then

$$\begin{aligned} \mathbb{R}(pq,rs) &= \mathbb{R}[(p_2,p_3)(q_2,q_3),\ (r_2,r_3)(s_2,s_3)] \\ &= \mathbb{R}[(p_3,p_1)(q_3,q_1),\ (r_3,r_1)(s_3,s_1)] \\ &= \mathbb{R}[(p_1,p_2)(q_1,q_2),\ (r_1,r_2)(s_1,s_2)] \end{aligned}$$

whenever the cross ratios are meaningful, and in every case at least one of the cross ratios is meaningful.

Theorem 5

If $a:[a_1,a_2,a_3]$ and $b:[b_1,b_2,b_3]$ are distinct lines, and if $m_i = \lambda_i a + \mu_i b$, $i = 1, 2, 3, 4$, then

$$\mathbb{R}(m_1 m_2, m_3 m_4) = \mathbb{R}[(\lambda_1,\mu_1)(\lambda_2,\mu_2),\ (\lambda_3,\mu_3)(\lambda_4,\mu_4)]$$

Theorem 6

Every line-conic has an equation of the form

$$Al_1{}^2 + 2Hl_1 l_2 + Bl_2{}^2 + 2Gl_1 l_3 + 2Fl_2 l_3 + Cl_3{}^2 = 0$$

Theorem 7

The envelope of every equation of the form

$$Al_1{}^2 + 2Hl_1 l_2 + Bl_2{}^2 + 2Gl_1 l_3 + 2Fl_2 l_3 + Cl_3{}^2 = 0$$

is a line-conic.

Theorem 8

The equation

$$Al_1{}^2 + 2Hl_1 l_2 + Bl_2{}^2 + 2Gl_1 l_3 + 2Fl_2 l_3 + Cl_3{}^2 = 0$$

represents a singular line-conic if and only if

$$\begin{vmatrix} A & H & G \\ H & B & F \\ G & F & C \end{vmatrix} = 0$$

Theorem 9

If

$$Al_1{}^2 + 2Hl_1 l_2 + Bl_2{}^2 + 2Gl_1 l_3 + 2Fl_2 l_3 + Cl_3{}^2 = 0$$

is the equation of a nonsingular line-conic, Φ, then:

1. The equation of the pole of a line $[m_1,m_2,m_3]$ with respect to Φ is

$$(Am_1 + Hm_2 + Gm_3)l_1 + (Hm_1 + Bm_2 + Fm_3)l_2$$
$$+ (Gm_1 + Fm_2 + Cm_3)l_3 = 0$$

If $m:[m_1,m_2,m_3]$ is a line of Φ, this is the equation of the tangent-point of m.

2. The equation of the singular line-conic consisting of the two tangent points of Φ which lie on a line $[m_1,m_2,m_3]$ which is not on Φ is

$$(Al_1{}^2 + 2Hl_1l_2 + Bl_2{}^2 + 2Gl_1l_3 + 2Fl_2l_3 + Cl_3{}^2)$$
$$(Am_1{}^2 + 2Hm_1m_2 + Bm_2{}^2 + 2Gm_1m_3 + 2Fm_2m_3 + Cm_3{}^2)$$
$$-[(Am_1 + Hm_2 + Gm_3)l_1 + (Hm_1 + Bm_2 + Fm_3)l_2$$
$$+ (Gm_1 + Fm_2 + Cm_3)l_3]^2 = 0$$

So far, we have defined homogeneous point-coordinates and homogeneous line-coordinates, and we have derived the form of the equation of a line and the form of the equation of a point. However, we have not yet obtained the condition that a point (x_1,x_2,x_3) and a line $[l_1,l_2,l_3]$ should be incident. To do this, let us attempt to find an equation, $f(x_1,x_2,x_3,l_1,l_2,l_3) = 0$, which will be satisfied if and only if the point (x_1,x_2,x_3) lies on the line $[l_1,l_2,l_3]$. Clearly, such a condition must have the property that when the coordinates of a given line are substituted into it, the resulting equation is satisfied by the coordinates of every point on that line and by the coordinates of no other point. In other words, the resulting equation must be an equation of the line. Similarly, the required condition must be such that when the coordinates of a particular point are substituted into it, the resulting equation is satisfied by the coordinates of every line on that point and by the coordinates of no other line. In other words, it must be an equation of the given point. Thus, since the equations of both points and lines are linear, it follows that the condition we are seeking must be homogeneous and bilinear; i.e., it must be of the form

(1)
$$(a_1x_1 + b_1x_2 + c_1x_3)l_1 + (a_2x_1 + b_2x_2 + c_2x_3)l_2$$
$$+ (a_3x_1 + b_3x_2 + c_3x_3)l_3 = 0$$

Now suppose that the equations of the lines of the fundamental four-line are

$$x:[1,0,0]: a_1'x_1 + b_1'x_2 + c_1'x_3 = 0$$
$$y:[0,1,0]: a_2'x_1 + b_2'x_2 + c_2'x_3 = 0$$
$$z:[0,0,1]: a_3'x_1 + b_3'x_2 + c_3'x_3 = 0$$
$$i:[1,1,1]: a_4'x_1 + b_4'x_2 + c_4'x_3 = 0$$

Then, substituting the coordinates of the lines x, y, z into the bilinear relation (1), we must have

(2)
$$a_1x_1 + b_1x_2 + c_1x_3 = k_x(a_1'x_1 + b_1'x_2 + c_1'x_3)$$
$$a_2x_1 + b_2x_2 + c_2x_3 = k_y(a_2'x_1 + b_2'x_2 + c_2'x_3)$$
$$a_3x_1 + b_3x_2 + c_3x_3 = k_z(a_3'x_1 + b_3'x_2 + c_3'x_3)$$

and therefore Eq. (1) can be written in the form

$$k_x(a_1'x_1 + b_1'x_2 + c_1'x_3)l_1 + k_y(a_2'x_1 + b_2'x_2 + c_2'x_3)l_2$$
$$+ k_z(a_3'x_1 + b_3'x_2 + c_3'x_3)l_3 = 0$$

Finally, substituting the coordinates of the line i into the last equation and identifying the result with the equation of i, we obtain the equations

$$k_x a_1' + k_y a_2' + k_z a_3' = k_i a_4'$$
$$k_x b_1' + k_y b_2' + k_z b_3' = k_i b_4'$$
$$k_x c_1' + k_y c_2' + k_z c_3' = k_i c_4'$$

From these equations k_x, k_y, k_z can be found in terms of k_i and the known coefficients in the equations of x, y, z, i. Then the coefficients in Eq. (1) can be found by comparing like terms in Eqs. (2). Although only the lines x, y, z, i of the fundamental four-line were used in determining the coefficients in the incidence condition (1), it is not difficult to show that it is, in fact, the condition that an arbitrary point (x_1, x_2, x_3) and an arbitrary line $[l_1, l_2, l_3]$ should be incident (see Exercises 7 to 9).

If it turns out that $f(x_1, x_2, x_3, l_1, l_2, l_3) = 0$ is symmetric, that is, if it can be written in the form $\Lambda^T M X = 0$, where M is a symmetric matrix,

say $M = \begin{vmatrix} a & h & g \\ h & b & f \\ g & f & c \end{vmatrix}$, then the equation of the line $x:[1,0,0]$ is

$$ax_1 + hx_2 + gx_3 = 0$$

which is the equation of the polar of $X:(1,0,0)$ with respect to the conic $X^T M X = 0$. Similarly, y, z, and i are, respectively, the polars of $Y:(0,1,0)$, $Z:(0,0,1)$, and $I:(1,1,1)$ with respect to the conic $X^T M X = 0$. Conversely, it is clear that if x, y, z, i are the polars of X, Y, Z, I with respect to a conic $X^T M X = 0$, then the incidence condition is the bilinear form $\Lambda^T M X = 0$. Thus, summarizing, we have the following theorem.

Theorem 10

The condition that a point (x_1, x_2, x_3) should lie on a line $[l_1, l_2, l_3]$ is the symmetric relation $\Lambda^T M X = 0$ if and only if the complete four-line used to define the homogeneous line-coordinates is the polar with respect to the conic $X^T M X = 0$ of the complete four-point used to define the homogeneous point-coordinates.

Corollary 1

The condition that a point (x_1,x_2,x_3) should lie on a line $[l_1,l_2,l_3]$ is $l_1 x_1 + l_2 x_2 + l_3 x_3 = 0$ if and only if the complete four-point and the complete four-line used to define the homogeneous point- and line-coordinates, respectively, are polars with respect to the conic $x_1{}^2 + x_2{}^2 + x_3{}^2 = 0$.

Naturally, since we wish to have the simplest possible incidence condition, we shall suppose that the complete four-point and the complete four-line which are used to define point- and line-coordinates, respectively, are polars with respect to the conic $x_1{}^2 + x_2{}^2 + x_3{}^2 = X^T X = 0$. More specifically, this means that our two coordinate systems are so related that:

x is the line YZ,
y is the line ZX,
z is the line XY,
i is the polar of I with respect to $\triangle XYZ$.

With our coordinate systems now established so that the incidence condition is $l_1 x_1 + l_2 x_2 + l_3 x_3 = 0$, we can prove the following theorems, which we first encountered in our study of conics in Π_2 (Sec. 4.9).

Theorem 11

The equation of the envelope of the tangents to the point-conic $X^T M X = 0$ is $\Lambda^T M^{-1} \Lambda = 0$.

Theorem 12

The equation of the locus of the tangent points of the line-conic $\Lambda^T N \Lambda = 0$ is $X^T N^{-1} X = 0$.

Exercises

1. Prove Theorem 11.   2. Prove Theorem 12.

3. Find the equation of the envelope of the tangents to each of the following point-conics:

(a) $x_1{}^2 - 2x_1 x_2 + 4x_1 x_3 - 3x_3{}^2 = 0$
(b) $x_1{}^2 + 2x_1 x_2 - x_2{}^2 + 4x_2 x_3 + 2x_3{}^2 = 0$
(c) $-2x_1{}^2 + 2x_1 x_2 + x_1 x_3 + x_2 x_3 + x_3{}^2 = 0$

4. Under what conditions, if any, will the point equation and the line equation of a conic be identical in structure?

5. Determine the form of the incidence condition if the sides of the four-line used to determine the coordinates of a line are:

(a) $x: x_1 + x_2 = 0, y: x_2 + x_3 = 0, z: x_1 + x_3 = 0, i: x_1 + x_2 + x_3 = 0$

(b) $x: x_1 = 0, \ y: x_1 + x_2 = 0, \ z: x_1 + x_2 + x_3 = 0, \ i: x_1 - 2x_2 + x_3 = 0$

(c) $x: x_1 + x_2 = 0, y: x_1 - x_2 + x_3 = 0, z: x_2 + x_3 = 0,$
$\quad i: 2x_1 + x_2 + 2x_3 = 0$

6. Determine the equations of the sides of the four-line used to define the coordinates of a line if the incidence condition is:

(a) $l_2x_1 + l_3x_2 + l_1x_3 = 0$ $\qquad$ (b) $l_2x_1 + l_1x_2 + l_3x_3 = 0$

(c) $(l_1 + 2l_2)x_1 - (l_1 - l_3)x_2 + (2l_2 - l_3)x_3 = 0$

7. Letting (a_ix) denote the expression $a_ix_1 + b_ix_2 + c_ix_3$, show that if $(a_1x) = 0$, $(a_2x) = 0$, $(a_3x) = 0$, and $\lambda'(a_1x) + \mu'(a_2x) + \nu'(a_3x) = 0$ are, respectively, the equations of the lines x, y, z, i of the fundamental four-line, then the coordinates of the line whose equation is $\lambda(a_1x) + \mu(a_2x) + \nu(a_3x) = 0$ are $[\lambda/\lambda', \mu/\mu', \nu/\nu']$.

8. Show that if substituting the coordinates of an arbitrary line into Eq. (1) yields the equation of that line, then substituting the coordinates of an arbitrary point into Eq. (1) will yield the equation of that point.

9. Using the results of Exercises 7 and 8, show that Eq. (1) is the incidence condition for an arbitrary point and an arbitrary line.

9.9 Conclusion

In this chapter our axiomatic development of projective geometry culminated in the introduction of coordinates, both nonhomogeneous and homogeneous, into the projective plane. We accomplished this by first developing an algebra of collinear points and, dually, of concurrent lines. Then because we were able to verify that this algebra satisfied the postulates of a field, we were able to postulate an isomorphism between the points of an arbitrary line and the field of complex numbers, which led easily to a system of coordinates for the points of the entire plane and, dually, to a coordinate system for the lines of the plane. Because of the principle of duality, points have no priority over lines, or vice versa. Hence the introduction of co-ordinates for points and coordinates for lines involves two independent processes. However, in order to preserve the simple incidence condition $l_1x_1 + l_2x_2 + l_3x_3 = 0$ for a point (x_1,x_2,x_3) and a line $[l_1,l_2,l_3]$, we found that the complete four-point involved in the definition of point-coordinates and the complete four-line involved in the definition of line-coordinates had to be polars of each other with respect to the nonsingular point-conic $x_1{}^2 + x_2{}^2 + x_3{}^2 = 0$ or, dually, the nonsingular line-conic $l_1{}^2 + l_2{}^2 + l_3{}^2 = 0$.

  After the introduction of coordinates, we derived the equations describing projectivities, lines, and conics. Although for completeness we quoted a number of results on lines and conics, we did not go into them in detail because, with the introduction of homogeneous coordinates into the projective plane, it became clear that the system we have developed axiomatically is isomorphic to the algebraic representation, and the algebraic proofs and illustrations of the properties of lines and conics which we gave in Chaps. 4 and 5 are applicable without change in the context of the projective plane we have developed from Axioms 1 to 10.

ten

THE INTRODUC= TION OF A METRIC

10.1 Introduction

Up to this point, metrical ideas have been conspicuously absent from our work. Neither the measurement of distance nor that of angles has been considered, and, except incidentally in occasional applications to euclidean geometry, such related topics as betweenness, congruence, similarity, perpendicularity, and parallelism have had no place in our discussions. It is possible, however, without adding any new axioms, to introduce definitions of distance and angle measure into the projective plane in such a way that at least some of the familiar properties of these concepts are preserved. Moreover, if the application of these definitions is restricted to pairs of points and lines in suitable subsets of the projective plane, we obtain metrical geometries exhibiting *all* the properties of the euclidean plane or the elliptic and hyperbolic noneuclidean planes, as we choose. It was this which inspired the English mathematician Arthur Cayley

(1821–1895) to remark, with more enthusiasm than accuracy perhaps, "Projective geometry is all geometry."

In the sections which follow, we shall discuss, in turn, the measurement of distance and angle in the entire projective plane and the specializations of these processes which lead to elliptic, hyperbolic, and euclidean geometry. In doing this we shall need a few simple properties of the trigonometric, hyperbolic, and logarithmic functions of a complex variable, as well as certain matric identities; and since we are not presupposing a knowledge of these results, we shall interrupt our development, when necessary, to introduce them.

10.2 The Definition of Distance and Angle Measure

In elementary geometry the distance between two points, P and Q, denoted by the symbol (PQ), possesses, among others, the following properties:

D.1. $(PP) = 0$.
D.2. $(PQ) = -(QP)$.†
D.3. If P, Q, R are collinear, then $(PQ) + (QR) + (RP) = 0$.‡

Similarly, in elementary geometry the measure, in radians, of the angle between two directed lines, p and q, which we shall denote by the symbol $[pq]$, possesses the following properties:

A.1. $[pp] = 2n\pi$.
A.2. $[pq] = -[qp] + 2n\pi$.
A.3. If p, q, r are concurrent, then $[pq] + [qr] + [rp] = 2n\pi$.

Although properties A.1 to A.3 bear a striking resemblance to properties D.1 to D.3, they obviously differ because the term $2n\pi$ appears only in the three expressions relating to angle measurement. That there should be a difference of some sort is, of course, not surprising, since the principle of duality does not hold in elementary geometry. On the other hand, since the principle of duality is valid in projective geometry, one would expect that the original formulations of the measurement of distance and the measurement of angle over the entire projective plane would be duals of each other, and this is indeed the case.

---

† In current axiomatic developments of elementary geometry there is an increasing tendency to regard distance as a function which associates with every pair of points a *nonnegative* number which is zero if and only if the points of the pair are coincident. Under this assumption, one would, of course, have $(PQ) = (QP)$. However, for our purposes this is less convenient than the use of *directed distances* implied by property D.2.
‡ Unless $P = Q = R$, this is meaningful if and only if directed distances are used, i.e., if and only if property D.2 holds.

In defining the measurement of distance and angle in the projective plane, we naturally wish to retain as many as possible of the properties of measurement familiar to us from elementary geometry. The clue that suggests how this should be done is found in the multiplication theorem for cross ratios (Exercise 13, Sec. 4.7):

Theorem 1

If M_1, M_2, P, Q, and R are collinear points such that M_1 and M_2 are distinct and such that none of the points P, Q, R coincides with either M_1 or M_2, then

$$R(M_1M_2,PQ)R(M_1M_2,QR)R(M_1M_2,RP) = 1$$

If we take the natural logarithm of the expression appearing in the statement of this theorem, we have

(1) $\ln R(M_1M_2,PQ) + \ln R(M_1M_2,QR) + \ln R(M_1M_2,RP) = \ln 1 = 0$

which would be precisely property D.3 if $\ln R(M_1M_2,PQ)$ were defined to be the distance (PQ) and the other terms were similarly interpreted. Moreover, from the properties of the cross ratio which we developed in Sec. 4.7, we know that $R(M_1M_2,PP) = 1$. Hence

(2) $$\ln R(M_1M_2,PP) = \ln 1 = 0$$

which clearly corresponds to property D.1. Also

$$R(M_1M_2,PQ) = \frac{1}{R(M_1M_2,QP)}$$

Hence

(3) $$\ln R(M_1M_2,PQ) = -\ln R(M_1M_2,QP)$$

which, under our tentative definition of distance is just property D.2. Finally, we note that (1), (2), and (3) still express properties D.1, D.2, and D.3 if the distance (PQ) is defined to be any nonzero multiple of $R(M_1M_2,PQ)$. These observations we now formalize in the following definition.

Definition 1

Let M_1 and M_2 be distinct points chosen arbitrarily on a line m. Then the distance between any other points, P and Q, on m is

$$(PQ) = k_d \ln R(M_1M_2,PQ) = k_d \ln R(PQ,M_1M_2)$$

where k_d is an arbitrary nonzero constant.

The points M_1 and M_2, in terms of which distances on m are measured, are called the **metric gauge points** on m. The constant k_d is called the **scale constant.**

Dually, of course, we have the following definition for angle measure.

Definition 2

Let m_1 and m_2 be distinct lines chosen arbitrarily on a point M. Then the measure of the angle between any other lines, p and q, on m is $[pq] = k_a \ln \mathrm{R}\,(m_1 m_2, pq) = k_a \ln \mathrm{R}\,(pq, m_1 m_2)$, where k_a is an arbitrary nonzero constant.

The lines m_1 and m_2 in terms of which angles on M are measured are called the **metric gauge lines** on M. The constant k_a is also called the **scale constant.**

The distance between two points, P and Q, is not an intrinsic property of P and Q and the scale constant k_d, as it appears to be in elementary geometry, but depends also upon the gauge points, M_1 and M_2. Similarly, the measure of the angle between two lines, p and q, is not an intrinsic property of p and q and the scale constant k_a but depends also on the gauge lines m_1 and m_2. How the gauge points, M_1 and M_2, are to be chosen on the various lines of the projective plane, and how the gauge lines, m_1 and m_2, are to be chosen on the various points of the projective plane will be discussed in Sec. 10.4.

It should be noted that whereas in elementary geometry distance and angle measure are always real numbers, this is presumably not the case in projective geometry. In fact, in the projective plane cross ratios may be either real or complex, since they are functions of coordinates which themselves may be either real or complex. Hence, in applying our definitions of distance and angle measure we are immediately faced with a problem we do not encounter in elementary analysis: What do we mean by the logarithm of a negative number or of a complex number? To answer this question and certain related ones soon to confront us, we shall devote the next section to a brief investigation of the properties of the logarithmic, trigonometric, and hyperbolic functions of complex arguments. In particular, and perhaps surprisingly, this will introduce into the measure of distance, as well as into the measurement of angle, the periodic term $2n\pi$, which occurs in properties A.1 to A.3 but appears to be missing in properties D.1 to D.3.

Exercises

1. If the gauge points on a certain line, m, are $M_1:(2,2,1)$ and $M_2:(-4,0,1)$, if $k_d = 2$, and if P and Q are the points $(-5,1,2)$ and $(-1,1,1)$, what is (PQ)?

2. If the gauge lines on a certain point, M, are $m_1:[2,0,1]$ and $m_2:[1,2,-1]$, if $k_a = 3$, and if p and q are the lines $[3,2,0]$ and $[0,-4,3]$, what is $[pq]$?

3. Given Γ: $x_1{}^2 - 2x_2x_3 = 0$, P:(2,3,0), Q:(2,4,-1), and $k_d = 3$. What is (PQ) if M_1 and M_2 are the points in which PQ intersects Γ?

4. If p and q are, respectively, the lines $x_1 + x_2 + x_3 = 0$ and $x_1 - x_2 + x_3 = 0$, and if m_1 and m_2 are, respectively, the lines joining the point pq to the points (1,0,1) and (2,1,1), what is $[pq]$?

5. If M_1 and M_2 are interchanged, how is (PQ) affected?

6. Does (M_1M_2) have any meaning?

7. Does (PQ) have meaning if $M_1 = M_2$?

8. If, as in elementary geometry, we say that Q is between P and R if and only if P, Q, and R are collinear and $(PQ) + (QR) = (PR)$, show that each of three collinear points is between the other two.

9. What is (PQ) if one of the points P, Q coincides with either M_1 or M_2? What other name does this suggest for the points M_1 and M_2?

10. If P and Q are two points on a line m, and if d is a nonzero constant, show that there are infinitely many pairs of points (M_1,M_2) on m in terms of which, as gauge points, we have $(PQ) = d$. Show also that the points $\{M_1\}$ and the points $\{M_2\}$ for which this is the case are projectively related.

10.3 A Little Complex Analysis

As the starting point of our digression to discover the meaning of the logarithm of a general complex number, we take the famous formula of Euler

$$(1) \qquad e^{i\theta} = \cos \theta + i \sin \theta$$

This is usually developed in a nonrigorous fashion in calculus as a by-product of the discussion of the Maclaurin expansions of e^x, $\cos x$, and $\sin x$, and hence is familiar to most students. It is a well-known fact of algebra that any complex number, $z = x + iy$, can be written in the standard polar form $z = r(\cos \theta + i \sin \theta)$, where $r = |z|$ is the length, or absolute value, of z and $\theta = \arg z$ is the angle, or argument, of z (Fig. 10.1). Hence, using (1), it appears that the general complex number z can also be written in the exponential form $z = re^{i\theta}$. Furthermore, it also follows from (1) that

$$(2) \qquad e^z = e^{x+iy} = e^x e^{iy} = e^x(\cos y + i \sin y)\dagger$$

which shows that e^x is the length of e^z and that y is the angle of e^z.

† This is actually the definition of e^z, rather than a result to be derived or proved. For a careful motivation of this definition, based on the fundamental properties of analytic functions, see, for example, C. R. Wylie, Jr., "Advanced Engineering Mathematics," 3d ed., pp. 656–658, McGraw-Hill Book Company, New York, 1966.

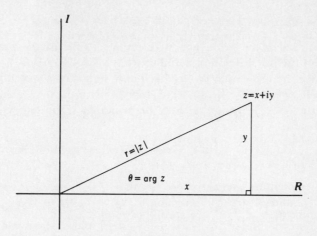

Fig. 10.1   The geometrical representation of the complex number $z = x + iy$.

The natural logarithm of $z = x + iy$ we now define implicitly as the function $w = \ln z$ which satisfies the equation

$$(3) \qquad\qquad e^w = z$$

If we let $w = u + iv$ and write z in the exponential form $re^{i\theta_1}$, where θ_1 is the particular argument of z which lies in the interval $-\pi < \theta \leq \pi$, we then must have, from (3), $e^{u+iv} = re^{i\theta_1}$, or

$$(4) \qquad\qquad e^u e^{iv} = re^{i\theta_1}$$

Since complex numbers which are equal must have equal lengths, it follows from (4) that $e^u = r$ or $u = \ln r$, where, since $r = |z|$ is a positive real number, only the familiar properties of the real logarithmic function are involved in the calculation of u. Moreover, since the angles of equal complex numbers must either be of equal measure or have measures which differ by an integral multiple of 2π, it also follows that $v = \theta_1 + 2n\pi$. Hence we have, finally,

$$(5) \qquad\qquad \ln z = w = u + iv = \ln |z| + i(\theta_1 + 2n\pi)$$

which is the formula for the logarithm of a general complex number. Because of the term $2n\pi$ in Eq. (4), it is clear that every number except zero has infinitely many logarithms. For each positive real number, one and only one of these logarithms is real, and it is upon these real values that all our elementary calculations with logarithms are based. For negative real numbers and numbers which are strictly complex, *none* of the logarithms is real.

Example 1

If p and q are harmonic conjugates with respect to the metric gauge lines, m_1 and m_2, on the intersection of p and q, what is $[pq]$?

Since the cross ratio of any harmonic tetrad is -1, it follows that $[pq] = k_a \ln \text{R}\,(pq,m_1m_2) = k_a \ln\,(-1)$. Now the length, or absolute value, of the number -1 is, of course, 1, and the principal argument, θ_1, of -1 is clearly π. Hence

$$[pq] = k_a \ln\,(-1) = k_a[\ln 1 + i(\pi + 2n\pi)] = k_a i(\pi + 2n\pi)$$

Example 2

If M_1:$(3 + 2i, -i, 1)$ and M_2:$(-2 + 6i, 1, 2i)$ are chosen as the metric gauge points on the line m: $x_1 + 2x_2 - 3x_3 = 0$, and if the scale constant is given the value $k_d = 1/2i$, what is the distance, (PQ), from P:$(3,0,1)$ to Q:$(-3 + i, i, -1 + i)$?

According to Definition 1, Sec. 10.2, $(PQ) = k_d \ln \text{R}\,(PQ,M_1M_2)$. Moreover, according to Theorem 4, Sec. 9.7, the cross ratio of four collinear points can be computed by using consistently any two of the three coordinates of the points, provided these lead to a meaningful value of the cross ratio. Hence, working with the second and third coordinates of the given points, since they are a little simpler than the first coordinates, we have

$$\text{R}\,(PQ,M_1M_2) = \frac{\begin{vmatrix} 0 & -i \\ 1 & 1 \end{vmatrix} \cdot \begin{vmatrix} i & 1 \\ -1+i & 2i \end{vmatrix}}{\begin{vmatrix} 0 & 1 \\ 1 & 2i \end{vmatrix} \cdot \begin{vmatrix} i & -i \\ -1+i & 1 \end{vmatrix}} = 1 - i$$

Now the absolute value of the complex number $z = 1 - i$ is $|z| = \sqrt{2}$, and the principal angle, θ_1, is the value of $\tan^{-1}\,(-1/1)$ which lies in the range $-\pi < \theta \le \pi$, that is, $\theta_1 = -\pi/4$. Therefore

$$\ln \text{R}\,(PQ,M_1M_2) = \ln\,(1 - i) = \ln \sqrt{2} + i\left(-\frac{\pi}{4} + 2n\pi\right)$$

and

$$(PQ) = k_d \ln \text{R}\,(PQ,M_1M_2) = \frac{1}{2i}\left[\frac{1}{2}\ln 2 + i\left(-\frac{\pi}{4} + 2n\pi\right)\right]$$

$$= \left(-\frac{\pi}{8} - \frac{i \ln 2}{4}\right) + n\pi$$

Let us now return to Eq. (1) and the equation obtained from it by replacing θ by $-\theta$:

$$\cos \theta + i \sin \theta = e^{i\theta}$$
$$\cos \theta - i \sin \theta = e^{-i\theta}$$

From these relations we obtain, by addition and subtraction,

$$\cos \theta = \frac{e^{i\theta} + e^{-i\theta}}{2}$$

(6)

$$\theta \text{ real}$$

$$\sin \theta = \frac{e^{i\theta} - e^{-i\theta}}{2i}$$

(7)

Since exponential functions of a complex variable have now been defined, it is possible to define $\cos z$ and $\sin z$ simply by allowing the symbol θ in Eq. (6) and Eq. (7) to take on complex values. Thus,

$$\cos z = \frac{e^{iz} + e^{-iz}}{2}$$

(8)

$$\sin z = \frac{e^{iz} - e^{-iz}}{2i}$$

(9)

From these definitions it is easy to establish the validity of such identities as

$$\cos^2 z + \sin^2 z = 1$$
$$\cos (z_1 \pm z_2) = \cos z_1 \cos z_2 \mp \sin z_1 \sin z_2$$
$$\sin (z_1 \pm z_2) = \sin z_1 \cos z_2 \pm \cos z_1 \sin z_2$$

If we expand the exponentials in (8), using Eq. (2), we find

$$\cos z = \frac{e^{i(x+iy)} + e^{-i(x+iy)}}{2}$$

$$= \frac{e^{-y}(\cos x + i \sin x) + e^{y}(\cos x - i \sin x)}{2}$$

$$= \cos x \frac{e^{y} + e^{-y}}{2} - i \sin x \frac{e^{y} - e^{-y}}{2}$$

or, using the usual definitions of the hyperbolic functions of real variables,

(10) $$\cos z = \cos x \cosh y - i \sin x \sinh y$$

In a similar fashion, it is easy to show that

(11) $$\sin z = \sin x \cosh y + i \cos x \sinh y$$

In particular, taking $x = 0$ in (10) and (11), we obtain

(12) $$\cos iy = \cosh y$$

(13) $$\sin iy = i \sinh y$$

Example 3

For what values of z, if any, is $\sin z = 2$?

Using Eq. (11), we must have $\sin x \cosh y + i \cos x \sinh y = 2$, which requires that, simultaneously, $\sin x \cosh y = 2$ and $\cos x \sinh y = 0$. From the second of these equations we infer that either $y = 0$ or $x = \pi/2 + n\pi$. If $y = 0$, the first equation leads to the condition $\sin x = 2$, which is impossible since x is a real quantity. Hence the possibility that $y = 0$ must be rejected. On the other hand, if $x = \pi/2 + n\pi$, then the first equation reduces to $(-1)^n \cosh y = 2$. Since the hyperbolic cosine is positive for all real values of its argument, it follows from the last equation that n must be even, say $n = 2m$. Hence $y = \cosh^{-1} 2$. Therefore the values of z for which $\sin z = 2$ are

$$z = x + iy = \left(\frac{\pi}{2} + 2m\pi\right) + i \cosh^{-1} 2$$

To define the hyperbolic functions $\cosh z$ and $\sinh z$, we simply let the symbol θ in the usual definitions when θ is real, namely,

$$\cosh \theta = \frac{e^\theta + e^{-\theta}}{2} \quad \text{and} \quad \sinh \theta = \frac{e^\theta - e^{-\theta}}{2}$$

take on complex values. Thus

$$\text{(14)} \qquad \cosh z = \frac{e^z + e^{-z}}{2}$$

$$\text{(15)} \qquad \sinh z = \frac{e^z - e^{-z}}{2}$$

From these definitions, the validity of such identities as

$$\cosh^2 z - \sinh^2 z = 1$$
$$\cosh (z_1 \pm z_2) = \cosh z_1 \cosh z_2 \pm \sinh z_1 \sinh z_2$$
$$\sinh (z_1 \pm z_2) = \sinh z_1 \cosh z_2 \pm \cosh z_1 \sinh z_2$$

can easily be established.

By expanding the exponentials and regrouping, as we did in deriving Eq. (10), we obtain without difficulty the formulas

$$\text{(16)} \qquad \cosh z = \cosh x \cos y + i \sinh x \sin y$$

$$\text{(17)} \qquad \sinh z = \sinh x \cos y + i \cosh x \sin y$$

In particular, setting $x = 0$, we obtain

$$\text{(18)} \qquad \cosh iy = \cos y$$

$$\text{(19)} \qquad \sinh iy = i \sin y$$

In later sections of this chapter we shall occasionally need to consider cosines and sines, as well as the hyperbolic cosine and hyperbolic sine, of complex quantities, and the formulas we have just obtained will be useful in this connection.

Exercises

1. Derive Eq. (11).

2. Derive Eq. (16).

3. Derive Eq. (17).

4. Show that $\cos^2 z + \sin^2 z = 1$.

5. For what values of z, if any, is $\sin z = -3$?

6. For what values of z, if any, is $\cosh z = -2$?

7. Show that the only zeros of $\cos z$ are the real values $z = \pi/2 + n\pi$.

8. Show that the only values of z for which $\sin z$ is zero are the real values $z = n\pi$.

9. Show that $\cosh z = 0$ if and only if $z = [(2m + 1)/2]\pi i$.

10. Show that e^z is never zero.

11. What are the values of:
(a) $\ln (2 + 3i)$ (b) $\ln (-3 + 4i)$ (c) $\ln (-1 - i)$

12. What are the values of:
(a) $\cos (3i)$ (b) $\cos (1 - 2i)$ (c) $\cos (-3 + 2i)$

13. What are the values of:
(a) $\sin (-2i)$ (b) $\sin (1 - i)$ (c) $\sin (3 + 2i)$

14. If a line m is parametrized in terms of the metric gauge points M_1 and M_2 as base points, what is the distance from $P:(1,1)$ to $Q:(1,i)$?

15. What is the principal value of $[pq]$ if p and q are, respectively, the lines $2x + z = 0$ and $x - iy + z = 0$, and the metric gauge lines, m_1 and m_2, on the intersection of p and q are, respectively, $x + iy = 0$ and $2iy - z = 0$.

16. Prove the following result. If p and q are, respectively, the lines $y = (\tan \theta)x$ and $y = (\tan \phi)x$, and if m_1 and m_2 are the lines joining the intersection of p and q to the circular points at infinity, $I:(1,i,0)$ and $J:(1,-i,0)$,

then the scale constant k_a can be chosen so that $[pq]$, as given by Definition 2, is equal to the measure of the angle between p and q in the euclidean sense.[1]

10.4 The Metric Gauge Conic

In Sec. 10.2 we saw that to measure distances on a line we needed to identify two points of the line as *metric gauge points* and that to measure angles in a pencil we needed to identify two lines of the pencil as *metric gauge lines*. Over the entire projective plane, this means that two points must be chosen on every line and that two lines must be chosen on every point. Theoretically, these points and lines can be selected in any way we choose, but practically it is desirable to identify them in some simple, systematic fashion. The most natural way is to choose the metric gauge elements so that, collectively, they comprise the points and the lines of a nonsingular conic, Γ. The metric gauge points on a general line are then the two points in which that line intersects Γ, and the metric gauge lines on a general point are the two tangents of Γ which pass through that point. These observations we now formalize in the following generalization of Definitions 1 and 2, Sec. 10.2.

Definition 1

Let Γ be an arbitrary nonsingular conic in the projective plane, and let k_d and k_a be arbitrary nonzero constants. Then:

1. If P and Q are any points in the projective plane and if M_1 and M_2 are the points of Γ collinear with P and Q, then the distance (PQ) is equal to any one of the values of $k_d \ln \text{R}(M_1M_2,PQ) = k_d \ln \text{R}(PQ,M_1M_2)$ provided this expression is meaningful.

2. If p and q are any lines in the projective plane and if m_1 and m_2 are the lines of Γ concurrent with p and q, then the angle $[pq]$ is equal to any one of the values of $k_a \ln \text{R}(m_1m_2,pq) = k_a \ln \text{R}(pq,m_1m_2)$ provided this expression is meaningful.

Neither distance nor angle measure is defined if the corresponding expression in 1 or 2 is undefined.

The conic Γ required by Definition 1 for the measurement of distance and angle is called the **metric gauge conic.**[2] We shall sometimes

---

[1] This result was discovered in 1853 by the French mathematician Edmond Laguerre (1834–1886), who was the first to recognize the relation between cross ratios and the measurement of angles. Arthur Cayley, in his famous "Sixth Memoir on Quantics," published in 1859, was the first to recognize and develop the relation between cross ratios and the measurement of distance.

[2] From the time of Cayley it has been customary to refer to the conic Γ as the **absolute conic** or simply the **absolute.** We shall avoid this name, however, since it suggests that there is something intrinsically different or special about Γ, whereas *any* nonsingular conic can be chosen as the metric gauge conic.

refer to the metric gauge conic more specifically as the **metric gauge locus** or the **metric gauge envelope,** according as we are concerned primarily with the measurement of distance or the measurement of angle.

As Definition 1 implies, there are pairs of points (P,Q) for which the distance (PQ) is undefined. Specifically, if both P and Q are points of Γ, the cross ratio $\mathrm{R}(PQ,M_1M_2)$ assumes the indeterminate form $\frac{0}{0}$, and the distance (PQ) can be assigned no meaning. Furthermore, if just one of the points P, Q is a point of Γ, then the distance (PQ) is infinite For if P coincides with M_1 or if Q coincides with M_2, then $\mathrm{R}(PQ,M_1M_2) = 0$; and if P coincides with M_2 or if Q coincides with M_1, then $\mathrm{R}(PQ,M_1M_2)$ is infinite; and in either case $(PQ) = k_d \ln \mathrm{R}(PQ,M_1M_2)$ is infinite. Thus, although there is no difference between the points of Γ and any other points in the projective plane, once Γ is chosen as the metric gauge conic, its points may with some justification be called "points at infinity" and Γ itself may be called the "conic at infinity."

In elementary geometry not only is the distance between a point and itself equal to zero, but conversely if $(PQ) = 0$, then P and Q are the same point. It is interesting to note that this is not the case in projective geometry. In fact, if P and Q are points on any line tangent to the metric gauge conic, then M_1 and M_2 coincide and $\mathrm{R}(PQ,M_1M_2) = \mathrm{R}(PQ,M_1M_1) = 1$. Hence, whether P and Q coincide or not, we have

$$(PQ) = k_d \ln \mathrm{R}(PQ,M_1M_1) = k_d \ln 1 = k_d 2n\pi i$$

which is zero if $n = 0$. Lines whose points have this seemingly unusual property, i.e., lines tangent to the metric gauge conic, are called **isotropic lines.** Lines which are not isotropic lines, i.e., all lines which are not tangent to the metric gauge conic, are called **ordinary lines.**

Dually, if both p and q are lines which are tangent to the metric gauge conic, then $\mathrm{R}(pq,m_1m_2)$ is an indeterminate of the form $\frac{0}{0}$, and the angle $[pq]$ can be assigned no meaning. If just one of the lines p, q is tangent to Γ, then $\mathrm{R}(pq,m_1m_2)$ is either 0 or ∞, and $[pq] = k_a \ln \mathrm{R}(pq,m_1m_2)$ is infinite. Also, if p and q are lines which intersect in a point, M, of Γ, then there is just one line of Γ concurrent with p and q, namely, the tangent to Γ at M. In this case, $m_1 = m_2$, and $\mathrm{R}(pq,m_1m_2) = \mathrm{R}(pq,m_1m_1) = 1$. Therefore, whether p and q are coincident or not, we have

$$[pq] = k_a \ln \mathrm{R}(pq,m_1m_1) = k_a \ln 1 = k_a 2n\pi i$$

which is zero if $n = 0$. This, of course, is consistent with the fact that in elementary geometry it is customary to say that the measure of the angle between two parallel lines is zero. It is an interesting commentary on the bias developed by our earlier work in elementary geometry that whereas it is quite natural that the angle between distinct lines can be zero, it seems strange to us that the distance between distinct points can be zero. Incidentally, the observations we have just been making can be rephrased by saying that distinct lines which determine an angle of zero measure necessarily intersect in a point of Γ, that is, a "point at infinity," which perhaps gives a little more meaning to the statement occasionally heard in discussions of elementary geometry that "parallel lines meet at infinity." These "points at infinity," i.e., the points of the metric gauge conic, Γ, we shall call **isotropic points.** The term *isotropic point* is, of course, the dual of the term *isotropic line*, which we introduced above. Points which are not isotropic, i.e., points which are not on the metric gauge conic, we shall refer to as **ordinary points.**

The choice of which one of the metric gauge points on a given line is to be called M_1 and which is to be called M_2 is, of course, completely arbitrary. Making a choice is equivalent to what in elementary geometry we refer to as "choosing a positive direction on a line," for clearly

$$k_d \ln \mathrm{R}(PQ,M_1M_2) = -k_d \ln \mathrm{R}(PQ,M_2M_1) + 2n\pi i k_d$$

In other words, interchanging the metric gauge points on a line, m, results in changing the sign of the principal value of the distance between any two points of m. It is conventional, however, that when more than one distance is to be measured on a given line, the gauge points on that line are taken in the same order for all measurements. Thus if P, Q, R, S, ... are points on a line m, and if $(PQ) = k_a \ln \mathrm{R}(PQ,M_1M_2)$, then (RS) is equal to $k_d \ln \mathrm{R}(RS,M_1M_2)$ and not $k_d \ln \mathrm{R}(RS,M_2M_1)$.

We now turn our attention to several theorems involving the metric gauge conic. The first deals with points which are conjugate with respect to the metric gauge conic.

Theorem 1

Two points, P and Q, are conjugate with respect to the metric gauge conic if and only if $(PQ) = (2n + 1)k_d\pi i$.

Proof From earlier results we know that two points, P and Q, are conjugate with respect to the metric gauge conic, Γ, if and only if P, Q and the points of Γ collinear with them, namely, M_1 and M_2, form a harmonic tetrad, i.e., if and only if $\mathrm{R}(PQ,M_1M_2) = -1$. This in turn is true if and only if $(PQ) = k_d \ln \mathrm{R}(PQ,M_1M_2) = k_d \ln(-1) = k_d(2n + 1)\pi i$, which proves the theorem.

By a similar argument we can establish the dual theorem:

Theorem 2

Two lines, p and q, are conjugate with respect to the metric gauge conic if and only if $[pq] = (2n + 1)k_a\pi i$.

Points which are conjugate with respect to the metric gauge conic are called **orthogonal points** and are said to be **orthogonal** to each other. Lines which are conjugate with respect to the metric gauge conic are called **orthogonal lines** and are said to be **orthogonal** to each other. As our development proceeds, we shall see that orthogonal lines have properties analogous to those of perpendicular lines in elementary geometry.

Our next theorem is a fundamental one which expresses the distance between two points in terms of the coordinates of the points and the equation of the metric gauge conic.

Theorem 3

In terms of a homogeneous coordinate system in the projective plane, let the metric gauge locus be defined by the equation $X^T A X = 0$. Then the distance between any two ordinary points, P and Q, is given by the formula

$$\cosh^2 \frac{(PQ)}{2k_d} = \frac{(P^T A Q)^2}{(P^T A P)(Q^T A Q)}$$

Proof   Let $P: \begin{Vmatrix} x_1 \\ y_1 \\ z_1 \end{Vmatrix}$ and $Q: \begin{Vmatrix} x_2 \\ y_2 \\ z_2 \end{Vmatrix}$ be any two ordinary points, let M_1 and M_2

be the points in which the line PQ intersects the metric gauge conic, $X^T A X = 0$, and let PQ be parametrized in terms of P and Q as base points. Then $M_1 = \lambda_1 P + \mu_1 Q$ and $M_2 = \lambda_2 P + \mu_2 Q$ where, since M_1 and M_2 are on the metric gauge conic, (λ_1, μ_1) and (λ_2, μ_2) satisfy the quadratic equation

$$(1) \qquad \lambda^2 P^T A P + 2\lambda\mu P^T A Q + \mu^2 Q^T A Q = 0$$

Now since the parameters of P are $(1,0)$ and the parameters of Q are $(0,1)$, it follows that

$$\text{R}(PQ, M_1 M_2) = \text{R}[(1,0)(0,1), (\lambda_1, \mu_1)(\lambda_2, \mu_2)] = \frac{\lambda_2 \mu_1}{\lambda_1 \mu_2}^\dagger$$

Hence

$$(2) \qquad (PQ) = k_d \ln \text{R}(PQ, M_1 M_2) = k_d \ln \frac{\lambda_2 \mu_1}{\lambda_1 \mu_2}$$

Therefore

$$\frac{\lambda_2 \mu_1}{\lambda_1 \mu_2} = e^{(PQ)/k_d} \qquad \text{and} \qquad \frac{\lambda_1 \mu_2}{\lambda_2 \mu_1} = e^{-(PQ)/k_d}$$

† Since P and Q are ordinary points, none of the numbers λ_1, λ_2, μ_1, μ_2 can be zero. Furthermore, neither $P^T A P$ nor $Q^T A Q$ can be zero.

Adding the last two equations, we have

$$e^{(PQ)/k_d} + e^{-(PQ)/k_d} = \frac{\lambda_2\mu_1}{\lambda_1\mu_2} + \frac{\lambda_1\mu_2}{\lambda_2\mu_1} = \frac{(\lambda_1\mu_2)^2 + (\lambda_2\mu_1)^2}{\lambda_1\lambda_2\mu_1\mu_2}$$

If we now add 2 to both members of the last equation and then divide by 4, we obtain

$$\frac{e^{(PQ)/k_d} + 2 + e^{-(PQ)/k_d}}{4} = \frac{1}{4}\left[\frac{(\lambda_1\mu_2)^2 + (\lambda_2\mu_1)^2 + 2\lambda_1\lambda_2\mu_1\mu_2}{\lambda_1\lambda_2\mu_1\mu_2}\right]$$

or

(3)
$$\left[\frac{e^{(PQ)/2k_d} + e^{-(PQ)/2k_d}}{2}\right]^2 = \frac{(\lambda_1\mu_2 + \lambda_2\mu_1)^2}{4\lambda_1\lambda_2\mu_1\mu_2}$$

Now the left member of Eq. (3) is precisely $\cosh^2[(PQ)/2k_d]$. Moreover, from the root-coefficient relations for the quadratic equation (1), we have

$$\frac{\lambda_1}{\mu_1} + \frac{\lambda_2}{\mu_2} = \frac{\lambda_1\mu_2 + \lambda_2\mu_1}{\mu_1\mu_2} = -\frac{2P^TAQ}{P^TAP} \quad\text{and}\quad \frac{\lambda_1}{\mu_1}\frac{\lambda_2}{\mu_2} = \frac{Q^TAQ}{P^TAP}$$

Hence the right-hand side of Eq. (3) reduces to $(P^TAQ)^2/(P^TAP)(Q^TAQ)$, and the theorem is proved.

Corollary 1

$$\sinh^2\frac{(PQ)}{2k_d} = \frac{(P^TAQ)^2 - (P^TAP)(Q^TAQ)}{(P^TAP)(Q^TAQ)}$$

Corollary 2

If P and Q are any two ordinary points, then

$$(PQ) = k_d\ln\left[\frac{-P^TAQ - \sqrt{(P^TAQ)^2 - (P^TAP)(Q^TAQ)}}{-P^TAQ + \sqrt{(P^TAQ)^2 - (P^TAP)(Q^TAQ)}}\right]$$

Proof From the quadratic equation (1), we have

$$\frac{\lambda_1}{\mu_1}, \frac{\lambda_2}{\mu_2} = \frac{-P^TAQ \pm \sqrt{(P^TAQ)^2 - (P^TAP)(Q^TAQ)}}{P^TAP}$$

If we now agree, for definiteness, that the ratio λ_1/μ_1 corresponds to the positive sign in the last expression, we have, from (2),

$$(PQ) = k_d\ln\left[\frac{\lambda_2/\mu_2}{\lambda_1/\mu_1}\right] = k_d\ln\left[\frac{-P^TAQ - \sqrt{(P^TAQ)^2 - (P^TAP)(Q^TAQ)}}{-P^TAQ + \sqrt{(P^TAQ)^2 - (P^TAP)(Q^TAQ)}}\right]$$

as asserted.

In exactly the same fashion we can establish the dual results:

Theorem 4

In terms of a homogeneous coordinate system in the projective plane, let the metric gauge envelope be defined by the equation $\Lambda^T B \Lambda = 0$. Then the angle between any two ordinary lines, p and q, is given by the formula

$$\cosh^2 \frac{[pq]}{2k_a} = \frac{(p^T Bq)^2}{(p^T Bp)(q^T Bq)}$$

Corollary 1

$$\sinh^2 \frac{[pq]}{2k_a} = \frac{(p^T Bq)^2 - (p^T Bp)(q^T Bq)}{(p^T Bp)(q^T Bq)}$$

Corollary 2

If p and q are any two ordinary lines, then

$$[pq] = k_a \ln \left[\frac{-(p^T Bq) - \sqrt{(p^T Bq)^2 - (p^T Bp)(q^T Bq)}}{-(p^T Bq) + \sqrt{(p^T Bq)^2 - (p^T Bp)(q^T Bq)}} \right]$$

With Theorem 1 available, we can now determine the locus of points whose distance from a given ordinary point is a given constant:

Theorem 5

If the equation of the metric gauge locus is $X^T AX = 0$, the locus of points which are at a given distance, $2k_d n\pi i + d$, from a given ordinary point, P, consists of all the ordinary points on the conic

$$(P^T AX)^2 = \left(\cosh^2 \frac{d}{2k_d} \right) (X^T AX)(P^T AP)$$

If $d = (2m + 1)k_d\pi i$, this conic is singular and consists of the polar of P with respect to the metric gauge locus. If $d = 0$, the conic is singular and consists of the two tangents from P to the metric gauge conic. If d is neither 0 nor $(2m + 1)\pi i$, the conic is nonsingular and is tangent to the metric gauge locus at the points common to the polar of P and the metric gauge locus.

Proof Using the result of Theorem 3, it is clear that the coordinates of a general point, X, whose distance from the given point P is $2k_d n\pi i + d$ must satisfy the equation

$$\frac{(P^T AX)^2}{(P^T AP)(X^T AX)} = \cosh^2 \left(\frac{2k_d n\pi i + d}{2k_d} \right) = \cosh^2 \left(n\pi i + \frac{d}{2k_d} \right)$$

$$= \left(\cosh n\pi i \cosh \frac{d}{2k_d} + \sinh 2\pi i \sinh \frac{d}{2k_d} \right)^2$$

Using Eqs. (18) and (19), Sec. 10.3, the right-hand side of the last equation reduces to $\cosh^2(d/2k_d)$, which proves the first assertion of the theorem.

To identify the equidistant locus more explicitly, we observe first that if $d = (2m + 1)k_d\pi i$, then

$$\cosh^2\left(\frac{d}{2k_d}\right) = \cosh^2\frac{(2m + 1)k_d\pi i}{2k_d} = \cosh^2\left(\frac{2m + 1}{2}\,\pi i\right)$$

$$= \cos^2\left(\frac{2m + 1}{2}\,\pi\right) = 0$$

Hence, in this case, the equation of the equidistant locus reduces to $P^T AX = 0$, which is precisely the equation of the polar of P with respect to the metric gauge locus.

If $d = 0$, then $\cosh^2(d/2k_d) = 1$. In this case, the equation of the equidistant locus becomes $(P^T AX)^2 = (P^T AP)(X^T AX)$, which is simply the condition that the discriminant of the quadratic equation $\lambda^2 P^T AP + 2\lambda\mu P^T AX + \mu^2 X^T AX = 0$ should be zero; and this we have already identified (Theorem 8, Sec. 9.7) as the equation of the singular conic consisting of the tangents from P to the conic $X^T AX = 0$.

Finally, if d is neither 0 nor $(2m + 1)k_d\pi i$, then the quantity $\alpha = \cosh^2(d/2k_d)$ is neither 0 (see Exercise 9, Sec. 10.3) nor 1, and the equation of the equidistant curve becomes

$$(P^T AX)^2 = \alpha(P^T AP)(X^T AX) \qquad \alpha \neq 0, 1$$

This is evidently the equation of a conic since it is of the second degree in X. Moreover, this conic clearly has two intersections with the metric gauge conic at each point where the polar of P, namely, $P^T AX = 0$, meets the metric gauge conic. Furthermore, except when $\alpha = 0$ or 1, this conic is nonsingular, since the only singular conics having coincident intersections with the metric gauge conic at its intersections with the polar of P are the polar of P, counted twice, and the pair of tangents to the metric gauge conic from P (Fig. 10.2). Thus the last assertion of the theorem is established.

In the same way we can prove the dual theorem:

Theorem 6

If the equation of the metric gauge envelope is $\Lambda^T B\Lambda = 0$, then the envelope of lines forming an angle of measure $2k_a n\pi i + a$ with a given line, p, consists of all the ordinary lines of the conic

$$(p^T B\Lambda)^2 = \left(\cosh^2\frac{a}{2k_a}\right)(p^T Bp)(\Lambda^T B\Lambda)$$

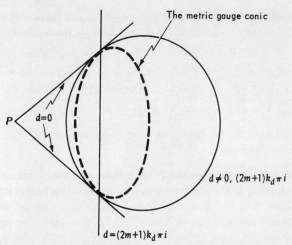

The metric gauge conic

P $d=0$

$d \neq 0,\ (2m+1)k_d\pi i$

$d=(2m+1)k_d\pi i$

Fig. 10.2   Singular and nonsingular equidistant curves.

If $a = (2m + 1)k_a\pi i$, this conic is singular and consists of the pole of p with respect to the metric gauge conic. If $a = 0$, the conic is singular and consists of the two points common to p and the metric gauge conic. If a is neither 0 nor $(2m + 1)k_a\pi i$, the conic is nonsingular and is tangent to the metric gauge conic at the lines common to the pole of p and the metric gauge envelope.

Exercises

1. Given the metric gauge conic $\Gamma: X^T A X = 0$ and a point P with real coordinates which is not on Γ. What is the locus of points whose distance from P is real if:

(a) k_d is real?   (b) k_d is pure imaginary?

2. Given the metric gauge conic $\Gamma: x_1{}^2 - 2x_2x_3 = 0$ and the points $P_1:(1,0,0)$, $P_2:(1,1,0)$, $P_3:(1,1,2)$. Determine the sides and angles of $\triangle P_1P_2P_3$ if:

(a) $k_d = k_a = \frac{1}{2}$ (b) $k_d = k_a = 1/2i$

3. Work Exercise 2 given $\Gamma: x_1{}^2 + x_2{}^2 + x_3{}^2 = 0$.

4. If P, Q, Q' are three collinear points such that $(PQ) = (PQ')$, show that $Q = Q'$.

5. If P, P', Q, Q' are four collinear points such that (P,P') is a pair of orthogonal points and $|(PQ)| = |(PQ')|$, show that $|(P'Q)| = |(P'Q')|$.

6. If P, P', Q, Q' are four ordinary points which are collinear on an ordinary line, if $(PQ) = (Q'P)$, and if $(P'Q) = (Q'P')$, show that P and P' are orthogonal.

7. If the vertices and sides of the triangle of reference are ordinary points and ordinary lines with respect to the metric gauge conic

$$ax_1{}^2 + 2hx_1x_2 + bx_2{}^2 + 2gx_1x_3 + 2fx_2x_3 + cx_3{}^2 = 0$$

determine the lengths of the sides and the measures of the angles of the triangle of reference.

8. The **distance from a point** P **to a line** p is said to be d if and only if $(PQ) = d$, where Q is the intersection of p and the line on P which is orthogonal to p. Show that the distance from a point to its polar with respect to the metric gauge conic is $(2n + 1)k_d\pi i$.

9. Show that the distance from a point $P:(p_1,p_2,p_3)$ to a line $l:[l_1,l_2,l_3]$ satisfies the equation

$$\cosh^2 \frac{d}{2k_d} = 1 - \frac{(P^T l)^2\, |A|}{(P^T AP)(l^T \text{ adj } A\, l)}$$

where $X^T AX = 0$ is the equation of the metric gauge conic.

10. Dualize the definition given in Exercise 8, and then, using the result of Exercise 9, show that the angle between a point and a line is equal to the distance between them.

11. Show that if two ordinary points are equidistant from an ordinary line, they are also equidistant from the pole of that line with respect to the metric gauge conic.

12. What is the locus of points which are equidistant from an ordinary line?

13. Let T be a projectivity between cobasal ranges on an ordinary line l. Show that the distance between any ordinary point, A, on l and its image under T depends only on the projectivity T and is independent of the particular point A.

14. Let Φ be any nonsingular conic distinct from the metric gauge conic Γ, and let A, B, C, D, P be five points on Φ such that each of the lines PA, PB, PC, PD is an ordinary line. Show that if $k_a = \frac{1}{2}$, then the value of the expression

$$\frac{\sinh^2 \angle APC \, \sinh^2 \angle BPD}{\sinh^2 \angle APD \, \sinh^2 \angle BPC}$$

is independent of P.

15. Show that if $k_a = k_{\bar{a}}$, the distance between two ordinary points is equal to the angle between their polars with respect to the metric gauge conic.

10.5 Matric Identities for a General Triangle

In this section we shall develop a number of matric identities relating to a general triangle. These will be essential in our work as we attempt to develop elliptic and hyperbolic geometry by suitably specializing the metric gauge conic.

Let $X = \left\| \begin{matrix} x \\ y \\ z \end{matrix} \right\|$ be the coordinate vector of a general point, and let

$\Lambda = \left\| \begin{matrix} l \\ m \\ n \end{matrix} \right\|$ be the coordinate vector of a general line. Let

$$ a = \left\| \begin{matrix} a_{11} & a_{12} & a_{13} \\ a_{21} & a_{22} & a_{23} \\ a_{31} & a_{32} & a_{33} \end{matrix} \right\| $$

be the matrix of a nonsingular point-conic, or conic-locus, $f: X^T a X = 0$. Then

$$ A = \operatorname{adj} a = \left\| \begin{matrix} A_{11} & A_{21} & A_{31} \\ A_{12} & A_{22} & A_{32} \\ A_{13} & A_{23} & A_{33} \end{matrix} \right\| $$

is the matrix of the nonsingular line-conic, or conic-envelope, $F: \Lambda^T A \Lambda = 0$, which corresponds to f. Now let $P_1 = \left\| \begin{matrix} x_1 \\ y_1 \\ z_1 \end{matrix} \right\|$, $P_2 = \left\| \begin{matrix} x_2 \\ y_2 \\ z_2 \end{matrix} \right\|$, $P_3 = \left\| \begin{matrix} x_3 \\ y_3 \\ z_3 \end{matrix} \right\|$ be

three noncollinear points, and let $t = \| P_1 \quad P_2 \quad P_3 \| = \left\| \begin{matrix} x_1 & x_2 & x_3 \\ y_1 & y_2 & y_3 \\ z_1 & z_2 & z_3 \end{matrix} \right\|$.

Obviously, $|t| \neq 0$, since P_1, P_2, P_3 are noncollinear.

Furthermore, let p_1, p_2, p_3 be, respectively, the coordinate vectors of the lines P_2P_3, P_3P_1, P_1P_2. This means that we have $P_2P_3 : p_1 = \left\| \begin{matrix} l_1 \\ m_1 \\ n_1 \end{matrix} \right\|$, where

l_1, m_1, n_1 are to be read from the matrix

$$ \left\| \begin{matrix} x_2 & y_2 & z_2 \\ x_3 & y_3 & z_3 \end{matrix} \right\| $$

$P_3P_1 : p_2 = \left\| \begin{matrix} l_2 \\ m_2 \\ n_2 \end{matrix} \right\|$, where l_2, m_2, n_2 are to be read from the matrix

$$ \left\| \begin{matrix} x_3 & y_3 & z_3 \\ x_1 & y_1 & z_1 \end{matrix} \right\| $$

and $P_1P_2 : p_3 = \left\| \begin{matrix} l_3 \\ m_3 \\ n_3 \end{matrix} \right\|$, where l_3, m_3, n_3 are to be read from the matrix

$$\left\| \begin{matrix} x_1 & y_1 & z_1 \\ x_2 & y_2 & z_2 \end{matrix} \right\|$$

Clearly, l_i, m_i, n_i are the cofactors of the respective elements in the ith column of the determinant $|t|$. Hence, if

$$T = \| p_1 \quad p_2 \quad p_3 \| = \left\| \begin{matrix} l_1 & l_2 & l_3 \\ m_1 & m_2 & m_3 \\ n_1 & n_2 & n_3 \end{matrix} \right\|, \text{ it follows that adj } t = T^T \text{ and}$$

(1) $$t^{-1} = \frac{\text{adj } t}{|t|} = \frac{T^T}{|t|}$$

Now using the first, or partitioned, form of the matrix t, we have

(2) $\quad t^T a t = \left\| \begin{matrix} P_1{}^T \\ P_2{}^T \\ P_3{}^T \end{matrix} \right\| \cdot a \cdot \| P_1 \quad P_2 \quad P_3 \| = \left\| \begin{matrix} P_1{}^T a P_1 & P_1{}^T a P_2 & P_1{}^T a P_3 \\ P_2{}^T a P_1 & P_2{}^T a P_2 & P_2{}^T a P_3 \\ P_3{}^T a P_1 & P_3{}^T a P_2 & P_3{}^T a P_3 \end{matrix} \right\|$

$$= \left\| \begin{matrix} f_{11} & f_{12} & f_{13} \\ f_{21} & f_{22} & f_{23} \\ f_{31} & f_{32} & f_{33} \end{matrix} \right\|$$

where, for notational convenience, we have put $f_{ij} = P_i{}^T a P_j$. Of course, since $P_i{}^T a P_j = P_j{}^T a P_i$, it follows that $f_{ij} = f_{ji}$. Since the determinant of a product of square matrices is equal to the product of the determinants of the individual matrices, and since the determinant of a square matrix and the determinant of its transpose are equal, Eq. (2) implies that

(3) $$|t^T a t| = |t|^2 |a| = |f_{ij}|$$

In particular, since $|a| \neq 0$, because the conic $X^T a X = 0$ is nonsingular, and since $|t| \neq 0$ because P_1, P_2, P_3 are noncollinear points, it follows that $|f_{ij}| \neq 0$.

Also, using the partitioned form of the matrix T, we have

(4) $\quad T^T A T = \left\| \begin{matrix} p_1{}^T \\ p_2{}^T \\ p_3{}^T \end{matrix} \right\| \cdot A \cdot \| p_1 \quad p_2 \quad p_3 \| = \left\| \begin{matrix} p_1{}^T A p_1 & p_1{}^T A p_2 & p_1{}^T A p_3 \\ p_2{}^T A p_1 & p_2{}^T A p_2 & p_2{}^T A p_3 \\ p_3{}^T A p_1 & p_3{}^T A p_2 & p_3{}^T A p_3 \end{matrix} \right\|$

$$= \left\| \begin{matrix} F_{11} & F_{12} & F_{13} \\ F_{21} & F_{22} & F_{23} \\ F_{31} & F_{32} & F_{33} \end{matrix} \right\|$$

where now, for notational convenience, we have put $F_{ij} = p_i{}^T A p_j$. Of course, since $p_i{}^T A p_j = p_j{}^T A p_i$, it follows that $F_{ij} = F_{ji}$. This result implies that $|T|^2 |A| = |F_{ij}|$. In particular, since $|T| \neq 0$ because p_1, p_2, p_3 are nonconcurrent lines, and since $|A| \neq 0$ because the determinant of the adjoint of a (3,3) matrix is equal to the square of the determinant of that matrix, it follows that $|F_{ij}| \neq 0$.

Now we have just verified that $t^T a t = \| f_{ij} \|$ is nonsingular. Hence it has an inverse, which in fact is equal to

$$(t^T a t)^{-1} = t^{-1} a^{-1} (t^T)^{-1} = t^{-1} a^{-1} (t^{-1})^T$$

$$= \frac{T^T}{|t|} \frac{A}{|a|} \left(\frac{T^T}{|t|} \right)^T \qquad \text{by (1)}$$

$$= \frac{T^T A T}{|t|^2 |a|}$$

$$= \frac{T^T A T}{|t^T a t|} \qquad \text{by (3)}$$

Therefore, from the definition of the inverse of a matrix, it follows that adj $(t^T a t) = T^T A T$. Thus, recalling (2) and (4), we have

$$(5) \qquad \text{adj} \begin{Vmatrix} f_{11} & f_{12} & f_{13} \\ f_{21} & f_{22} & f_{23} \\ f_{31} & f_{32} & f_{33} \end{Vmatrix} = \begin{Vmatrix} F_{11} & F_{12} & F_{13} \\ F_{21} & F_{22} & F_{23} \\ F_{31} & F_{32} & F_{33} \end{Vmatrix}$$

and from this, by comparing corresponding elements, we obtain the important identities

$$(6) \qquad F_{ii} = f_{jj} f_{kk} - f_{jk}^2$$

$$(i,j,k) \text{ any permutation of } (1,2,3)$$

$$(7) \qquad F_{ij} = f_{ik} f_{jk} - f_{ij} f_{kk}$$

Useful results analogous to (6) and (7), with the f's and F's interchanged, can be obtained as follows. Obviously

$$I = \| f_{ij} \|^{-1} \cdot \| f_{ij} \| = \frac{\text{adj} \| f_{ij} \|}{|f_{ij}|} \| f_{ij} \|$$

$$= \| F_{ij} \| \cdot \frac{\| f_{ij} \|}{|f_{ij}|} \qquad \text{by (5)}$$

From this we conclude that

$$(8) \qquad \frac{\| f_{ij} \|}{|f_{ij}|} = \| F_{ij} \|^{-1}$$

Therefore, since

$$\| F_{ij} \|^{-1} = \frac{\text{adj} \| F_{ij} \|}{|F_{ij}|}$$

and since, from (5), $|F_{ij}| = |f_{ij}|^2$, it follows from (8) that

$$\text{adj } \|F_{ij}\| = |F_{ij}| \cdot \|F_{ij}\|^{-1} = |f_{ij}|^2 \frac{\|f_{ij}\|}{|f_{ij}|} = |f_{ij}| \cdot \|f_{ij}\|$$

Finally, by comparing corresponding elements, we infer from this that

$$(9) \quad f_{ii} = \frac{(F_{jj}F_{kk} - F_{jk}{}^2)}{|f_{ij}|}$$

(i,j,k) any permutation of $(1,2,3)$

$$(10) \quad f_{ij} = \frac{(F_{ik}F_{jk} - F_{ij}F_{kk})}{|f_{ij}|}$$

Exercises

1. Given $a = \begin{Vmatrix} 1 & 0 & 1 \\ 0 & 1 & 0 \\ 1 & 0 & 2 \end{Vmatrix}$, $P_1 = \begin{Vmatrix} 1 \\ 1 \\ 0 \end{Vmatrix}$, $P_2 = \begin{Vmatrix} 0 \\ 0 \\ 1 \end{Vmatrix}$, $P_3 = \begin{Vmatrix} 1 \\ -1 \\ 1 \end{Vmatrix}$. What is A? t? T? $\|f_{ij}\|$? $\|F_{ij}\|$?

2. Using the data of Exercise 1, verify Eqs. (1), (5), and (8).

3. Do any of the formulas derived in this section require modification if one or more of the points P_1, P_2, P_3 is on $X^T aX = 0$?

4. Dualize the development of this section; i.e., repeat the discussion beginning with a line-conic and three lines.

10.6 Elliptic Geometry

We have repeatedly stressed the fact that with respect to their geometric properties, the points of the projective plane are all identical. On the other hand, once a coordinate system is established in the projective plane, two classes of points become distinguishable, namely, those whose coordinates are all real, or can be made real by multiplying them by a suitable nonzero proportionality constant, and those whose coordinates do not have this property. *With respect to a particular coordinate system*, points of the first class we shall call **real points** and points of the second class we shall call **complex points.** Similar remarks hold, of course, for lines.

Likewise, with respect to their geometrical properties, all non-singular conics in the projective plane are identical. However, once a coordinate system is established, several classes of nonsingular conics become distinguishable. In particular, some conics will have equations which are real, or can be made real by multiplying them by a suitable nonzero proportionality constant, while others will have equations which are not real and cannot be made real. Furthermore, conics with real equations, or **real conics** as we shall call them, may have the property that they contain no real points, or they may contain some real points.

As we have already observed, any nonsingular conic can be chosen as the metric gauge conic. However, to obtain the classical noneuclidean geometries by introducing measurement into the projective plane, it is necessary that the metric gauge conic be real. If this conic contains some real points, the resulting metrical geometry is identical to the one discovered by Lobachevski (1793–1856) and Bolyai (1802–1860) in the early years of the nineteenth century, and now known as hyperbolic geometry. If the conic contains no real points, the resulting metrical geometry is identical to the one discovered by Riemann (1826–1866) around the middle of the nineteenth century and now known as elliptic geometry. In this section we shall investigate the fundamental properties of elliptic geometry. In a later section we shall make a similar study of hyperbolic geometry.

Although it is possible, once a metric gauge conic is identified, to assign a distance to *every* pair of ordinary points and an angle measure to *every* pair of ordinary lines, elliptic geometry is not this extensive. It is, in fact, defined only over the real points in the projective plane. More specifically, *given a particular coordinate system, elliptic geometry is the study of the metrical relations imposed upon the real points and real lines of the coordinatized projective plane when the metric gauge conic is chosen to be a real conic which contains no real points.*

Let $X = \begin{Vmatrix} x \\ y \\ z \end{Vmatrix}$ be the coordinate vector of a general point in the

coordinatized projective plane, and let $a = \begin{Vmatrix} a_{11} & a_{12} & a_{13} \\ a_{21} & a_{22} & a_{23} \\ a_{31} & a_{32} & a_{33} \end{Vmatrix}$ be a real,

symmetric, nonsingular matrix. Then $f: X^T a X = 0$ is clearly the equation of a nonsingular real conic. If this conic is to have the further property that it contains no real points, then the only real matrix, X, which can satisfy the

equation $X^T a X = 0$ is the null matrix $X = \begin{Vmatrix} 0 \\ 0 \\ 0 \end{Vmatrix}$, which of course does

satisfy this equation, though it is not the coordinate vector of any point. Thus if $X^T a X = 0$ is to be the equation of a real conic which contains no real points, a must be a definite matrix (see Sec. 3.7), which without loss of generality we can suppose to be positive definite.

Preparatory to our discussion of elliptic geometry, we must establish the following properties of definite matrices.

Theorem 1

If M is a nonsingular square matrix which is conformable to a square matrix a, then $M^T a M$ is definite if and only if a is definite.

Proof Let $M^T a M = b$. Then for any column matrix, Q, we have

(1) $$Q^T b Q = Q^T (M^T a M) Q = (MQ)^T a (MQ)$$

Moreover, since $|M| \neq 0$, it follows that $MQ = 0$ if and only if $Q = 0$. Hence it is clear from (1) that there exists a nonzero column matrix for which $X^T b X = 0$ if and only if there is a nonzero column matrix for which $X^T a X = 0$. In other words, either a and b are both definite or neither is definite, as asserted.

Corollary 1

If a is a definite matrix, and if M is a nonsingular square matrix which is conformable to a, then a and $M^T a M$ are both positive definite or both negative definite.

Theorem 2

The adjoint of a square matrix, a, is definite if and only if a is definite.

Proof Let us suppose first that a is a definite matrix, and let A be its adjoint. If A is not definite, then there exists a nonzero column matrix, Q, such that $Q^T A Q = 0$. Furthermore, by dividing this equation by $|a|$, which is different from zero since a is definite and hence nonsingular, we have

(2) $$Q^T \frac{A}{|a|} Q = Q^T \frac{\text{adj } a}{|a|} Q = Q^T a^{-1} Q = 0$$

Now consider the equation $aP = Q$. Since $|a| \neq 0$, this nonhomogeneous equation has a nonzero solution vector P. Hence, substituting for Q in the last equation in (2), we have

$$(aP)^T a^{-1}(aP) = P^T a^T a^{-1} aP = P^T a a^{-1} aP = P^T aP = 0$$

But this contradicts the hypothesis that a is definite. Hence the assumption that A is not definite must be abandoned, and the "if" assertion of the theorem is established. The "only if" assertion of the theorem can be established in exactly the same fashion, beginning with the fact that when a is not definite, there exists a nonzero column matrix, Q, such that $Q^T a Q = 0$ and using the relation $a/|a| = A^{-1}$.

Corollary 1

If a is a definite matrix, then a and its adjoint are both positive definite or both negative definite.

The two basic assumptions of elliptic geometry are:

1. The metric gauge conic is a real, nonsingular conic, $f: X^T a X = 0$, which contains no real points.

2. Only the properties of real points are to be studied.

Bearing them in mind, let us consider a general triangle (Fig. 10.3) and the identities relating to it which we derived in Sec. 10.5. Obviously the matrix

$$t = \begin{Vmatrix} x_1 & x_2 & x_3 \\ y_1 & y_2 & y_3 \\ z_1 & z_2 & z_3 \end{Vmatrix}$$

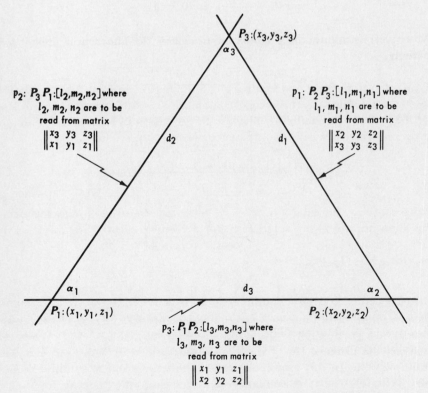

Fig. 10.3   The coordinates of the sides and vertices of a general triangle.

is nonsingular because $P_1 = \begin{Vmatrix} x_1 \\ y_1 \\ z_1 \end{Vmatrix}$, $P_2 = \begin{Vmatrix} x_2 \\ y_2 \\ z_2 \end{Vmatrix}$, $P_3 = \begin{Vmatrix} x_3 \\ y_3 \\ z_3 \end{Vmatrix}$, being the vertices of a triangle, are noncollinear. Moreover, the nature of the metric gauge conic assumed in (1) guarantees that the matrix a is definite and, as we observed above, it is no specialization to suppose further that it is positive definite. Hence $P_i^T a P_i = f_{ii} > 0$, and by Corollary 1, Theorem 1, the matrix

$$\begin{Vmatrix} f_{11} & f_{12} & f_{13} \\ f_{21} & f_{22} & f_{23} \\ f_{31} & f_{32} & f_{33} \end{Vmatrix} = t^T a t \text{ is positive definite. But if this is the case, then by}$$

Theorem 1, Sec. 3.7, every second-order minor of $t^T a t$ is positive. Thus, in particular, $f_{ii} f_{jj} - f_{ij}{}^2 > 0$, $i, j = 1, 2, 3$, $i \neq j$, which implies that

$$\frac{f_{ij}{}^2}{f_{ii} f_{jj}} < 1 \qquad \text{or} \qquad -1 < \frac{f_{ij}}{\sqrt{f_{ii} f_{jj}}} < 1$$

The last inequality now permits us to define a quantity θ by the equation

$$\cos \theta = \frac{f_{ij}}{\sqrt{f_{ii} f_{jj}}} \qquad 0 < \theta < \pi$$

Moreover, recalling the formula of Corollary 2, Theorem 3, Sec. 10.4, namely,

$$(P_i P_j) = k_d \ln \left[\frac{-f_{ij} - \sqrt{f_{ij}{}^2 - f_{ii} f_{jj}}}{-f_{ij} + \sqrt{f_{ij}{}^2 - f_{ii} f_{jj}}} \right]$$

we have, on dividing numerator and denominator by $\sqrt{f_{ii} f_{jj}}$,

$$(P_i P_j) = k_d \ln \left[\frac{-\dfrac{f_{ij}}{\sqrt{f_{ii} f_{jj}}} - \sqrt{\dfrac{f_{ij}{}^2}{f_{ii} f_{jj}} - 1}}{-\dfrac{f_{ij}}{\sqrt{f_{ii} f_{jj}}} + \sqrt{\dfrac{f_{ij}{}^2}{f_{ii} f_{jj}} - 1}} \right]$$

$$= k_d \ln \left(\frac{-\cos \theta - i \sin \theta}{-\cos \theta + i \sin \theta} \right)$$

or, using Eq. (1), Sec. 10.3,

$$(P_i P_j) = k_d \ln \left(\frac{-e^{i\theta}}{-e^{-i\theta}} \right) = k_d \ln e^{2i\theta} = k_d i (2\theta + 2n\pi)$$

In order that the distance between two real points shall be real, it is necessary that k_d be a pure imaginary, and it is customary to take $k_d = 1/2i$. When this is done, the distance $(P_i P_j)$ is uniquely determined to within an arbitrary multiple of π. In this respect the measurement of distance in elliptic geometry bears a striking resemblance to the measurement of angles in elementary geometry. Summarizing, then, we have the following theorem.

Theorem 3

In elliptic geometry, with metric gauge locus $f: X^T aX = 0$, the distance between any two points $P_i = \begin{Vmatrix} x_i \\ y_i \\ z_i \end{Vmatrix}$ and $P_j = \begin{Vmatrix} x_j \\ y_j \\ z_j \end{Vmatrix}$ is equal to $\theta + n\pi$, where

$$\cos \theta = \frac{f_{ij}}{\sqrt{f_{ii}f_{jj}}} \qquad \sin \theta = \sqrt{\frac{f_{ii}f_{jj} - f_{ij}{}^2}{f_{ii}f_{jj}}} \qquad f_{ij} = P_i{}^T a P_j$$

The formula for the measure of the angle between two lines cannot be obtained by duality from Theorem 3 because, since elliptic geometry is a geometry restricted to a proper subset of the projective plane, we do not know that the principle of duality is valid for this subsystem. Instead, we must reason as follows. By Corollary 1, Theorem 2, since a and $t^T at = \| f_{ij} \|$ are positive definite, so too are their adjoints, A and $T^T A T = \| F_{ij} \|$. But if A and $\| F_{ij} \|$ are positive definite, then $F_{ii} = p_i{}^T A p_i$ is positive, and every second-order minor in $\| F_{ij} \|$ is also positive. Hence, in particular,

$$F_{ii}F_{jj} - F_{ij}{}^2 > 0 \qquad \text{or} \qquad -1 < \frac{F_{ij}}{\sqrt{F_{ii}F_{jj}}} < 1$$

We can therefore define a quantity ϕ by the equation

$$(3) \qquad \cos \phi = \frac{F_{ij}}{\sqrt{F_{ii}F_{jj}}} \qquad 0 < \phi < \pi$$

Moreover, from the formula of Corollary 2, Theorem 4, Sec. 10.4, namely,

$$[p_i p_j] = k_a \ln \left[\frac{-F_{ij} - \sqrt{F_{ij}{}^2 - F_{ii}F_{jj}}}{-F_{ij} + \sqrt{F_{ij}{}^2 - F_{ii}F_{jj}}} \right]$$

we have, on dividing numerator and denominator by $\sqrt{F_{ii}F_{jj}}$ and then using (3),

$$[p_i p_j] = k_a \ln \left[\frac{-\cos \phi - i \sin \phi}{-\cos \phi + i \sin \phi} \right]$$

$$= k_a \ln \left(\frac{-e^{i\phi}}{-e^{-i\phi}} \right) = k_a \ln e^{2i\phi} = k_a i (2\phi + 2n\pi)$$

Taking $k_a = 1/2i$, we have thus established the following theorem.

Theorem 4

In elliptic geometry, with metric gauge envelope $F: \Lambda^T A \Lambda = 0$, the measure of the angle between any two lines $p_i = \begin{Vmatrix} l_i \\ m_i \\ n_i \end{Vmatrix}$ and $p_j = \begin{Vmatrix} l_j \\ m_j \\ n_j \end{Vmatrix}$

is equal to $\phi + n\pi$, where

$$\cos\phi = \frac{F_{ij}}{\sqrt{F_{ii}F_{jj}}} \qquad \sin\phi = \sqrt{\frac{F_{ii}F_{jj} - F_{ij}^2}{F_{ii}F_{jj}}} \qquad F_{ij} = p_i^T A p_j$$

It is important to note that in our analysis of the general triangle in elliptic geometry, points and lines do not play quite the same roles. In particular, the coordinates of the vertices of the triangle are *given* as initial data, and then the coordinates of the sides are computed from the coordinates of the, appropriate pairs of vertices. From the definition of the coordinates of the side $p_i: P_jP_k$ in terms of the second-order determinants in the matrix $\left\| \begin{matrix} x_j & y_j & z_j \\ x_k & y_k & z_k \end{matrix} \right\|$, it is clear that the coordinates of the line P_jP_k are the negatives of the coordinates of the line P_kP_j. With the definitions implied by the labeling in Fig. 10.3 and used in the derivation of the identities in Sec. 10.5, it is evident that the formulas of Theorem 4 define the measure of the angle between the lines $p_i: P_jP_k$ and $p_j: P_kP_i$, whereas the angle in $\triangle P_1P_2P_3$ whose vertex is P_k is the angle between P_kP_j and P_kP_i, that is, the angle between the lines whose coordinate-vectors are $-p_i$ and p_j. Now since F_{ij} is a bilinear function of p_i and p_j, it follows when one or the other of these coordinate vectors changes sign, i.e., when the direction of one or the other of these lines is reversed, that F_{ij} changes sign also. Hence, with the cyclic definitions of the coordinates of the sides of a general triangle on which the identities of Sec. 10.5 are based, the measures of the angles of a triangle are defined not by the formula of Theorem 4 but rather by the formula

$$(4) \qquad\qquad \cos\alpha_k = -\frac{F_{ij}}{\sqrt{F_{ii}F_{jj}}}$$

It is convenient now to collect the formulas relating to a general triangle, $\triangle P_1P_2P_3 = \triangle p_1p_2p_3$, with sides d_1, d_2, d_3 and angles α_1, α_2, α_3, which we need in the rest of our work:

$$\cos d_i = \frac{f_{jk}}{\sqrt{f_{jj}f_{kk}}} \qquad \sin d_i = \sqrt{\frac{f_{jj}f_{kk} - f_{jk}^2}{f_{jj}f_{kk}}}$$

$$= \sqrt{\frac{F_{ii}}{f_{jj}f_{kk}}} \qquad \text{by (6), Sec. 10.5}$$

$$\cos\alpha_i = \frac{-F_{jk}}{\sqrt{F_{jj}F_{kk}}} \qquad \sin\alpha_i = \sqrt{\frac{F_{jj}F_{kk} - F_{jk}^2}{F_{jj}F_{kk}}}$$

$$= \sqrt{\frac{|f_{ii}|f_{ii}}{F_{jj}F_{kk}}} \qquad \text{by (9) Sec. 10.5}$$

The basic laws for the general triangle in elliptic geometry are given in the next three theorems.

Theorem 5

In any triangle in elliptic geometry, with sides d_1, d_2, d_3 and angles α_1, α_2, α_3,

$$\cos d_i = \cos d_j \cos d_k + \sin d_j \sin d_k \cos \alpha_i$$

Proof Using the appropriate formulas, and their obvious permutations, from the list we have just compiled, we have for the right-hand side of the relation given in the theorem

$$\frac{f_{ik}}{\sqrt{f_{ii}f_{kk}}} \frac{f_{ij}}{\sqrt{f_{ii}f_{jj}}} + \sqrt{\frac{F_{jj}}{f_{ii}f_{kk}}} \sqrt{\frac{F_{kk}}{f_{ii}f_{jj}}} \left(-\frac{F_{jk}}{\sqrt{F_{jj}F_{kk}}}\right)$$

$$= \frac{f_{ik}f_{ij} - F_{jk}}{f_{ii}\sqrt{f_{jj}f_{kk}}} = \frac{f_{ik}f_{ij} - (f_{ij}f_{ik} - f_{jk}f_{ii})}{f_{ii}\sqrt{f_{jj}f_{kk}}} \qquad \text{by (7), Sec. 10.5}$$

$$= \frac{f_{jk}}{\sqrt{f_{jj}f_{kk}}} = \cos d_i$$

as asserted.

Theorem 6

In any triangle in elliptic geometry, with sides d_1, d_2, d_3 and angles α_1, α_2, α_3,

$$\cos \alpha_i = -\cos \alpha_j \cos \alpha_k + \sin \alpha_j \sin \alpha_k \cos d_i$$

Proof Working, again, with the right-hand side of the relation given in the theorem, we have

$$-\left(\frac{-F_{ik}}{\sqrt{F_{ii}F_{kk}}}\right)\left(\frac{-F_{ij}}{\sqrt{F_{ii}F_{jj}}}\right) + \sqrt{\frac{|f_{ij}|f_{jj}}{F_{ii}F_{kk}}} \sqrt{\frac{|f_{ij}|f_{kk}}{F_{ii}F_{jj}}} \frac{f_{jk}}{\sqrt{f_{jj}f_{kk}}}$$

$$= \frac{-F_{ik}F_{ij} + |f_{ij}|f_{jk}}{F_{ii}\sqrt{F_{jj}F_{kk}}} = \frac{-F_{ik}F_{ij} + (F_{ij}F_{ik} - F_{jk}F_{ii})}{F_{ii}\sqrt{F_{jj}F_{kk}}} \qquad \text{by (10), Sec. 10.5}$$

$$= \frac{-F_{jk}}{\sqrt{F_{jj}F_{kk}}} = \cos \alpha_i$$

as asserted.

Theorem 7

In any triangle in elliptic geometry, with sides d_1, d_2, d_3 and angles α_1, α_2, α_3,

$$\frac{\sin \alpha_1}{\sin d_1} = \frac{\sin \alpha_2}{\sin d_2} = \frac{\sin \alpha_3}{\sin d_3} = \sqrt{\frac{|f_{ij}|f_{11}f_{22}f_{33}}{F_{11}F_{22}F_{33}}}$$

Proof Evaluating the typical fraction in the relation given in the theorem, we have

$$\frac{\sin \alpha_i}{\sin d_i} = \frac{\sqrt{\dfrac{|f_{ij}| f_{ii}}{F_{jj}F_{kk}}}}{\sqrt{\dfrac{F_{ii}}{f_{jj}f_{kk}}}} = \sqrt{\frac{|f_{ij}| f_{ii} f_{jj} f_{kk}}{F_{ii}F_{jj}F_{kk}}}$$

as asserted.

Using Theorems 5 and 6, we can now prove the famous angle-sum theorem of elliptic geometry.

Theorem 8

In elliptic geometry, the sum of the measures of the angles of any triangle is greater than π.

Proof In a triangle with sides d_1, d_2, d_3 and angles α_1, α_2, α_3, let the notation be chosen so that $\pi > \alpha_1 \geq \alpha_2 \geq \alpha_3 > 0$, and therefore

(5)
$$-\pi < \alpha_1 - \alpha_2 - \alpha_3 < \pi$$

Now, by Theorem 6,

$$\cos \alpha_1 = - \cos \alpha_2 \cos \alpha_3 + \sin \alpha_2 \sin \alpha_3 \cos d_1$$

and, from trigonometry,

$$\cos (\alpha_2 + \alpha_3) = \cos \alpha_2 \cos \alpha_3 - \sin \alpha_2 \sin \alpha_3{}^\dagger$$

Adding the last two equations, we have

$$\cos \alpha_1 + \cos (\alpha_2 + \alpha_3) = - \sin \alpha_2 \sin \alpha_3 (1 - \cos d_1)$$

or, converting the sum on the left to a product,

$$2 \cos \frac{\alpha_1 + \alpha_2 + \alpha_3}{2} \cos \frac{\alpha_1 - \alpha_2 - \alpha_3}{2} = - \sin \alpha_2 \sin \alpha_3 (1 - \cos d_1)$$

The right-hand side of this equation is clearly negative. Moreover, it follows from (5) that

$$-\frac{\pi}{2} < \frac{\alpha_1 - \alpha_2 - \alpha_3}{2} < \frac{\pi}{2}$$

---

† Although the usual derivation of this formula involves an argument based upon the properties of euclidean geometry, this result is actually a fact of analytical trigonometry, independent of any geometrical considerations. See, for instance, E. T. Whittaker and G. N. Watson, "Modern Analysis," pp. 584–585, The Macmillan Company, New York, 1943.

and hence $\cos\left[(\alpha_1 - \alpha_2 - \alpha_3)/2\right] > 0$. Therefore $\cos\left[(\alpha_1 + \alpha_2 + \alpha_3)/2\right]$ must be negative, and hence $(\alpha_1 + \alpha_2 + \alpha_3)/2 > \pi/2$, or $\alpha_1 + \alpha_2 + \alpha_3 > \pi$, as asserted.

It is a matter of some interest to investigate the assertions of Theorems 5 to 7 for infinitesimal triangles, i.e., for triangles whose sides, d_1, d_2, d_3, are arbitrarily small. This we do by replacing the trigonometric functions of the sides which appear in the formulas of Theorems 5 to 7 by their Maclaurin expansions and then neglecting all but the leading terms. From Theorem 5 we thus obtain

$$\left(1 - \frac{d_i^2}{2} + \cdots\right) = \left(1 - \frac{d_j^2}{2} + \cdots\right)\left(1 - \frac{d_k^2}{2} + \cdots\right)$$
$$+ (d_j - \cdots)(d_k - \cdots)\cos\alpha_i$$

or, retaining only the lowest powers of the infinitesimal quantities d_i, d_j, d_k,

$$d_i^2 = d_j^2 + d_k^2 - 2d_jd_k\cos\alpha_i$$

which is just the familiar law of cosines of elementary euclidean geometry. Similarly, from Theorem 6 we obtain

$$\cos\alpha_i = -\cos\alpha_j\cos\alpha_k + \sin\alpha_j\sin\alpha_k\left(1 - \frac{d_i^2}{2} + \cdots\right)$$

or, again, retaining only the lowest powers of d_i, d_j, d_k,

$$\cos\alpha_i = -\cos\alpha_j\cos\alpha_k + \sin\alpha_j\sin\alpha_k$$
$$= -\cos(\alpha_j + \alpha_k) = \cos(\pi - \alpha_j - \alpha_k)$$

This implies that $\alpha_i = \pi - \alpha_j - \alpha_k$ or $\alpha_i + \alpha_j + \alpha_k = \pi$, which is precisely the angle-sum theorem of euclidean geometry. Finally, from Theorem 7 we have

$$\frac{\sin\alpha_1}{d_1 - d_1^3/6 + \cdots} = \frac{\sin\alpha_2}{d_2 - d_2^3/6 + \cdots} = \frac{\sin\alpha_3}{d_3 - d_3^3/6 + \cdots}$$

or, retaining only the leading terms in the infinite series in the denominators,

$$\frac{\sin\alpha_1}{d_1} = \frac{\sin\alpha_2}{d_2} = \frac{\sin\alpha_3}{d_3}$$

which is the euclidean law of sines.

The preceding discussion has established the following important fact: *for arbitrarily small triangles, the formulas of elliptic geometry are arbitrarily close to the corresponding formulas of euclidean geometry*, or to put it in another way, *the geometry of infinitesimal regions in the elliptic plane is approximately euclidean*. Using this fact, we shall now conclude our brief exploration of

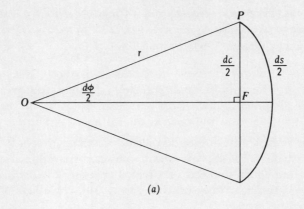

(a)

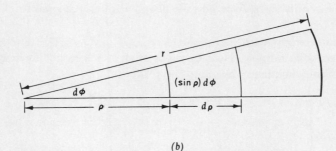

(b)

Fig. 10.4   Figures used in the proofs of Theorems 9 and 10.

elliptic geometry by computing the length of a circular arc, the area of a circular sector, and the area of a general triangle.

Theorem 9

In elliptic geometry, the length of the arc of a sector of central angle θ in a circle of radius r is $(\sin r)\theta$.

Proof   Consider first the infinitesimal sector of a circle shown in Fig. 10.4a. Applying the formula of Theorem 7 to the right triangle OPF, we have

$$\frac{\sin r}{1} = \frac{\sin (dc/2)}{\sin (d\phi/2)}$$

Now since the central angle, $d\phi$, and the chord, dc, are both infinitesimals, we have to an arbitrary degree of approximation,

$$\sin \frac{d\phi}{2} = \frac{d\phi}{2} \qquad \sin \frac{dc}{2} = \frac{dc}{2} \qquad dc = ds$$

Hence $ds = (\sin r)\, d\phi$ and, integrating with respect to ϕ from $\phi = 0$ to $\phi = \theta$, we have $s = (\sin r)\theta$, as asserted. If r is also infinitesimal, this reduces to the euclidean arc-length formula, $s = r\theta$.

Theorem 10

In elliptic geometry, the area of a sector of central angle θ in a circle of radius r is $(1 - \cos r)\theta$.

Proof Consider first the infinitesimal sector of a circle shown in Fig. 10.4*b*. To an arbitrary degree of approximation, the area of one of the infinitesimal "rectangles" into which the sector is divided is, using Theorem 9, $(\sin \rho\, d\phi)$ $d\rho$. Hence, integrating with respect to ρ from 0 to r, we have for the area of the infinitesimal sector

$$d\phi \int_0^r \sin \rho\, d\rho = d\phi \, [-\cos \rho]_0^r = (1 - \cos r)\, d\phi$$

Finally, integrating with respect to ϕ from 0 to θ, we find the area of a general sector to be $(1 - \cos r)\theta$, as asserted. If r is also infinitesimal, this reduces to the euclidean formula, $A = r^2\theta/2$.

Theorem 11

In elliptic geometry, the area of a general triangle with angles $\alpha_1,\, \alpha_2,\, \alpha_3$ is $\alpha_1 + \alpha_2 + \alpha_3 - \pi$.

Proof We shall establish this theorem by first proving it for a general right triangle. To do this, let us consider a general right triangle and subdivide it into infinitesimal sectors, as shown in Fig. 10.5*a*. By Theorem 10, the area of the typical sector in this subdivision is approximately $(1 - \cos r)\, d\theta$. Hence the area of the triangle is

$$\int_0^{\alpha_1} (1 - \cos r)\, d\theta = \alpha_1 - \int_0^{\alpha_1} \cos r\, d\theta$$

Now from Theorem 5, applied to $\triangle P_1 P_2' P_3$, we have

$$\cos r = \cos d_2 \cos d_1' - 0 \qquad \text{or} \qquad \cos d_1' = \frac{\cos r}{\cos d_2}$$

and from Theorem 7, applied to $\triangle P_1 P_2' P_3$, we have

$$\frac{\sin r}{1} = \frac{\sin d_1'}{\sin \theta} \qquad \text{or} \qquad \sin d_1' = \sin r \sin \theta$$

Therefore, $\sin^2 d_1' + \cos^2 d_1' = \sin^2 r \sin^2 \theta + \dfrac{\cos^2 r}{\cos^2 d_2} = 1$

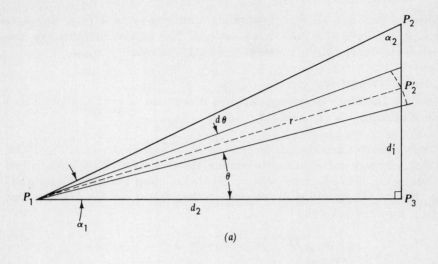

(a)

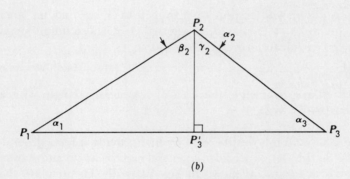

(b)

Fig. 10.5 Figures used in the proof of Theorem 11.

Hence, substituting $1 - \cos^2 r$ for $\sin^2 r$ and solving for $\cos^2 r$, we find

$$\cos^2 r = \frac{\cos^2 d_2(1 - \sin^2 \theta)}{1 - \sin^2 \theta \cos^2 d_2} = \frac{\cos^2 d_2 \cos^2 \theta}{\cos^2 d_2 + \sin^2 d_2 - \sin^2 \theta \cos^2 d_2}$$

$$= \frac{\cos^2 d_2 \cos^2 \theta}{\cos^2 d_2 \cos^2 \theta + \sin^2 d_2}$$

$$= \frac{\cos^2 \theta}{\cos^2 \theta + \tan^2 d_2}$$

$$= \frac{\cos^2 \theta}{\sec^2 d_2 - \sin^2 \theta}$$

Thus

$$\int_0^{a_1} \cos r \, d\theta = \int_0^{a_1} \frac{\cos \theta \, d\theta}{\sqrt{\sec^2 d_2 - \sin^2 \theta}}$$

$$= \text{Sin}^{-1} \left(\frac{\sin \theta}{\sec d_2} \right) \Big|_0^{a_1}$$

$$= \text{Sin}^{-1} (\cos d_2 \sin \alpha_1)$$

$$= \text{Sin}^{-1} (\cos \alpha_2) \qquad \text{by Theorem 6}$$

$$= \frac{\pi}{2} - \alpha_2$$

Hence the area of the given right triangle is equal to

$$\alpha_1 - \left(\frac{\pi}{2} - \alpha_2 \right) = \alpha_1 + \alpha_2 - \frac{\pi}{2} = \alpha_1 + \alpha_2 + \alpha_3 - \pi$$

since, for the right triangle, $\alpha_3 = \pi/2$.

To prove the theorem for a general triangle, we merely divide the triangle into two right triangles by drawing the altitude from some convenient vertex to the opposite side and then applying to each of the right triangles thus formed the result we have just established. Referring to Fig. 10.5b, this gives us

$$\text{Area } \triangle P_1 P_2 P_3 = \text{area } \triangle P_1 P_2 P_3' + \text{area } \triangle P_2 P_3' P_3$$

$$= \left(\alpha_1 + \beta_2 - \frac{\pi}{2} \right) + \left(\gamma_2 + \alpha_3 - \frac{\pi}{2} \right)$$

$$= \alpha_1 + (\beta_2 + \gamma_2) + \alpha_3 - \pi$$

$$= \alpha_1 + \alpha_2 + \alpha_3 - \pi$$

as asserted.

Other properties of elliptic geometry will be found among the exercises.

Exercises

1. Prove Corollary 1, Theorem 1.

2. Prove Corollary 1, Theorem 2.

3. Prove that the s, s, s congruence theorem for triangles is valid in elliptic geometry.

4. Is the a, s, a congruence theorem for triangles valid in elliptic geometry?

5. Is the s, a, s congruence theorem for triangles valid in elliptic geometry?

6. Prove that in elliptic geometry if two sides of a triangle are equal, the angles opposite these sides are equal, and conversely.

7. Given $P_1:(1,1,0)$, $P_2:(0,2,1)$, $P_3:(1,-1,2)$. Determine the sides and angles of $\triangle P_1 P_2 P_3$ if the equation of the metric gauge conic is:

(a) $x_1{}^2 + x_2{}^2 + x_3{}^2 = 0$ (b) $x_1{}^2 + x_2{}^2 + 2x_2 x_3 + 2x_3{}^2 = 0$
(c) $x_1{}^2 - 2x_1 x_2 + 3x_2{}^2 - 2x_2 x_3 + x_3{}^2 = 0$
(d) $2x_1{}^2 - 2x_1 x_2 + 5x_2{}^2 + 2x_2 x_3 + x_3{}^2 = 0$

8. Complete the proof of Theorem 2 by supplying the details of the "only if" part of the argument.

9. Prove that in any triangle in elliptic geometry, if $\alpha_i > \alpha_j$ then $d_i > d_j$, and conversely.

10. In the elliptic plane, let α and β be the angles opposite the sides a and b of a right triangle whose hypotenuse is h. Prove the following formulas:

(a) $\sin b = \sin h \sin \beta$ (b) $\cos h = \cos a \cos b$
(c) $\cos \alpha = \sin \beta \cos a$
(d) $\cos \alpha \cos \beta = \sin \alpha \sin \beta \cos h$

11. Solve the following right triangles:

(a) $\alpha = \pi/6$, $h = 1$ (b) $a = 1$, $b = 2$
(c) $a = 1$, $h = 2$ (d) $a = 1$, $\beta = \pi/3$

12. Solve the following oblique triangles:

(a) $d_1 = d_2 = 2$, $d_3 = 1$ (b) $d_1 = 1$, $d_2 = 2$, $\alpha_3 = 3\pi/4$

13. What theorems about circles were implicitly assumed in the proofs of Theorems 9 and 10? Prove as many of these as you can.

14. How are distances and angles in the elliptic plane affected by a transformation, $T: X' = MX$, which leaves the metric gauge conic invariant?

15. Prove that in the elliptic plane if two triangles are similar, they are congruent.

10.7 The Geometry on a Euclidean Sphere

Although it is not in the mainstream of our work in projective geometry, it is interesting to note that elliptic geometry, as developed in the last section, is essentially the same as the geometry on the surface of a euclidean sphere.

To verify this, let us recall that on the sphere, great circles, i.e., circles cut from the sphere by planes containing the center of the sphere, play the roles of lines. Triangles on the sphere are then three-sided figures whose sides are arcs of great circles.

We begin by considering on a sphere of radius r a general right spherical triangle, $\triangle ABC$, with right angle at C, as shown in Fig. 10.6a. Next we pass a plane perpendicular to OA at some convenient point, A', and intersecting OB in B' and OC in C', as shown in Fig. 10.6b. Clearly, the plane angle $\angle B'A'C'$ measures the dihedral angle between the plane OAB and the plane OAC, and hence measures the angle α in the given right spherical triangle, $\triangle ACB$. By construction, the following plane triangles are right triangles:

$\triangle OA'B'$, right angle at A'

$\triangle OA'C'$, right angle at A'

$\triangle A'C'B'$, right angle at C'

$\triangle OC'B'$, right angle at C'

Therefore, from plane trigonometry,

$$\sin \alpha = \sin (\angle B'A'C') = \frac{B'C'}{A'B'} = \frac{B'C'/OB'}{A'B'/OB'} = \frac{\sin (\angle B'OC')}{\sin (\angle B'OA')}$$

Hence, since $a = r\angle B'OC'$ and $c = r\angle B'OA'$, we have

(1)
$$\sin \alpha = \frac{\sin (a/r)}{\sin (c/r)}$$

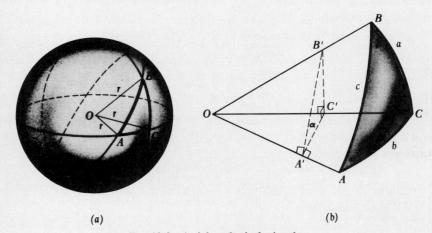

(a)   (b)

Fig. 10.6   A right spherical triangle.

Similarly,

(2) $\cos \alpha = \cos (\angle B'A'C') = \dfrac{A'C'}{A'B'} = \dfrac{A'C'/OA'}{A'B'/OA'} = \dfrac{\tan (\angle A'OC')}{\tan (\angle A'OB')}$

$= \dfrac{\tan (b/r)}{\tan (c/r)}$

and

(3) $\tan \alpha = \tan (\angle B'A'C') = \dfrac{B'C'}{A'C'} = \dfrac{B'C'/OC'}{A'C'/OC'} = \dfrac{\tan (\angle B'OC')}{\sin (\angle A'OC')}$

$= \dfrac{\tan (a/r)}{\sin (b/r)}$

If we now divide Eq. (1) by Eq. (2), we obtain

$$\tan \alpha = \frac{\sin (a/r) \tan (c/r)}{\sin (c/r) \tan (b/r)} = \frac{\sin (a/r) \cos (b/r)}{\cos (c/r) \sin (b/r)}$$

or, substituting for $\tan \alpha$ from (3)

$$\frac{\tan (a/r)}{\sin (b/r)} = \frac{\sin (a/r)}{\cos (a/r) \sin (b/r)} = \frac{\sin (a/r) \cos (b/r)}{\cos (c/r) \sin (b/r)}$$

whence

(4) $$\cos \frac{c}{r} = \cos \frac{a}{r} \cos \frac{b}{r}$$

Finally, from (3) we have

$$\frac{\sin \alpha}{\cos \alpha} = \frac{\sin (a/r)}{\cos (a/r) \sin (b/r)}$$

from which we find

$$\cos \alpha = \frac{\sin \alpha \sin (b/r) \cos (a/r)}{\sin (a/r)}$$

or, substituting for $\sin \alpha$ from (1) and substituting for $\sin (b/r)$ from an obvious permutation of (1),

$$\cos \alpha = \frac{\sin (a/r)}{\sin (c/r)} \left[\sin \beta \sin \left(\frac{c}{r} \right) \right] \frac{\cos (a/r)}{\sin (a/r)}$$

Hence

(5) $$\cos \alpha = \cos \frac{a}{r} \sin \beta$$

Formulas (1) to (5) and their obvious permutations can all be read from the diagram shown in Fig. 10.7 by using what are known as **Napier's rules**:[1]

Letting the prefix "co-" denote "complement of,"

1. The sine of any term in the figure is equal to the product of the tangents of the two adjacent terms.

[1] Named for the Scottish mathematician John Napier (1550–1617), best known as the inventor of logarithms.

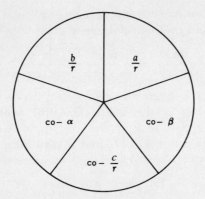

Fig. 10.7 Figure to be used with Napier's rules.

2. The sine of any term in the figure is equal to the product of the cosines of the two nonadjacent terms.

Let us now consider a general spherical triangle, subdivided into two right spherical triangles, as shown in Fig. 10.8. By Eq. (1) applied to $\triangle ADC$ and $\triangle BDC$, we have

$$\sin \frac{p}{r} = \sin \frac{b}{r} \sin \alpha \qquad \text{and} \qquad \sin \frac{p}{r} = \sin \frac{a}{r} \sin \beta$$

Hence, eliminating $\sin (p/r)$, we have

$$\frac{\sin \alpha}{\sin (a/r)} = \frac{\sin \beta}{\sin (b/r)}$$

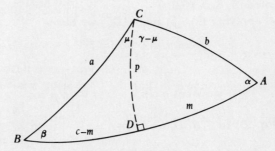

Fig. 10.8 A general spherical triangle decomposed into two right spherical triangles.

Repeating this argument, after dropping a perpendicular from either A or B to the opposite side, and combining the results, we obtain

$$(6) \qquad \frac{\sin \alpha}{\sin (a/r)} = \frac{\sin \beta}{\sin (b/r)} = \frac{\sin \gamma}{\sin (c/r)}$$

If we choose to work on a sphere for which $r = 1$, which is equivalent to our choice of $k_d = k_a = 1/2i$ in the last section, (6) becomes precisely the assertion of Theorem 7, Sec. 10.5.

Referring again to Fig. 10.8, and applying Eq. (4) to $\triangle BDC$, we find that $\cos (a/r) = \cos [(c - m)/r] \cos (p/r)$, or, expanding $\cos [(c - m)/r]$,

$$(7) \qquad \cos \frac{a}{r} = \cos \frac{c}{r} \cos \frac{m}{r} \cos \frac{p}{r} + \sin \frac{c}{r} \sin \frac{m}{r} \cos \frac{p}{r}$$

Also, from Eq. (4) applied to $\triangle ADC$, we have

$$(8) \qquad \cos \frac{m}{r} \cos \frac{p}{r} = \cos \frac{b}{r}$$

and from Eq. (1) and Eq. (5) applied to $\triangle ADC$, we have

$$\sin \frac{m}{r} = \sin \frac{b}{r} \sin (\gamma - \mu) \qquad \text{and} \qquad \cos \alpha = \cos \frac{p}{r} \sin (\gamma - \mu)$$

or, eliminating $\sin (\gamma - \mu)$ between these two equations,

$$(9) \qquad \sin \frac{m}{r} \cos \frac{p}{r} = \sin \frac{b}{r} \cos \alpha$$

Now substituting from (8) and (9) into (7), we obtain

$$(10) \qquad \cos \frac{a}{r} = \cos \frac{b}{r} \cos \frac{c}{r} + \sin \frac{b}{r} \sin \frac{c}{r} \cos \alpha$$

which is the law of cosines for spherical triangles. If $r = 1$, it is precisely the assertion of Theorem 5, Sec. 10.5.

Finally, applying Eq. (5) to $\triangle ADC$, we have

$$(11) \qquad \cos \alpha = \sin (\gamma - \mu) \cos \frac{p}{r}$$

$$= \sin \gamma \cos \mu \cos \frac{p}{r} - \cos \gamma \sin \mu \cos \frac{p}{r}$$

Also, applying Eq. (5) to $\triangle BCD$, we have

$$(12) \qquad \sin \mu \cos \frac{p}{r} = \cos \beta$$

Furthermore, from Eqs. (4) and (5) applied to $\triangle BCD$, we have

$$\cos \frac{a}{r} = \cos \frac{p}{r} \cos \frac{c-m}{r} \qquad \text{and} \qquad \cos \mu = \cos \frac{c-m}{r} \sin \beta$$

or, eliminating $\cos [(c-m)/r]$,

(13) $$\cos \mu \cos \frac{p}{r} = \cos \frac{a}{r} \sin \beta$$

Hence substituting from Eqs. (12) and (13) into Eq. (11), we obtain

(14) $$\cos \alpha = -\cos \beta \cos \gamma + \sin \beta \sin \gamma \cos \frac{a}{r}$$

If $r = 1$, this is the assertion of Theorem 6, Sec. 10.5.

It is easy to verify that as r becomes infinite, Eqs. (6), (10), and (14) reduce to the corresponding euclidean laws. In other words, *as r becomes infinite, the geometry of any bounded portion of a sphere of radius r approaches euclidean geometry.*

Exercises

1. Determine the limiting forms of Eqs. (6), (10), and (14) as r becomes infinite.

2. Verify that Eqs. (2) and (3) are valid for right triangles in the elliptic plane.

10.8 Hyperbolic Geometry

As a specialization of the projective plane, hyperbolic geometry differs from elliptic geometry in two significant respects. In the first place, in hyperbolic geometry the metric gauge conic is a real, nonsingular conic containing real points, whereas in elliptic geometry it is a real, nonsingular conic containing *no* real points. Second, hyperbolic geometry involves only a proper subset of the real points of the coordinatized projective plane, whereas elliptic geometry is concerned with *all* the real points in the projective plane.

Let $f \colon X^T aX = 0$ be the equation of a real, nonsingular conic which contains at least one real point. Since f is nonsingular, we know that $|a| \neq 0$, and, to be specific, we shall suppose that $|a| > 0$. This is no restriction, because if $|a| < 0$, multiplying the equation of f by -1 will make $|a| > 0$, since the matrix a is of odd order. Now it is not difficult to show (see Exercise 15) that if $X^T aX = 0$ is a real, nonsingular conic containing at least one real point, then there exist points at which $X^T aX$ is positive and also points at which $X^T aX$ is negative. When $|a| > 0$, a real point, P_i, will be called an **interior point** or an **exterior point** according as $f_{ii} = P_i^T a P_i$ is positive or negative. Similarly, if $\Lambda^T A \Lambda = 0$ is the equation of the metric gauge conic in line-coordinates, and if $|a|$, and hence $|A|$, is positive, we shall call a real line, p_i, an **interior line** or an **exterior line** according as $F_{ii} = p_i^T A p_i$ is negative or positive.[1] In hyperbolic geometry, we are concerned exclusively with interior points and interior lines. In other words, *hyperbolic geometry is the study of the metrical relations imposed upon the interior points and interior lines of the coordinatized projective plane when the metric gauge conic is chosen to be a real, nonsingular conic containing at least one real point.*

Preparatory to defining the measurement of distance and angle in the hyperbolic plane, we must establish a number of preliminary results describing how the identities derived in Sec. 10.5 for a general triangle and a general conic are affected by our current choice of metric gauge conic.

Theorem 1

If at least one of the points P_1, P_2 is an interior point, then $F_{33} = f_{11}f_{22} - f_{12}^2 < 0$.

Proof Since $\|F_{ij}\|$ is the adjoint of $\|f_{ij}\|$, and since we know from Eq. (3), Sec. 10.5, that $|f_{ij}| = |t|^2 \cdot |a|$, it follows that $|F_{ij}| = |t|^4 \cdot |a|^2 > 0$. Also, assuming that of the two given points, P_1 is an interior point, we have $f_{11} > 0$ and, by Eq. (9), Sec. 10.5,

$$\begin{vmatrix} F_{22} & F_{23} \\ F_{32} & F_{33} \end{vmatrix} = |f_{ij}|f_{11} = |t|^2 \cdot |a|f_{11} > 0$$

Therefore, if F_{33} were positive, the conditions of Theorem 1, Sec. 3.7, would be met, and the matrix $\|F_{ij}\|$ would be positive definite. However, the matrix, a, of our current metric gauge conic is not definite, since the quadratic form $X^T a X$ vanishes for at least one real nonzero matrix $X = P$. Therefore, by Theorem 1, Sec. 10.6, the matrix $\|f_{ij}\| = t^T a t$ is not definite, and by Theorem 2, Sec. 10.6, its adjoint, the matrix $\|F_{ij}\| = T^T A T$ is not definite.

[1] The appropriateness of these definitions can be seen by considering the euclidean circle $-x^2 - y^2 + 1 = 0$ or, in homogeneous coordinates, $-x_1^2 - x_2^2 + x_3^2 = 0$. For this conic, the matrix a is $\begin{Vmatrix} -1 & 0 & 0 \\ 0 & -1 & 0 \\ 0 & 0 & 1 \end{Vmatrix}$, the adjoint matrix, A, is also $\begin{Vmatrix} -1 & 0 & 0 \\ 0 & -1 & 0 \\ 0 & 0 & 1 \end{Vmatrix}$, and both $|a|$ and $|A|$ are clearly positive. Now the origin, i.e., the point $P_1:(0,0,1)$, obviously is in the interior of the circle, and a trivial calculation shows that $P_1^T a P_1 > 0$. On the other hand, the point $P_2:(2,0,1)$ is clearly outside the circle, and for it we find at once that $P_2^T a P_2 < 0$. Moreover, although no line can lie entirely in the interior of the given circle, the line $p_1:[1,-1,0]$, that is, the line $x - y = 0$, does pass through the interior; and for it we have $p_1^T A p_1 < 0$. On the other hand, the line $p_2:[0,1,-2]$, that is, the line $y = 2$, certainly lies entirely outside the circle; and for it we have $p_2^T A p_2 > 0$. Thus in the euclidean plane, our abstract definitions of interior and exterior points and interior and exterior lines appear to check our intuitive understanding of these terms.

Hence F_{33} cannot be positive. Moreover, it is clear from Eq. (9), Sec. 10.5, that $F_{33} \neq 0$. Therefore $F_{33} < 0$, and the theorem is proved.

Corollary 1

Any line on an interior point is an interior line.

Proof Let P_1 be an interior point, so that $f_{11} > 0$, and let P_2 be any other point. Then for the line P_1P_2, whose coordinates determine F_{33}, we have $F_{33} = f_{11}f_{22} - f_{12}{}^2$. By Theorem 1, this quantity is negative, and therefore P_1P_2 is an interior line, as asserted.

Corollary 2

An exterior line contains only exterior points.

Proof If an exterior line contained an interior point, it would, by Corollary 1, have to be an interior line, contrary to hypothesis.

Theorem 2

The polar of an interior point with respect to the metric gauge conic is an exterior line.

Proof Let P_1 be an interior point, so that $P_1{}^T a P_1 > 0$, and let p_1 be its polar line with respect to the metric gauge conic. Then since the equation of the polar of P_1 can be written in either of the forms $P_1{}^T a X = 0$ or $p_1{}^T X = 0$, it follows that $p_1{}^T = k P_1{}^T a$, where k is a nonzero scalar constant. Therefore $p_1{}^T A p_1 = (k P_1{}^T a) A (k a P_1)$ and, since $a A = I |a|$, $p_1{}^T A p_1 = k^2 |a| P_1{}^T a P_1$. Hence, since both $|a|$ and $P_1{}^T a P_1$ are positive, it follows that $p_1{}^T A p_1 > 0$. Thus p_1 is an exterior line, and the theorem is proved.

Theorem 3

If P_1, P_2, P_3 are interior points, then of the three quantities $f_{12} = P_1{}^T a P_2$, $f_{23} = P_2{}^T a P_3$, $f_{13} = P_1{}^T a P_3$, either one or three must be positive.

Proof We observe first that none of the quantities $P_1{}^T a P_2$, $P_2{}^T a P_3$, $P_1{}^T a P_3$ can be zero, since each is the evaluation of the polar of an interior point for a (different) interior point, and, according to Theorem 2 and Corollary 2, Theorem 1, the polar of an interior point is an exterior line, necessarily containing no interior points.

Now, as we observed in the proof of Theorem 1,

$$|f_{ij}| = \begin{vmatrix} f_{11} & f_{12} & f_{13} \\ f_{21} & f_{22} & f_{23} \\ f_{31} & f_{32} & f_{33} \end{vmatrix} > 0$$

Hence, expanding in terms of the elements of the first row,

$$f_{11}(f_{22}f_{33} - f_{23}{}^2) + f_{12}(f_{23}f_{31} - f_{21}f_{33}) + f_{13}(f_{21}f_{32} - f_{31}f_{22})$$
$$= f_{11}f_{22}f_{33} - f_{11}f_{23}{}^2 + 2f_{12}f_{13}f_{23} - f_{12}{}^2f_{33} - f_{13}{}^2f_{22} > 0$$

Moreover, since P_2P_3, P_3P_1, P_1P_2 are interior lines, it follows that

$$F_{11} = f_{22}f_{33} - f_{23}{}^2 < 0 \quad F_{22} = f_{11}f_{33} - f_{13}{}^2 < 0 \quad F_{33} = f_{11}f_{22} - f_{12}{}^2 < 0$$

Therefore, if we replace $f_{23}{}^2, f_{13}{}^2$, and $f_{12}{}^2$ in the expansion of $|f_{ij}|$ by $f_{22}f_{33}$, $f_{11}f_{33}$, and $f_{11}f_{22}$, respectively, we are surely overestimating $|f_{ij}|$. Thus $0 < |f_{ij}| < 2f_{12}f_{13}f_{23} - 2f_{11}f_{22}f_{33}$. The last term is surely negative, since f_{11}, f_{22}, f_{33} are all positive because P_1, P_2, P_3 are interior points. Hence we have a contradiction unless the first term is positive, which is possible if and only if either one or three of the quantities f_{12}, f_{13}, f_{23} is positive, as asserted.

Theorem 4

Coordinates may be chosen so that $f_{ij} = P_i{}^T a P_j > 0$ for all pairs of interior points.

Proof   Let $P_1{}^T a X = 0$ be the polar line of an arbitrary interior point, P_i. Then since $P_1{}^T a X = 0$ is an exterior line, its equation can be satisfied by no interior point; that is, $P_1{}^T a P_i \neq 0$. Thus $P_1{}^T a P_i$ is either positive or can be made positive by multiplying the coordinates of P_i by an arbitrary negative proportionality constant. Thus for all interior points, $P_1{}^T a P_i > 0$. Now consider any two interior points, P_i and P_j. Since $P_1{}^T a P_i$ and $P_1{}^T a P_j$ are both positive, as we have just shown, it follows from Theorem 3 that $P_i{}^T a P_j$ is also positive, and our proof is complete.

In the rest of our work in hyperbolic geometry, we shall always suppose that the coordinates of all interior points have been adjusted so that for any two such points, P_i and P_j, we have $P_i{}^T a P_j > 0$.

Theorem 5

If p_1 and p_2 are two lines which intersect in an interior point, then $F_{11}F_{22} - F_{12}{}^2 > 0$.

Proof   Let $p_1: P_1P_3$ and $p_2: P_2P_3$ be two lines intersecting in the interior point P_3. Then by Eq. (9), Sec. 10.5, we have $F_{11}F_{22} - F_{12}{}^2 = |f_{ij}|f_{33}$. Since $|f_{ij}| = |t|^2|a|$ and $|a| > 0$, it follows that $|f_{ij}| > 0$. Moreover, $f_{33} > 0$, since P_3 is an interior point. Hence $F_{11}F_{22} - F_{12}{}^2 > 0$, as asserted.

Using Theorems 1 and 4, we can now define the distance between any two interior points, P_i and P_j, in the hyperbolic plane. For from Theorem

1, we have

$$f_{ii}f_{jj} - f_{ij}{}^2 < 0 \qquad \text{or} \qquad \frac{f_{ij}{}^2}{f_{ii}f_{jj}} > 1$$

Also, from Theorem 4, we know that $f_{ij} > 0$, and therefore

$$\frac{f_{ij}}{\sqrt{f_{ii}f_{jj}}} > 1$$

It is thus possible to define a positive number θ by the equation

$$\cosh \theta = \frac{f_{ij}}{\sqrt{f_{ii}f_{jj}}}$$

Moreover, from Corollary 2, Theorem 3, Sec. 10.4, we have

$$(P_iP_j) = k_d \ln \left[\frac{-f_{ij} - \sqrt{f_{ij}{}^2 - f_{ii}f_{jj}}}{-f_{ij} + \sqrt{f_{ij}{}^2 - f_{ii}f_{jj}}} \right]$$

$$= k_d \ln \left[\frac{\dfrac{-f_{ij}}{\sqrt{f_{ii}f_{jj}}} - \sqrt{\dfrac{f_{ij}{}^2}{f_{ii}f_{jj}} - 1}}{\dfrac{-f_{ij}}{\sqrt{f_{ii}f_{jj}}} + \sqrt{\dfrac{f_{ij}{}^2}{f_{ii}f_{jj}} - 1}} \right]$$

$$= k_d \ln \left[\frac{- \cosh \theta - \sinh \theta}{- \cosh \theta + \sinh \theta} \right]$$

$$= k_d \ln \left(\frac{-e^{\theta}}{-e^{-\theta}} \right) = k_d \ln e^{2\theta} = k_d(2\theta + 2ni\pi)$$

or, choosing $k_d = \frac{1}{2}$, we have $(P_iP_j) = \theta + ni\pi$. Of the various values of the distance (P_iP_j), only one, corresponding to $n = 0$, is real. This is in sharp contrast to the case in elliptic geometry, where (P_iP_j) has infinitely many real values. Summarizing, then, we have the following theorem.

Theorem 6

In hyperbolic geometry, with metric gauge conic $X^TaX = 0$, the distance between any two (interior) points

$$P_i = \left\| \begin{matrix} x_i \\ y_i \\ z_i \end{matrix} \right\| \qquad \text{and} \qquad P_j = \left\| \begin{matrix} x_j \\ y_j \\ z_j \end{matrix} \right\|$$

is equal to θ, where

$$\cosh \theta = \frac{f_{ij}}{\sqrt{f_{ii}f_{jj}}} \qquad \sinh \theta = \sqrt{\frac{f_{ij}{}^2 - f_{ii}f_{jj}}{f_{ii}f_{jj}}} \qquad f_{ij} = P_i{}^TaP_j$$

To obtain the formula for measuring the angle between two lines, p_i and p_j, we use the result of Theorem 5, namely, $F_{ii}F_{jj} - F_{ij}^2 > 0$, from which it follows that

$$0 < \frac{F_{ij}^2}{F_{ii}F_{jj}} < 1 \qquad \text{or} \qquad -1 < \frac{F_{ij}}{\sqrt{F_{ii}F_{jj}}} < 1$$

We can therefore define a quantity ϕ by the equation

(1)
$$\cos \phi = \frac{-F_{ij}}{\sqrt{F_{ii}F_{jj}}} \qquad -0 < \phi < \pi$$

Moreover, from Corollary 2, Theorem 4, Sec. 10.4, we have

$$[p_i p_j] = k_a \ln \left[\frac{-F_{ij} - \sqrt{F_{ij}^2 - F_{ii}F_{jj}}}{-F_{ij} + \sqrt{F_{ij}^2 - F_{jj}}} \right]$$

Hence, dividing numerator and denominator of the argument of the logarithm by $\sqrt{F_{ii}F_{jj}}$, and then using (1), we find

$$[p_i p_j] = k_a \ln \frac{\cos \phi - i \sin \phi}{\cos \phi + i \sin \phi}$$

$$= k_a \ln \left(\frac{e^{-i\phi}}{e^{i\phi}} \right) = k_a \ln e^{-2i\phi} = k_a i(-2\phi + 2n\pi)$$

In order for the angle $[p_i p_j]$ to be real, k_a must be a pure imaginary, and for convenience we take it to be $-1/2i$, giving us

$$[p_i p_j] = \phi - n\pi = \phi + m\pi$$

Here, as in both elliptic geometry and elementary geometry, the angle between two lines has infinitely many values. Summarizing, then, we have the following theorem.

Theorem 7

In hyperbolic geometry, with metric gauge conic $X^T a X = 0$, the angle between any two (interior) lines,

$$p_i = \left\| \begin{array}{c} l_i \\ m_i \\ n_i \end{array} \right\| \qquad \text{and} \qquad p_j = \left\| \begin{array}{c} l_j \\ m_j \\ n_j \end{array} \right\|$$

is equal to $\phi + m\pi$, where

$$\cos \phi = \frac{-F_{ij}}{\sqrt{F_{ii}F_{jj}}} \qquad \sin \phi = \sqrt{\frac{F_{ii}F_{jj} - F_{ij}^2}{F_{ii}F_{jj}}} \qquad F_{ij} = p_i^T A p_j$$

and A is the adjoint of the matrix of the metric gauge conic.

Again, exactly as in our discussion of the general triangle in elliptic geometry in Sec. 10.6, we note that if p_i and p_j are the sides of a triangle determining the angle α_k, and if the coordinates of p_i and p_j are defined cyclically from the coordinates of the vertices P_i, P_j, P_k, as shown in Fig. 10.3, then α_k is defined not by the formula of Theorem 7 but rather by the formula

$$\cos \alpha_k = \frac{F_{ij}}{\sqrt{F_{ii}F_{jj}}}$$

It is convenient now before proving the fundamental laws for the general triangle, $\triangle P_1P_2P_3$ with sides d_1, d_2, d_3 and angles α_1, α_2, α_3, in the hyperbolic plane, to collect the formulas we shall need for this purpose:

$$\cosh d_i = \frac{f_{jk}}{\sqrt{f_{jj}f_{kk}}} \qquad \sinh d_i = \sqrt{\frac{f_{jk}^2 - f_{jj}f_{kk}}{f_{jj}f_{kk}}}$$

$$= \sqrt{\frac{-F_{ii}}{f_{jj}f_{kk}}} \qquad \text{by (6) Sec. 10.5}$$

$$\cos \alpha_i = \frac{F_{jk}}{\sqrt{F_{jj}F_{kk}}} \qquad \sin \alpha_i = \sqrt{\frac{F_{jj}F_{kk} - F_{jk}^2}{F_{jj}F_{kk}}}$$

$$= \sqrt{\frac{f_{ii}\,|f_{ij}|}{F_{jj}F_{kk}}} \qquad \text{by (9) Sec. 10.5}$$

With these, we can now prove the following fundamental results very much as we proved the corresponding theorems in elliptic geometry.

Theorem 8

In a triangle in the hyperbolic plane with sides d_1, d_2, d_3 and angles α_1, α_2, α_3,

$$\cosh d_i = \cosh d_j \cosh d_k - \sinh d_j \sinh d_k \cos \alpha_i$$

Proof Using the formulas (and their obvious permutations) from the list we have just compiled, we have for the right-hand side of the formula given in the theorem

$$\frac{f_{ik}}{\sqrt{f_{ii}f_{kk}}} \frac{f_{ij}}{\sqrt{f_{ii}f_{jj}}} - \sqrt{\frac{-F_{jj}}{f_{ii}f_{kk}}} \sqrt{\frac{-F_{kk}}{f_{ii}f_{jj}}} \frac{F_{jk}}{\sqrt{F_{jj}F_{kk}}}$$

$$= \frac{f_{ij}f_{ik} - F_{jk}}{f_{ii}\sqrt{f_{jj}f_{kk}}} = \frac{f_{ii}f_{ik} - (f_{ij}f_{ik} - f_{jk}f_{ii})}{f_{ii}\sqrt{f_{jj}f_{kk}}}$$

$$= \frac{f_{jk}}{\sqrt{f_{jj}f_{kk}}} = \cosh d_i$$

as asserted.

Theorem 9

In a triangle in the hyperbolic plane with sides d_1, d_2, d_3 and angles α_1, α_2, α_3,

$$\cos \alpha_i = - \cos \alpha_j \cos \alpha_k + \sin \alpha_j \sin \alpha_k \cosh d_i$$

Proof Working with the right-hand side of the formula in the theorem, just as in the proof of the preceding theorem, we have

$$- \frac{F_{ik}}{\sqrt{F_{ii}F_{kk}}} \frac{F_{ij}}{\sqrt{F_{ii}F_{jj}}} + \sqrt{\frac{f_{jj}\,|f_{ij}|}{F_{ii}F_{kk}}} \sqrt{\frac{f_{kk}\,|f_{ij}|}{F_{ii}F_{jj}}} \frac{f_{jk}}{\sqrt{f_{jj}f_{kk}}}$$

$$= \frac{-F_{ik}F_{ij} + f_{jk}\,|f_{ij}|}{-F_{ii}\sqrt{F_{jj}F_{kk}}} \qquad \text{since } F_{ii} < 0$$

$$= \frac{-F_{ik}F_{ij} + (F_{ik}F_{ij} - F_{jk}F_{ii})}{-F_{ii}\sqrt{F_{jj}F_{kk}}}$$

$$= \frac{F_{jk}}{\sqrt{F_{jj}F_{kk}}} = \cos \alpha_i$$

as asserted.

Theorem 10

In a triangle in the hyperbolic plane with sides d_1, d_2, d_3 and angles α_1, α_2, α_3,

$$\frac{\sin \alpha_1}{\sinh d_1} = \frac{\sin \alpha_2}{\sinh d_2} = \frac{\sin \alpha_3}{\sinh d_3} = \sqrt{\frac{f_{11}f_{22}f_{33}\,|f_{ij}|}{-F_{11}F_{22}F_{33}}}$$

Proof Evaluating the typical fraction in the formula of the theorem, we have

$$\frac{\sin \alpha_i}{\sinh d_i} = \frac{\sqrt{\dfrac{f_{ii}\,|f_{ij}|}{F_{jj}F_{kk}}}}{\sqrt{\dfrac{-F_{ii}}{f_{jj}f_{kk}}}}$$

$$= \sqrt{\frac{f_{ii}f_{jj}f_{kk}\,|f_{ij}|}{-F_{ii}F_{jj}F_{kk}}}$$

as asserted.

Theorem 11

In hyperbolic geometry, the sum of the measures of the angles of any triangle is less than π.

Proof In a triangle with sides d_1, d_2, d_3 and angles α_1, α_2, α_3, let the notation be chosen so that $\pi > \alpha_1 \geqq \alpha_2 \geqq \alpha_3 > 0$ and therefore

$$(2) \qquad\qquad -\pi < \alpha_1 - \alpha_2 - \alpha_3 < \pi$$

Now by Theorem 9, we have

$$\cos \alpha_1 = - \cos \alpha_2 \cos \alpha_3 + \sin \alpha_2 \sin \alpha_3 \cosh d_1$$

and, from trigonometry,

$$\cos (\alpha_2 + \alpha_3) = \cos \alpha_2 \cos \alpha_3 - \sin \alpha_2 \sin \alpha_3$$

Adding the last two equations, we obtain

$$\cos \alpha_1 + \cos (\alpha_2 + \alpha_3) = \sin \alpha_2 \sin \alpha_3 (\cosh d_1 - 1)$$

or, converting the left-hand side to a product,

$$2 \cos \frac{\alpha_1 + \alpha_2 + \alpha_3}{2} \cos \frac{\alpha_1 - \alpha_2 - \alpha_3}{2} = \sin \alpha_2 \sin \alpha_3 (\cosh d_1 - 1)$$

The right-hand side of the last equation is clearly positive. Moreover, from (2) it is clear that $\cos [(\alpha_1 - \alpha_2 - \alpha_3)/2] > 0$. Hence $\cos [(\alpha_1 + \alpha_2 + \alpha_3)/2]$ must be positive, which implies that

$$0 < \frac{\alpha_1 + \alpha_2 + \alpha_3}{2} < \frac{\pi}{2} \qquad \text{or} \qquad \alpha_1 + \alpha_2 + \alpha_3 < \pi$$

as asserted.

It is interesting to note that in hyperbolic geometry, the assertions of Theorems 8 to 10 reduce, respectively, to the euclidean law of cosines, the euclidean angle-sum theorem, and the euclidean law of sines as the lengths d_1, d_2, and d_3 all approach zero. Thus, *for arbitrarily small triangles, the formulas of hyperbolic geometry are arbitrarily close to the corresponding laws of euclidean geometry*, or, to put it another way, *the geometry of infinitesimal regions of the hyperbolic plane is approximately euclidean.*

Analogous to Theorems 9 to 11 of Sec. 10.6, we have the following results, whose proofs we leave as exercises.

Theorem 12

In hyperbolic geometry, the length of the arc of a sector of central angle ϕ in a circle of radius r is equal to $(\sinh r)\phi$.

Theorem 13

In hyperbolic geometry, the area of a sector of central angle ϕ in a circle of radius r is equal to $(\cosh r - 1)\phi$.

Theorem 14

In hyperbolic geometry, the area of a triangle whose angles are α_1, α_2, α_3 is equal to $\pi - \alpha_1 - \alpha_2 - \alpha_3$.

In the last section, we found that the formulas of elliptic geometry were also formulas describing the geometry on the surface of a euclidean sphere, once lines on the surface of the sphere were suitably defined. Similarly, there is a surface on which the geometry, once lines have been suitably defined, is described by the formulas of hyperbolic geometry. Unfortunately, this surface, the so-called **pseudo sphere** defined by the cartesian equation

$$z = r\left(\text{sech}^{-1} \sqrt{\frac{x^2 + y^2}{r^2}} - \sqrt{1 - \frac{x^2 + y^2}{r^2}} \right)$$

is quite unfamiliar to us, and its properties can be explored only by methods beyond the scope of this book.

Exercises

1. Prove the following formulas for a right triangle in the hyperbolic plane if α and β are the angles opposite the sides a and b and if h is the hypotenuse of the triangle:

(a) $\sinh b = \sinh h \sin \beta$
(c) $\cos \alpha = \sin \beta \cosh a$
(b) $\cosh h = \cosh a \cosh b$
(d) $\cos \alpha \cos \beta = \sin \alpha \sin \beta \cosh h$

2. Solve the following right triangles:

(a) $\alpha = \pi/6$, $h = 1$
(c) $a = 1$, $h = 2$
(b) $a = 1$, $b = 2$
(d) $a = 1$, $\beta = \pi/3$

3. Solve the following oblique triangles:

(a) $d_1 = d_2 = 2$, $d_3 = 1$
(b) $d_1 = 1$, $d_2 = 2$, $\alpha_3 = 3\pi/4$

4. Prove that the s,s,s congruence theorem for triangles is valid in hyperbolic geometry.

5. Is the a,s,a congruence theorem for triangles valid in hyperbolic geometry?

6. Is the s,a,s congruence theorem for triangles valid in hyperbolic geometry?

7. Prove that in the hyperbolic plane if two triangles are similar, they are congruent.

8. Prove that in hyperbolic geometry if two sides of a triangle are equal, the angles opposite these sides are equal, and conversely.

9. Prove that in any triangle in the hyperbolic plane if $\alpha_i > \alpha_j$, then $d_i > d_j$, and conversely.

10. Given $P_1:(1,1,0)$, $P_2:(0,2,1)$, $P_3:(1,-1,2)$. Determine the sides and angles of $\triangle P_1 P_2 P_3$ if the equation of the metric gauge conic is:

(a) $x_1{}^2 + x_2{}^2 - x_3{}^2 = 0$

(b) $x_1{}^2 - 2x_2 x_3 = 0$

(c) $x_2 x_3 + x_3 x_1 + x_1 x_2 = 0$

(d) $x_1{}^2 - x_2{}^2 + 2x_1 x_3 = 0$

11. Prove Theorem 12.

12. Prove Theorem 13.

13. Prove Theorem 14.

14. How are distances and angles in the hyperbolic plane affected by a transformation, $T: X' = MX$, which leaves the metric gauge conic invariant?

15. If $X^T aX = 0$ is a nonsingular real conic which contains at least one real point, show that there are real points at which $X^T aX$ is positive and real points at which $X^T aX$ is negative. *Hint:* Let P be a point at which $X^T aX = 0$, let Q be a point at which $X^T aX \neq 0$, and consider the possible values of $(\lambda P + \mu Q)^T a(\lambda P \overset{\text{\tiny .}}{+} \mu Q)$.

10.9 Conclusion

In reflecting on the derivation of elliptic and hyperbolic geometry from projective geometry, it is important to bear in mind that as far as their geometric properties are concerned, the points of the projective plane are all alike, and so too are its lines. Distinctions between real and complex elements, or between interior and exterior elements, are quite arbitrary and reflect no intrinsic differences. By changing from one coordinate system to another, points that were real may be made complex, and vice versa. And depending on the choice of metric gauge conic, a point may be either interior or exterior, or the distinction may be meaningless; two lines may or may not be perpendicular, a triangle may or may not be isosceles, or may or not have its angle sum less than π.

In the development of elliptic geometry, a metric gauge conic containing no real points is required. As a consequence, the isotropic points on every line in the elliptic plane are strictly complex; i.e., they do not exist in the elliptic plane itself. Thus there are no infinitely distant real points, and, as we saw in Sec. 10.6, there is an upper bound to the principal value of the distance between two points. Because of this, together with the fact that to the principal value of any distance one may add any integral multiple of π,[†] it is sometimes said that *in the elliptic plane distance is finite but unbounded.*

† Or some other "natural" constant, depending on the choice of the scale constant k_d.

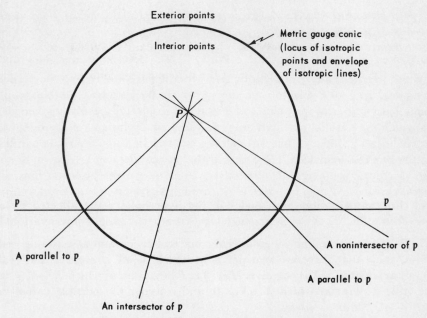

Exterior points

Interior points

Metric gauge conic
(locus of isotropic
points and envelope
of isotropic lines)

P

p

p

A nonintersector of p

A parallel to p

A parallel to p

An intersector of p

Fig. 10.9   A suggestive representation of the hyperbolic plane.

On the other hand, the development of hyperbolic geometry requires a metric gauge conic containing real points. Moreover, it is easy to show that every interior line, i.e., every line in the hyperbolic plane, intersects the metric gauge conic in distinct real points. Hence every line in the hyperbolic plane contains two distinct infinitely distant points and, consequently, pairs of points whose distance apart is arbitrarily large. In hyperbolic geometry the (real) distance between two points is unique rather than infinitely multiple-valued, as in elliptic geometry.

In the projective plane, two lines always have a unique intersection. If the lines are real, it is clear from the algebra of simultaneous linear equations that their intersection is also real. Hence, since the elliptic plane consists, by definition, of *all* real elements in the projective plane, it follows that in elliptic geometry two lines always intersect. In other words, *in elliptic geometry there are no parallel lines.*

On the other hand, not all real points of the coordinatized projective plane belong to the hyperbolic plane, but only those which are interior points. Hence the (real) intersection of two lines of the hyperbolic plane may or may not be a point of that plane. If such an intersection is an isotropic point, i.e., a point of the metric gauge conic, the lines are said to be parallel since they "intersect at infinity." If such an intersection is an exterior point, the lines are said to be nonintersectors. Since, as we noted above, every hyperbolic line contains two real isotropic points, it follows that *in hyperbolic geometry, two distinct lines can be drawn parallel to a given line through a point not on that line*. These properties of the hyperbolic plane are illustrated in Fig. 10.9.

Historically, both elliptic and hyperbolic geometry were developed independently of projective geometry by mathematicians working with modified versions of the axioms of euclidean geometry, specifically, versions in which the parallel postulate was contradicted. If, instead of assuming, as in euclidean geometry, that through a point not on a line a unique parallel to the line can be drawn, one assumed that *no* parallel can be drawn, one is led to elliptic geometry. If, alternatively, one adopts the assertion that *two* parallels can be drawn to a given line through a given point not on that line, one is led to hyperbolic geometry.

eleven

SINGULAR METRIC GAUGES

11.1 Introduction

In the last chapter we observed that in both elliptic and hyperbolic geometry, the properties of infinitesimal triangles were arbitrarily close to those of triangles in euclidean geometry. This suggests that in some sense, euclidean geometry is the limiting form of both elliptic geometry and hyperbolic geometry, and in this chapter we shall explore this possibility.

Naturally, if we are to obtain a specialization of the projective plane which will have the familiar euclidean properties, our analysis must involve only real elements. However, the only possibilities for a real, nonsingular metric gauge conic have already been explored. For such a conic either contains no real points or at least one real point, and, as we have seen, the first case leads to elliptic geometry and the second to hyperbolic geometry. It appears, therefore, that as our metric gauge conic we must choose a real conic which is singular. In fact, guided by the euclidean parallel postulate, which implies that there is just one isotropic point on each line, it seems that we must choose a conic consisting of a repeated real line as our metric gauge locus. However, this presents one obvious problem, because if the metric gauge points, M_1 and M_2, on a general line are coincident, then for any points, P and Q, we have $\mathrm{R}(PQ, M_1M_2) = \mathrm{R}(PQ, M_1M_1) = 1$, and $(PQ) = k_d \ln \mathrm{R}(PQ, M_1M_2) = 0$, regardless of whether the points P and Q are distinct or coincident!

To avoid this difficulty, we shall use instead of a single metric gauge conic, a pencil of such conics, $C_1 + \lambda C_2 = 0$, so constructed that the general member of the pencil is real and nonsingular, while the particular member, $C_1 = 0$, corresponding to $\lambda = 0$, consists of a repeated real line. Hopefully, letting $\lambda \to 0$ at the appropriate point in our work will then give us the euclidean structure we are seeking.

This is still not the whole story, however, because it is necessary to distinguish two cases, according as the intersections of the repeated line $C_1 = 0$ and the conic $C_2 = 0$ are or are not real. In the latter case, the metric gauge envelope consists of two pencils of *complex* lines on the (complex) intersections of $C_1 = 0$ and $C_2 = 0$, and the analysis does indeed lead to euclidean geometry. In the former case, the metric gauge envelope consists of two pencils of *real* lines on the (real) intersections of $C_1 = 0$ and $C_2 = 0$, and the analysis leads to a curious variation of euclidean geometry which has been of some interest in the theory of relativity.

11.2 The Measurement of Distance

Let $X^T a X = 0$ be the equation of a real conic consisting of a line $\rho : \alpha x + \beta y + \gamma z = 0$ counted twice. Then, perforce, since $X^T a X \equiv \rho^2 = 0$, we have

$$(1) \qquad a = \begin{Vmatrix} \alpha^2 & \alpha\beta & \alpha\gamma \\ \beta\alpha & \beta^2 & \beta\gamma \\ \gamma\alpha & \gamma\beta & \gamma^2 \end{Vmatrix}$$

Furthermore, let $X^T b X = 0$ be the equation of a real conic which intersects the line ρ in two conjugate complex points, I and J. Finally, let us take $X^T(a + \lambda b)X = 0$ as the metric gauge conic. We propose to investigate the geometry imposed on the real elements in the projective plane by this choice of metric when $\lambda \to 0$ through either positive values or negative values. Preparatory to this we must first establish several preliminary results.

Lemma 1

If $X^T a X = 0$ is the equation of a conic consisting of a line $\rho : \alpha x + \beta y + \gamma z = 0$ counted twice, and if $P_1 = \begin{Vmatrix} x_1 \\ y_1 \\ z_1 \end{Vmatrix}$ and $P_2 = \begin{Vmatrix} x_2 \\ y_2 \\ z_2 \end{Vmatrix}$ are any two points, then $P_i^T a P_i = \rho_i^2$, $i = 1,\ 2$, and $P_1^T a P_2 = \rho_1 \rho_2$, where $\rho_i = \alpha x_i + \beta y_i + \gamma z_i$.

Proof The first assertion of the lemma follows immediately from the fact that, by hypothesis, $X^T a X = \rho^2$. The second assertion follows similarly from the fact, easily verified by a trivial calculation, that $P_i^T a X = \rho_i \rho$.

Lemma 2

If $X^T a X = 0$ is the equation of a conic consisting of a line $\rho : \alpha x + \beta y + \gamma z = 0$ counted twice, and if $P_1 = \left\| \begin{matrix} x_1 \\ y_1 \\ z_1 \end{matrix} \right\|$ and $P_2 = \left\| \begin{matrix} x_2 \\ y_2 \\ z_2 \end{matrix} \right\|$ are any two points, then $(P_1^T a P_2)^2 = (P_1^T a P_1)(P_2^T a P_2)$.

Proof The assertion of the lemma follows immediately from Lemma 1, since $P_1^T a P_2 = \rho_1 \rho_2$, $P_1^T a P_1 = \rho_1^2$, and $P_2^T a P_2 = \rho_2^2$.

Lemma 3

If $P_1 = \left\| \begin{matrix} x_1 \\ y_1 \\ z_1 \end{matrix} \right\|$, $P_2 = \left\| \begin{matrix} x_2 \\ y_2 \\ z_2 \end{matrix} \right\|$, and $P_3 = \left\| \begin{matrix} x_3 \\ y_3 \\ z_3 \end{matrix} \right\|$ are the vertices of a triangle whose sides are $p_1 = \left\| \begin{matrix} l_1 \\ m_1 \\ n_1 \end{matrix} \right\|$, $p_2 = \left\| \begin{matrix} l_2 \\ m_2 \\ n_2 \end{matrix} \right\|$, and $p_3 = \left\| \begin{matrix} l_3 \\ m_3 \\ n_3 \end{matrix} \right\|$, then $\rho_j P_i - \rho_i P_j = D p_k$, $i,\ j,\ k$ a cyclic permutation of 1, 2, 3, where $\rho_i = \alpha x_i + \beta y_i + \gamma z_i$ and $D = \left\| \begin{matrix} 0 & -\gamma & \beta \\ \gamma & 0 & -\alpha \\ -\beta & \alpha & 0 \end{matrix} \right\|$.

Proof If $P_1 = \left\| \begin{matrix} x_1 \\ y_1 \\ z_1 \end{matrix} \right\|$, $P_2 = \left\| \begin{matrix} x_2 \\ y_2 \\ z_2 \end{matrix} \right\|$, and $P_3 = \left\| \begin{matrix} x_3 \\ y_3 \\ z_3 \end{matrix} \right\|$ are the vertices of a triangle, then under the cyclic convention we adopted in Sec. 10.6, the coordinate-vector of the line $p_k = P_i P_j$ is

$$p_k = \left\| \begin{matrix} l_k \\ m_k \\ n_k \end{matrix} \right\| = \left\| \begin{matrix} y_i z_j - y_j z_i \\ z_i x_j - z_j x_i \\ x_i y_j - x_j y_i \end{matrix} \right\|$$

Hence

$$Dp_k = \begin{Vmatrix} 0 & -\gamma & \beta \\ \gamma & 0 & -\alpha \\ -\beta & \alpha & 0 \end{Vmatrix} \cdot \begin{Vmatrix} y_i z_j - y_j z_i \\ z_i x_j - z_j x_i \\ x_i y_j - x_j y_i \end{Vmatrix}$$

$$= \begin{Vmatrix} -\gamma(z_i x_j - z_j x_i) + \beta(x_i y_j - x_j y_i) \\ \gamma(y_i z_j - y_j z_i) - \alpha(x_i y_j - x_j y_i) \\ -\beta(y_i z_j - y_j z_i) + \alpha(z_i x_j - z_j x_i) \end{Vmatrix}$$

$$= \begin{Vmatrix} x_i(\alpha x_j + \beta y_j + \gamma z_j) - x_j(\alpha x_i + \beta y_i + \gamma z_i) \\ y_i(\alpha x_j + \beta y_j + \gamma z_j) - y_j(\alpha x_i + \beta y_i + \gamma z_i) \\ z_i(\alpha x_j + \beta y_j + \gamma z_j) - z_j(\alpha x_i + \beta y_i + \gamma z_i) \end{Vmatrix}$$

$$= \rho_j P_i - \rho_i P_j$$

as asserted.

Lemma 4

If $P_1 = \begin{Vmatrix} x_1 \\ y_1 \\ z_1 \end{Vmatrix}$, $P_2 = \begin{Vmatrix} x_2 \\ y_2 \\ z_2 \end{Vmatrix}$, $P_3 = \begin{Vmatrix} x_3 \\ y_3 \\ z_3 \end{Vmatrix}$ are the vertices of a triangle

whose sides are $p_1 = \begin{Vmatrix} l_1 \\ m_1 \\ n_1 \end{Vmatrix}$, $p_2 = \begin{Vmatrix} l_2 \\ m_2 \\ n_2 \end{Vmatrix}$, $p_3 = \begin{Vmatrix} l_3 \\ m_3 \\ n_3 \end{Vmatrix}$, then the coordinate-

vector of the intersection of $\rho: \alpha x + \beta y + \gamma z = 0$ and the line $P_i P_j$ is Dp_k,

where $D = \begin{Vmatrix} 0 & -\gamma & \beta \\ \gamma & 0 & -\alpha \\ -\beta & \alpha & 0 \end{Vmatrix}$.

Proof Since the coordinates of $P_i P_j = p_k$ are $[l_k, m_k, n_k]$, the equation of $P_i P_j$ can be written $l_k x + m_k y + n_k z = 0$. Hence the coordinate-vector of the intersection of ρ and the line $P_i P_j$ is

$$\begin{Vmatrix} -\gamma m_k + \beta n_k \\ \gamma l_k - \alpha n_k \\ -\beta l_k + \alpha m_k \end{Vmatrix} = \begin{Vmatrix} 0 & -\gamma & \beta \\ \gamma & 0 & -\alpha \\ -\beta & \alpha & 0 \end{Vmatrix} \cdot \begin{Vmatrix} l_k \\ m_k \\ n_k \end{Vmatrix} = Dp_k$$

as asserted.

Now let $P_1 = \begin{Vmatrix} x_1 \\ y_1 \\ z_1 \end{Vmatrix}$ and $P_2 = \begin{Vmatrix} x_2 \\ y_2 \\ z_2 \end{Vmatrix}$ be any two ordinary real

points, i.e., any real points not on the metric gauge conic, $X^T(a + \lambda b)X = 0$. Then from Corollary 1, Theorem 3, Sec. 10.4, we have

$$\sinh^2 \frac{(P_1 P_2)}{2k_d} = \frac{[P_1{}^T(a + \lambda b)P_2]^2 - [P_1{}^T(a + \lambda b)P_1][P_2{}^T(a + \lambda b)P_2]}{[P_1{}^T(a + \lambda b)P_1][P_2{}^T(a + \lambda b)P_2]}$$

or, expanding, simplifying by means of Lemmas 1 and 2, and then dividing by $-\lambda$,

$$\frac{\sinh^2\left[(P_1P_2)/2k_d\right]}{-\lambda} = \frac{-2\rho_1\rho_2P_1{}^Tb P_2 + \rho_1{}^2P_2{}^Tb P_2 + \rho_2{}^2P_1{}^Tb P_1 + \lambda(\cdot\cdot\cdot)}{\rho_1{}^2\rho_2{}^2 + \lambda(\cdot\cdot\cdot)}$$

If $\lambda > 0$, say $\lambda = R^2$, we take $k_d = 1/2iR$. Then

$$\frac{\sinh^2\left[(P_1P_2)/2k_d\right]}{-\lambda} = (P_1P_2)^2\frac{\sinh^2\left[iR(P_1P_2)\right]}{-R^2(P_1P_2)^2} = (P_1P_2)^2\left[\frac{i\sin R(P_1P_2)}{iR(P_1P_2)}\right]^2$$

and this approaches $(P_1P_2)^2$ as R, and therefore λ, approaches 0. On the other hand, if $\lambda < 0$, say $\lambda = -R^2$, we take $k_d = 1/2R$. Then

$$\frac{\sinh^2\left[(P_1P_2)/2k_d\right]}{-\lambda} = (P_1P_2)^2\left[\frac{\sinh R(P_1P_2)}{R(P_1P_2)}\right]^2$$

and again this approaches $(P_1P_2)^2$ as R, and therefore λ, approaches 0. Hence, whether λ is positive or negative, it follows that as $\lambda \to 0$,

$$(2) \qquad (P_1P_2)^2 = \frac{\rho_1{}^2P_2{}^Tb P_2 - 2\rho_1\rho_2P_1{}^Tb P_2 + \rho_2{}^2P_1{}^Tb P_1}{\rho_1{}^2\rho_2{}^2}$$

where $\rho_i = \alpha x_i + \beta y_i + \gamma z_i$.

Now if we remember that ρ_i is just a scalar, the expression for $(P_1P_2)^2$ can be written

$$(3) \qquad (P_1P_2)^2 = \frac{(\rho_2P_1 - \rho_1P_2)^Tb(\rho_2P_1 - \rho_1P_2)}{\rho_1{}^2\rho_2{}^2}$$

Furthermore, if we introduce the notation $p_3 = P_1P_2$ in anticipation of our real interest in P_1P_2 as one side of a general triangle, then by Lemma 3 the last expression becomes

$$(4) \qquad (P_1P_2)^2 = \frac{(Dp_3)^Tb(Dp_3)}{\rho_1{}^2\rho_2{}^2} = \frac{p_3{}^T(D^Tb D)p_3}{\rho_1{}^2\rho_2{}^2}$$

Equations (2), (3), and (4) all confront us with the following apparent paradox. Although the matrix b disappears from the equation of the metric gauge conic as $\lambda \to 0$, it nonetheless appears to play a permanent and decisive role in the calculation of distance! In other words, although it is our intuitive feeling that distances should be defined to within an arbitrary scale constant by the metric gauge conic, which in the limit consists of the line ρ and the pencils on the points I and J independent of which conic $X^TbX = 0$ determines I and J on ρ, the matrix b still seems to enter into the formula for distance in a nonproportional way.

To resolve this paradox, it is convenient to think of the line ρ as being parametrized in terms of I and J as base points. Then the intersection,

Q, of $p_3 = P_1P_2$ and ρ can be written in the form $Q = \mu I + \nu J$. Moreover, by Lemma 4 the coordinate-vector of this intersection is precisely the matrix Dp_3; that is, $\mu I + \nu J = Dp_3$. Hence, substituting into Eq. (4), we have

$$(5) \qquad (P_1P_2)^2 = \frac{(\mu I + \nu J)^T b(\mu I + \nu J)}{\rho_1{}^2\rho_2{}^2}$$

$$= \frac{2\mu\nu I^T bJ}{\rho_1{}^2\rho_2{}^2}$$

since $I^T bI = J^T bJ = 0$ because both I and J are points of $X^T bX = 0$. Now I, J, and $I^T bJ$ are clearly independent of P_1 and P_2. Hence from Eq. (5) it follows that the matrix b does indeed appear in the formula for distance only as a matric factor in the scalar constant $I^T bJ$.

Since we want all distances to be real, it is necessary to verify that the right-hand side of Eq. (5) is nonnegative. To do this, we observe first that since I and J are the intersections of the real line ρ and the real conic $X^T bX = 0$, therefore the coordinates of J are, respectively, the complex conjugates of the coordinates of I; that is, $J = \bar{I}$. Furthermore, since the intersection, Q, of the real line P_1P_2 and the real line ρ is necessarily real, it follows that the expression $Q = \mu I + \nu J = \mu I + \nu \bar{I}$ is equal to its complex conjugate; that is, $\mu I + \nu \bar{I} = \bar{\mu}\bar{I} + \bar{\nu}I$. This equation can always be satisfied by taking $\nu = \bar{\mu}$, and this we shall suppose done. Finally, since $X^T bX = 0$ is either a nonsingular conic or a singular conic consisting of distinct lines, it follows that J cannot lie on the polar of I with respect to $X^T bX = 0$, and hence $I^T bJ \neq 0$. Therefore, being real, $I^T bJ$ is either positive or can be made positive by multiplying b by -1. Thus since $I^T bJ$ can be supposed positive, and since $\mu\nu = \mu\bar{\mu}$ is always nonnegative, it follows that with suitable conventions the right-hand side of Eq. (5) is always nonnegative, and all distances are real, as required.

Exercises

1. Explain in detail why J cannot lie on the polar of I with respect to $X^T bX = 0$.

2. Why is $I^T bJ$ real?

3. Show that assuming $\nu = \bar{\mu}$ implies no loss of generality.

4. If P_1, P_2, P_3 are, respectively, $(1,0,1)$, $(1,1,0)$, $(1,-2,1)$, find the sides of $\triangle P_1P_2P_3$, given $\rho: y + z = 0$ and $X^T bX = 0: x^2 = yz$.

11.3 The Measurement of Angle

Before angle measurement can be defined, it is necessary that in addition to the equation of the metric gauge locus, $X^T(a + \lambda b)X = 0$, we also have the equation of the metric gauge envelope, $\Lambda^T \operatorname{adj}(a + \lambda b)\Lambda = 0$. Hence we must now compute the adjoint of the matrix

$$a + \lambda b = \begin{Vmatrix} \alpha^2 + \lambda b_{11} & \alpha\beta + \lambda b_{12} & \alpha\gamma + \lambda b_{13} \\ \alpha\beta + \lambda b_{21} & \beta^2 + \lambda b_{22} & \beta\gamma + \lambda b_{23} \\ \alpha\gamma + \lambda b_{31} & \beta\gamma + \lambda b_{32} & \gamma^2 + \lambda b_{33} \end{Vmatrix} \qquad b_{ij} = b_{ji}$$

The·evaluation of the cofactors of this matrix, though tedious, is completely straightforward. In each cofactor, the terms which do not involve λ drop out, and what remains consists of a group of terms multiplied by λ and a second group multiplied by λ^2. The scalar factor λ can therefore be removed from the adjoint matrix and then divided from the equation of the metric gauge envelope. Doing this, and then letting $\lambda \to 0$, we find that the equation of the metric gauge envelope becomes $\Lambda^T M\Lambda = 0$, where M is the matrix

$$\begin{Vmatrix} \beta^2 b_{33} - 2\beta\gamma b_{23} + \gamma^2 b_{22} & -\alpha\beta b_{33} + \alpha\gamma b_{23} + \beta\gamma b_{13} - \gamma^2 b_{12} & \alpha\beta b_{23} - \alpha\gamma b_{22} + \beta\gamma b_{12} - \beta^2 b_{13} \\ -\alpha\beta b_{33} + \alpha\gamma b_{23} + \beta\gamma b_{13} - \gamma^2 b_{12} & \alpha^2 b_{33} - 2\alpha\gamma b_{13} + \gamma^2 b_{11} & \alpha\beta b_{13} + \alpha\gamma b_{12} - \beta\gamma b_{11} - \alpha^2 b_{23} \\ \alpha\beta b_{23} - \alpha\gamma b_{22} + \beta\gamma b_{12} - \beta^2 b_{13} & \alpha\beta b_{13} + \alpha\gamma b_{12} - \beta\gamma b_{11} - \alpha^2 b_{23} & \alpha^2 b_{22} - 2\alpha\beta b_{12} + \beta^2 b_{11} \end{Vmatrix}$$

It is important to note, as direct calculation will verify at once, that

$$M = \begin{Vmatrix} 0 & \gamma & -\beta \\ -\gamma & 0 & \alpha \\ \beta & -\alpha & 0 \end{Vmatrix} \cdot \begin{Vmatrix} b_{11} & b_{12} & b_{13} \\ b_{21} & b_{22} & b_{23} \\ b_{31} & b_{32} & b_{33} \end{Vmatrix} \cdot \begin{Vmatrix} 0 & -\gamma & \beta \\ \gamma & 0 & -\alpha \\ -\beta & \alpha & 0 \end{Vmatrix} \qquad b_{ij} = b_{ji}$$

$$= D^T b D$$

The matrix, M, of the metric gauge envelope is thus the same matrix which appeared in the final form of the distance formula, Eq. (4), Sec. 11.2.

Now the metric gauge lines on the intersection of two arbitrary lines, $p_1 = \begin{Vmatrix} l_1 \\ m_1 \\ n_1 \end{Vmatrix}$ and $p_2 = \begin{Vmatrix} l_2 \\ m_2 \\ n_2 \end{Vmatrix}$, are the lines of the pencil $\mu_1 p_1 + \mu_2 p_2$ whose parameters satisfy the equation

$$(\mu_1 p_1 + \mu_2 p_2)^T M(\mu_1 p_1 + \mu_2 p_2) = 0$$

or

$$\mu_1{}^2 p_1{}^T M p_1 + 2\mu_1\mu_2 p_1{}^T M p_2 + \mu_2{}^2 p_2{}^T M p_2 = 0$$

Since the gauge lines on any real point are clearly complex, it follows that

$$(p_1{}^T M p_2)^2 - (p_1{}^T M p_1)(p_2{}^T M p_2) < 0$$

or

$$0 < \frac{(p_1{}^T M p_2)^2}{(p_1{}^T M p_1)(p_2{}^T M p_2)} < 1$$

It is therefore possible to define a quantity ϕ by the equation

$$(1) \qquad \cos \phi = \frac{p_1^T M p_2}{\sqrt{(p_1^T M p_1)(p_2^T M p_2)}}$$

Moreover, from Corollary 2, Theorem 4, Sec. 10.4, we have

$$[p_1 p_2] = k_a \ln \left[\frac{-p_1^T M p_2 - \sqrt{(p_1^T M p_2)^2 - (p_1^T M p_1)(p_2^T M p_2)}}{-p_1^T M p_2 + \sqrt{(p_1^T M p_2)^2 - (p_1^T M p_1)(p_2^T M p_2)}} \right]$$

Hence, dividing numerator and denominator by $\sqrt{(p_1^T M p_1)(p_2^T M p_2)}$ and then simplifying the result by means of Eq. (1), we have

$$[p_1 p_2] = k_a \ln \left(\frac{-\cos \phi - i \sin \phi}{-\cos \phi + i \sin \phi} \right) = k_a \ln e^{2i\phi} = k_a(2i\phi + 2n\pi i)$$

Thus, if we take $k_a = 1/2i$, the quantity ϕ defined by Eq. (1) is precisely the measure of the angle determined by the lines p_1 and p_2.

Exercises

1. Find the measures of the angles of $\triangle P_1 P_2 P_3$ in Exercise 4, Sec. 11.3.

2. Show that $p^T M p$ has the same sign for all ordinary lines, p, and that without loss of generality it can be taken positive.

11.4 The Laws for a General Triangle

We shall now prove that the geometry arising from the singular metric gauge we have been considering is euclidean by showing that the euclidean cosine law and the euclidean sine law hold for a general triangle.

Let P_1, P_2, P_3 be the vertices of an arbitrary triangle, let $p_1 = P_2 P_3$, $p_2 = P_3 P_1$, $p_3 = P_1 P_2$ be the sides of $\triangle P_1 P_2 P_3$, let $(P_i P_j) = d_k$, and let $[p_i p_j] = \alpha_k$. Then, using Eq. (4), Sec. 11.2, and Eq. (1), Sec. 11.3, we have

$$(1) \quad d_i^2 + d_j^2 - 2 d_i d_j \cos \alpha_k$$

$$= \frac{p_i^T D^T b D p_i}{\rho_j^2 \rho_k^2} + \frac{p_j^T D^T b D p_j}{\rho_i^2 \rho_k^2} - 2 \sqrt{\frac{p_i^T D^T b D p_i}{\rho_j^2 \rho_k^2}} \sqrt{\frac{p_j^T D^T b D p_j}{\rho_i^2 \rho_k^2}}$$

$$\cdot \left(\frac{-p_i^T M p_j}{\sqrt{(p_i^T M p_i)(p_j^T M p_j)}} \right)^{\dagger}$$

$$= \frac{(\rho_i D p_i)^T b (\rho_i D p_i) + (\rho_j D p_j)^T b (\rho_j D p_j) + 2(\rho_i D p_i)^T b (\rho_j D p_j)}{\rho_i^2 \rho_j^2 \rho_k^2}$$

$$= \frac{(\rho_i D p_i + \rho_j D p_j)^T b (\rho_i D p_i + \rho_j D p_j)}{\rho_i^2 \rho_j^2 \rho_k^2}$$

† We must remember, as we observed in Sec. 10.6, that to obtain $\cos \alpha_k$, the sign of $\cos [p_i p_j]$, as given by Eq. (1), Sec. 11.3, must be reversed.

Now using Lemma 3, Sec. 11.2, we have

$$\begin{aligned}
(\rho_i D p_i + \rho_j D p_j) &= \rho_i(\rho_k P_j - \rho_j P_k) + \rho_j(\rho_i P_k - \rho_k P_i) \\
&= -\rho_k(\rho_j P_i - \rho_i P_j) \\
&= -\rho_k D p_k
\end{aligned}$$

Therefore the fraction in the last line of (1) can be written

$$\frac{(-\rho_k D p_k)^T b(-\rho_k D p_k)}{\rho_i{}^2 \rho_j{}^2 \rho_k{}^2} = \frac{p_k{}^T D^T b D p_k}{\rho_i{}^2 \rho_j{}^2} = (P_i P_j)^2 = d_k{}^2$$

which verifies the euclidean law of cosines for $\triangle P_1 P_2 P_3$.

The verification of the law of sines follows immediately from the law of cosines, for

$$\begin{aligned}
\frac{\sin \alpha_i}{d_i} &= \frac{1}{d_i} \sqrt{1 - \left(\frac{d_j{}^2 + d_k{}^2 - d_i{}^2}{2 d_j d_k}\right)^2} \\
&= \sqrt{\frac{-(d_i{}^4 + d_j{}^4 + d_k{}^4) + 2(d_j{}^2 d_k{}^2 + d_i{}^2 d_k{}^2 + d_i{}^2 d_j{}^2)}{4 d_i{}^2 d_j{}^2 d_k{}^2}}
\end{aligned}$$

Since the last fraction is a symmetric function of d_i, d_j, d_k, it follows that

$$\frac{\sin \alpha_i}{d_i} = \frac{\sin \alpha_j}{d_j} = \frac{\sin \alpha_k}{d_k}$$

and the law of sines is established.

The euclidean law of cosines and the euclidean law of sines suffice to prove the euclidean angle-sum theorem. Thus, using the law of cosines and the law of sines, we have

$$\begin{aligned}
- \cos \alpha_j \cos \alpha_k + \sin \alpha_j \sin \alpha_k &= - \cos \alpha_j \cos \alpha_k + d_j d_k \frac{\sin \alpha_j}{d_j} \frac{\sin \alpha_k}{d_k} \\
&= - \frac{d_i{}^2 + d_k{}^2 - d_j{}^2}{2 d_i d_k} \cdot \frac{d_i{}^2 + d_j{}^2 - d_k{}^2}{2 d_i d_j} \\
&\quad + \frac{2(d_j{}^2 d_k{}^2 + d_i{}^2 d_k{}^2 + d_i{}^2 d_j{}^2) - (d_i{}^4 + d_j{}^4 + d_k{}^4)}{4 d_i{}^2 d_j d_k} \\
&= \frac{-d_i{}^2 + d_j{}^2 + d_k{}^2}{2 d_j d_k} = \cos \alpha_i
\end{aligned}$$

Hence $\cos \alpha_i = - \cos(\alpha_j + \alpha_k) = \cos(\pi - \alpha_j - \alpha_k)$, which implies that $\alpha_i = \pi - \alpha_j - \alpha_k$, or $\alpha_i + \alpha_j + \alpha_k = \pi$.

We can now see the possibility of many different euclidean structures. For instance, if we take the metric gauge line, ρ, to be the line $z = 0$ and $X^T b X = 0$ to be the conic $x^2 + y^2 = 0$, then I and J are the points

$(1,i,0)$ and $(1, -i, 0)$, and for the distance formula we have, from Eq. (2), Sec. 11.2,

$$(P_1P_2)^2 = \frac{z_1{}^2(x_2{}^2 + y_2{}^2) - 2z_1z_2(x_1x_2 + y_1y_2) + z_2{}^2(x_1{}^2 + y_1{}^2)}{z_1{}^2z_2{}^2}$$

$$= \frac{(x_1z_2 - x_2z_1)^2 + (y_1z_2 - y_2z_1)^2}{z_1{}^2z_2{}^2}$$

If, for convenience, we let $z_1 = z_2 = 1$, as we can since the euclidean plane consists only of real ordinary points, i.e., real points which do not lie on $z = 0$, we obtain the familiar euclidean distance formula

$$(P_1P_2)^2 = (x_1 - x_2)^2 + (y_1 - y_2)^2$$

Similarly, since $b = \begin{Vmatrix} 1 & 0 & 0 \\ 0 & 1 & 0 \\ 0 & 0 & 0 \end{Vmatrix}$ and $D = \begin{Vmatrix} 0 & -1 & 0 \\ 1 & 0 & 0 \\ 0 & 0 & 0 \end{Vmatrix}$, we

have $M = D^T b D = \begin{Vmatrix} 1 & 0 & 0 \\ 0 & 1 & 0 \\ 0 & 0 & 0 \end{Vmatrix}$. Hence for the measurement of angles, we

have, from Eq. (1), Sec. 11.3, the formula

$$\cos \phi = \frac{l_1l_2 + m_1m_2}{\sqrt{l_1{}^2 + m_1{}^2}\,\sqrt{l_2{}^2 + m_2{}^2}}$$

which we recognize as the euclidean formula for the angle between the lines whose direction numbers are $[l_1, m_1]$ and $[l_2, m_2]$.

On the other hand, if we keep $x^2 + y^2 = 0$ as the conic $X^T b X = 0$, but take $x + y + z = 0$ to be the metric gauge line, ρ, then I and J are the points $(1, i, -1 - i)$ and $(1, -i, -1 + i)$. Since the only points we are concerned with are the real points which are not on $x + y + z = 0$, it is clearly no specialization to assume (analogous to the simplifying assumption $z = 1$ in the preceding discussion) that the coordinates of all points in the new euclidean plane are adjusted so that $x + y + z = 1$. Then from Eq. (2), Sec. 11.2, we obtain again the distance formula

$$(P_1P_2)^2 = (x_1 - x_2)^2 + (y_1 - y_2)^2$$

However, since now

$$D = \begin{Vmatrix} 0 & -1 & 1 \\ 1 & 0 & -1 \\ -1 & 1 & 0 \end{Vmatrix}$$

we have

$$M = D^T b D = \begin{Vmatrix} 1 & 0 & -1 \\ 0 & 1 & -1 \\ -1 & -1 & 2 \end{Vmatrix}$$

and therefore, from Eq. (1), Sec. 11.3, the measure of the angle between two lines is given by the formula

$$\cos \phi = \frac{(l_1 - n_1)(l_2 - n_2) + (m_1 - n_1)(m_2 - n_2)}{\sqrt{(l_1 - n_1)^2 + (m_1 - n_1)^2}\,\sqrt{(l_2 - n_2)^2 + (m_2 - n_2)^2}}$$

It is interesting to analyze in each of these euclidean geometries the triangle whose vertices are $P_1:(0,0,1)$, $P_2:(0,2,1)$, $P_3:(1,0,1)$. In the first of our geometries, the sides of this triangle are obviously

$$(P_1P_2) = 1 \qquad (P_2P_3) = \sqrt{5} \qquad (P_1P_3) = 2$$

and its angles are

$$\angle P_2P_1P_3 = 90° \qquad \angle P_1P_3P_2 = \text{Cos}^{-1}\frac{1}{\sqrt{5}} \qquad \angle P_3P_2P_1 = \text{Cos}^{-1}\frac{2}{\sqrt{5}}$$

Preparatory to using the formulas in our second geometry, we must first "normalize" the coordinates of P_1, P_2, and P_3 by making their sum equal to 1 in each case. This gives us $P_1:(0,0,1)$, $P_2:(0,\frac{2}{3},\frac{1}{3})$, $P_3:(\frac{1}{2},0,\frac{1}{2})$, and then

$$(P_1P_2) = \tfrac{2}{3} \qquad (P_2P_3) = \tfrac{5}{6} \qquad (P_1P_3) = \tfrac{1}{2}$$
$$\angle P_2P_1P_3 = 90° \qquad \angle P_1P_3P_2 = \text{Cos}^{-1}\tfrac{4}{5} \qquad \angle P_3P_2P_1 = \text{Cos}^{-1}\tfrac{3}{5}$$

Thus, even within the framework of euclidean geometry, the lengths and angles of a triangle are not determined solely by the vertices of the triangle but depend also upon the particular euclidean metric gauge being used.

Exercises

1. Find the distance formula and angle-measurement formula, given

(a) $\rho: y + z = 0$, $X^T bX = 0: x^2 = yz$
(b) $\rho: x + y + z = 0$, $X^T bX = 0: x^2 = yz$
(c) $\rho: x - y = 0$, $X^T bX = 0: x^2 + z^2 = 0$
(d) $\rho: x + y - z = 0$, $X^T bX = 0: yz + zx - xy = 0$

2. Find the sides and the angles of the triangle whose vertices are $P_1:(1,0,2)$, $P_2:(1,2,0)$, $P_3:(2,1,1)$ using the formulas in each part of Exercise 1.

11.5 The Geometry When I and J Are Real

When the points I and J in which the metric gauge line $\rho: \alpha x + \beta y + \gamma z = 0$ and the conic $X^T bX = 0$ intersect are real rather than conjugate complex, the resulting geometry, though akin to euclidean geometry, nonetheless differs from it in several striking ways. In the first place, we note that in the work of Sec. 11.2 up to and including Eq. (5), no assumptions were made about the nature of I and J. Hence these results are valid whether I and J are real or complex. However, when I and J are real, the parameters (μ, ν) of the intersection of the line P_1P_2 and the gauge line ρ are real and may be of the same or of opposite sign. The product $\mu\nu$ may therefore be either positive or negative, and it is impossible to establish a single convention for the matrix b which will ensure that in all cases $(P_1P_2)^2$ is nonnegative and (P_1P_2) is real.

For convenience in our subsequent discussions of distance and angle, let us agree to say that a point $Q:(\mu,\nu)$ is "between" I and J if $\mu\nu > 0$ and is "not between" I and J if $\mu\nu < 0$.† Let us also agree to say that a line which intersects ρ in a point between I and J is a line of type 1 and a line which intersects ρ in a point which is not between I and J is a line of type 2. Then to calculate the distance (P_1P_2), we must first determine whether the point in which P_1P_2 intersects ρ is or is not between I and J; that is, we must determine whether P_1P_2 is a line of type 1 or a line of type 2. Having done this and having adjusted the sign of the matrix b so that $I^T bJ > 0$, we then have from Eq. (5), Sec. 11.2,

$$(P_1P_2)^2 = \begin{cases} \dfrac{2\mu\nu I^T bJ}{\rho_1{}^2\rho_2{}^2} & P_1P_2 \text{ a line of type 1} \\[3mm] -\dfrac{2\mu\nu I^T bJ}{\rho_1{}^2\rho_2{}^2} & P_1P_2 \text{ a line of type 2} \end{cases}$$

The two cases can, of course, be included in one formula by writing

$$(1) \qquad (P_1P_2)^2 = \frac{2\,|\mu\nu I^T bJ|}{\rho_1{}^2\rho_2{}^2} = \frac{|(\rho_2P_1 - \rho_1P_2)^T b(\rho_2P_1 - \rho_1P_2)|}{\rho_1{}^2\rho_2{}^2}$$

In any event, it is always possible to measure the distance between any real points.

The corresponding result for angles is not true, however, for it turns out that angle measures can be assigned in a consistent way only when the lines involved are of the same type. To verify this, we observe that the gauge lines on the intersection of two ordinary lines, p_1 and p_2, are the lines of the pencil $\eta_1 p_1 + \eta_2 p_2$ whose parameters satisfy the equation

$$\eta_1{}^2 p_1{}^T Mp_1 + 2\eta_1\eta_2 p_1{}^T Mp_2 + \eta_2{}^2 p_2{}^T Mp_2 = 0$$

† The motivation for this terminology is found in the familiar fact of elementary analytic geometry that the point

$$P:\left(\frac{\mu x_1 + \nu x_2}{\mu + \nu}, \frac{\mu y_1 + \nu y_2}{\mu + \nu}\right)$$

is between the points $P_1:(x_1, y_1)$ and $P_2:(x_2, y_2)$ if and only if $\mu\nu > 0$.

Moreover, when I and J are both real, the gauge lines on any real point must also be real. Hence the roots of the last equation must be real, which implies that $(p_1{}^T M p_2)^2 > (p_1{}^T M p_1)(p_2{}^T M p_2)$. If p_1 and p_2 are lines of the same type, then distances on p_1 and p_2 are measured by the same formula, and therefore it follows from Eq. (4), Sec. 11.2, that $p_1{}^T M p_1$ and $p_2{}^T M p_2$ have the same sign. Hence in this case

$$\frac{(p_1{}^T M p_2)^2}{(p_1{}^T M p_1)(p_2{}^T M p_2)} > 1$$

and it is therefore possible to define a positive quantity, ϕ, by writing

$$(2) \qquad \cosh \phi = \frac{|p_1{}^T M p_2|}{\sqrt{(p_1{}^T M p_1)(p_2{}^T M p_2)}}$$

By an argument identical to the one used in Secs. 10.6, 10.7, and 11.3, it can be shown that the quantity ϕ defined by Eq. (2) is in fact $[p_1 p_2]$, provided k_a is taken to be $\frac{1}{2}$.

On the other hand, if the lines p_1 and p_2 are of different types, then $p_1{}^T M p_1$ and $p_2{}^T M p_2$ are of opposite sign and, regardless of the sign of $p_1{}^T M p_2$, the argument of the logarithm in the formula

$$[p_1 p_2] = k_a \ln \left[\frac{-p_1{}^T M p_2 - \sqrt{(p_1{}^T M p_2)^2 - (p_1{}^T M p_1)(p_2{}^T M p_2)}}{-p_1{}^T M p_2 + \sqrt{(p_1{}^T M p_2)^2 - (p_1{}^T M p_1)(p_2{}^T M p_2)}} \right]$$

is negative. The logarithm itself is therefore imaginary, and hence to obtain a real value for $[p_1 p_2]$ it is necessary that k_a be imaginary. But this is inadmissible, for the validity of the fundamental formula $[pq] + [qr] = [pr]$ requires that the same value of k_a be used in computing each angle; and this will not be the case unless p, q, and r are all lines of the same type.

It follows from the preceding discussion that the only triangles which can be studied in real terms are those whose sides are all of the same type. The principal properties of such triangles are described in the following theorems.

Theorem 1

If the sides p_1, p_2, p_3 of $\triangle P_1 P_2 P_3$ are all of type 1, then of the three quantities $p_1{}^T M p_2$, $p_2{}^T M p_3$, $p_1{}^T M p_3$ either one or three must be positive. If p_1, p_2, p_3 are all of type 2, then of the three quantities $p_1{}^T M p_2$, $p_2{}^T M p_3$, $p_1{}^T M p_3$, either one or three must be negative.

Proof   For convenience, let us denote $p_i{}^T M p_j$ by the symbol p_{ij}. Then since $\|p_{ij}\| = \|p_i p_j p_k\|^T M \|p_i p_j p_k\| = T^T M T = T^T (D^T b D) T$, it follows that $|p_{ij}| = |T|^2 |D|^2 |b| = 0$, since $|D| = 0$. Therefore, expanding $|p_{ij}|$ in terms of the elements of the first row, we have

$$0 = p_{11}(p_{22}\,p_{33} - p_{23}{}^2) - p_{12}(p_{12}\,p_{33} - p_{13}\,p_{23}) + p_{13}(p_{12}\,p_{23} - p_{13}\,p_{22})$$

$$(3) \qquad 0 = -p_{11}(p_{23}{}^2 - p_{22}\,p_{33}) - p_{12}{}^2\,p_{33} - p_{13}{}^2\,p_{22} + 2p_{12}\,p_{23}\,p_{13}$$

Now, as we observed above, $p_{ij}{}^2 > p_{ii}\,p_{jj}$. Hence when p_1, p_2, p_3 are all of type 1, so that $p_{ii} > 0$, it follows that the right-hand side of Eq. (3) is increased if we replace $p_{ij}{}^2$ by the smaller positive quantity $p_{ii}\,p_{jj}$; that is, $0 < 2p_{12}\,p_{23}\,p_{13} - 2p_{11}\,p_{22}\,p_{33}$. Moreover, when p_1, p_2, p_3 are lines of type 1, the second term in this inequality is clearly negative. Therefore the first term must be positive, which, in turn, requires that either one or three of the quantities p_{ij} be positive, as asserted.

On the other hand, if p_1, p_2, p_3 are all of type 2, so that $p_{ii} < 0$, then the right-hand side of Eq. (3) is decreased if we replace $p_{ij}{}^2$ by the smaller positive quantity $p_{ii}\,p_{jj}$; that is, $0 > 2p_{12}\,p_{23}\,p_{13} - 2p_{11}\,p_{22}\,p_{33}$. In this case, the second term is clearly positive. Hence the inequality can hold if and only if the first term is negative, which requires that either one or three of the quantities p_{ij} be negative, as asserted.

Theorem 2

If $\triangle P_1 P_2 P_3$ has sides of length d_1, d_2, d_3 and angles of measure α_1, α_2, α_3, then:

1. If the sides p_1, p_2, p_3 are all of type 1,

$$d_i{}^2 = d_j{}^2 + d_k{}^2 - 2d_j d_k \cdot \begin{cases} \cosh \alpha_i & p_j{}^T M p_k < 0 \\ -\cosh \alpha_i & p_j{}^T M p_k > 0 \end{cases}$$

2. If the sides p_1, p_2, p_3 are all of type 2,

$$d_i{}^2 = d_j{}^2 + d_k{}^2 + 2d_j d_k \cdot \begin{cases} \cosh \alpha_i & p_j{}^T M p_k < 0 \\ -\cosh \alpha_i & p_j{}^T M p_k > 0 \end{cases}$$

Proof Exactly as in the proof of the euclidean law of cosines in Sec. 11.4, we have the identity

$$\frac{p_i{}^T M p_i}{\rho_j{}^2 \rho_k{}^2} = \frac{p_j{}^T M p_j}{\rho_i{}^2 \rho_k{}^2} + \frac{p_k{}^T M p_k}{\rho_i{}^2 \rho_j{}^2}$$

$$- 2\sqrt{\frac{p_j{}^T M p_j}{\rho_i{}^2 \rho_k{}^2} \frac{p_k{}^T M p_k}{\rho_i{}^2 \rho_j{}^2}} \left(-\frac{p_j{}^T M p_k}{\sqrt{(p_j{}^T M p_j)(p_k{}^T M p_k)}} \right)$$

If the sides of the given triangle are all of type 1, then the first three terms in this equation are, respectively, $d_i{}^2$, $d_j{}^2$, $d_k{}^2$ and we have, using Eq. (2),

$$d_i{}^2 = d_j{}^2 + d_k{}^2 - 2d_jd_k \cdot \begin{cases} \cosh \alpha_i & p_j{}^T Mp_k < 0 \\ -\cosh \alpha_i & p_j{}^T Mp_k > 0 \end{cases}$$

On the other hand, if the sides p_1, p_2, p_3 are all of type 2, then the first three terms in the identity are, respectively, $-d_i{}^2$, $-d_j{}^2$, $-d_k{}^2$; and after multiplication by -1 we have

$$d_i{}^2 = d_j{}^2 + d_k{}^2 + 2d_jd_k \cdot \begin{cases} \cosh \alpha_i & p_j{}^T Mp_k < 0 \\ -\cosh \alpha_i & p_j{}^T Mp_k > 0 \end{cases}$$

as asserted.

Corollary 1

In any triangle whose sides are lines of the same type, either $d_i > d_j + d_k$ or $d_i < |d_j - d_k|$.

Proof Since $\cosh \alpha_i > 1$, we have, from Theorem 2,

$$\left| \frac{d_i{}^2 - d_j{}^2 - d_k{}^2}{2d_jd_k} \right| > 1$$

Hence $\qquad \dfrac{d_i{}^2 - d_j{}^2 - d_k{}^2}{2d_jd_k} > 1 \qquad$ or $\qquad \dfrac{d_i{}^2 - d_j{}^2 - d_k{}^2}{2d_jd_k} < -1$

In the first case, we have $d_i{}^2 > d_j{}^2 + 2d_jd_k + d_k{}^2 = (d_j + d_k)^2$. In the second case, we have $d_i{}^2 < d_j{}^2 - 2d_jd_k + d_k{}^2 = (d_j - d_k)^2$. Hence, taking square roots, we have either $d_i > d_j + d_k$ or $d_i < |d_j - d_k|$, as asserted.

Theorem 3

In any triangle whose sides are lines of the same type,

$$\frac{\sinh \alpha_1}{d_1} = \frac{\sinh \alpha_2}{d_2} = \frac{\sinh \alpha_3}{d_3}$$

Proof In any case, regardless of the type of the lines forming the given triangle and regardless of the sign of $p_j{}^T Mp_k$, we have

$$\cosh \alpha_i = \frac{d_i{}^2 - d_j{}^2 - d_k{}^2}{\pm 2d_jd_k}$$

Hence

$$\sinh^2 \alpha_i = \frac{(d_i{}^2 - d_j{}^2 - d_k{}^2)^2}{4d_j{}^2 d_k{}^2} - 1$$

$$= \frac{d_i{}^4 + d_j{}^4 + d_k{}^4 - 2(d_i{}^2 d_j{}^2 + d_j{}^2 d_k{}^2 + d_k{}^2 d_i{}^2)}{4d_j{}^2 d_k{}^2}$$

and therefore

$$\frac{\sinh \alpha_i}{d_i} = \sqrt{\frac{d_i^4 + d_j^4 + d_k^4 - 2(d_i^2 d_j^2 + d_j^2 d_k^2 + d_k^2 d_i^2)}{4 d_i^2 d_j^2 d_k^2}}$$

Since the expression on the right is symmetric in d_i, d_j, d_k, the theorem is proved.

Corollary 1

In any triangle whose sides are lines of the same type, the side opposite the greater angle is the greater.

Theorem 4

In any triangle whose sides are lines of the same type, the largest angle is equal to the sum of the other two angles.

Proof Consider first a triangle whose sides are all of type 1. Then by Theorem 1, at least one of the quantities $p_1^T M p_2$, $p_1^T M p_3$, $p_2^T M p_3$ must be positive. To be specific, let us suppose that $p_2^T M p_3 > 0$. Then $p_1^T M p_2$ and $p_1^T M p_3$ are either both positive or both negative, and in either case we can write

$$-\cosh \alpha_2 \cosh \alpha_3 - \sinh \alpha_2 \sinh \alpha_3$$
$$= -\cosh \alpha_2 \cosh \alpha_3 - d_2 d_3 \frac{\sinh \alpha_2}{d_2} \frac{\sinh \alpha_3}{d_3}$$
$$= -\frac{(d_2^2 - d_1^2 - d_3^2)(d_3^2 - d_1^2 - d_2^2)}{4 d_1^2 d_2 d_3}$$
$$\qquad - \frac{d_1^4 + d_2^4 + d_3^4 - 2(d_1^2 d_2^2 + d_2^2 d_3^2 + d_1^2 d_3^2)}{4 d_1^2 d_2 d_3}$$
$$= \frac{d_1^2 - d_2^2 - d_3^2}{-2 d_2 d_3}$$

By the first part of Theorem 2 and our assumption that $p_2^T M p_3 > 0$, it follows that the last fraction is equal to $-\cosh \alpha_1$. Hence

$$-\cosh \alpha_2 \cosh \alpha_3 - \sinh \alpha_2 \sinh \alpha_3 = -\cosh \alpha_1$$

or $\cosh \alpha_1 = \cosh (\alpha_2 + \alpha_3)$, which implies that $\alpha_1 = \alpha_2 + \alpha_3$.

Similarly, if the sides of the triangle are all of type 2, then at least one of the quantities $p_1^T M p_2$, $p_1^T M p_3$, $p_2^T M p_3$, say $p_2^T M p_3$, must be negative and the other two must be either both positive or both negative. Hence, just as before, we are led to the relation

$$-\cosh \alpha_2 \cosh \alpha_3 - \sinh \alpha_2 \sinh \alpha_3 = \frac{d_1^2 - d_2^2 - d_3^2}{-2 d_2 d_3}$$

By the second part of Theorem 2 and our assumption that $p_2^T M p_3 < 0$, it follows that

$$\frac{d_1^2 - d_2^2 - d_3^2}{-2d_2 d_3} = -\cosh \alpha_1$$

and therefore, as before, $\cosh(\alpha_2 + \alpha_3) = \cosh \alpha_1$ and $\alpha_1 = \alpha_2 + \alpha_3$.

Corollary 1

For triangles whose sides are all of type 1, exactly one of the quantities $p_1^T M p_2$, $p_2^T M p_3$, $p_1^T M p_3$ is positive; and for triangles whose sides are all of type 2, exactly one of the quantities $p_1^T M p_2$, $p_2^T M p_3$, $p_1^T M p_3$ is negative.

Example 1

If we take the metric gauge line, ρ, to be the line $z = 0$ and the conic $X^T b X = 0$ to be the conic $x^2 - y^2 = 0$, then I and J are the points $(1,1,0)$ and $(1,-1,0)$ and from Eq. (1) the distance formula is

$$(P_1 P_2)^2 = \frac{|z_1^2(x_2^2 - y_2^2) - 2z_1 z_2(x_1 x_2 - y_1 y_2) + z_2^2(x_1^2 - y_1^2)|}{z_1^2 z_2^2}$$

or

$$(P_1 P_2) = \sqrt{\frac{|(z_1 x_2 - x_1 z_2)^2 - (z_1 y_2 - y_1 z_2)^2|}{z_1^2 z_2^2}}$$

If, for convenience, we take $z_1 = z_2 = 1$, as we always can since we are concerned only with the metric relations between real points which are not on the line $z = 0$, the distance formula becomes simply

$$(P_1 P_2) = \sqrt{|(x_2 - x_1)^2 - (y_2 - y_1)^2|}$$

Similarly, since $b = \begin{Vmatrix} 1 & 0 & 0 \\ 0 & -1 & 0 \\ 0 & 0 & 0 \end{Vmatrix}$ and $D = \begin{Vmatrix} 0 & -1 & 0 \\ 1 & 0 & 0 \\ 0 & 0 & 0 \end{Vmatrix}$,

it follows that $M = D^T b D = \begin{Vmatrix} -1 & 0 & 0 \\ 0 & 1 & 0 \\ 0 & 0 & 0 \end{Vmatrix}$. Hence for the measurement of

angles we have, from Eq. (2), the formulas

$$\cosh \phi = \begin{cases} \dfrac{-l_1 l_2 + m_1 m_2}{\sqrt{(-l_1^2 + m_1^2)(-l_2^2 + m_2^2)}} & p_1^T M p_2 > 0 \\[4mm] \dfrac{l_1 l_2 - m_1 m_2}{\sqrt{(-l_1^2 + m_1^2)(-l_2^2 + m_2^2)}} & p_1^T M p_2 < 0 \end{cases}$$

In particular, for the triangle whose vertices are $P_1:(0,4,1)$, $P_2:(1,2,1)$, $P_3:(2,8,1)$ and whose sides are therefore

$$p_1 = P_2P_3:[-6, 1, 4] \qquad p_2 = P_3P_1:[4, -2, 8] \qquad p_3 = P_1P_2:[2, 1, -4]$$

we have $p_1{}^T M p_1 = -35$, $p_2{}^T M p_2 = -12$, $p_3{}^T M p_3 = -3$, $p_2{}^T M p_3 = -10$, $p_1{}^T M p_3 = 13$, $p_1{}^T M p_2 = 22$. Hence, each side of $\triangle P_1P_2P_3$ is a line of type 2, and it is possible to study the triangle numerically. Using the formulas we developed above, we find immediately that $d_1 = (P_2P_3) = \sqrt{35}$, $d_2 = (P_3P_1) = \sqrt{12}$, $d_3 = (P_1P_2) = \sqrt{3}$, and $\cosh \alpha_1 = \cosh \angle P_2P_1P_3 = \frac{5}{3}$, $\cosh \alpha_2 = \cosh \angle P_3P_2P_1 = 13/\sqrt{105}$, $\cosh \alpha_3 = \cosh \angle P_1P_3P_2 = 11/\sqrt{105}$. Using these values, we have in verification of Theorem 2,

$$d_1{}^2 = d_2{}^2 + d_3{}^2 + 2d_2d_3 \cosh \alpha_1 = 12 + 3 + 2 \cdot \sqrt{12} \cdot \sqrt{3}\,\tfrac{5}{3} = 35$$

$$d_2{}^2 = d_1{}^2 + d_3{}^2 - 2d_1d_3 \cosh \alpha_2 = 35 + 3 - 2 \cdot \sqrt{35} \cdot \sqrt{3}\,\frac{13}{\sqrt{105}} = 12$$

$$d_3{}^2 = d_1{}^2 + d_2{}^2 - 2d_1d_2 \cosh \alpha_3 = 35 + 12 - 2 \cdot \sqrt{35} \cdot \sqrt{12}\,\frac{11}{\sqrt{105}} = 3$$

Furthermore, $\sinh \alpha_1 = \sqrt{\tfrac{25}{9} - 1} = \tfrac{4}{3}$, $\sinh \alpha_2 = \sqrt{\tfrac{169}{105} - 1} = 8/\sqrt{105}$, $\sinh \alpha_3 = \sqrt{\tfrac{121}{105} - 1} = 4/\sqrt{105}$. Hence, in verification of Theorem 3, we have

$$\frac{\sinh \alpha_1}{d_1} = \frac{\tfrac{4}{3}}{\sqrt{35}} = \frac{4}{3\sqrt{35}} \qquad \frac{\sinh \alpha_2}{d_2} = \frac{8/\sqrt{105}}{\sqrt{12}} = \frac{4}{3\sqrt{35}}$$

$$\frac{\sinh \alpha_3}{d_3} = \frac{4/\sqrt{105}}{\sqrt{3}} = \frac{4}{3\sqrt{35}}$$

Finally we note that

$$\cosh (\alpha_2 + \alpha_3) = \cosh \alpha_2 \cosh \alpha_3 + \sinh \alpha_2 \sinh \alpha_3$$

$$= \frac{13}{\sqrt{105}} \frac{11}{\sqrt{105}} + \frac{8}{\sqrt{105}} \frac{4}{\sqrt{105}} = \frac{175}{105} = \frac{5}{3} = \cosh \alpha_1$$

which verifies Theorem 4.

Incidentally, it is easy to check that $d_1 > d_2 + d_3$, in verification of Corollary 1, Theorem 1.

Exercises

1. Prove Corollary 1, Theorem 4.

2. Find the distance- and angle-measurement formulas, given

(a) $\rho: x = 0$, $X^T b X = 0: x^2 - yz = 0$

(b) $\rho: x + y + z = 0$, $X^T b X = 0: x^2 - z^2 = 0$

3. Given $P_1:(1,0,0)$, $P_2:(1,1,0)$, $P_3:(1,0,2)$. Find the sides and angles of $\triangle P_1 P_2 P_3$ using the formulas of each part of Exercise 2.

4. What congruence theorems, if any, are there when I and J are real?

11.6 Conclusion

In this chapter we have been primarily concerned with the question of how euclidean geometry can be obtained as a specialization of the projective plane, and our discussion made it clear that to do this it is necessary to use a singular metric gauge conic introduced via a certain limiting process. More specifically, we found that the metric gauge locus must consist of a repeated real line, ρ, and the metric gauge envelope must consist of a pair of pencils whose vertices are conjugate complex points on ρ. If the line ρ is one of the coordinate axes in the underlying projective plane, the geometry imposed on the real elements in the plane is not only euclidean but is, in fact, the familiar geometry of the cartesian plane. In other words, not only does it satisfy the basic euclidean laws, but the formulas for the distances and angles appearing in these laws are the familiar formulas of elementary analytic geometry. On the other hand, if ρ is not one of the coordinate axes, then although the resulting geometry is still euclidean, the formulas for measuring distances and angles are not necessarily those of cartesian geometry. More-over, when ρ is not one of the coordinate axes of the projective plane, a point always has three, rather than two, coordinates, regardless of whether the coordinate system in the resulting euclidean plane is homogeneous or non-homogeneous.

We also considered in some detail the geometry imposed on the real elements of the projective plane when the vertices of the pencils forming the singular metric gauge envelope are real points on ρ. This geometry is of no great historical or practical interest, but it provides an example, more striking even than elliptic or hyperbolic geometry, of a geometrical system in which the "immutable" and "eternal" properties of euclidean geometry are violated. In it, for instance, the distance between two distinct real points may be zero, the triangle inequality may or may not hold, and the shortest distance between two points is not the length of the segment they determine. Moreover, in this geometry there are two classes of lines, and angles can be assigned real measures only when their sides are lines belonging to the same class. Furthermore, although there are properties analogous to the cosine law and the law of sines for triangles whose sides are lines belonging to the same class, the euclidean angle-sum theorem is replaced by the surprising requirement that the largest angle of a triangle is always equal to the sum of the other two angles.

Discussions of noneuclidean geometry often close with the question: What *is* the geometry of the "real" world? It is said that Gauss once attempted to answer this question by making very accurate measurements of the angles of a triangle whose vertices were points on three widely separated peaks in the Alps. Had the difference between the angle sum and 180° exceeded his expected errors of measurement, Gauss would have concluded that the geometry of the universe was not euclidean and was therefore, presumably, either elliptic or hyperbolic. However, the angle sum differed from 180° by an amount within the range of experimental error, and so the result was inconclusive.

It is conceivable that as man probes more and more deeply into space, he will someday be able to make observations on a scale so large that he can convince himself by actual measurement that the geometry of the "real" world is either elliptic or hyperbolic, if such indeed is the case. On the other hand, since the properties of sufficiently small configurations in both elliptic and hyperbolic geometry are arbitrarily close to the corresponding euclidean properties, and since in a system the size of the universe, "sufficiently small" may still be so large that such regions will be all that man can ever explore, it may be that no observations can ever settle the question. It may also be that all our measuring procedures have an intrinsic euclidean bias, equivalent to the singular metric gauge conic which gives rise to euclidean geometry, so that no matter on what scale we make our measurements, our observations will always make the universe appear euclidean.

Finally, it may well be that the question we have been considering is entirely meaningless. In fact, our work in the last two chapters has made it clear that a scheme of measurement is not necessarily inherent in a space but is instead superimposed on the space by an observer who chooses certain elements of the space to play a special role in his scheme of measurement. Perhaps the "real" world can be simultaneously elliptic, hyperbolic, and euclidean, according as its observers theorize from one point of view or another or measure it with instruments biased in favor of one geometry or another.

appendix 1

A REVIEW OF DETERMI= NANTS

A determinant of order n is a certain expression involving n^2 quantities which is defined in Definition 2, below. The usual symbol for a determinant of order n is formed by displaying between vertical bars the n^2 quantities upon which the determinant depends:

$$(1) \quad |A| = |a_{ij}| = \begin{vmatrix} a_{11} & a_{12} & \cdots & a_{1n} \\ a_{21} & a_{22} & \cdots & a_{2n} \\ \cdot & \cdot & \cdots & \cdot \\ a_{n1} & a_{n2} & \cdots & a_{nn} \end{vmatrix}$$

The word *determinant* is commonly used to denote this symbol as well as the expression which it represents.

The quantities a_{ij} are called the **elements** of the determinant (1). The first subscript associated with an element in the double-subscript notation illustrated in (1) tells us the row in which the element lies; the second subscript tells us the column in which the element lies. From this convent on it is clear that a_{ij} and a_{ji} are the names of different elements and that in general $a_{ij} \neq a_{ji}$.

The determinant, M_{ij}, formed by

the $(n-1)^2$ elements which remain when the ith row and jth column are deleted from an nth-order determinant $|A|$ is called the **minor** of the element a_{ij} which is common to the ith row and jth column of $|A|$. The quantity

$$A_{ij} = (-1)^{i+j} M_{ij}$$

is called the **cofactor** of the element a_{ij}.

The determinant, $M_{ij,kl}$, formed by the $(n-2)^2$ elements which remain when the ith and jth rows and the kth and lth columns are deleted from an nth-order determinant, $|A|$, is called the **complementary minor** of the second-order determinant

$$\begin{vmatrix} a_{ik} & a_{il} \\ a_{jk} & a_{jl} \end{vmatrix}$$

common to the two deleted rows and the two deleted columns. The quantity

$$A_{ij,kl} = (-1)^{i+j+k+l} M_{ij,kl}$$

is called the **algebraic complement** of the determinant

$$\begin{vmatrix} a_{ik} & a_{il} \\ a_{jk} & a_{jl} \end{vmatrix}$$

The ideas of complementary minors and algebraic complements can be generalized to the case where more than two rows and more than two columns are deleted from a determinant, but we shall not need to make this extension in our work.

Example 1

In the third-order determinant $\begin{vmatrix} y^2 & 0 & b-5 \\ 1 & a & 7 \\ 6 & c & x \end{vmatrix}$ the minor of the element c in the third row and second column is the second-order determinant

$$M_{32} = \begin{vmatrix} y^2 & b-5 \\ 1 & 7 \end{vmatrix}$$

which remains when the third row and second column are deleted. The cofactor of the element c is

$$A_{32} = (-1)^{3+2} M_{32} = -\begin{vmatrix} y^2 & b-5 \\ 1 & 7 \end{vmatrix}$$

Example 2

In the fourth-order determinant

$$\begin{vmatrix} a_{11} & a_{12} & a_{13} & a_{14} \\ a_{21} & a_{22} & a_{23} & a_{24} \\ a_{31} & a_{32} & a_{33} & a_{34} \\ a_{41} & a_{42} & a_{43} & a_{44} \end{vmatrix}$$

the complementary minor of the determinant $\begin{vmatrix} a_{11} & a_{14} \\ a_{21} & a_{24} \end{vmatrix}$ whose elements are common to the first and second rows and the first and fourth columns is the determinant $M_{12,14} = \begin{vmatrix} a_{32} & a_{33} \\ a_{42} & a_{43} \end{vmatrix}$ which remains when the first and second rows and the first and fourth columns are deleted. The algebraic complement of the determinant $\begin{vmatrix} a_{11} & a_{14} \\ a_{21} & a_{24} \end{vmatrix}$ is

$$A_{12,14} = (-1)^{1+2+1+4} M_{12,14} = \begin{vmatrix} a_{32} & a_{33} \\ a_{42} & a_{43} \end{vmatrix}$$

We now define determinants inductively, beginning with the familiar definition of a determinant of the second order.

Definition 1

$$\begin{vmatrix} a_{11} & a_{12} \\ a_{21} & a_{22} \end{vmatrix} = a_{11}a_{22} - a_{12}a_{21}$$

Definition 2

A determinant of order n is equal to the sum of the products of the elements of any row or column and their respective cofactors.

According to Definition 2, a determinant of order n depends upon n determinants of order $n - 1$, each of which in turn depends upon $n - 1$ determinants of order $n - 2$, and so on until, finally, the determinant is expressed in terms of second-order determinants, each of which can be expanded by Definition 1. However, before Definition 2 can be accepted, it is necessary to prove that the same expression is obtained regardless of which row or which column is used for the expansion. Fortunately, this is the case,[1] and Definition 2 is unambiguous.

Example 3

Using the elements of the first row of the determinant

$$|A| = \begin{vmatrix} a_{11} & a_{12} & a_{13} \\ a_{21} & a_{22} & a_{23} \\ a_{31} & a_{32} & a_{33} \end{vmatrix}$$

[1] Proofs of the theorems cited in this section can be found in C. R. Wylie, Jr., "Advanced Engineering Mathematics," 3d ed., pp. 400–414, McGraw-Hill Book Company, New York, 1966.

and their cofactors, we have

$$|A| = a_{11}\begin{vmatrix} a_{22} & a_{23} \\ a_{32} & a_{33} \end{vmatrix} + a_{12}\left(-\begin{vmatrix} a_{21} & a_{23} \\ a_{31} & a_{33} \end{vmatrix}\right) + a_{13}\begin{vmatrix} a_{21} & a_{22} \\ a_{31} & a_{32} \end{vmatrix}$$

$$= a_{11}(a_{22}a_{33} - a_{23}a_{32}) - a_{12}(a_{21}a_{33} - a_{23}a_{31}) + a_{13}(a_{21}a_{32} - a_{22}a_{31})$$

$$= a_{11}a_{22}a_{33} + a_{12}a_{23}a_{31} + a_{13}a_{21}a_{32} - a_{11}a_{23}a_{32} - a_{12}a_{21}a_{33} - a_{13}a_{22}a_{31}$$

It is easy to verify that the expression derived for $|A|$ in Example 3 is obtained no matter which row or column is used. It is convenient to note that in this case the expansion of $|A|$ can also be obtained by copying the first two columns of $|A|$ on the right and then adding the signed products of the elements on the various diagonals in the resulting array, as indicated in the following diagram:

$$
\begin{array}{ccc}
(+) & (+) & (+) \\
a_{11} & a_{12} & a_{13} & a_{11} & a_{12} \\
a_{21} & a_{22} & a_{23} & a_{21} & a_{22} \\
a_{31} & a_{32} & a_{33} & a_{31} & a_{32} \\
(-) & (-) & (-)
\end{array}
$$

This method of diagonal expansion is incorrect for determinants of order higher than 3.

The basic properties of determinants are contained in the following theorems.

Theorem 1

If $|A|$ is any determinant, and if $|B|$ is the determinant whose rows are the respective columns of $|A|$, then $|B| = |A|$.

Theorem 2

Let any two rows (or columns) be selected from a determinant $|A|$. Then $|A|$ is equal to the sum of the products of all the second-order minors contained in the chosen pair of rows (or columns) each multiplied by its algebraic complement.

This method of expanding a determinant is known as **Laplace's expansion.**

Theorem 3

If all the elements in any row or in any column of a determinant are zero, the value of the determinant is zero.

Theorem 4

If each element in one row or in one column of a determinant is multiplied by c, the value of the determinant is multiplied by c.

Theorem 5

If $|A|$ is any determinant, and if $|B|$ is the determinant obtained from $|A|$ by interchanging any two rows or any two columns of $|A|$, then $|B| = -|A|$.

Theorem 6

If corresponding elements of two rows or of two columns of a determinant are proportional, the value of the determinant is zero.

Theorem 7

If the elements in one column of a determinant are expressed as binomials, the determinant can be written as the sum of two determinants, according to the formula

$$
\begin{vmatrix}
a_{11} & \cdots & a_{1j} + \alpha_{1j} & \cdots & a_{1n} \\
a_{21} & \cdots & a_{2j} + \alpha_{2j} & \cdots & a_{2n} \\
\cdot & \cdots & \cdot & \cdots & \cdot \\
a_{n1} & \cdots & a_{nj} + \alpha_{nj} & \cdots & a_{nn}
\end{vmatrix}
$$

$$
=
\begin{vmatrix}
a_{11} & \cdots & a_{1j} & \cdots & a_{1n} \\
a_{21} & \cdots & a_{2j} & \cdots & a_{2n} \\
\cdot & \cdots & \cdot & \cdots & \cdot \\
a_{n1} & \cdots & a_{nj} & \cdots & a_{nn}
\end{vmatrix}
+
\begin{vmatrix}
a_{11} & \cdots & \alpha_{1j} & \cdots & a_{1n} \\
a_{21} & \cdots & \alpha_{2j} & \cdots & a_{2n} \\
\cdot & \cdots & \cdot & \cdots & \cdot \\
a_{n1} & \cdots & \alpha_{nj} & \cdots & a_{nn}
\end{vmatrix}
$$

A similar result holds for a determinant containing a row of elements which are binomials.

Theorem 8

The value of a determinant is unaltered if the elements of any row (or column) are modified by adding to them the same multiple of the corresponding elements in any other row (or column).

Theorem 8 is very useful in the practical expansion of determinants, for by its repeated application one can reduce to zero a number of the elements in a chosen row (or column) of the given determinant. Then, when the determinant is expanded in terms of this row (or column) most of the products involved will be zero and the computation will be appreciably shortened.

Example 4

Find the value of the determinant

$$
\begin{vmatrix}
2 & 1 & -1 & 3 & 1 \\
0 & 2 & 1 & 1 & 2 \\
4 & 1 & 0 & 2 & 3 \\
1 & 5 & 1 & 3 & -1 \\
2 & 0 & 2 & -1 & 2
\end{vmatrix}
$$

Here, in an attempt to introduce as many zeros as possible into some row, let us add the third column to the second and to the fifth, and let us add twice the third column to the first and three times the third column to the fourth. This gives us the new but equal determinant

$$\begin{vmatrix} 0 & 0 & -1 & 0 & 0 \\ 2 & 3 & 1 & 4 & 3 \\ 4 & 1 & 0 & 2 & 3 \\ 3 & 6 & 1 & 6 & 0 \\ 6 & 2 & 2 & 5 & 4 \end{vmatrix}$$

Expanding this in terms of the first row, according to Definition 2, we have

$$(-1)(-1)^{1+3} \begin{vmatrix} 2 & 3 & 4 & 3 \\ 4 & 1 & 2 & 3 \\ 3 & 6 & 6 & 0 \\ 6 & 2 & 5 & 4 \end{vmatrix}$$

Now, subtracting twice the first column from both the second column and the third column, we obtain the equal determinant

$$- \begin{vmatrix} 2 & -1 & 0 & 3 \\ 4 & -7 & -6 & 3 \\ 3 & 0 & 0 & 0 \\ 6 & -10 & -7 & 4 \end{vmatrix}$$

or, expanding in terms of the third row,

$$-(3)(-1)^{3+1} \begin{vmatrix} -1 & 0 & 3 \\ -7 & -6 & 3 \\ -10 & -7 & 4 \end{vmatrix}$$

We can now simplify this by further row or column manipulations, or, since it is of the third order, we can expand it by the diagonal method. The result is 90.

Theorem 9

The sum of the products formed by multiplying each element of any row (or column) of a determinant by the cofactors of the corresponding elements of any other row (or column) is zero.

Corollary 1

If A_{ij} is the cofactor of the element a_{ij} in the determinant $|A| = |a_{ij}|$, then

$$\sum_{k=1}^{n} a_{ik}A_{jk} = \begin{cases} 0 & i \neq j \\ |A| & i = j \end{cases} \quad \text{and} \quad \sum_{i=1}^{n} a_{ik}A_{il} = \begin{cases} 0 & k \neq l \\ |A| & k = l \end{cases}$$

Example 5

If we take the elements of the first row of the determinant

$$\begin{vmatrix} a_{11} & a_{12} & a_{13} \\ a_{21} & a_{22} & a_{23} \\ a_{31} & a_{32} & a_{33} \end{vmatrix}$$

and multiply them by the cofactors of the corresponding elements in the third row, say, we obtain the sum

$$a_{11}\begin{vmatrix} a_{12} & a_{13} \\ a_{22} & a_{23} \end{vmatrix} - a_{12}\begin{vmatrix} a_{11} & a_{13} \\ a_{21} & a_{23} \end{vmatrix} + a_{13}\begin{vmatrix} a_{11} & a_{12} \\ a_{21} & a_{22} \end{vmatrix}$$

and this is clearly the expansion of the determinant

$$\begin{vmatrix} a_{11} & a_{12} & a_{13} \\ a_{21} & a_{22} & a_{23} \\ a_{11} & a_{12} & a_{13} \end{vmatrix}$$

according to the third row. Since this determinant has two identical rows, it vanishes identically.

Theorem 10

The product of two determinants of the same order is a determinant of the same order in which the element in the ith row and jth column is the sum of the products of corresponding elements in the ith row of the first determinant and the jth column of the second determinant.

Example 6

If $|A| = \begin{vmatrix} 1 & 2 & -1 \\ 0 & 3 & 1 \\ 2 & 2 & 3 \end{vmatrix}$ and $|B| = \begin{vmatrix} 4 & 0 & 1 \\ 3 & 1 & 2 \\ 1 & 1 & -1 \end{vmatrix}$ it is easy to

verify that $|A| = 17$ and $|B| = -10$. Hence $|A| \cdot |B| = -170$. On the other hand, $|A| \cdot |B|$ can also be computed by using Theorem 10 to construct a new determinant whose value is equal to $|A| \cdot |B|$:

$$\begin{vmatrix} 1 & 2 & -1 \\ 0 & 3 & 1 \\ 2 & 2 & 3 \end{vmatrix} \cdot \begin{vmatrix} 4 & 0 & 1 \\ 3 & 1 & 2 \\ 1 & 1 & -1 \end{vmatrix} = \begin{vmatrix} 4+6-1 & 0+2-1 & 1+4+1 \\ 0+9+1 & 0+3+1 & 0+6-1 \\ 8+6+3 & 0+2+3 & 2+4-3 \end{vmatrix}$$

$$= \begin{vmatrix} 9 & 1 & 6 \\ 10 & 4 & 5 \\ 17 & 5 & 3 \end{vmatrix} = -170$$

The same result is obtained, of course, if Theorem 10 is used to multiply $|A|$ and $|B|$ in the order $|B| \cdot |A|$.

Exercises

1. Use Definition 2 to expand each of the following determinants. In each case, use two different rows or two different columns, and verify that they yield the same answer.

(a) $\begin{vmatrix} 1 & 1 & 1 \\ 1 & 2 & 3 \\ 3 & 4 & 6 \end{vmatrix}$
(b) $\begin{vmatrix} 1 & 2 & -1 \\ 3 & 0 & 2 \\ 1 & 4 & 1 \end{vmatrix}$
(c) $\begin{vmatrix} 1 & 2 & 3 \\ 4 & 5 & 6 \\ 1 & 1 & -2 \end{vmatrix}$

2. Expand each of the following determinants, using Theorem 8 to simplify the work:

(a) $\begin{vmatrix} 1 & 1 & 1 & -1 & -1 \\ 1 & 2 & 4 & -1 & 5 \\ 3 & 4 & 5 & -1 & -2 \\ 1 & 3 & 4 & 5 & -1 \\ 1 & -1 & -2 & 7 & -5 \end{vmatrix}$
(b) $\begin{vmatrix} 1 & 2 & 1 & 2 & 0 \\ 1 & -1 & 0 & 2 & 3 \\ 2 & 3 & 1 & -1 & 1 \\ 1 & 4 & 2 & 0 & -2 \\ 3 & 0 & -1 & -3 & 1 \end{vmatrix}$

3. Using Theorem 7, express each of the following determinants as a polynomial in λ:

(a) $\begin{vmatrix} 1 - \lambda & 1 & 2 \\ 2 & 2 - \lambda & 3 \\ 3 & 4 & 3 - \lambda \end{vmatrix}$
(b) $\begin{vmatrix} 2\lambda + 3 & \lambda + 2 & \lambda \\ \lambda - 1 & -\lambda + 2 & 2 \\ \lambda + 1 & 3\lambda + 1 & \lambda + 3 \end{vmatrix}$

(c) $\begin{vmatrix} \lambda + 2 & \lambda + 3 & \lambda - 1 \\ 2\lambda + 1 & \lambda - 3 & \lambda + 1 \\ \lambda & 3\lambda - 4 & 1 \end{vmatrix}$

appendix 2

A FINITE NONDESAR= GUESIAN GEOMETRY

In Sec. 4.3, after we had established the validity of Desargues' triangle theorem in E_2^+ and Π_2, we noted that there were geometries satisfying the fundamental axioms of incidence and extension (Axioms 1 to 6) in which this theorem was false. In this section we shall describe a geometry in which Desargues' theorem does not hold and illustrate by particular examples some of the peculiar properties of such a system.

The geometry we shall consider is a finite geometry containing 91 points, P_1, $P_2, \ldots, P_{91}$, and 91 lines, $l_1, l_2, \ldots, l_{91}$, in which 10 points lie on every line and 10 lines pass through every point. As with the finite geometries we considered in Sec. 6.2, this system is completely specified when we know which points lie on which lines or, equivalently, which lines pass through which points, and this information is given in Tables A.1 and A.2. Table A.1 lists the names of the 91 lines and after each name lists the names of the 10 points which lie on that line. Table A.2 gives us the same

Table A.1

Line	Points on line										Line	Points on line									
1	1	2	3	4	5	6	7	8	9	10	47	6	23	36	44	48	58	69	80	82	84
2	1	11	12	13	14	15	16	17	18	82	48	6	11	24	34	45	46	59	67	79	85
3	1	19	20	21	22	23	24	25	26	27	49	6	12	22	35	43	47	60	68	81	83
4	1	28	29	30	31	32	33	34	35	36	50	6	13	26	30	38	51	61	71	73	87
5	1	37	38	39	40	41	42	43	44	45	51	6	14	27	28	39	49	62	72	74	88
6	1	46	47	48	49	50	51	52	53	54	52	6	15	25	29	37	50	63	70	75	86
7	1	55	56	57	58	59	60	61	62	63	53	6	16	20	33	41	54	55	66	76	90
8	1	64	65	66	67	68	69	70	71	72	54	6	17	21	31	42	52	56	64	77	91
9	1	73	74	75	76	77	78	79	80	81	55	6	18	19	32	40	53	57	65	78	89
10	1	83	84	85	86	87	88	89	90	91	56	7	24	35	41	52	57	71	74	82	86
11	2	19	28	37	46	55	64	73	82	83	57	7	11	22	36	42	53	55	72	75	87
12	2	11	20	29	38	47	56	65	74	84	58	7	12	23	34	40	54	56	70	73	88
13	2	12	21	30	39	48	57	66	75	85	59	7	13	27	29	44	46	60	66	77	89
14	2	13	22	31	40	49	58	67	76	86	60	7	14	25	30	45	47	58	64	78	90
15	2	14	23	32	41	50	59	68	77	87	61	7	15	26	28	43	48	59	65	76	91
16	2	15	24	33	42	51	60	69	78	88	62	7	16	21	32	38	49	63	69	79	83
17	2	16	25	34	43	52	61	72	80	89	63	7	17	19	33	39	50	61	67	81	84
18	2	17	26	35	44	53	62	70	79	90	64	7	18	20	31	37	51	62	68	80	85
19	2	18	27	36	45	54	63	71	81	91	65	8	25	31	38	53	59	66	81	82	88
20	3	20	30	40	50	60	72	79	82	91	66	8	11	26	32	39	54	60	64	80	86
21	3	11	21	28	41	51	58	70	81	89	67	8	12	27	33	37	52	58	65	79	87
22	3	12	19	29	42	49	59	71	80	90	68	8	13	19	34	41	47	62	69	75	91
23	3	13	23	33	43	53	63	64	74	85	69	8	14	20	35	42	48	63	67	73	89
24	3	14	24	31	44	54	61	65	75	83	70	8	15	21	36	40	46	61	68	74	90
25	3	15	22	32	45	52	62	66	73	84	71	8	16	22	28	44	50	56	71	78	85
26	3	16	26	36	37	47	57	67	77	88	72	8	17	23	29	45	51	57	72	76	83
27	3	17	27	34	38	48	55	68	78	86	73	8	18	24	30	43	49	55	70	77	84
28	3	18	25	35	39	46	56	69	76	87	74	9	26	33	45	49	56	68	75	82	89
29	4	21	29	43	54	62	67	78	82	87	75	9	11	27	31	43	50	57	69	73	90
30	4	11	19	30	44	52	63	68	76	88	76	9	12	25	32	44	51	55	67	74	91
31	4	12	20	28	45	53	61	69	77	86	77	9	13	20	36	39	52	59	70	78	83
32	4	13	24	32	37	48	56	72	81	90	78	9	14	21	34	37	53	60	71	76	84
33	4	14	22	33	38	46	57	70	80	91	79	9	15	19	35	38	54	58	72	77	85
34	4	15	23	31	39	47	55	71	79	89	80	9	16	23	30	42	46	62	65	81	86
35	4	16	27	35	40	51	59	64	75	84	81	9	17	24	28	40	47	63	66	80	87
36	4	17	25	36	41	49	60	65	73	85	82	9	18	22	29	41	48	61	64	79	88
37	4	18	26	34	42	50	58	66	74	83	83	10	27	32	42	47	61	70	76	82	85
38	5	22	34	39	51	63	65	77	82	90	84	10	11	25	33	40	48	62	71	77	83
39	5	11	23	35	37	49	61	66	78	91	85	10	12	26	31	41	46	63	72	78	84
40	5	12	24	36	38	50	62	64	76	89	86	10	13	21	35	45	50	55	65	80	88
41	5	13	25	28	42	54	57	68	79	84	87	10	14	19	36	43	51	56	66	79	86
42	5	14	26	29	40	52	55	69	81	85	88	10	15	20	34	44	49	57	64	81	87
43	5	15	27	30	41	53	56	67	80	83	89	10	16	24	29	39	53	58	68	73	91
44	5	16	19	31	45	48	60	70	74	87	90	10	17	22	30	37	54	59	69	74	89
45	5	17	20	32	43	46	58	71	75	88	91	10	18	23	28	38	52	60	67	75	90
46	5	18	21	33	44	47	59	72	73	86											

Table A.2

Point	Lines on point										Point	Lines on point									
1	1	2	3	4	5	6	7	8	9	10	47	6	12	26	34	46	49	60	68	81	83
2	1	11	12	13	14	15	16	17	18	19	48	6	13	27	32	44	47	61	69	82	84
3	1	20	21	22	23	24	25	26	27	28	49	6	14	22	36	39	51	62	73	74	88
4	1	29	30	31	32	33	34	35	36	37	50	6	15	20	37	40	52	63	71	75	86
5	1	38	39	40	41	42	43	44	45	46	51	6	16	21	35	38	50	64	72	76	87
6	1	47	48	49	50	51	52	53	54	55	52	6	17	25	30	42	54	56	67	77	91
7	1	56	57	58	59	60	61	62	63	64	53	6	18	23	31	43	55	57	65	78	89
8	1	65	66	67	68	69	70	71	72	73	54	6	19	24	29	41	53	58	66	79	90
9	1	74	75	76	77	78	79	80	81	82	55	7	11	27	34	42	53	57	73	76	86
10	1	83	84	85	86	87	88	89	90	91	56	7	12	28	32	43	54	58	71	74	87
11	2	12	21	30	39	48	57	66	75	84	57	7	13	26	33	41	55	56	72	75	88
12	2	13	22	31	40	49	58	67	76	85	58	7	14	21	37	45	47	60	67	79	89
13	2	14	23	32	41	50	59	68	77	86	59	7	15	22	35	46	48	61	65	77	90
14	2	15	24	33	42	51	60	69	78	87	60	7	16	20	36	44	49	59	66	78	91
15	2	16	25	34	43	52	61	70	79	88	61	7	17	24	31	39	50	63	70	82	83
16	2	17	26	35	44	53	62	71	80	89	62	7	18	25	29	40	51	64	68	80	84
17	2	18	27	36	45	54	63	72	81	90	63	7	19	23	30	38	52	62	69	81	85
18	2	19	28	37	46	55	64	73	82	91	64	8	11	23	35	40	54	60	66	82	88
19	3	11	22	30	44	55	63	68	79	87	65	8	12	24	36	38	55	61	67	80	86
20	3	12	20	31	45	53	64	69	77	88	66	8	13	25	37	39	53	59	65	81	87
21	3	13	21	29	46	54	62	70	78	86	67	8	14	26	29	43	48	63	69	76	91
22	3	14	25	33	38	49	57	71	82	90	68	8	15	27	30	41	49	64	70	74	89
23	3	15	23	34	39	47	58	72	80	91	69	8	16	28	31	42	47	62	68	75	90
24	3	16	24	32	40	48	56	73	81	89	70	8	18	21	33	44	52	58	73	77	83
25	3	17	28	36	41	52	60	65	76	84	71	8	19	22	34	45	50	56	71	78	84
26	3	18	26	37	42	50	61	66	74	85	72	8	17	20	32	46	51	57	72	79	85
27	3	19	27	35	43	51	59	67	75	83	73	9	11	25	36	46	50	58	69	75	89
28	4	11	21	31	41	51	61	71	81	91	74	9	12	23	37	44	51	56	70	76	90
29	4	12	22	29	42	52	59	72	82	89	75	9	13	24	35	45	52	57	68	74	91
30	4	13	20	30	43	50	60	73	80	90	76	9	14	28	30	40	53	61	72	78	83
31	4	14	24	34	44	54	64	65	75	85	77	9	15	26	31	38	54	59	73	79	84
32	4	15	25	32	45	55	62	66	76	83	78	9	16	27	29	39	55	60	71	77	85
33	4	16	23	33	46	53	63	67	74	84	79	9	18	20	34	41	48	62	67	82	87
34	4	17	27	37	38	48	58	68	78	88	80	9	17	22	33	43	47	64	66	81	86
35	4	18	28	35	39	49	56	69	79	86	81	9	19	21	32	42	49	63	65	80	88
36	4	19	26	36	40	47	57	70	77	87	82	2	11	20	29	38	47	56	65	74	83
37	5	11	26	32	39	52	64	67	78	90	83	10	11	24	37	43	49	62	72	77	84
38	5	12	27	33	40	50	62	65	79	91	84	10	12	25	35	41	47	63	73	78	85
39	5	13	28	34	38	51	63	66	77	89	85	10	13	23	36	42	48	64	71	79	83
40	5	14	20	35	42	55	58	70	81	84	86	10	14	27	31	46	52	56	66	80	87
41	5	15	21	36	43	53	56	68	82	85	87	10	15	28	29	44	50	57	67	81	88
42	5	16	22	37	41	54	57	69	80	83	88	10	16	26	30	45	51	58	65	82	86
43	5	17	23	29	45	49	61	73	75	87	89	10	17	21	34	40	55	59	69	74	90
44	5	18	24	30	46	47	59	71	76	88	90	10	18	22	32	38	53	60	70	75	91
45	5	19	25	31	44	48	60	72	74	86	91	10	19	20	33	39	54	61	68	76	89
46	6	11	28	33	45	48	59	70	80	85											

information in dual form; i.e., it lists the names of the 91 points, and after each name lists the names of the 10 lines which pass through that point. Limitations of space make it impossible to list the full names of the points and lines in these tables. As a result the designations P and l are not included, and only the number identifying a given point or a given line is listed. This should cause no confusion, however, since the headings of the columns in each table make it clear whether a particular number, say 25, refers to the point P_{25} or to the line l_{25}. Thus in Table A.1, the entry 25 in the left-hand column denotes the line l_{25}, and the numbers following this entry tell us that the points P_3, P_{15}, P_{22}, P_{32}, P_{45}, P_{52}, P_{62}, P_{66}, P_{73}, and P_{84}, and only these points, lie on l_{25}. Similarly, the entry 25 in the left-hand column of Table A.2 identifies P_{25}, and the numbers following this entry tell us that the lines l_3, l_{17}, l_{28}, l_{36}, l_{41}, l_{52}, l_{60}, l_{65}, l_{76}, and l_{84}, and no others, pass through this point. Though it is a very tedious undertaking, it can be shown that the system described by Tables A.1 and A.2 satisfies Axioms 1 to 6, Sec. 6.2.

Our first example[1] shows that the geometry described by Tables A.1 and A.2 is indeed nondesarguesian.

Example 1 A counterexample to the theorem of Desargues

Using Tables A.1 and A.2, it is easy to verify that for the centrally perspective triangles, $\triangle P_{46}P_{61}P_{83}$ and $\triangle P_{54}P_{59}P_{90}$ (which are perspective from P_1),

$P_{46}P_{61} = l_{70}$, $P_{54}P_{59} = l_{90}$, and the intersection of l_{70} and l_{90} is P_{74}.

$P_{46}P_{83} = l_{11}$, $P_{54}P_{90} = l_{53}$, and the intersection of l_{11} and l_{53} is P_{55}.

$P_{61}P_{83} = l_{24}$, $P_{59}P_{90} = l_{22}$, and the intersection of l_{24} and l_{22} is P_3.

If Desargues' theorem were true, $\triangle P_{46}P_{61}P_{83}$ and $\triangle P_{54}P_{59}P_{90}$ would be axially perspective; i.e., the intersections of pairs of corresponding sides, P_3, P_{55}, P_{74}, would be collinear. However,

P_3P_{55} is the line l_{27}.

P_3P_{74} is the line l_{23}.

$P_{55}P_{74}$ is the line l_{76}.

Thus, instead of being collinear, P_3, P_{55}, and P_{74} are the vertices of a proper triangle, and the conclusion of Desargues' theorem is not fulfilled!

---

[1] Table A.1 and a number of the examples contained in this section are taken from Steven H. Heath, "An Introduction to Finite Projective Geometry," a master's thesis written at the University of Utah in 1967.

In Sec. 6.2 we found that Desargues' theorem could be proved if, in addition to Axioms 1 to 6, we assumed the truth of any one of the following propositions:

1. The theorem of Pappus.

2. A projectivity between ranges on distinct lines in which the intersection of the lines is self-corresponding is a perspectivity.

3. A projectivity between two ranges is completely determined when three pairs of corresponding points are given.

Obviously, then, since Desargues' theorem is false in the system we are discussing, none of these propositions can hold. The following examples illustrate this fact.

Example 2   A counterexample to the theorem of Pappus

For the two collinear triples (P_{37}, P_{40}, P_{45}) and (P_{73}, P_{74}, P_{81}) (collinear on l_5 and l_9, respectively),

$P_{37}P_{74} = l_{90}$, $P_{40}P_{73} = l_{58}$, and the intersection of l_{58} and l_{90} is P_{54}.

$P_{37}P_{81} = l_{32}$, $P_{45}P_{73} = l_{25}$, and the intersection of l_{25} and l_{32} is P_{32}.

$P_{40}P_{81} = l_{42}$, $P_{45}P_{74} = l_{44}$, and the intersection of l_{42} and l_{44} is P_5.

If the theorem of Pappus were true, the three intersections, P_5, P_{32}, and P_{54}, would be collinear. However,

P_5P_{32} is the line l_{45}.
P_5P_{54} is the line l_{41}.
$P_{32}P_{54}$ is the line l_{66}.

Thus instead of being collinear, P_5, P_{32}, and P_{54} are the vertices of a proper triangle, and therefore the theorem of Pappus fails to hold.

Example 3   A counterexample to the perspectivity theorem

If we combine the two perspectivities

$$l_{50}(P_6P_{13}P_{26}P_{30}P_{38}P_{51}P_{61}P_{71}P_{73}P_{87}) \overset{P_1}{\wedge} l_{51}(P_6P_{14}P_{27}P_{28}P_{39}P_{49}P_{62}P_{72}P_{74}P_{88})$$
$$l_{51}(P_6P_{14}P_{27}P_{28}P_{39}P_{49}P_{62}P_{72}P_{74}P_{88}) \overset{P_{70}}{\wedge} l_{54}(P_6P_{91}P_{42}P_{21}P_{52}P_{77}P_{17}P_{64}P_{31}P_{56})$$

we obtain the projectivity

$$l_{50}(P_6P_{13}P_{26}P_{30}P_{38}P_{51}P_{61}P_{71}P_{73}P_{87}) \sim l_{54}(P_6P_{91}P_{42}P_{21}P_{52}P_{77}P_{17}P_{64}P_{31}P_{56})$$

between l_{50} and l_{54} in which the intersection, P_6, is a self-corresponding point. However, the projectivity is *not* a perspectivity because, for instance, the lines joining the corresponding points

P_{13} and P_{91}, namely, l_{68},
P_{26} and P_{42}, namely, l_{37},
P_{30} and P_{21}, namely, l_{13},

are not concurrent since

l_{68} and l_{37} intersect in P_{34},
l_{68} and l_{13} intersect in P_{75},
l_{13} and l_{37} intersect in P_{66}.

Example 4   A counterexample to Axiom 7

From the perspectivities

$$l_1(P_1P_2P_3P_4P_5P_6P_7P_8P_9P_{10}) \overset{P_{12}}{\barwedge} l_{20}(P_{82}P_{30}P_3P_{20}P_{60}P_{50}P_{40}P_{79}P_{91}P_{72})$$

and

$$l_{20}(P_{82}P_{30}P_3P_{20}P_{60}P_{50}P_{40}P_{79}P_{91}P_{72}) \overset{P_{15}}{\barwedge} l_{11}(P_{82}P_{83}P_{73}P_{64}P_{37}P_2P_{46}P_{55}P_{28}P_{19})$$

we have the projectivity

(1) $l_1(P_1P_2P_3P_4P_5P_6P_7P_8P_9P_{10}) \sim l_{11}(P_{82}P_{83}P_{73}P_{64}P_{37}P_2P_{46}P_{55}P_{28}P_{19})$

On the other hand, from the perspectivities

$$l_1(P_1P_2P_3P_4P_5P_6P_7P_8P_9P_{10}) \overset{P_{11}}{\barwedge} l_{83}(P_{82}P_{47}P_{70}P_{76}P_{61}P_{85}P_{42}P_{32}P_{27}P_{10})$$

and

$$l_{83}(P_{82}P_{47}P_{70}P_{76}P_{61}P_{85}P_{42}P_{32}P_{27}P_{10}) \overset{P_{12}}{\barwedge} l_{11}(P_{82}P_{83}P_{73}P_{64}P_{28}P_2P_{19}P_{55}P_{37}P_{46})$$

we have the projectivity

(2) $l_1(P_1P_2P_3P_4P_5P_6P_7P_8P_9P_{10}) \sim l_{11}(P_{82}P_{83}P_{73}P_{64}P_{28}P_2P_{19}P_{55}P_{37}P_{46}$

Obviously, the projectivity (1) and the projectivity (2) are different since they assign different images to the point P_{10}, for instance However, they have *six* pairs of corresponding points in common, namely, (P_1,P_{82}), (P_2,P_{83}), (P_3,P_{73}), (P_4,P_{64}), (P_6,P_2), (P_8,P_{55}), whereas if Axiom 7 were valid, they could have no more than *two* pairs of corresponding points in common without being identical.

In view of the fact that Axiom 7 is not valid, it is not surprising that conics, too, have unusual properties in the geometry we are here considering. Some of these are illustrated in the next two examples.

Example 5

The perspectivities

$$P_5(l_1 l_{38} l_{39} l_{40} l_{41} l_{42} l_{43} l_{44} l_{45} l_{46}) \overset{l_{22}}{\wedge} P_{10}(l_1 l_{91} l_{88} l_{85} l_{83} l_{89} l_{86} l_{87} l_{84} l_{90})$$

and

$$P_{10}(l_1 l_{91} l_{88} l_{85} l_{83} l_{89} l_{86} l_{87} l_{84} l_{90}) \overset{l_{75}}{\wedge} P_{20}(l_{77} l_{53} l_{88} l_{64} l_3 l_{69} l_{20} l_{45} l_{12} l_{31})$$

determine the projectivity

$$P_5(l_1 l_{38} l_{39} l_{40} l_{41} l_{42} l_{43} l_{44} l_{45} l_{46}) \sim P_{20}(l_{77} l_{53} l_{88} l_{64} l_3 l_{69} l_{20} l_{45} l_{12} l_{31})$$

The locus of the intersections of corresponding lines of these two projective pencils is the conic

$$\Gamma : \{P_9, P_{90}, P_{49}, P_{62}, P_{25}, P_{14}, P_{30}, P_5, P_{20}, P_{86}\}$$

Contrary to the expectations we have built up during our study of projective planes in which Desargues' theorem holds, there are lines which meet Γ in more than two points, and there are points of Γ at which there are more than one tangent. In particular

2 lines, l_{60} and l_{80}, meet Γ in 4 points,
1 line, l_{51}, meets Γ in 3 points,
30 lines meet Γ in 2 points,
29 lines meet Γ in 1 point; i.e., are tangent to Γ,
29 lines have no point in common with Γ,

and

At 1 point, P_{30}, there are 5 tangents to Γ,
At 2 points, P_{14} and P_{62}, there are 4 tangents to Γ,
At 4 points, P_9, P_{25}, P_{86}, and P_{90}, there are 3 tangents to Γ,
At 1 point, P_{49}, there are 2 tangents to Γ,
At 2 points, P_5 and P_{20}, there is a single tangent to Γ.

These facts are illustrated in Fig. A.1, where the multisecants of Γ and the tangents at each point of Γ are indicated.

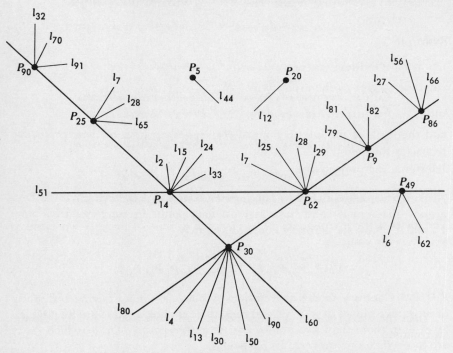

Fig. A.1   The multisecants and tangents of the conic Γ of Example 5.

Example 6

The perspectivities

$$P_{15}(l_2l_{16}l_{25}l_{34}l_{43}l_{52}l_{61}l_{70}l_{79}l_{88}) \overset{l_1}{\wedge} P_{33}(l_4l_{16}l_{23}l_{33}l_{46}l_{53}l_{63}l_{67}l_{74}l_{84})$$

and

$$P_{33}(l_4l_{16}l_{23}l_{33}l_{46}l_{53}l_{63}l_{67}l_{74}l_{84}) \overset{l_{75}}{\wedge} P_{52}(l_{54}l_{42}l_{17}l_{56}l_{25}l_{91}l_6l_{67}l_{77}l_{30})$$

combine to give us the projectivity

$$P_{15}(l_2l_{16}l_{25}l_{34}l_{43}l_{52}l_{61}l_{70}l_{79}l_{88}) \sim P_{52}(l_{54}l_{42}l_{17}l_{56}l_{25}l_{91}l_6l_{67}l_{77}l_{30})$$

and the locus of the intersections of corresponding lines of these two projective pencils is the conic

$$\Gamma_1:\{P_{17},P_{69},P_{52},P_{71},P_{15},P_{75},P_{48},P_8,P_9,P_{44}\}$$

Similarly, the perspectivities

$$P_{15}(l_2 l_{16} l_{25} l_{34} l_{43} l_{52} l_{61} l_{70} l_{79} l_{88}) \overset{l_1}{\wedge} P_{20}(l_3 l_{12} l_{20} l_{31} l_{45} l_{53} l_{64} l_{69} l_{77} l_{88})$$

$$P_{20}(l_3 l_{12} l_{20} l_{31} l_{45} l_{53} l_{64} l_{69} l_{77} l_{88}) \overset{l_{10}}{\wedge} P_7(l_1 l_{63} l_{61} l_{56} l_{58} l_{60} l_{64} l_{59} l_{62} l_{57})$$

and

$$P_7(l_1 l_{63} l_{61} l_{56} l_{58} l_{60} l_{64} l_{59} l_{62} l_{57}) \overset{l_{75}}{\wedge} P_{52}(l_{77} l_6 l_{17} l_{56} l_{25} l_{91} l_{54} l_{67} l_{42} l_{30})$$

combine to give us the projectivity

$$P_{15}(l_2 l_{16} l_{25} l_{34} l_{43} l_{52} l_{61} l_{70} l_{79} l_{88}) \sim P_{52}(l_{77} l_6 l_{17} l_{56} l_{25} l_{91} l_{54} l_{67} l_{42} l_{30})$$

and the locus of the intersections of corresponding lines of these projective pencils is the conic

$$\Gamma_2 : \{P_{13}, P_{51}, P_{52}, P_{71}, P_{15}, P_{75}, P_{91}, P_8, P_{85}, P_{44}\}$$

Although we know that two nonsingular conics in a system in which Desargues' theorem is valid have at most four points in common, the conics Γ_1 and Γ_2 have six points in common, namely,

$$P_8, \ P_{15}, \ P_{44}, \ P_{52}, \ P_{71}, \ P_{75} \quad (!)$$

Exercises

1. Does there exist a pair of projectivities with more than six pairs of mates in common?

2. Does there exist a pair of nonsingular conics with more than six points in common?

ANSWERS TO ODD=NUMBERED EXERCISES

CHAPTER 1

Sec. 1.2, p. 10

1. No. The center of the given circle is the point $X = 0$, $Y = 2$, and its image is the point $X' = 0$, $Z' = \frac{4}{5}$. On the other hand, the center of the image circle is the point $X' = 0$, $Z' = \frac{3}{2}$.

3. Ellipse $4(X')^2 + 3(Z' - 1)^2 = 3$. Hyperbola $2(X')^2 - 2(Z' - \frac{5}{4})^2 = -\frac{9}{8}$.

5. The equations of the transformation are $X' = 4X/(Y + 3)$, $Z' = (2Y - 2)/(Y + 3)$, where X' and Z' are now coordinates in the plane $y = 1$.

7. Yes. The points on the line of intersection of the object plane and the picture plane are invariant.

9. No. The line in the object plane which is the intersection of the object plane and the plane through the viewing point parallel to the picture plane has no image. All other lines in the object plane are represented by lines. Yes.

11. Yes. For instance, in Example 1 the image of the circle $X^2 + Y^2 + aX + bY + c = 0$ is a circle provided $a = 0$ and $c = 3b - 5$.

13. Both parabolas and hyperbolas may have ellipses, parabolas, or hyperbolas as their images.

15. No. This will be the case if and only if the segment $\overline{QR}$ contains no point of the line of intersection of the object plane and the plane through the viewing point parallel to the picture plane.

17. No.

19. Yes, provided the line itself has an image. (See the answer to Exercise 9.)
23. Yes. Let l be the line of intersection of the object plane and the image plane, let $\triangle ABC$ be a triangle in the object plane, let A_1, B_1, C_1 be, respectively, the intersections of l and BC, CA, AB, let A' and B' be, respectively, points in the picture plane such that the angles inscribed in the circular arcs $\overparen{A'B_1C_1}$ and $\overparen{A_1B'C_1}$ are of measure $\alpha \neq 90,^{\circ}$ and let S_a and S_b be the cones whose vertices are A and B and whose directrices are the circles containing $\overparen{A'B_1C_1}$ and $\overparen{A_1B'C_1}$, respectively. Then any point, D, on the curve of intersection of S_a and S_b is a viewing point from which $\triangle ABC$ will appear on the picture plane as an isosceles triangle provided only that each point of the triangle has an image from D. (See the answer to Exercise 9.)

Sec. 1.3, p. 15

1. No.
3. Let p_1 and p_2 be two parallel lines which are not parallel to the axis, l, of the plane perspectivity, and let P be the point of intersection of their images. Then the line through P which is parallel to l is the line into which the vanishing line in the picture plane is rotated. This line has no preimage in the plane of the plane perspective.
5. No. Any line on the intersection of GH and $G'H'$ which does not pass through C is the axis of a plane perspective in which $G \rightarrow G'$ and $H \rightarrow H'$.

9. $x = \dfrac{3a}{3 - 2b} \qquad y = \dfrac{-b}{3 - 2b} \qquad 3 - 2b \neq 0 \qquad v: y = \tfrac{3}{2}$

11. $x = \dfrac{4a - b}{a - b + 3} \qquad y = \dfrac{3b}{a - b + 3} \qquad a - b + 3 \neq 0$

    $v: x - y + 3 = 0$

13. $(0, 4, -3)$ 15. $(0, -3)$
17. Let V be an arbitrary point on the given vanishing line, let t be the line through G' parallel to VC, and let L be the intersection of VG and t. Then the axis is the line on L which is parallel to v.

19. $x = \dfrac{3a}{b} \qquad y = \dfrac{5b - 6}{b} \qquad b \neq 0 \qquad y = 0$

21. $x' = \dfrac{x}{y} \qquad y' = \dfrac{1}{y} \qquad y = 1 \qquad$ 23. $ab = 1$

Sec. 1.4, p. 23

1. Use the transformation described in this section for transforming a general quadrilateral into a parallelogram whose diagonals are perpendicular.

3. By an obvious modification of the construction for locating the center of a plane perspective which will transform a given angle into a right angle, a center can be found which can be used to transform a given angle into an angle of arbitrary measure. Applying this to two angles of a given triangle gives two possible centers, located as the intersections of two circles.

5. Use the construction of Exercise 3 in the particular case when the base angles are transformed so that the common measure of their images is $45°$.

7. Yes. Let $\overline{AB}$ be the given segment and $\overline{A'B'}$ be an image segment of the required length. Then any perspective whose center is the intersection of AA' and BB' and whose axis passes through the intersection of AB and $A'B'$ will map $\overline{AB}$ into $\overline{A'B'}$.

9. Yes.

11. By choosing v so that it does not intersect, intersects in two points, or is tangent to the parabola.

13. If the image is a parabola, the center is the focus of the parabola. If the image is a hyperbola or an ellipse, the center is one of the foci.

15. Let C and v be, respectively, the center and vanishing line of the given perspective, let V_1 and V_2 be the intersections of v and the given circle, Γ, let A be the intersection of the tangents to Γ at V_1 and V_2, and let V_b be the intersection of v and the bisector of $\angle V_1 C V_2$. Then the intersection of Γ and AV_b are the preimages of the vertices of the image of Γ.

17. Every point in the plane is the vertex of a unique right angle which is transformed into a right angle by a given plane perspective.

19. Yes. Let P_1P_2, P_1P_3, P_1P_4 intersect the chosen vanishing line, v, in the points V_2, V_3, V_4, respectively, and determine the circle which is the locus of points, C, such that $\angle V_2 C V_4 = \angle V_3 C V_4$. Let P_2P_1, P_2P_3, P_2P_4 intersect v in the points $\bar{V}_1$, $\bar{V}_3$, $\bar{V}_4$, respectively, and determine the locus of points, C, such that $\angle \bar{V}_1 C \bar{V}_4 = \angle \bar{V}_3 C \bar{V}_4$. Then either of the intersections of these two circles is a center which with v will accomplish the required mapping.

21. Yes. Choose C to be one focus of the given ellipse. Then if the line, l, through C perpendicular to the major axis of the ellipse intersects the ellipse in the points T_1 and T_2, the required vanishing line is the line parallel to l through the intersection of the tangents to the ellipse at T_1 and T_2.

CHAPTER 2

Sec. 2.2, p. 35

1. (*a*) (2,3,1), (4,6,2), (−2,−3,−1)
   (*b*) (−1,4,1), (−2,4,2), (3,−12,−3)

(c) $(-3,-5,1)$, $(3,5,-1)$, $(6,10,-2)$

(d) $(0,2,1)$, $(0,4,2)$, $(0,6,3)$

(e) $(\frac{2}{3},\frac{4}{3},1)$, $(2,4,3)$, $(-4,-8,-6)$

(f) $(-\frac{1}{2},\frac{3}{2},1)$, $(-1,3,2)$, $(1,-3,-2)$

(g) $(\frac{1}{2},-\frac{2}{3},1)$, $(3,-4,6)$, $(6,-8,12)$

(h) $(-3,0,1)$, $(3,0,-1)$, $(6,0,-2)$

3. (a) $x_2x_3 = x_1^2$ (b) $x_1x_2 = 2x_3^2$

    (c) $x_1^2 + x_2^2 = r^2x_3^2$ (d) $2x_1x_2 + 3x_1x_3 + x_2x_3 = x_3^2$

    (e) $x_2^2x_3 = x_1^3$ (f) $x_2x_3^2 = x_1^3 - 2x_1^2x_3$

    (g) $x_1^2 - 2x_1x_2 + 3x_2x_3 = 2x_3^2$ (h) $x_1x_2^2x_3 = x_1x_3^3 + x_2^4$

5. (a) $[1,-2,-3]$ (b) $[3,2,-1]$ (c) $[2,-1,0]$

    (d) $[1,0,-3]$ (e) $[0,1,0]$

11. (a) $(-4,6,5)$, $-4l_1 + 6l_2 + 5l_3 = 0$

    (b) $(2,5,-8)$, $2l_1 + 5l_2 - 8l_3 = 0$

    (c) $(0,0,1)$, $l_3 = 0$

    (d) $(4,-1,0)$, $4l_1 - l_2 = 0$

    (e) $(4,-3,2)$, $4l_1 - 3l_2 + 2l_3 = 0$

    (f) $(1,5,-2)$, $l_1 + 5l_2 - 2l_3 = 0$

13. $\dfrac{x_2'x_3 - x_2x_3'}{x_1'x_3 - x_1x_3'}$

15. (a) $[0,1,1]$ (b) $[1,4,2]$, $[1,1,-1]$ (c) $[1,1,1]$, $[1,1,-1]$

    (d) $[0,1,0]$, $[1,1,1]$

19. (a) $l_1^2 + l_2^2 = l_3^2$ (b) $4l_2l_3 = l_1^2$

    (c) $8l_1l_2 = l_3^2$ (d) $4l_1^3 + 27l_2l_3^2 = 0$

    (e) $l_1^2 - 8l_1l_2 + 2l_1l_3 + l_3^2 = 0$ (f) $l_1^2 - 6l_1l_2 + l_2^2 - 4l_2l_3 = 0$

    (g) $l_1^2 - l_2^2 = l_3^2$ (h) $27l_2^2l_3 + 4l_1^3 = 0$

23. No, since $a \not\parallel a$ under the usual definition of parallel lines. No. Yes.

Sec. 2.3, p. 42

1. (a) $(1,2,0)$ (b) $(4,3,0)$ (c) $(-2,1,0)$

    (d) $(0,1,0)$ (e) $(1,0,0)$ (f) $(3,1,0)$

3. (a) $(2,1,0)$ (b) $(5,-2,0)$ (c) $(1,1,0)$

5. (a) $(1,i,0)$, $(1,-i,0)$ (b) $(1,1,0)$, $(1,-1,0)$ (c) $(2,1,0)$, $(1,1,0)$

    (d) $(0,1,0)$ (e) $(1,-2,0)$

    (f) $(-1 + i, 1, 0)$, $(-1 - i, 1, 0)$

7. (a) No   (b) Yes, the ideal line itself has this property.

9. (a) $[1,0,1]$, $[0,1,1]$, both in E_2

    (b) $[\sqrt{2},\sqrt{2},2]$, $[\sqrt{2},\sqrt{2},-2]$, both in E_2

    (c) $[1,1,1]$, $[0,1,0]$ in E_2; $[-1 \pm i\sqrt{3}, -1 \mp i\sqrt{3}, 2]$ in Π_2

    (d) $[0,0,1]$ in E_2^+; $[-2,\pm i,1]$ in Π_2

11. $(a,b,0)$, $(a,-b,0)$ 13. (b) Yes

Sec. 2.4, p. 47

1. $y = x^2$ does not contain the point $(0,1,0)$; $x_2 x_3 = x_1{}^2$ does.

3. (a) Parabola, hyperbola, ellipse
   (b) Hyperbola, hyperbola, parabola
   (c) Parabola, parabola, ellipse
   (d) Circle, hyperbola, hyperbola

5. (a) Two nonintersecting hyperbolas, two nonintersecting hyperbolas, two concentric circles
   (b) Two nonintersecting hyperbolas, a circle and a hyperbola tangent at $(0,0)$ and at $(0,1)$, two parabolas tangent at $(0,1)$
   (c) Two hyperbolas tangent at $(0,1)$ and at $(0,-1)$, two nonintersecting hyperbolas, a circle and an ellipse tangent at $(1,0)$ and at $(-1,0)$
   (d) Two nonintersecting hyperbolas, a circle and a hyperbola tangent at $(0,1)$ and at $(0,-1)$, a circle and a hyperbola tangent at $(0,1)$ and at $(0,-1)$.

7. (a) $\frac{1}{2}$, 1, $\frac{1}{3}$
   (b) 1, there is no angle since $[0, 1, 0]$ is the ideal line, $-\frac{1}{2}$.
   (c) The lines are parallel, -1, 1.
   (d) 1, $\frac{1}{2}$, there is no angle since $[0,0,2]$ is the ideal line.

9.

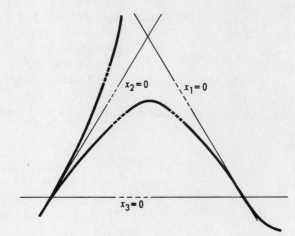

CHAPTER 3

Sec. 3.2, p. 60

1. $AB = \begin{Vmatrix} 17 & 24 \\ 8 & 10 \end{Vmatrix}$ $BA = \begin{Vmatrix} -1 & 2 & 10 \\ -1 & 4 & 6 \\ -1 & 6 & 24 \end{Vmatrix}$

$ABA = \begin{Vmatrix} -7 & 34 & 140 \\ -2 & 16 & 62 \end{Vmatrix}$ $BAB = \begin{Vmatrix} 33 & 44 \\ 58 & 78 \\ 83 & 112 \end{Vmatrix}$

3. $A^2 = \begin{Vmatrix} -1 & 8 \\ -4 & 7 \end{Vmatrix}$ $A^2 - 3A + 2I = \begin{Vmatrix} -2 & 2 \\ -1 & 0 \end{Vmatrix}$

5. (a) $A^2 = \begin{Vmatrix} -5 & -6 \\ 9 & -2 \end{Vmatrix}$ $A^2 - 3A + 2I = \begin{Vmatrix} -6 & 0 \\ 0 & -6 \end{Vmatrix}$

(b) $A^2 = \begin{Vmatrix} 1 & 2 & 2 \\ -2 & 7 & 2 \\ -1 & 2 & 6 \end{Vmatrix}$ $A^2 - 3A + 2I = \begin{Vmatrix} 0 & -1 & 2 \\ -2 & 6 & -4 \\ 2 & -7 & 8 \end{Vmatrix}$

9. AB is symmetric if and only if $AB = BA$.

13. $x_1^2 - 2x_1x_2 + 4x_1x_3 - x_2^2 + 6x_2x_3 + 4x_3^2$

15. (a) $X = -\dfrac{1}{2} \begin{Vmatrix} 6 & 6 \\ -5 & -4 \end{Vmatrix}$ (b) $X = -\dfrac{1}{3} \begin{Vmatrix} -4 & -6 \\ 5 & 6 \end{Vmatrix}$

(c) $X = \dfrac{1}{2} \begin{Vmatrix} 0 & 0 \\ 1 & 2 \end{Vmatrix}$ (d) $X = \begin{Vmatrix} 1 & 2 \\ -1 & -2 \end{Vmatrix}$ No

Sec. 3.3, p. 67

1. (a) adj $A = \begin{Vmatrix} 3 & -2 \\ -2 & 1 \end{Vmatrix}$ $A^{-1} = \begin{Vmatrix} -3 & 2 \\ 2 & -1 \end{Vmatrix}$

(b) adj $A = \begin{Vmatrix} 4 & -2 \\ -2 & 1 \end{Vmatrix}$ A^{-1} does not exist since $|A| = 0$

(c) adj $A = \begin{Vmatrix} 0 & -3 \\ 4 & 2 \end{Vmatrix}$ $A^{-1} = \dfrac{1}{12} \begin{Vmatrix} 0 & -3 \\ 4 & 2 \end{Vmatrix}$

(d) adj $A = \begin{Vmatrix} 1 & 1 \\ -1 & 2 \end{Vmatrix}$ $A^{-1} = \dfrac{1}{3} \begin{Vmatrix} 1 & 1 \\ -1 & 2 \end{Vmatrix}$

3. (a) All values except $-2, 3$ (b) All values
   (c) All values except $1, 2$ (d) No values
   (e) All values

5. (a) $|A| = 1$, $A^{-1} = \begin{Vmatrix} 3 & -1 \\ -2 & 1 \end{Vmatrix}$

   (b) $|A| = -2$, $A^{-1} = -\dfrac{1}{2} \begin{Vmatrix} 5 & -3 \\ -4 & 2 \end{Vmatrix}$

   (c) $|A| = -1$, $A^{-1} = \begin{Vmatrix} -4 & 3 & -1 \\ 5 & -3 & 1 \\ -2 & 1 & 0 \end{Vmatrix}$

   (d) $|A| = 2$, $A^{-1} = \dfrac{1}{2} \begin{Vmatrix} -2 & 6 & 2 \\ -1 & 3 & 2 \\ 2 & -4 & -2 \end{Vmatrix}$

7. Yes   9. Yes, if $|A| \neq 0$.

Sec. 3.4, p. 73

3. The identity transformation, $Z = X$. The transformation, $X = AY$.

5. If $ad - bc \neq 0$, and if neither (x_1, x_2) nor (y_1, y_2) is the pair $(0,0)$, then if these values satisfy the bilinear equation, there exists a value of k for which they will also satisfy the matric equation.

7. (b) $10y_1{}^2 - 27y_1 y_2 + 18y_2{}^2 + 26y_1 y_3 - 35y_2 y_3 + 17y_3{}^2 = 0$,
$l_1' = l_1 - 6l_2 + 2l_3$

9. $l_2' = -l_1 + 9l_2 - 3l_3$,
$l_3' = l_1 - 8l_2 + 3l_3$

Sec. 3.5, p. 78

7. Linearly independent

9. (a) No   (b) Yes   (c) No   15. (b) No

Sec. 3.6, p. 89

1. (a) 3   (b) 2   (c) 1

3. (a) $x = -\frac{11}{5}$, $y = 3$, $z = -\frac{7}{5}$ (b) $x = \frac{14}{5}$, $y = -\frac{6}{5}$, $z = \frac{9}{5}$
   (c) $x = 0$, $y = -1$, $z = 0$

7. $\lambda = -8$, $x_1 = 3$, $x_2 = 2$; $\lambda = 2$, $x_1 = 1$, $x_2 = 4$

9. $\lambda = -1$, $x_1 = 1$, $x_2 = 1$, $x_3 = -2$
$\lambda = 1$, $x_1 = 1$, $x_2 = -1$, $x_3 = 0$
$\lambda = 5$, $x_1 = 1$, $x_2 = 3$, $x_3 = 4$

Sec. 3.7, p. 95

1. (a) Indefinite   (b) Positive definite   (c) Positive definite
   (d) Indefinite   (e) Positive definite

3. $\begin{Vmatrix} -3 + i\sqrt{3} \\ -3 + i\sqrt{3} \\ 12 - i\sqrt{3} \end{Vmatrix}$, $\begin{Vmatrix} -3 - i\sqrt{3} \\ -3 - i\sqrt{3} \\ 12 + i\sqrt{3} \end{Vmatrix}$

9. The quadratic forms corresponding to the pairs $(\lambda,\mu) = (-1,1)$, $(5,1)$, $(6,1)$ are singular.

CHAPTER 4

Sec. 4.2, p. 107

1. If the points in each set are named P_1, P_2, P_3, then
    (a) $P_1 - P_2 + P_3 = 0$ (b) $-P_1 + 2P_2 + P_3 = 0$
    (c) $3P_1 + P_2 - P_3 = 0$ (d) $-P_1 + P_2 + P_3 = 0$

3. If the points in each set are named P_1, P_2, P_3, P_4, then
    (a) $P_4 = -3P_2 + 2P_3$ (b) $P_4 = \frac{3}{2}P_1 - \frac{7}{2}P_2 + P_3$
    (c) $P_4 = \frac{1}{3}P_2 - \frac{1}{3}P_3$ (d) $P_4 = -\frac{5}{3}P_1 + \frac{7}{3}P_2 + \frac{8}{3}P_3$

5. $\lambda = y_1 + y_2 + y_3,\ \mu = -2$ 9. No

Sec. 4.3, p. 114

1. (a) $A'':(3,-4,2), B'':(1,4,2), C'':(1,0,1)$ are collinear on $4x_1 + x_2 - 4x_3 = 0$.
   (b) $A'':(3,2,2), B'':(2,5,1), C'':(7,1,5)$ are collinear on $8x_1 - x_2 - 11x_3 = 0$. (c) $A'':(3,2,2), B'':(1,4,2), C'':(4,1,2)$ are collinear on $2x_1 + 2x_2 - 5x_3 = 0$. (d) $A'':(0,2,-3), B'':(1,1,0), C'':(2,0,3)$ are collinear on $3x_1 - 3x_2 - 2x_3 = 0$.

3. Let a, b, c be three lines on a point L, and let a', b', c' be three lines on a second point L', none of these lines coinciding with the line determined by L and L'. Then the join of bc' and $b'c$, the join of ac' and $a'c$, and the join of ab' and $a'b$ are collinear.

7. (a) If two triangles are centrally perspective in E_2, then

    1. If the triangles contain no pair of corresponding sides which are parallel, the intersections of corresponding sides are collinear.

    2. If the triangles contain exactly one pair of corresponding sides which are parallel, these sides are parallel to the line determined by the intersections of the two pairs of nonparallel corresponding sides.

    3. If the triangles contain two pairs of corresponding sides which are parallel, the remaining sides are also parallel.

(*b*) If two triangles in E_2 are so related that the lines joining corresponding vertices are parallel, then 1, 2, 3, as in part (*a*).

11. $(a\beta)(b\gamma)(c\alpha) - (a\gamma)(b\alpha)(c\beta) = 0$

13. No

Sec. 4.4, p. 121

1. $\lambda\mu' - \lambda'\mu - 3\mu\mu' = 0$ 3. (*a*) (2,3,0)   (*b*) (1,−2,1)

5. $3\alpha - 5\gamma = 0$ 7. (*a*) No   (*b*) Sometimes

Sec. 4.5, p. 132

1. The correspondence is not a perspectivity.

3. $U{:}(1,0,1)$, $V{:}(3,2,3)$

5. The locus of U is $x_1 + 3x_2 - x_3 = 0$, minus the points $(0,1,3)$ and $(1,0,1)$. The locus of V is $x_1 - x_2 + x_3 = 0$, minus the points $(0,1,1)$ and $(1,0,-1)$.

7. (*a*) $2x_1 - 2x_2 + x_3 = 0$, minus the points $(0,1,2)$ and $(1,1,0)$
   (*b*) $8x_1 - 5x_2 + 2x_3 = 0$, minus the points $(0,2,5)$ and $(1,2,1)$
   (*c*) $2x_1 + 4x_2 - x_3 = 0$, minus the points $(0,1,4)$ and $(-1,1,2)$

11. $(0,1,1)$ and $(0,1,0)$

13. (*a*) $\left\|\begin{matrix}\lambda_2\\\mu_2\end{matrix}\right\| = \left\|\begin{matrix}1 & 0\\-1 & -1\end{matrix}\right\| \cdot \left\|\begin{matrix}\lambda_1\\\mu_1\end{matrix}\right\|$ (*b*) $\left\|\begin{matrix}\lambda_2\\\mu_2\end{matrix}\right\| = \left\|\begin{matrix}-1 & -1\\1 & -1\end{matrix}\right\| \cdot \left\|\begin{matrix}\lambda_1\\\mu_1\end{matrix}\right\|$

15. (*a*) $\left\|\begin{matrix}\lambda_2\\\mu_2\end{matrix}\right\| = \left\|\begin{matrix}1 & 0\\0 & -1\end{matrix}\right\| \cdot \left\|\begin{matrix}\lambda_1\\\mu_1\end{matrix}\right\|$ (*b*) $\left\|\begin{matrix}\lambda_2\\\mu_2\end{matrix}\right\| = \left\|\begin{matrix}1 & -1\\0 & -1\end{matrix}\right\| \cdot \left\|\begin{matrix}\lambda_1\\\mu_1\end{matrix}\right\|$

17. (*a*) $\left\|\begin{matrix}\lambda_2\\\mu_2\end{matrix}\right\| = \left\|\begin{matrix}0 & 1\\1 & 2\end{matrix}\right\| \cdot \left\|\begin{matrix}\lambda_1\\\mu_1\end{matrix}\right\|$ (*b*) $\left\|\begin{matrix}\lambda_2\\\mu_2\end{matrix}\right\| = \left\|\begin{matrix}0 & 1\\1 & 1\end{matrix}\right\| \cdot \left\|\begin{matrix}\lambda_1\\\mu_1\end{matrix}\right\|$

19. (*a*) $\left\|\begin{matrix}\lambda_2\\\mu_2\end{matrix}\right\| = \left\|\begin{matrix}0 & 1\\-1 & 0\end{matrix}\right\| \cdot \left\|\begin{matrix}\lambda_1\\\mu_1\end{matrix}\right\|$ (*b*) $\left\|\begin{matrix}\lambda_2\\\mu_2\end{matrix}\right\| = \left\|\begin{matrix}0 & 1\\-1 & -1\end{matrix}\right\| \cdot \left\|\begin{matrix}\lambda_1\\\mu_1\end{matrix}\right\|$

Sec. 4.6, p. 142

1. (*a*) $\lambda_1\lambda_2 - \lambda_1\mu_2 + 3\mu_1\lambda_2 - 2\mu_1\mu_2 = 0$
   (*b*) $2\lambda_1\lambda_2 - \lambda_1\mu_2 + \mu_1\lambda_2 + \mu_1\mu_2 = 0$
   (*c*) $-3\lambda_1\lambda_2 + \lambda_1\mu_2 - 4\mu_1\lambda_2 + 2\mu_1\mu_2 = 0$
   (*d*) $3\lambda_1\lambda_2 - \lambda_1\mu_2 + 2\mu_1\lambda_2 - 2\mu_1\mu_2 = 0$
   (*e*) $\lambda_1\lambda_2 - \mu_1\mu_2 = 0$

3. Taking P_1 and Q_1 as the base points on the first line and P_2 and Q_2 as the base points on the second line, the equations of the projectivities are:

(a) $\lambda_1\mu_2 - \mu_1\lambda_2 = 0$ (b) $2\lambda_1\mu_2 - \mu_1\lambda_2 = 0$

(c) $2\lambda_1\mu_2 + 3\mu_1\lambda_2 = 0$

Sec. 4.7, p. 150

1. (a) $\frac{10}{9}$ (b) $-\frac{5}{7}$ (c) $-\frac{5}{4}$ (d) $\frac{2}{3}$

7. 2

17. (a) $\frac{8}{15}$ or $\frac{15}{8}$, depending on how the parameters are paired

(b) $a_0b_2 - 2a_1b_1 + a_2b_0 = 0$

19. Yes

21. Let $\triangle p_1p_2p_3$ be an arbitrary triangle, and let u be an arbitrary line. If q_i is the join of the point p_jp_k and the harmonic conjugate of p_iu with respect to p_ip_j and p_ip_k, then q_1, q_2, q_3 are concurrent.

Sec. 4.8, p. 162

1. (a) $(1,1)$, $(2,-3)$ (b) $(1,2)$, $(3,-1)$ (c) $(2,-1)$, $(1,1)$

(d) $(3,-1)$, $(1,1)$ (e) $(3,1)$, $(2,-1)$ (f) $(1,1)$

5. $2(a_{11} + a_{22})^2 = a_{11}a_{22} - a_{12}a_{21}$

7. (a) 2   (b) 4   (c) 5   (d) 3   (e) 3   (f) 6   (g) 3   (h) 4

17. (a) The equation of the mapping is $\left\|\begin{matrix}\lambda_2\\\lambda_2'\end{matrix}\right\| = \left\|\begin{matrix}1 & 0\\-1 & -1\end{matrix}\right\| \cdot \left\|\begin{matrix}\lambda_1\\\lambda_1'\end{matrix}\right\|$.

(b) The equation of the mapping is $\left\|\begin{matrix}\lambda_2\\\lambda_2'\end{matrix}\right\| = \left\|\begin{matrix}1 & 0\\0 & -1\end{matrix}\right\| \cdot \left\|\begin{matrix}\lambda_1\\\lambda_1'\end{matrix}\right\|$.

19. The equation of the transformation is $\left\|\begin{matrix}\lambda_2\\\lambda_2'\end{matrix}\right\| = \left\|\begin{matrix}-v_3 & v_1\\-v_2 & v_3\end{matrix}\right\| \cdot \left\|\begin{matrix}\lambda_1\\\lambda_1'\end{matrix}\right\|$.

21. $(0,1,0)$, $(0,1,-3)$

23. (a) $x_1 + x_3 = 0$

(b) $(3x_1 - 2x_2 + x_3)^2 + 72(x_2^2 + x_2x_3) = 0$

Sec. 4.9, p. 175

1. (a) $(1,1,2)$, $(2,2,5)$ (b) $(2,0,1)$

(c) $(2, 2, -1 - \sqrt{6})$, $(2, 2, -1 + \sqrt{6})$

(d) $(i, -1, -1 + i)$, $(-i, -1, -1 - i)$

3. (a) $a_{23}x_2x_3 + a_{13}x_1x_3 + a_{12}x_1x_2 = 0$
   (b) $a_{11}x_1{}^2 + a_{23}x_2x_3 = 0$
   (c) $a_{11}x_1{}^2 + a_{22}x_2{}^2 + a_{33}x_3{}^2 = 0$

5. (a) $(1,-5,2)$ (b) $(5,36,32)$ (c) $(1,2,3)$

7. (a) $(0,0,1), (0,1,0), (2,-4,-1), (1,1,1)$
   (b) $(0,0,1), (0,1,0), (1,1,1)$

9. (a) $(r-1)x_1x_2 - rx_1x_3 + x_2x_3 = 0$
   (b) $rx_1x_2 + (2-r)x_1x_3 + rx_2{}^2 + (r-1)x_2x_3 + (1-2r)x_3{}^2 = 0$

11. (a) $(0,0,1), (1,-1,1), (1,1,0)$ (b) $(1,-3,-1), (2,3,0), (1,-4,1)$
    (c) $(2,3,0), (1,-3,-1), (1,-4,1)$

13. $-a_1a_2 + a_1a_3 + a_2{}^2 + 2a_2a_3 + a_3{}^2 = 0$
    $-8x_1x_2 + 8x_1x_3 + x_2{}^2 + 2x_2x_3 + x_3{}^2 = 0$

17. (a) $y_1' = y_1y_3, y_2' = y_2y_3, y_3' = 2y_1y_2$
    (b) $y_1' = y_1y_3, y_2' = y_2y_3, y_3' = -y_1{}^2 + 2y_1y_2$
    (c) $y_1' = -y_1y_2 - y_1y_3 + y_3{}^2, y_2' = y_2{}^2 - y_2y_3, y_3' = y_2y_3 - 2y_1y_2$
    (d) $y_1' = -y_1y_2 - y_1y_3 + y_3{}^2, y_2' = -y_1y_2 + y_2{}^2 - y_2y_3,$
    $y_3' = y_1{}^2 - 2y_1y_2 - y_1y_3 + y_2y_3$

CHAPTER 5

Sec. 5.2, p. 185

1. Yes

5. C_1: $9x^2 - 6xy + y^2 - 20x + 4y + 12 = 0$, a parabola
        $12x^2 + 4xy + y^2 - 20x - 6y + 9 = 0$, an ellipse
        $12x^2 - 20xy + 9y^2 + 4x - 6y + 1 = 0$, an ellipse
   C_2: $4x^2 + xy - 8x + 4 = 0$, a hyperbola
        $4x^2 - 8x + y + 4 = 0$, a parabola
        $4x^2 - 8xy + 4y^2 + y = 0$, a parabola
   C_3: $x^2 + 4xy - 3y^2 + 4x - 12y - 8 = 0$, a hyperbola
        $8x^2 + 12xy + 3y^2 - 4x - 4y - 1 = 0$, a hyperbola
        $8x^2 - 4xy - y^2 + 12x - 4y + 3 = 0$, a hyperbola

7. (a) $x_1' = 12x_1 + 12x_2 - 6x_3$ (b) $x_1' = \quad\;\; - 3x_2 + 3x_3$
       $x_2' = \;\; 2x_1 + \;\; 2x_2 + 4x_3$ $x_2' = 2x_1 - 4x_2 + 2x_3$
       $x_3' = \;\; 2x_1 - \;\; 3x_2 - \;\; x_3$ $x_3' = 6x_1 - 6x_2 + 6x_3$

9. (a) The new ideal line ($x_3' = 0$) can be any line of the envelope $a_1 - 2a_2 = 0$, provided that the new unit point is chosen on the (singular) conic $(2a_2 - a_3)x_1{}^2 + a_2(2x_1x_2 + x_2{}^2) = 0$.

(*b*) The new ideal line ($x_3' = 0$) can be any line of the envelope $2a_1a_2 - 2a_2{}^2 - a_3{}^2 = 0$, provided the new unit point is chosen on the (singular) conic $a_1{}^2x_1{}^2 = a_2{}^2(2x_1{}^2 + 2x_1x_2 + x_2{}^2)$.

Sec. 5.3, p. 194

1. $y_1' = -y_1 - y_2 + y_3$, $y_2' = 3y_1 - 5y_2 + y_3$, $y_3' = 6y_1 - 2y_2 - 2y_3$. The transformation is singular. Any point on l is transformed into itself. The point O does not have a unique image since it is transformed into every point on l.

3. (*a*) $y_1' = -3y_1, y_2' = 3y_2, y_3' = -y_1 + y_2$. The transformation is singular. The invariant points are $(0,3,1)$ and $(3,0,1)$. The point O does not have a unique image since it is transformed into the entire line l.
   (*b*) $y_1' = 0$, $y_2' = 0$, $y_3' = y_1 + 2y_2$. The transformation is singular. There are no invariant points. Every point on the line $x_1 + 2x_2 = 0$ is transformed into the entire line l.

7. $P' = (A_2{}^T)^{-1}A_1{}^T P$

9. The characteristic values of a nonsingular matrix and the characteristic values of its inverse are reciprocals.

13. The conic $X^T A X = 0$. The envelope consisting of the lines tangent to $X^T A X = 0$.

15. (*a*) No   (*b*) No

Sec. 5.4, p. 199

1. No

3. The only characteristic value of A, and therefore of BAB^{-1}, is $k = 1$. For this root both A and BAB^{-1} are of rank 1. Two linearly independent characteristic vectors of BAB^{-1} are

$$X_1 = \begin{Vmatrix} 1 \\ -2 \\ 0 \end{Vmatrix} \qquad X_2 = \begin{Vmatrix} 1 \\ -2 \\ 1 \end{Vmatrix}$$

5. The characteristic values of A, and therefore of BAB^{-1}, are -1, 2, 3. For each of these roots, both A and BAB^{-1} are of rank 2. The characteristic vectors of BAB^{-1} are

$$k = -1: X_1 = \begin{Vmatrix} 5 \\ 6 \\ 6 \end{Vmatrix} \qquad k = 2: X_2 = \begin{Vmatrix} 1 \\ 6 \\ 3 \end{Vmatrix} \qquad k = 3: X_3 = \begin{Vmatrix} 1 \\ 2 \\ 2 \end{Vmatrix}$$

Sec. 5.5, p. 203

1. (a) $k_1 = -1$, $k_2 = 2$, $k_3 = -2$; $F_1:(2,-1,2)$, $F_2:(1,0,1)$, $F_3:(2,-1,1)$
   (b) $k_1 = 1$, $k_2 = -1$, $k_3 = 2$; $F_1:(1,1,-1)$, $F_2:(2,1,0)$, $F_3:(0,1,-1)$
   (c) $k_1 = 1$, $k_2 = 2$, $k_3 = 3$; $F_1:(1,1,-1)$, $F_2:(2,1,0)$, $F_3:(0,1,-1)$
   (d) $k_1 = 1$, $k_2 = i$, $k_3 = -i$; $F_1:(2,-1,1)$, $F_2:(-1 - 3i, 2i\ -1 - 3i)$,
   $F_3:(-1 + 3i, -2i, -1 + 3i)$

3. If k_1, k_2, k_3 are distinct cube roots of 1. If k_1, k_2, k_3 are distinct fourth roots of 1. If k_1, k_2, k_3 are distinct nth roots of 1, provided they are not also all pth roots of 1 for some $p < n$.

5. Rotation about the origin through an angle of measure $-\theta$.

7. (a) $x_1' = k_1x_1 \qquad + (k_3 - k_1)x_3$
   $x_2' = \qquad k_2x_2 + (k_3 - k_2)x_3$
   $x_3' = \qquad\qquad k_3x_3$
   (b) $x_1' = 2k_1x_1$
   $x_2' = \qquad (k_2 + k_3)x_2 + (k_2 - k_3)x_3$
   $x_3' = \qquad (k_2 - k_3)x_2 + (k_2 + k_3)x_3$
   (c) $x_1' = \quad (k_2 + k_3)x_1 - (k_2 - k_3)x_2 + (k_2 - k_3)x_3$
   $x_2' = \quad (k_3 - k_1)x_1 + (k_3 + k_1)x_2 - (k_3 - k_1)x_3$
   $x_3' = -(k_1 - k_2)x_1 + (k_1 - k_2)x_2 + (k_1 + k_2)x_3$

9. If the coordinate system is chosen so that the equations of the collineation are given by (1), the equations of the required loci are

$$a_1(k_2 - k_3)x_2x_3 + a_2(k_3 - k_1)x_3x_1 + a_3(k_1 - k_2)x_1x_2 = 0$$

and

$$k_1a_1(k_2 - k_3)x_2'x_3' + k_2a_2(k_3 - k_1)x_3'x_1' + k_3a_3(k_1 - k_2)x_1'x_2' = 0$$

11. $\left(\dfrac{a}{k_2 - k_3}, \dfrac{b}{k_3 - k_1}, \dfrac{c}{k_1 - k_2}\right)$

Sec. 5.6, p. 208

5. (a) The collineation is of type III. $F_1:(-1,3,2)$ is the fixed point corresponding to the simple root $k_1 = 2$; $f_1: 2x_1 + x_2 - x_3 = 0$ is the line of fixed points corresponding to the double root $k_2 = 1$.
   (b) The collineation is of type II. The fixed points are $F_1:(1,2,0)$ and $F_2:(0,1,1)$, corresponding, respectively, to the simple root $k_1 = 2$ and the double root $k_2 = -1$.
   (c) The collineation is of type II. The fixed points are $F_1:(1,-1,1)$ and $F_2:(1,2,2)$, corresponding, respectively, to the simple root $k_1 = -1$ and the double root $k_2 = 1$.

(*d*) The collineation is of type III. $F_1:(-1,3,2)$ is the fixed point corresponding to the simple root $k_1 = -1$; $f_1: 2x_1 + x_2 - x_3 = 0$ is the line of invariant points corresponding to the double root $k_2 = 1$.

7. (*a*) $x_1' = k_1 x_1$

$x_2' = \qquad\qquad a_{22}x_2 + (a_{22} - k_2)\ x_3 \qquad a_{22} \neq k_2$

$x_3' = \quad (k_2 - a_{22})x_2 + (2k_2 - a_{22})x_3$

(*b*) $x_1' = \qquad k_2 x_1 \qquad\qquad + a_{13}x_3$

$x_2' = (k_2 - k_1)x_1 + k_1 x_2 + a_{13}x_3 \qquad a_{13} \neq 0$

$x_3' = \qquad\qquad\qquad k_2 x_3$

9. $T_1:\ x_1' = x_2,\ x_2' = x_1,\ x_3' = x_3$

$T_2:\ x_1' = x_1 x_3,\ x_2' = x_2 x_3,\ x_3' = x_1 x_2$

Sec. 5.7, p. 212

5. (*a*) $x_1' = k_1 x_1 \qquad\qquad + a_{13}x_3$

$x_2' = a_{21}x_1 + k_1 x_2 + a_{23}x_3 \qquad a_{21},\, a_{13},\, k_1 \neq 0$

$x_3' = \qquad\qquad k_1 x_3$

(*b*) $x_1' = (k_1 + a_{31})x_1 + a_{12}x_2 \qquad - a_{31}x_3 \qquad k,a_{31} \neq 0$

$x_2' = \qquad\qquad k_1 x_2 \qquad\qquad\qquad a_{12} \neq a_{32}$

$x_3' = \qquad a_{31}x_1 + a_{32}x_2 + (k - a_{31})x_3$

(*c*) $x_1' = k_1 x_1 + \quad a_{12}x_2 \qquad + a_{13}x_3 \qquad a_{12} + a_{13} \neq 0$

$x_2' = \qquad\qquad a_{22}x_2 + (k_1 - a_{22})\ x_3 \qquad k_1 \neq a_{22}$

$x_3' = -(k_1 - a_{22})x_2 + (2k_1 - a_{22})x_3$

7. If and only if they have either the same axis or the same center.

9. Each locus consists of the axis of the transformation and the line determined by the given point and the center of the transformation.

Sec. 5.8, p. 216

3. For all values of a and b the following transformations will accomplish the given mappings. The lack of uniqueness follows from the fact that in each set, three of the points and their images are collinear.

(*a*) $x_1' = 2ax_1 + (4a - 6b)x_2 - (4a - 4b)x_3$

$x_2' = 2bx_1 - \qquad 2bx_2$

$x_3' = \ bx_1 + \qquad bx_2 - \qquad 2bx_3$

(*b*) $x_1' = 2ax_1 - \ (2a + b)x_2 + (6a + 2b)x_3$

$x_2' = \qquad\qquad - \qquad 2bx_3$

$x_3' = \qquad\qquad bx_2 + \qquad 2bx_3$

5. (*a*) If P is the point $(1,1,1)$, the transformation is of type VI. If P is on one of the lines $x_1 = x_2$, $x_2 = x_3$, $x_3 = x_1$ but is not the point $(1,1,1)$, the transformation is of type III. If P is not on one of these three lines, the transformation is of type I.

(b) If P is a point of the cubic curve $(x_1 + x_2 + x_3)^3 = 27x_1x_2x_3$, the transformation is of type II. If P is not a point of this curve, the transformation is of type I.

Sec. 5.9, p. 227

5. No   11. (a) No   (b) No   (c) No   (d) No

23. For each triangle the axes of the six multiple perspectivities are the six sides of the other two triangles.

Sec. 5.10, p. 234

3. No

7. $x_1' = \quad a_{11}a_{12}x_1 + \quad a_{12}{}^2x_2 - \quad\quad ka_{12}{}^2x_3$
$x_2' = (1 - a_{11}{}^2)x_1 - a_{11}a_{12}x_2 - ka_{12}(1 + a_{11})x_3$
$x_3' = \quad\quad\quad\quad\quad\quad\quad\quad\quad\quad a_{12}x_3$

11. $b^2 - ac = (a_{11}a_{22} - a_{12}a_{21})^2\{(b')^2 - a'c'\}$. The discriminant of Γ is equal to $|A|^2$ times the discriminant of Γ'.

15. Yes

Sec. 5.11, p. 237

1. There is no similarity transformation of type IV. Similarity transformations of the other five types are defined by the following matrices

$$\text{Type I:} \quad \begin{Vmatrix} a_{11} & 0 & 0 \\ 0 & ka_{11} & 0 \\ 0 & 0 & a_{33} \end{Vmatrix} \quad \begin{matrix} k \neq 1 \\ a_{11} \neq a_{33} \end{matrix}$$

$$\text{Type II:} \quad \begin{Vmatrix} a_{33} & 0 & a_{13} \\ 0 & ka_{11} & a_{23} \\ 0 & 0 & a_{33} \end{Vmatrix} \quad \begin{matrix} ka_{11} \neq a_{33} \\ a_{13} \neq 0 \end{matrix}$$

$$\text{Type III:} \quad \begin{Vmatrix} a_{11} & 0 & 0 \\ 0 & a_{11} & 0 \\ 0 & 0 & a_{33} \end{Vmatrix} \quad a_{11} \neq a_{33}$$

$$\text{Type V:} \quad \begin{Vmatrix} a_{11} & 0 & a_{13} \\ 0 & a_{11} & a_{23} \\ 0 & 0 & a_{11} \end{Vmatrix} \quad a_{13}, a_{23} \text{ not both zero}$$

Type VI: the identity transformation

3. No

7. (a) $x_1' = \quad x_2 - 2x_3$ (b) No   9. $\sqrt{a_{11}{}^2 + a_{12}{}^2}$
$\quad\quad x_2' = x_1$
$\quad\quad x_3' = \quad\quad 2x_3$

Sec. 5.12, p. 240

3. (a) $x_1' = x_1 - 3x_2$ (b) Impossible   (c) Impossible
$\quad\quad x_2' = \quad\quad x_2$
$\quad\quad x_3' = \quad\quad x_3$

Sec. 5.13, p. 242

7. A translation in the direction of the line determined by the two centers of rotation

9. If and only if it is a rotation through 120° or 240°

CHAPTER 6

Sec. 6.3, p. 262

3. Yes

5. (a) $l_7(P_4P_6P_9P_{15}P_{18}) \overset{P_1}{\wedge} l_2(P_4P_{11}P_5P_{17}P_{14}) \overset{P_{19}}{\wedge} l_7(P_4P_6P_{15}P_{18}P_9)$
The transformation is not an involution.

(b) $l_4(P_2P_7P_9P_{17}P_{16}) \overset{P_1}{\wedge} l_{20}(P_{21}P_{14}P_{13}P_{15}P_{16}) \overset{P_5}{\wedge} l_{15}(P_7P_{11}P_{20}P_{15}P_3)$
$\overset{P_{18}}{\wedge} l_4(P_7P_2P_{17}P_9P_{16})$ The transformation is an involution.

7. No, because if $l_3 = l_1$, then $A_1A_3 = B_1B_3$, and W is indeterminate. The correspondence can be made one to one, however, by defining the image of l_3 to be the intersection of l_1 and UV. Given W on UV, the corresponding line, l_3, is then determined by P and the intersection of VA_2 and WA_1.

15. In a projectivity with distinct fixed points, the cross ratio of the fixed points and any point P and its image is independent of P.

Sec. 6.5, p. 274

3. If p is the pappus line of the projectivity, the image of P is the intersection of l' and p, and the preimage of P is the intersection of l and p.

CHAPTER 7

Sec. 7.2, p. 283

3. $(n^2 - n + 1)(n - 1)^3 n/6$

13. If A_1, A_2, A_3 are the finite vertices of the four-point, and if l is the line whose direction identifies the ideal vertex, A_0, then the diagonal point D_i is the intersection of A_jA_k and the line on A_i which is parallel to l.

15. If the finite sides of the four-line are a_1, a_2, a_3, then the diagonal line d_i is the line on a_ja_k which is parallel to a_i.

Sec. 7.3, p. 294

1. (a) D_1:(0,1,1), D_2:(1,0,1), D_3:(1,1,0), H_1:(2,1,1), H_1':(0,1,−1),
   H_2:(1,2,1), H_2':(1,0,−1), H_3:(1,1,2), H_3':(1,−1,0)
   (b) D_1:(2,1,1), D_2:(1,2,1), D_3:(1,1,2), H_1:(2,3,3), H_1':(0,1,−1),
   H_2:(3,2,3), H_2':(1,0,−1), H_3:(3,3,2), H_3':(1,−1,0)
   (c) D_1:(1,0,0), D_2:(0,1,0), D_3:(0,0,1), H_1:(0,1,1), H_1':(0,1,−1),
   H_2:(1,0,1), H_2':(1,0,−1), H_3:(1,1,0), H_3':(1,−1,0)

15. There is a unique four-point meeting the given requirements.

Sec. 7.4, p. 304

3. U and V must be harmonic conjugates with respect to the points in which their join intersects x and y.

7. If P is the point (p_1,p_2,p_3), then P' is the point

$$p_1(-p_1 + p_2 + p_3), p_2(p_1 - p_2 + p_3), p_3(p_1 + p_2 - p_3)$$

CHAPTER 8

Sec. 8.2, p. 330

3. $2n - 1$ if the conic consists of two lines, n if the conic consists of a single line.

9. $-\lambda\lambda' - 4\lambda\mu' + 2\lambda'\mu + 2\mu\mu' = 0$, $2x^2 + 2xy - y^2 - 4x + y = 0$

11. $rx^2 + (2r - 6)xy + 3y^2 - 3rx + 3y = 0$

15. If the locus of V is tangent to H, there is a single parabola in the pencil of conics on P_0, P_1, P_2, P_3. If the locus of V does not intersect H, there are no parabolas in the pencil.

Sec. 8.3, p. 335

9. $\lambda\lambda' - 2\lambda\mu' + \lambda'\mu + \mu\mu' = 0$, $x^2 + xy - 2x - y = 0$

Sec. 8.4, p. 341

9. Yes   17. Yes.

Sec. 8.5, p. 350

9. Axiom 8 does not hold in PG_5.

11. If $\triangle ABC$ and $\triangle A'B'C'$ are axially perspective, then the six lines determined by the vertices of one triangle and the noncorresponding vertices of the other triangle all belong to a nonsingular line-conic.

Sec. 8.6, p. 361

3. Under the same definition used for the polar of a point P with respect to a nonsingular conic, the polar of a point with respect to a singular conic is a line, coinciding with the line of the conic if the latter consists of a single line and passing through the vertex V of the conic if the conic consists of a pair of lines. In the latter case the line is the harmonic conjugate of the line PV with respect to the two components of the conic.

Sec. 8.7, p. 367

15. No

CHAPTER 9

Sec. 9.2, p. 385

11. The elements of the extension field are the quadratic polynomials $ax^2 + bx + c$, where a, b, $c = 0$, 1. If we let $0 = e_0$, $1 = e_1$, $x = e_2$, $x + 1 = e_3$, $x^2 = e_4$, $x^2 + 1 = e_5$, $x^2 + x = e_6$, $x^2 + x + 1 = e_7$, then:
(a) The addition table is

$+$	e_0	e_1	e_2	e_3	e_4	e_5	e_6	e_7
e_0	e_0	e_1	e_2	e_3	e_4	e_5	e_6	e_7
e_1	e_1	e_0	e_3	e_2	e_5	e_4	e_7	e_6
e_2	e_2	e_3	e_0	e_1	e_6	e_7	e_4	e_5
e_3	e_3	e_2	e_1	e_0	e_7	e_6	e_5	e_4
e_4	e_4	e_5	e_6	e_7	e_0	e_1	e_2	e_3
e_5	e_5	e_4	e_7	e_6	e_1	e_0	e_3	e_2
e_6	e_6	e_7	e_4	e_5	e_2	e_3	e_0	e_1
e_7	e_7	e_6	e_5	e_4	e_3	e_2	e_1	e_0

$\cdot$	e_0	e_1	e_2	e_3	e_4	e_5	e_6	e_7
e_0	e_0	e_0	e_0	e_0	e_0	e_0	e_0	e_0
e_1	e_0	e_1	e_2	e_3	e_4	e_5	e_6	e_7
e_2	e_0	e_2	e_4	e_6	e_3	e_1	e_7	e_5
e_3	e_0	e_3	e_6	e_5	e_7	e_4	e_1	e_2
e_4	e_0	e_4	e_3	e_7	e_6	e_2	e_5	e_1
e_5	e_0	e_5	e_1	e_4	e_2	e_7	e_3	e_6
e_6	e_0	e_6	e_7	e_1	e_5	e_3	e_2	e_4
e_7	e_0	e_7	e_5	e_2	e_1	e_6	e_4	e_3

(*b*) The addition table is the same as in part (*a*). The multiplication table is

$\cdot$	e_0	e_1	e_2	e_3	e_4	e_5	e_6	e_7
e_0	e_0	e_0	e_0	e_0	e_0	e_0	e_0	e_0
e_1	e_0	e_1	e_2	e_3	e_4	e_5	e_6	e_7
e_2	e_0	e_2	e_4	e_6	e_5	e_7	e_1	e_3
e_3	e_0	e_3	e_6	e_5	e_1	e_2	e_7	e_4
e_4	e_0	e_4	e_5	e_1	e_7	e_3	e_2	e_6
e_5	e_0	e_5	e_7	e_2	e_3	e_6	e_4	e_1
e_6	e_0	e_6	e_1	e_7	e_2	e_4	e_3	e_5
e_7	e_0	e_7	e_3	e_4	e_6	e_1	e_5	e_2

(*c*) The fields in parts *a* and *b* are isomorphic under the following one-to-one correspondence:

Part *a*	e_0	e_1	e_2	e_3	e_4	e_5	e_6	e_7
Part *b*	e_0	e_1	e_5	e_4	e_6	e_7	e_3	e_2

(*d*) No. The polynomial $x^3 + x^2 + x + 1$ is reducible over the field $\{0,1\}$.

Sec. 9.3, p. 396

1. Yes

5. Yes. Simply construct the point $(-B \pm \sqrt{B^2 - 4AC})/2A$, which is possible if and only if $A \neq L_0$ and $B^2 - 4AC > L_0$. The latter condition simply means that $B^2 - 4AC$ is a point L_x such that $x > 0$; the geometric criterion for this is the reality of the fixed points of the involution used in finding $\sqrt{B^2 - 4AC}$.

Sec. 9.4, p. 402

3. $L_0 \to L_0,\ L_1 \to L_1,\ L_\omega \to L_\omega;\ T_1 = I\colon u' = u$

$L_0 \to L_0,\ L_1 \to L_\omega,\ L_\omega \to L_1;\ T_2\colon u' = \dfrac{u}{u-1}$

$L_0 \to L_1,\ L_1 \to L_\omega,\ L_\omega \to L_0;\ T_3\colon u' = \dfrac{1}{-u+1}$

$L_0 \to L_1,\ L_1 \to L_0,\ L_\omega \to L_\omega;\ T_4\colon u' = -u+1$

$L_0 \to L_\omega,\ L_1 \to L_0,\ L_\omega \to L_1;\ T_5\colon u' = \dfrac{u-1}{u}$

$L_0 \to L_\omega,\ L_1 \to L_1,\ L_\omega \to L_0;\ T_6\colon u' = \dfrac{1}{u}$

Yes. These transformations form a group isomorphic with the cross-ratio group.

5. There is a single fixed point distinct from L_ω if and only if $c \neq 0$ and $(a-d)^2 + 4bc = 0$. L_ω is a single fixed point if and only if $c = 0$ and $a = d$.

Sec. 9.7, p. 421

1. Yes   7. $\mathbb{R}(P_1P_2, P_3P_4) = \tfrac{2}{3}$ using either set of base points.

9. Taking $(0,1,0)$ and $(0,0,1)$ as base points, $\left\| \begin{matrix} \lambda' \\ \mu' \end{matrix} \right\| = \left\| \begin{matrix} 1 & 0 \\ 0 & -2 \end{matrix} \right\| \cdot \left\| \begin{matrix} \lambda \\ \mu \end{matrix} \right\|$ is the equation of the projectivity.

11. (*a*) $a_{11}x_1{}^2 + a_{22}x_2{}^2 + a_{33}x_3{}^2 = 0$
    (*b*) $a_{23}x_2x_3 + a_{13}x_1x_3 + a_{12}x_1x_2 = 0$
    (*c*) $a_{11}x_1{}^2 + 2a_{23}x_2x_3 = 0$

13. If the four points are taken to be $(1,0,0)$, $(0,1,0)$, $(0,0,1)$, $(1,1,1)$, and if the coordinates of the given line are $[p_1,p_2,p_3]$, the equation of the required locus is

$$-p_1x_1{}^2 - p_2x_2{}^2 - p_3x_3{}^2 + (p_1 + p_2)x_1x_2 + (p_1 + p_3)x_1x_3$$
$$+ (p_2 + p_3)x_2x_3 = 0$$

Sec. 9.8, p. 428

3. (a) $6l_1l_2 + 7l_2{}^2 + 4l_2l_3 + l_3{}^2 = 0$
   (b) $3l_1{}^2 + 2l_1l_2 - l_2{}^2 - 2l_1l_3 + 2l_2l_3 + l_3{}^2 = 0$
   (c) $l_1{}^2 + 6l_1l_2 + 9l_2{}^2 - 4l_1l_3 - 12l_2l_3 + 4l_3{}^2 = 0$

5. (a) $(x_1 + x_2)l_1 + (x_2 + x_3)l_2 + (x_1 + x_2)l_3 = 0$
   (b) $x_1l_1 - 3(x_1 + x_2)l_2 + (x_1 + x_2 + x_3)l_3 = 0$
   (c) $(x_1 + x_2)l_1 + (x_1 - x_2 + x_3)l_2 + (x_2 + x_3)l_3 = 0$

CHAPTER 10

Sec. 10.2, p. 434

1. $2 \ln 3$ 3. $3 \ln 2$

5. The sign of (PQ) is changed.   7. $(PQ) = 0$

9. $(PQ) = \infty$. Thus M_1 and M_2 might be called "points at infinity."

Sec. 10.3, p. 440

5. $z = (4n + 1)\pi/2 + i \cosh^{-1} 3$

11. (a) $\ln \sqrt{13} + i(\tan^{-1} \frac{3}{2} + 2n\pi)$
    (b) $\ln 5 + i[\tan^{-1} (4/-3) + 2n\pi]$
    (c) $\ln \sqrt{2} + i(-3\pi/4 + 2n\pi)$

13. (a) $-i \sinh 2$ (b) $\sin 1 \cosh 1 - i \cos 1 \sinh 1$
    (c) $\sin 2 \cosh 3 + i \cos 2 \sinh 3$

15. $-k_a \ln 2$

Sec. 10.4, p. 448

1. (a) $(P^T A X)^2 - (P^T A P)(X^T A X) \geq 0$

   (b) $0 \leq \dfrac{(P^T A X)^2}{(P^T A P)(X^T A X)} \leq 1$

3. (a) $(P_2 P_3) = i \cos^{-1} \dfrac{1}{\sqrt{3}}$, $(P_1 P_3) = i \cos^{-1} \dfrac{1}{\sqrt{6}}$

   $(P_1 P_2) = \dfrac{i\pi}{4}$, $[p_2 p_3] = i \cos^{-1} \sqrt{\frac{2}{5}}$, $[p_1 p_3] = i \cos^{-1} \dfrac{1}{\sqrt{5}}$, $[p_1 p_2] = \dfrac{i\pi}{2}$

(b) $(P_2P_3) = \cos^{-1}\dfrac{1}{\sqrt{3}}$, $(P_1P_3) = \cos^{-1}\dfrac{1}{\sqrt{6}}$, $(P_1P_2) = \dfrac{\pi}{4}$,

$[p_2p_3] = \cos^{-1}\sqrt{\tfrac{2}{5}}$, $[p_1p_3] = \cos^{-1}\dfrac{1}{\sqrt{5}}$, $[p_1p_2] = \dfrac{\pi}{2}$

7. $(P_2P_3) = 2k_d \cosh^{-1}\dfrac{f}{\sqrt{ab}}$, $(P_1P_3) = 2k_d \cosh^{-1}\dfrac{g}{\sqrt{ab}}$,

$(P_1P_2) = 2k_d \cosh^{-1}\dfrac{h}{\sqrt{ab}}$, $[p_2p_3] = 2k_a \cosh^{-1}\dfrac{F}{\sqrt{AB}}$,

$[p_1p_3] = 2k_a \cosh^{-1}\dfrac{G}{\sqrt{AB}}$, $[p_1p_2] = 2k_a \cosh^{-1}\dfrac{H}{\sqrt{AB}}$,

where $A, B, \ldots$ are the cofactors of $a, b, \ldots$ in the discriminant of the given conic.

Sec. 10.5, p. 453

1.
$$A = \begin{Vmatrix} 2 & 0 & -1 \\ 0 & 1 & 0 \\ -1 & 0 & 1 \end{Vmatrix} \qquad t = \begin{Vmatrix} 1 & 0 & 1 \\ 1 & 0 & -1 \\ 0 & 1 & 0 \end{Vmatrix}$$

$$T = \begin{Vmatrix} 1 & -1 & 1 \\ 1 & 1 & -1 \\ 0 & 2 & 0 \end{Vmatrix} \qquad \|f_{ij}\| = \begin{Vmatrix} 2 & 1 & 1 \\ 1 & 2 & 3 \\ 1 & 3 & 6 \end{Vmatrix}$$

$$\|F_{ij}\| = \begin{Vmatrix} 3 & -3 & 1 \\ -3 & 11 & -5 \\ 1 & -5 & 3 \end{Vmatrix}$$

3. No

Sec. 10.6, p. 466

5. Yes

7. (a) $d_1 = \dfrac{\pi}{2}$, $d_2 = \dfrac{\pi}{2}$, $d_3 = \cos^{-1}\dfrac{\sqrt{10}}{5}$, $\alpha_1 = \dfrac{\pi}{2}$, $\alpha_2 = \dfrac{\pi}{2}$, $\alpha_3 = \cos^{-1}\dfrac{\sqrt{10}}{5}$

(b) $d_1 = \cos^{-1}\dfrac{\sqrt{15}}{6}$, $d_2 = \cos^{-1}\dfrac{\sqrt{3}}{3}$, $d_3 = \cos^{-1}\dfrac{3\sqrt{5}}{10}$,

$\alpha_1 = \cos^{-1}\dfrac{\sqrt{22}}{11}$, $\alpha_2 = \cos^{-1}\sqrt{\dfrac{5}{77}}$, $\alpha_3 = \cos^{-1}\dfrac{2\sqrt{2}}{35}$

(c) $d_1 = \cos^{-1}\left(\dfrac{-3\sqrt{14}}{14}\right)$, $d_2 = \cos^{-1}\left(\dfrac{-2\sqrt{7}}{7}\right)$, $d_3 = \cos^{-1}\dfrac{\sqrt{2}}{2}$,

$\alpha_1 = \cos^{-1}\left(\dfrac{-\sqrt{3}}{3}\right)$, $\alpha_2 = \cos^{-1}\left(\dfrac{-\sqrt{5}}{5}\right)$, $\alpha_3 = \cos^{-1}\dfrac{\sqrt{15}}{15}$

(d) $d_1 = \cos^{-1}\left(\dfrac{-7}{15}\right)$, $d_2 = \cos^{-1}\left(\dfrac{-\sqrt{5}}{15}\right)$, $d_3 = \cos^{-1}\frac{3}{5}$,

$\alpha_1 = \cos^{-1}\left(\dfrac{-13}{22}\right)$, $\alpha_2 = \cos^{-1}\frac{19}{44}$, $\alpha_3 = \cos^{-1}\frac{37}{44}$

11. (a) $a = 0.434$, $b = 0.933$, $\beta = 1.27$
   (b) $h = 1.80$, $\alpha = 1.04$, $\beta = 1.94$
   (c) $b = 2.45$, $\alpha = 1.18$, $\beta = 2.37$
   (d) $h = 1.26$, $b = 0.97$, $\alpha = 1.08$

Sec. 10.7, p. 472

1. Eq. (6): $\dfrac{\sin \alpha}{a} = \dfrac{\sin \beta}{b} = \dfrac{\sin \gamma}{c}$
  Eq. (10): $a^2 = b^2 + c^2 - 2bc \cos \alpha$
  Eq. (14): $\cos \alpha = \cos(\pi - \beta - \gamma)$

Sec. 10.8, p. 481

3. (a) $\alpha_1 = \alpha_2 = 1.07$, $\alpha_3 = 0.289$
   (b) $d_3 = 2.87$, $\alpha_1 = 0.095$, $\alpha_2 = 0.296$

5. Yes

CHAPTER 11

Sec. 11.3, p. 492

1. $[p_1 p_2] = 18°27'$, $[p_1 p_3] = 116°33'$, $[p_2 p_3] = 45°$

Sec. 11.4, p. 495

1. (a) $(P_1 P_2)^2 = (x_1 - x_2)^2 - (y_1 - y_2)(z_1 - z_2)$

$$\cos \phi = \frac{l_1 l_2 + (m_1 - n_1)(m_2 - n_2)}{\sqrt{\{l_1{}^2 + (m_1 - n_1)^2\}\{l_2{}^2 + (m_2 - n_2)^2\}}}$$

(b) $(P_1 P_2)^2 = (x_1 - x_2)^2 - (y_1 - y_2)(z_1 - z_2)$

$$\cos \phi = \frac{(l_1 - m_1)(l_2 - m_2) + (l_1 - n_1)(l_2 - n_2) + (m_1 - n_1)(m_2 - n_2)}{\sqrt{\{(l_1 - m_1)^2 + (l_1 - n_1)^2 + (m_1 - n_1)^2\}\{(l_2 - m_2)^2 + (l_2 - n_2)^2 + (m_2 - n_2)^2\}}}$$

(c) $(P_1P_2)^2 = (x_1 - x_2)^2 + (z_1 - z_2)^2$

$$\cos \phi = \frac{(l_1 + m_1)(l_2 + m_2) + n_1 n_2}{\sqrt{\{(l_1 + m_1)^2 + n_1^2\}\{(l_2 + m_2)^2 + n_2^2\}}}$$

(d) $(P_1P_2)^2 = -(x_1 - x_2)(y_1 - y_2) + (x_1 - x_2)(z_1 - z_2)$
$$+ (y_1 - y_2)(z_1 - z_2)$$

$$\cos \phi = \frac{\begin{aligned}(l_1 - m_1)(l_2 - m_2) + (l_1 + n_1)(l_2 + n_2)\\ + (m_1 + n_1)(m_2 + n_2)\end{aligned}}{\sqrt{\{(l_1 - m_1)^2 + (l_1 + n_1)^2 + (m_1 + n_1)^2\}}}$$
$$\cdot \{(l_2 - m_2)^2 + (l_2 + n_2)^2 + (m_2 + n_2)^2\}$$

Sec. 11.5, p. 502

3. (a) $(P_1P_2) = 1$, $(P_2P_3) = \sqrt{2}$, $(P_1P_3) = \sqrt{6}$, $[p_1p_2] = \cosh^{-1} \frac{7\sqrt{3}}{12}$,

$[p_2p_3] = \cosh^{-1} \frac{3\sqrt{2}}{4}$, $[p_1p_3] = \cosh^{-1} \frac{5\sqrt{6}}{12}$

(b) $(P_1P_2) = \frac{1}{2}$, $(P_2P_3) = \frac{\sqrt{15}}{6}$, $(P_1P_3) = \frac{2\sqrt{3}}{3}$,

$[p_1p_2] = \cosh^{-1} \frac{9\sqrt{5}}{20}$, $[p_2p_3] = \cosh^{-1} \frac{7\sqrt{3}}{12}$, $[p_1p_3] = \cosh^{-1} \frac{4\sqrt{15}}{15}$

APPENDIX 1

Page 512

1. (a) 1   (b) -22 (c) 9
3. (a) $-\lambda^3 + 6\lambda^2 + 9\lambda + 1$ (b) $\lambda^3 - 22\lambda^2 - 11\lambda + 22$
   (c) $3\lambda^3 - 9\lambda^2 - \lambda + 3$

APPENDIX 2

Page 521

1. The answer to this question appears to be unknown.

INDEX

Page references in italic indicate Exercises.

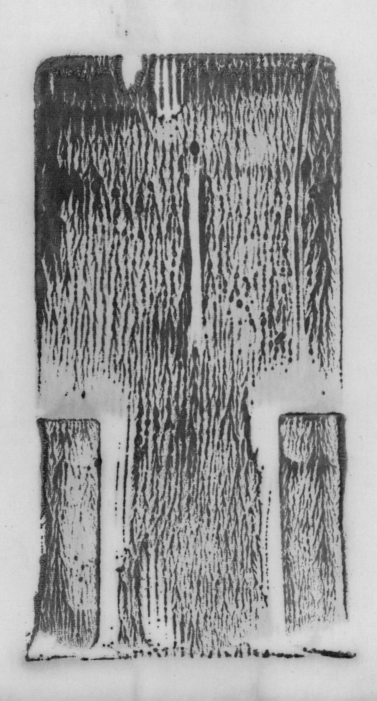